BASIC PRINCIPLES OF NUTRITION SCIENCE

Kruger Brentt
Publishers

BASIC PRINCIPLES OF NUTRITION SCIENCE

Jordan Moore
Barbara Kelly
Philip Carter

Kruger Brentt
Publishers
2024

Kruger Brentt Publishers UK. LTD.
Company Number 9728962

Regd. Office: 68 St Margarets Road, Edgware, Middlesex HA8 9UU

ISBN: 978-1-78715-167-3

For information on all our publications visit our website at http://krugerbrentt.com/

PREFACE

Nutrition science is the study of the relationship between food and health. It involves understanding how nutrients are digested, absorbed, and utilized by the body, as well as how diet can affect the risk of chronic diseases such as heart disease, diabetes, and cancer. In this essay, we will explore the principles of nutrition science and how they can help individuals make informed choices about their diet and health. One of the primary principles of nutrition science is that a healthy diet should be balanced and include a variety of foods from all food groups. This includes fruits, vegetables, whole grains, lean proteins, and healthy fats. Each of these food groups provides essential nutrients that the body needs to function properly, such as vitamins, minerals, and amino acids. A balanced diet can help prevent nutrient deficiencies and reduce the risk of chronic diseases. Consuming too many calories can lead to weight gain, which can increase the risk of chronic diseases such as diabetes and heart disease. On the other hand, consuming too few calories can lead to malnutrition and other health problems. It is important to find the right balance of calories to maintain a healthy weight and meet the body's nutrient needs.

Nutrition science also emphasizes the importance of choosing nutrient-dense foods. Nutrient-dense foods are those that provide a high amount of nutrients relative to their calorie content. Examples include fruits, vegetables, whole grains, lean proteins, and low-fat dairy products. Choosing these foods can help ensure that the body is getting the nutrients it needs without consuming too many calories. Principle of nutrition science is that dietary fat should be consumed in moderation and include healthy fats such as monounsaturated and polyunsaturated fats. These types of fats are found in foods such as nuts, seeds, avocado, and fish. They can help reduce the risk of heart disease and other chronic diseases when consumed in moderation. Nutrition science emphasizes the importance of avoiding excess consumption of sodium, added sugars, and saturated fats. These components are found in many processed and packaged foods and can increase the risk of chronic diseases when consumed in excess. It is important to read food labels and choose foods that are low in sodium, added sugars, and saturated fats. Nutrition science are essential for maintaining a healthy diet and preventing chronic diseases. These principles include consuming a balanced diet, appropriate portion sizes and calorie intake, choosing nutrient-dense foods, consuming healthy fats in moderation, and avoiding excess consumption of sodium, added sugars, and saturated fats. By following these principles, individuals can make informed choices about their diet and health and improve their overall well-being.

The present book contains 11 unit namely how to design a healthy diet; science and literacy in the area of nutrition; photosynthesis, digestion, and metabolism: molecules of life; carbohydrates; lipids; protein; health and body weight; vitamins and minerals-i; vitamins and minerals-ii; nutrition and physical activity & the impact of nutrition over the lifespan. This book describes how nutrition is related to health, including risk of chronic disease. Provides basic information about individual nutrient of high importance along with the latest recommended dietary allowances (RDAs) and dietary guidelines. Focuses on malnutrition problems through nutritional deficiency disorders. Helpful for the students and teachers of nutrition, dietetics, nursing, health science, and medicine.

We are grateful to all those persons as well as various books, manuals, periodicals, magazines, journals etc. that helped in the preparation of this book. In spite of the best efforts, it is possible that some errors may have occurred into the compilation and editing of the book. Further queries, constructive suggestions and criticisms for the improvement of the book are always welcomed and shall be thankfully acknowledged.

Jordan Moore

Barbara Kelly

Philip Carter

CONTENTS

UNIT-1

HOW TO DESIGN A HEALTHY DIET

DESIGNING A HEALTHY DIET: AN INTRODUCTION

What makes a diet "healthy"? What does the word "healthy" even mean? Each of us might picture something different when we think of a healthy diet, and if you travel around the world, you'll find even more variation in how people define this term.

Indeed, humans are incredibly flexible when it comes to food. We are omnivores, and we can survive and thrive on a wide variety of different foods. The foods that nourish our bodies are often the same foods that nourish our souls, bringing us together with friends and family, celebrating traditions and conjuring memories of meals past.

We'll begin our study of nutrition by zooming in on nutrients—the molecules in food that nourish us—to begin to understand what each gives us. Then, we'll zoom back out to consider some tools for choosing foods that will together provide us with all the nutrients we need. Because whatever the deep and complex meanings that food brings to our lives and our culture, we also want to choose foods that will enable us to be well, to fuel our activities, to prevent disease, and to live long, healthy lives.

HEALTH AND NUTRITION

Nutrition

Simply put, food is the plants and animals that we eat, and nutrition is how food affects the health of the body. According to the Academy of Nutrition and Dietetics, "Food is essential—it provides vital nutrients for survival, and helps the body function and stay healthy. Food is comprised of macronutrients including protein, carbohydrate and fat that not only offer calories to fuel the body and give it energy but play specific roles in maintaining health. Food also supplies micronutrients (vitamins and minerals) and phytochemicals that don't provide calories but serve a variety of critical functions to ensure the body operates optimally."[1] (Phytochemicals are compounds found in plants that give them their smell, taste, and color. They are not technically nutrients, but many have been shown to affect human health.)

The study of nutrition goes beyond just a discussion of food and the nutrients needed by the body. It includes how those nutrients are digested, absorbed, and used by the cells of the body. It examines how food provides energy for daily activities and how our food intake and choices impact body weight and risk for chronic diseases such as heart disease and type 2 diabetes. It also provides insight on behavioral, social, and environmental factors that influence what, how, when, and why we eat.[2] Thus, nutrition is an important part of the overall discussion of health and wellness.

HOW NUTRITION AFFECTS HEALTH

The World Health Organization (WHO) defines health as "a state of complete physical, mental and social well-being and not merely the absence of disease or infirmity."[3] The WHO recognizes nutrition as a critical part of health and development, noting that better nutrition is related to:[4]

- Improved infant, child and maternal health
- Stronger immune systems
- Safer pregnancy and childbirth
- Lower risk of non-communicable diseases (such as type 2 diabetes and cardiovascular disease)
- Greater longevity
- Greater productivity, creating opportunities to break cycles of poverty and hunger

Malnutrition, including both under-nutrition and over-nutrition, is a significant threat to human health. In fact, nutrition is associated with four of the top ten leading causes of death in the United States, including heart disease, cancer, diabetes, and stroke.[5]

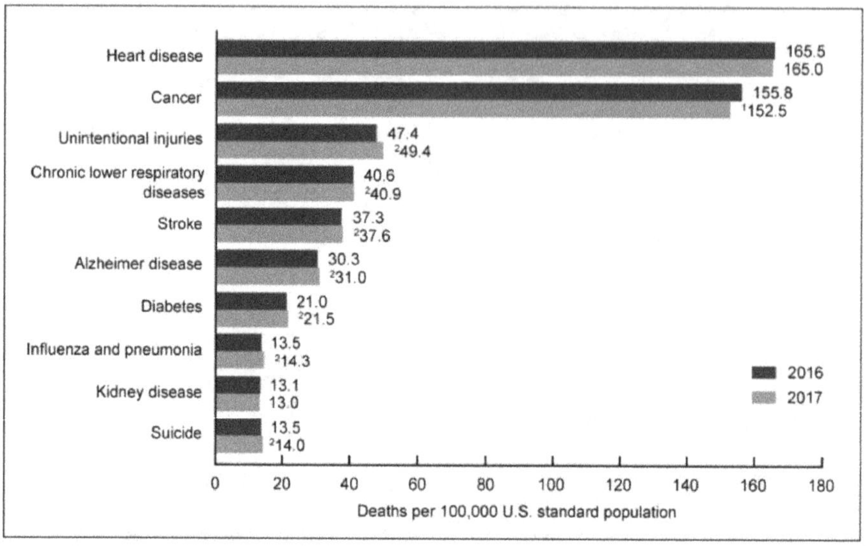

[1]Statistically significant decrease in age-adjusted death rate from 2016 to 2017 ($p < 0.05$).
[2]Statistically significant increase in age-adjusted death rate from 2016 to 2017 ($p < 0.05$).
NOTES: A total of 2,813,503 resident deaths were registered in the United States in 2017. The 10 leading causes accounted for 74.0% of all deaths in the United States in 2017. Causes of death are ranked according to number of deaths. Rankings for 2016 data are not shown. Data table for Figure 4 includes the number of deaths for leading causes. Access data table for Figure 4 at: https://www.cdc.gov/nchs/data/databriefs/db328_tables-508.pdf#4.
SOURCE: NCHS, National Vital Statistics System, Mortality.

Figure 1.1. Age-adjusted death rates for the 10 leading causes of deaths: United States, 2016 and 2017

Nutrition can affect the health of the mind as well as the body. For example, some research suggests that the foods people eat can influence their mood. A 2019 study of moderately-depressed people aged 17 to 35 years old found that when half of them shifted towards a Mediterranean-style eating pattern for 3 weeks—emphasizing more fruits and vegetables, whole grains, lean protein sources, unsweetened dairy, fish, nuts and seeds, olive oil, and spices—their depression levels decreased compared to participants who continued their usual eating habits. Some (but not all) other studies have also found links between healthier diets and decreased risk of depression. It's not clear why this might be, but researchers speculate that decreased inflammation or changes in the body's microbiome caused by these dietary patterns may play a role in brain functioning and mental health.[6] This is an area that requires much more research, but as you're thinking about dietary choices, it's worth thinking about how foods make you feel.

In addition to nutrition, health is affected by genetics, the environment, life cycle, and lifestyle. One important facet of lifestyle is personal dietary habits. Dietary habitsinclude what a person eats, how much a person eats during a meal, how frequently meals are consumed, and how often a person eats out. Other aspects of lifestyle include physical activity level, recreational drug use, and sleeping patterns, all of which play a role in health and impact food choices and nutrition status. Following a healthy lifestyle improves your overall health and well-being.

PERSONAL CHOICE: THE CHALLENGE OF CHOOSING FOODS

There are other factors besides environment and lifestyle that influence the foods you choose to eat. Food itself can regulate your appetite and how you feel. Multiple studies have demonstrated that some high-fiber foods and high-protein foods decrease appetite by slowing the digestive process and prolonging the feeling of being full (also called satiety). Making food choices that maximize nutrient intake and satiety can help manage how much you eat and how long before you eat again.

Food also has social, cultural, and religious significance, all of which impact the foods we choose to eat. The social meanings of food affect what people eat, as well as how and when. Special events in our lives—from birthdays to funerals—are commemorated with equally special foods. Cultural influences and upbringing can also shape an individual's food habits. Being aware of these factors can help people make healthier food choices, and still honor the traditions and ties they hold dear.

Factors that Drive Food Choices

A number of other factors affect the dietary choices individuals make, including:

- Taste, texture, and appearance. Individuals have a wide range of taste preferences, which influence their food choices. For example, some people dislike milk and others hate raw vegetables. Foods that may be unappealing at first to some people, like vegetables or tofu, can often be adapted to meet most taste preferences, and people can learn to like foods over time with repeated exposures.
- Economics. Access to fresh fruits and vegetables may be limited, particularly for those who live in economically disadvantaged or remote areas, where affordable food options are limited to convenience stores and fast food.
- Early food experiences. People who were not exposed to different foods as children, or who were forced to swallow every last bite of overcooked vegetables, may make limited food choices or experience food aversions as adults. On the other hand, those exposed to a variety of foods in the setting of pleasant family meals, are more likely to maintain those same eating habits in adulthood.
- Habits. It's common to establish eating routines, which can work both for and against optimal health. Habitually grabbing a fast food sandwich for breakfast can seem convenient, but might not offer substantial nutrition. Yet getting in the habit of drinking an ample amount of water each day can yield multiple benefits.
- Culture. The culture in which one grows up affects how one sees food in daily life and on special occasions.
- Geography. Where a person lives influences food choices. For instance, people who live in Midwestern US states have less access to seafood than those living along the coasts.
- Advertising. The media greatly influences food choices by persuading consumers to eat certain foods.
- Social factors. Any school lunchroom observer can testify to the impact of peer pressure on eating habits, and this influence lasts through adulthood. People make food choices based on how they see others and want others to see them. For example, individuals who are surrounded by others who consume fast food are more likely to do the same.
- Health concerns. Some people have food allergies or intolerances and need to avoid certain foods. Others may have developed health issues which require them to follow a low salt diet. In addition, people who have never

worried about their weight have a very different approach to eating than those who have long struggled to change their weight.

- Emotions. There is a wide range in how emotional issues affect eating habits. Food can be a source of comfort, such as the taste of a favorite dish from childhood. Or, for people with a history of disordered eating, it may also be a source of anxiety. When faced with a great deal of stress, some people tend to overeat, while others find it hard to eat at all.
- Green food/Sustainability choices. Based on a growing understanding of diet as a public and personal issue, more and more people are starting to make food choices based on their environmental impact. Realizing that their food choices help shape the world, many individuals are opting for a vegetarian diet, or, if they do eat animal products, striving to consider animal welfare and sustainability in their choices. Purchasing local and organic food products and items grown through sustainable products can help to shrink the environmental impact of one's food choices.

REFERENCES

[1]Academy of Nutrition and Dietetics. (2019). How to Explain Basic Nutrition Concepts. Retrieved December 18, 2019, from https://www.eatrightpro.org/practice/practice-resources/international-nutrition-pilot-project/how-to-explain-basic-nutrition-concepts

[2]Medline Plus. (2019). Definitions of Health Terms. Retrieved from https://medlineplus.gov/definitions/nutritiondefinitions.html

[3]World Health Organization. (n.d.) Constitution. Retrieved from https://www.who.int/about/who-we-are/constitution

[4]World Health Organization. (2018, February 22). Nutrition. Retrieved from https://www.who.int/news-room/facts-in-pictures/detail/nutrition

[5]Centers for Disease Control and Prevention. (2017, March 17). Leading Causes of Death. Retrieved from https://www.cdc.gov/nchs/fastats/leading-causes-of-death.htm

[6]Aubrey, A. & Chatterjee, R. (2019, October 19). Changing Your Diet Can Help Tamp Down Depression, Boost Mood. Retrieved from https://www.npr.org/sections/thesalt/2019/10/09/768665411/changing-your-diet-can-help-tamp-down-depression-boost-mood

DESCRIPTION OF MOLECULES

In order to understand the chemical structure of nutrients and how they function in the body and provide energy to cells of the body, you must first understand the basic chemical structure of molecules. Nutrients are chemical molecules that are found in foods and required by our bodies to maintain life and support growth and health. On this page, we'll zoom in to understand how atoms bond together to form these chemical molecules.

Matter is anything that has mass and takes up space. All living and nonliving things are composed of matter. Atoms are the fundamental unit of matter. The chair you are sitting in is made of atoms. The food you ate for lunch was built from atoms. Even the air you breathe is made of atoms. An atom is the smallest unit of an element, just like a blade is the smallest unit of grass. An element is made entirely from one specific type of atom. There are more than 100 elements that make up the world we live in, however hydrogen, carbon, nitrogen, and oxygen make up the bulk of all living things. Many elements are found in the foods we eat and all of them are found on the Periodic Table of Elements.

Atoms are unimaginably small. Even within a single microscopic cell, there is room for not just billions, but trillions or even hundreds of trillions of atoms. The atoms themselves are made of even smaller particles called protons, neutrons, and electrons. Protons and neutrons are found in the nucleus (center) of the atom, while electrons are found outside the nucleus in regions called shells. Protons have a positive charge, neutrons have no charge, and electrons are negatively charged. Because protons and neutrons are contained in the dense nucleus of the atom, the nucleus has a positive charge. And since opposites attract, electrons are attracted to this nucleus and move around it in an electron cloud surrounding the nucleus. This attraction keeps the atom together, much like the force of gravity keeps the moon in orbit around Earth.

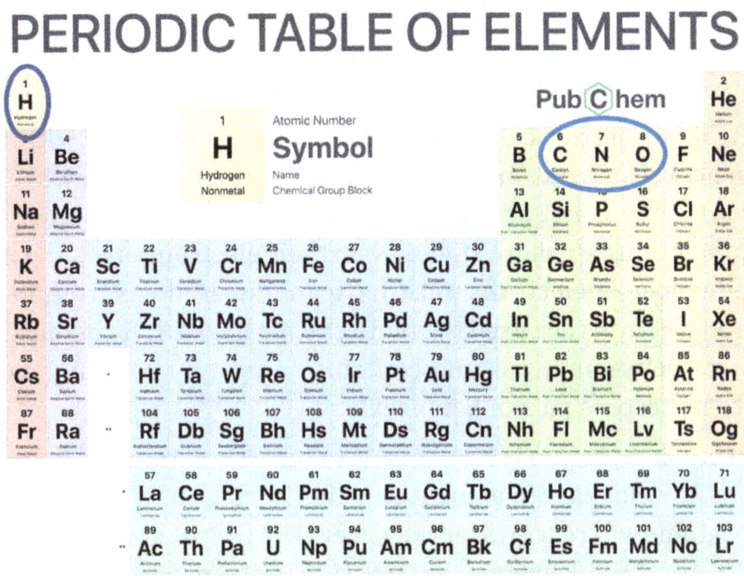

Figure 1.2. The Periodic Table of Elements. Note the four elements circled in blue (hydrogen, carbon, nitrogen, and oxygen). These four elements make up the bulk of all living things.

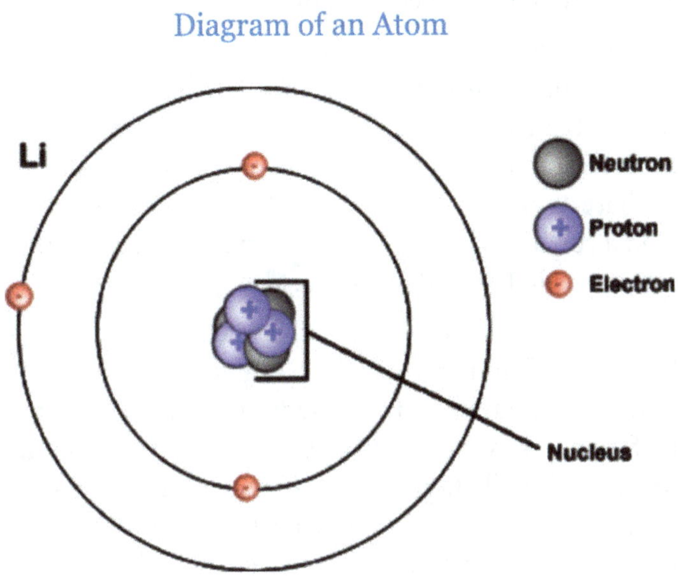

Figure 1.3. Diagram of a lithium atom, showing th placement of protons, neutrons, and electrons. Protons and neutrons are found in the nucleus (center) of the atom. Electrons are found outside the nucleus.

Molecules and Covalent Bonding

Atoms combine to form a larger and more complex entity called a molecule. Molecules are composed of two or more atoms held together by chemical bonds.

The electrons of an atom contain energy. This energy is stored within the charge and movement of electrons and the bonds that atoms make with one another. However, this energy is not always stable, depending on the number of electrons within an atom. Atoms are more stable when their electrons orbit in pairs. An atom with an odd number of electrons must have an unpaired electron. In most cases, these unpaired electrons are used to create chemical bonds. A chemical bond is the attractive force between atoms and contains energy. By bonding, electrons find pairs, and atoms become part of a molecule.

The most stable situation for an atom is to have its outer shell completely filled with electrons. It is not easy to explain why this is true, but it's a rule of thumb that predicts how atoms will react with each other. The first electron shell of an atom is considered full (or stable) when it contains two electrons, and the second and third shells are full (stable) with eight electrons. Atoms tend to bond to other atoms in such a way that both atoms have filled outer shells as a result of the interaction. While some elements may be able to hold more electrons in their third shell, most of the important elements in biology (e.g. hydrogen, carbon, nitrogen, and oxygen) are considered stable with eight electrons in this outer shell.

Instead of transferring their electrons completely, atoms typically remain in very close contact and share electrons so that their outer shells are filled. In essence, a shared electron is counted "twice" and participates in a larger shell that joins the two atoms. A single pair of shared electrons makes a single covalent bond. Atoms can also share two pairs of electrons (in a double bond). This sharing of electrons is called a covalent bond. Covalent bonds are the strongest, most stable types of chemical bonds in the biological world.

One example of covalent bonding to form a molecule is the formation of methane, a colorless and flammable gas that results from burning gasoline or fossil fuels (Fig. 1.4). One carbon atom and four hydrogen atoms react to form methane. The outer shell of carbon has four electrons, so carbon can share an electron with four other atoms (which will then give carbon a full outer shell of 8 electrons). Hydrogen has a single electron in its outermost shell and can share this electron with one other atom. The carbon atom forms a covalent bond with four hydrogen atoms to form a molecule of methane. In a methane molecule, carbon effectively has a "full" second shell (eight electrons) and each hydrogen has a "full" first shell (two electrons). Each hydrogen requires one covalent bond to fill its first shell. Each carbon requires four covalent bonds to fill its second shell.

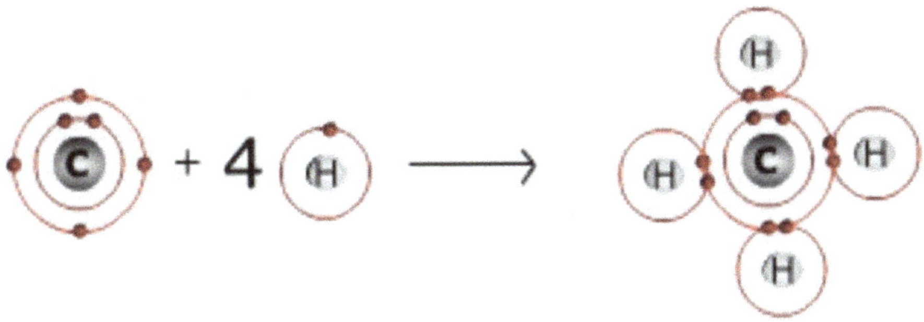

Figure 1.4. A molecule of methane. Carbon and hydrogen react to form methane by sharing electrons through a covalent bond.

In food and in components of the human body, energy resides in the chemical bonds of specific molecules. Bond formation and bond breaking are chemical reactions that involve the movement of electrons between atoms. These chemical reactions occur continuously in the body. When the chemical bonds of nutrients in the foods we eat are broken, energy is released. That energy is used by cells of the body to perform daily functions and tasks such as breathing, walking up a flight of steps, and studying for a test.

Biological Macromolecules

As we noted earlier, atoms are the building blocks of all matter. Biological macromolecules are formed when atoms of carbon, hydrogen, oxygen, and nitrogen bond with each other in unique and varied ways. Biological macromolecules are the raw materials used to build living organisms. They are special molecules that contain carbon atoms covalently bonded with hydrogen atoms.

There are three classes of biological macromolecules (or macronutrients) that we will study in this course: carbohydrates, lipids, proteins. These macronutrients are probably already familiar to you, because they make up the nutrients you ingest every time you eat. In this way, you provide your cells with the building materials and energy necessary to sustain life. The next section will take a closer look at these important macronutrients and the role they play in our diet and in providing energy to cells.

NUTRIENT CLASSIFICATION

Food is one of life's greatest pleasures. It offers amazing flavors, aromas, and textures. Food also provides our body with essential nutrients and non-nutrients like phytochemicals, both of which are vital to health. This section will discuss the six classes of nutrients and how these nutrients can be classified.

What Are Nutrients?

Nutrients are chemical substances found in food that are required by the body to provide energy, give the body structure, and help regulate chemical processes. There are six classes of nutrients:

- Carbohydrates
- Lipids
- Proteins
- Water
- Vitamins
- Minerals

Nutrients can be further classified as either macronutrients or micronutrients and either organic or inorganic, as well as whether or not they provide energy to the body (energy-yielding). We'll discuss these different ways of classifying nutrients in the following sections.

Macronutrients

Nutrients that are needed in large amounts are called macronutrients. There are three classes of macronutrients: carbohydrates, lipids, and proteins. Water is also a macronutrient in the sense that you require a large amount of it, but unlike the other macronutrients, it does not yield energy.se that you require a large amount of it, but unlike the other macronut

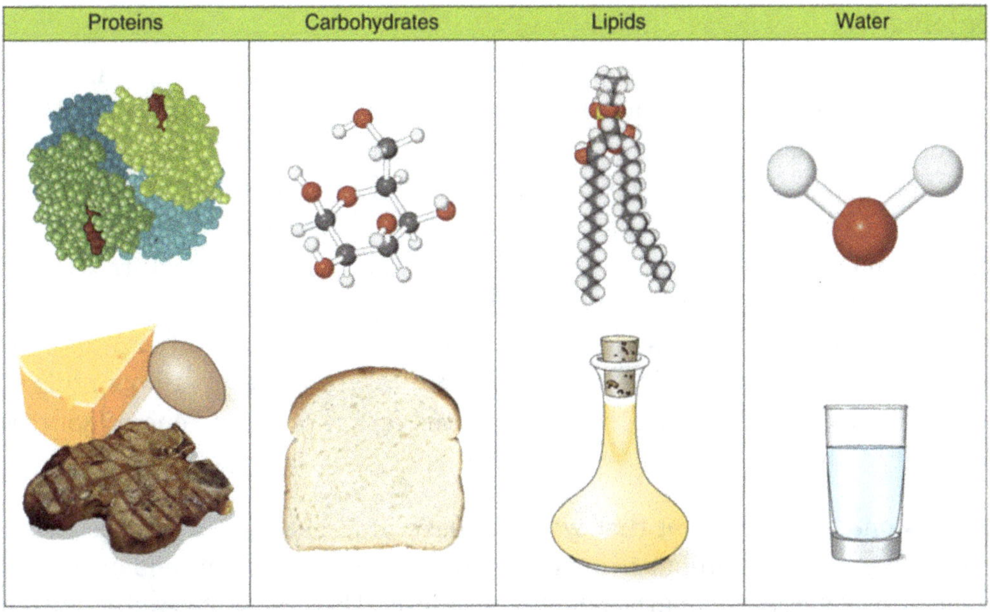

Figure 1.5. Macronutrients include proteins, carbohydrates, lipids, and water. This figure illustrates each nutrient's chemical structure and examples of food sources.

Carbohydrates

Carbohydrates are molecules composed of carbon, hydrogen, and oxygen. The major food sources of carbohydrates are grains, dairy products, fruits, legumes, and starchy vegetables, like potatoes. Non-starchy vegetables, like carrots, also contain carbohydrates, but in lesser quantities.

Carbohydrates are broadly classified into two groups based on their chemical structure: simple carbohydrates (often called simple sugars) and complex carbohydrates, which include fiber, starch, and glycogen. Carbohydrates are a major fuel source for all cells of the body, and certain cells, like cells of the central nervous system and red blood cells, rely solely on carbohydrates for energy.

Lipids

Lipids are also a family of molecules composed of carbon, hydrogen, and oxygen, but unlike carbohydrates, they are insoluble in water. Lipids are found predominantly in butter, oils, meats, dairy products, nuts and seeds, and in many processed foods. The three main types of lipids are triglycerides, phospholipids, and sterols. The main job of lipids is to provide or store energy. In addition to energy storage, lipids serve as major components of cell membranes, surround and protect organs, provide insulation to aid in temperature regulation, and regulate many other functions in the body.

Proteins

Proteins are large molecules composed of chains of amino acids, which are simple subunits made of carbon, oxygen, hydrogen, and nitrogen. Food sources of proteins include meats, dairy products, seafood, and a variety of plant-based foods, like beans, nuts, and seeds. The word protein comes from a Greek word meaning "of primary importance," which is an apt description of these macronutrients as they are also known as the "workhorses" of life. Proteins provide structure to bones, muscles, and skin, and they play a role in conducting most of the chemical reactions occurring in the body. Scientists estimate that more than 100,000 different proteins exist within the human body. Proteins can also provide energy, though this is a relatively minor function, as carbohydrates and fat are preferred energy sources.

Water

There is one other nutrient that we must have in large quantities: water. Water does not contain carbon but is composed of two hydrogens and one oxygen per molecule of water. More than 60 percent of your total body weight is water. Without it, nothing could be transported in or out of the body, chemical reactions would not occur, organs would not be cushioned, and body temperature would fluctuate widely. On average, an adult consumes just over two liters of water per day from food and drink combined. Since water is so critical for life's basic processes, we can only survive a few days without it, making it one of the most vital nutrients.

Micronutrients

Micronutrients are nutrients required by the body in smaller amounts, but they're still essential for carrying out bodily functions. Micronutrients include all of the essential minerals and vitamins. There are 16 essential minerals and 13 essential vitamins (Table 1.1 and Table 1.2). In contrast to carbohydrates, lipids, and proteins, micronutrients are not a source of energy, but they assist in the process of energy metabolism as cofactors or components of enzymes (known as coenzymes). Enzymes are proteins that catalyze (or accelerate) chemical reactions in the body; they're involved in all aspects of body functions, including producing energy, digesting nutrients, and building macromolecules.

Minerals

Minerals are inorganic substances that are classified depending on how much the body requires. Trace minerals, such as molybdenum, selenium, zinc, iron, and iodine, are only required in amounts of a few milligrams or less per day. Major minerals, such as calcium, magnesium, potassium, sodium, and phosphorus, are required in amounts of hundreds of milligrams or more per day. Many minerals are critical for enzyme function, and others are used to maintain fluid balance, build bone tissue, synthesize hormones, transmit nerve impulses, contract and relax muscles, and protect against harmful free radicals in the body. To give you an appreciation of the many functions of minerals, the table below has a complete list of all the minerals and their major functions. (Note: There is no need to memorize these minerals and functions at this point in the course.)

Table 1.1. Minerals and their major functions

Major Minerals	Major Function
Sodium	Fluid balance, nerve transmission, muscle contraction
Chloride	Fluid balance, stomach acid production
Potassium	Fluid balance, nerve transmission, muscle contraction
Calcium	Bone and teeth health maintenance, nerve transmission, muscle contraction, blood clotting
Phosphorus	Bone and teeth health maintenance, acid-base balance
Magnesium	Protein production, nerve transmission, muscle contraction
Sulfur	Protein production
Trace Minerals	**Function**
Iron	Carries oxygen, assists in energy production
Zinc	Protein and DNA production, wound healing, growth, immune system function
Iodine	Thyroid hormone production, growth, metabolism
Selenium	Antioxidant
Copper	Coenzyme, iron metabolism
Manganese	Coenzyme
Fluoride	Bone and teeth health maintenance, tooth decay prevention
Chromium	Assists insulin in glucose metabolism
Molybdenum	Coenzyme

Vitamins

Vitamins are organic nutrients that are categorized based on their solubility in water. The water-soluble vitamins are vitamin C and all of the B vitamins. The fat-soluble vitamins are vitamins A, D, E, and K. Vitamins are required to perform many functions in the body, such as making red blood cells, synthesizing bone tissue, and playing a role in normal vision, nervous system function, and immune function. To give you an appreciation of the many functions of vitamins, the table below lists the 13 essential vitamins and their major functions. (Note: There is no need to memorize these vitamins and functions at this point in the course.)

Table 1.2. Vitamins and their major functions

Water-Soluble Vitamins	Major Functions
Thiamin (B1)	Coenzyme, energy metabolism assistance
Riboflavin (B2)	Coenzyme, energy metabolism assistance
Niacin (B3)	Coenzyme, energy metabolism assistance
Pantothenic acid (B5)	Coenzyme, energy metabolism assistance
Pyridoxine (B6)	Coenzyme, energy metabolism assistance
Biotin (B7)	Coenzyme, amino acid and fatty acid metabolism
Folate (B9)	Coenzyme, essential for growth
Cobalamin (B12)	Coenzyme, red blood cell synthesis
C (ascorbic acid)	Collagen synthesis, antioxidant
Fat-Soluble Vitamins	**Major Functions**
A	Vision, reproduction, immune system function
D	Bone and teeth health maintenance, immune system function
E	Antioxidant, cell membrane protection
K	Bone and teeth health maintenance, blood clotting

As you might suspect based on the major functions of vitamins listed above, vitamin deficiencies can cause severe health problems and even death. For example, a deficiency in niacin causes a disease called pellagra, which was common in the early twentieth century in some parts of the United States. The common signs and symptoms of pellagra are known as the "4D's—diarrhea, dermatitis, dementia, and death." Until scientists discovered that better diets relieved the signs and symptoms of pellagra, many people with the disease ended up hospitalized and in asylums awaiting death. The following video gives an overview of pellagra and how its cure was discovered through a change in diet.

Energy-yielding nutrients

The macronutrients—carbohydrate, protein, and fat—are the only nutrients that provide energy to the body. The energy from macronutrients comes from their chemical bonds. This chemical energy is converted into cellular energy that can be utilized to perform work, allowing cells to conduct their basic functions. Although vitamins also have energy in their chemical bonds, our bodies do not make the enzymes to break these bonds and release this energy. (This is fortunate, as we need vitamins for their specific functions, and breaking them down to use for energy would be a waste.) Food energy is measured in kilocalories (kcals). A kilocalorie is the amount of energy needed to raise 1 kilogram of water by 1 degree Celsius. The kilocalories stored in food can be determined by putting the food into a bomb calorimeter and measuring the energy output (energy = heat produced).

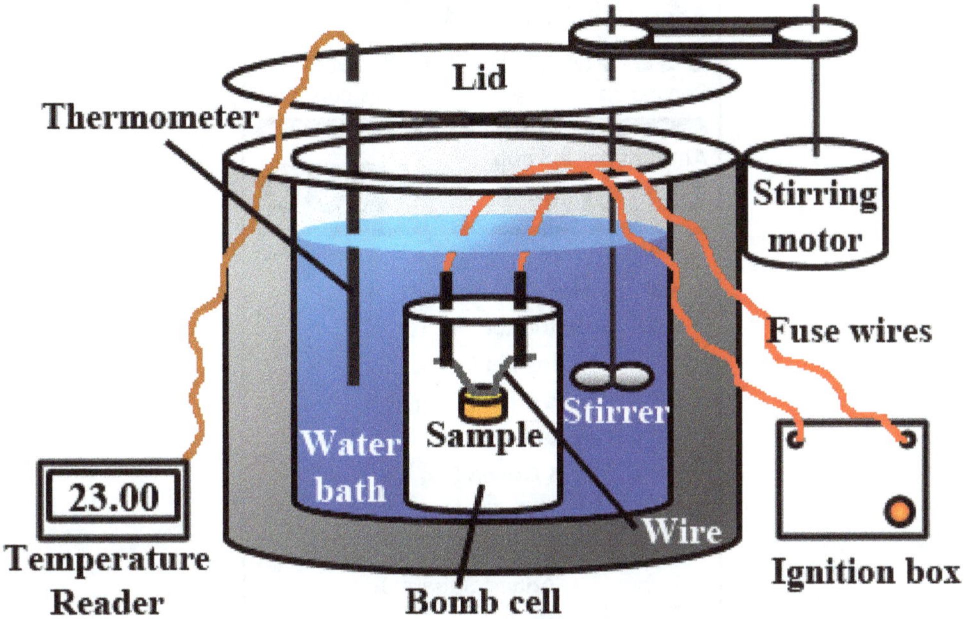

Figure 1.6. A Bomb calorimeter

In the US, the kilocalorie (kcal) is the most commonly used unit of energy and is often just referred to as a calorie. Strictly speaking, a kcal is 1000 calories. In nutrition, the term calories almost always refers to kcals. Sometimes the kcal is indicated by capitalizing calories as "Calories." For the sake of simplicity, we'll use the terms "calories" and "kilocalories" interchangeably in this book.

Below is a list of energy sources in the diet from lowest to highest calories per gram (a gram is about the weight of a paperclip). Notice the addition of alcohol. Although alcohol does provide energy, it isn't a nutrient, because it isn't required as a source of nourishment to the body.

Energy Sources (kcal/g)

- Carbohydrates 4
- Protein 4
- Alcohol 7
- Lipids 9

Carbohydrates and proteins provide 4 calories per gram, and fats provide 9 calories per gram. Fat is the most energy-dense nutrient, because it provides the most calories per gram (more than double carbohydrates and protein).

Nutrition Facts

8 servings per container

Serving size 2/3 cup (55g)

Amount per 2/3 cup

Calories 230

% DV*	
12%	**Total Fat** 8g
5%	Saturated Fat 1g
	Trans Fat 0g
0%	**Cholesterol** 0mg
7%	**Sodium** 160mg
12%	**Total Carbs** 37g
14%	Dietary Fiber 4g
	Sugars 1g
	Added Sugars 0g
	Protein 3g
10%	**Vitamin D** 2mcg
20%	**Calcium** 260mg
45%	**Iron** 8mg
5%	**Potassium** 235mg

* Footnote on Daily Values (DV) and calories reference to be inserted here.

Figure 1.7. Nutrition Facts

When you look at the Nutrition Facts panel on a food label, you'll see that it lists calories, as well as grams of total fat, total carbohydrates, and protein per serving. From these values, you can estimate the amount of calories coming from the different macronutrients.

Looking at the values in the Nutrition Facts label, you can convert grams into calories by doing the following calculations:

- 8 grams of fat x 9 kcal/g = 72 kcals
- 37 grams of carbohydrate x 4 kcal/g = 148 kcals
- 3 grams of protein x 4 kcal/g = 12 kcals

You can double check your math by adding the calories per serving provided from fat, carbohydrate, and protein (232 calories for the example above). This number should come close to the total calories per serving listed on the Nutrition Facts. It will not always match up exactly (like in the example above) due to rounding.

Organic and Inorganic Nutrients

So far, we've categorized nutrients as macronutrients or micronutrients and based on whether or not they're energy-yielding. There is one more way to categorize nutrients: organic or inorganic. When you think of the word "organic," you might think of how foods are produced (with or without synthetic fertilizers and pesticides), but in this case we are referring to the chemical structure of a nutrient.

Organic Nutrients

The organic nutrients include the macronutrients (carbohydrate, protein, and fat) and vitamins. An organic nutrient contains both carbon and hydrogen. Organic nutrients can be made by living organisms and are complex, made up of many elements (carbon, hydrogen, oxygen, and sometimes nitrogen) bonded together. In a sense, they are "alive," and therefore can be destroyed or broken down.

Vitamin E (shown below) is an organic molecule, because it contains both carbon and hydrogen atoms. Vitamin E is synthesized by plants and can be destroyed by heat during cooking.

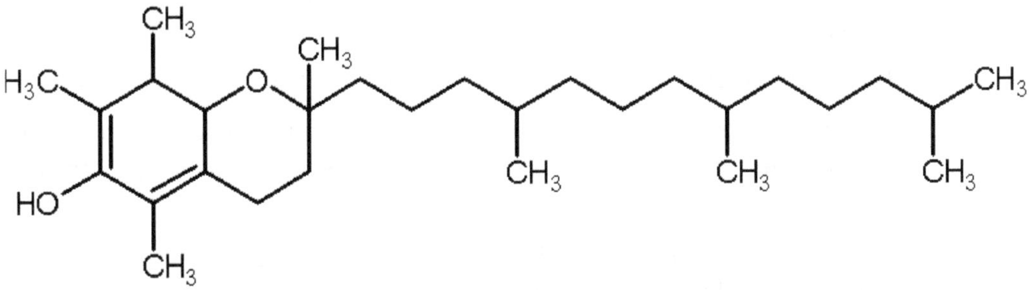

Figure 1.8. Chemical structure of Vitamin E

Inorganic Nutrients

Inorganic nutrients include both water and minerals. Inorganic nutrients do not contain both carbon and hydrogen, and they are not created or destroyed. Minerals can't be destroyed, so they are the ash left when a food is burned to completion. Minerals are also not digested or broken down, as they are already in their simplest form. They are absorbed as-is, then shuttled around the body for their different functions, and then excreted.

Summary

The different categories of nutrients are summarized in the following table.

Table 1.3. Summary of Nutrient Classifications

Classification	Nutrient
Macronutrient	Carbohydrate, protein, lipids, water
Micronutrient	Vitamins, minerals
Energy-Yielding	Carbohydrate, protein, fat
Organic	Carbohydrate, protein, lipids, vitamins
Inorganic	Minerals, water

DETERMINING NUTRIENT NEEDS: DIETARY REFERENCE INTAKES

How do we know how much of a given nutrient people should eat, or how much is too much? For this information, we can turn to the Dietary Reference Intakes (DRI)—a set of recommendations developed by the National Academies of Sciences, Engineering, and Medicine to describe the amounts of specific nutrients and energy that people should consume in order to stay healthy. They are developed by groups of nutrition scientists, who together evaluate the research to determine how much of a nutrient is required to prevent deficiencies and chronic disease, as well as how much is excessive and could cause toxicity. The DRI standards are specific to people living in the United States and Canada, and they're meant to be used by people who are generally healthy, because those with specific health conditions may have different nutrient requirements.

The DRI standards can be divided into two main categories:

- Recommendations for energy intake – How many calories are required, and how much energy should proportionately come from carbohydrate, fat, and protein?
- Recommendations for nutrient intake – How much of each nutrient should be consumed, and how much is excessive?

We'll discuss each of these categories, and then we'll discuss some of the ways that the DRI standards are used. Be prepared to learn a lot of acronyms!

DRI recommendations for energy intake

The DRIs include two types of recommendations related to energy intake:

Estimated Energy Requirement (EER). The EER is an estimate of how many calories a person needs to consume, on average, each day to stay healthy, based on their age, sex, height, weight, and physical activity level. For adults, the EER is meant to be a caloric intake that maintains energy balance, meaning that it won't cause weight loss or gain. For children, the EER includes the energy needed for normal growth. For pregnant or lactating women, it includes energy needed for development of the fetus and other pregnancy requirements or for milk production. Different EER values were also developed for different physical activity levels, because greater physical activity requires more energy.[1] The EER should be considered a "ballpark" estimate of a person's caloric needs. As we'll learn later in the term, the way that people process and utilize energy is highly variable, and two people can have the same sex, weight, height, and level of physical activity but different caloric needs.

Acceptable Macronutrient Distribution Ranges (AMDR). The AMDR is the calculated range of how much energy from carbohydrate, fat, and protein is recommended for a healthy diet. People who do not meet the AMDR may have increased risk of developing health complications—although these are also ballpark recommendations, not absolute requirements for health. Keep in mind that the percentage of daily caloric intake from the three energy-yielding macronutrients will add up to 100 percent, so the proportion of each influences the other two. For example, someone consuming a very low carbohydrate diet, with just 5 to 10 percent of calories coming from carbohydrates would not only fall short of the AMDR for carbohydrate but also exceed the recommended amounts of fat and/or protein, because the rest of daily calories must come from these macronutrients. The AMDR recommendations are based on balancing carbohydrate, fat, and protein to allow for adequate amounts of all three, and they are wide enough ranges that many different types of diets can fit within them.

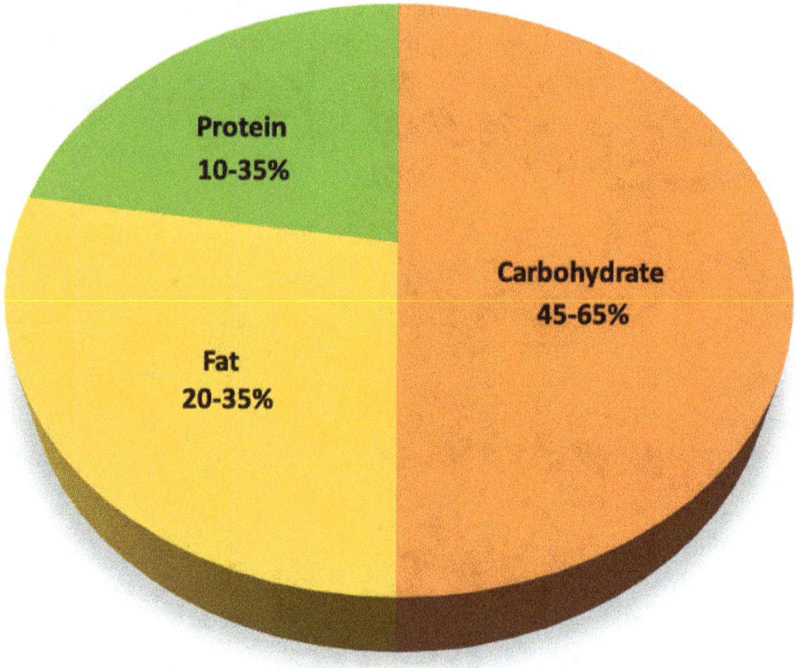

Figure 1.9. Acceptable Macronutrient Distribution Ranges (AMDR) for the three energy-yielding macronutrients.

Daily Recommendations for Nutrient Intake

There are four different types of DRI values used to describe recommendations for intake of individual nutrients:

- Estimated Average Requirements (EAR)
- Recommended Dietary Allowances (RDA)
- Adequate Intakes (AI)
- Tolerable Upper Intake Levels (UL)

DRI values are summarized in tables to make it easy to find a specific value for a person based on their life stage and sex. For example, part of a table of EAR values for macronutrients and vitamins is shown below.

Dietary Reference Intakes (DRIs): Estimated Average Requirements
Food and Nutrition Board, Institute of Medicine, National Academies

Life Stage Group	CHO (g/d)	Protein (g/kg/d)	Vit A (µg/d)[a]	Vit C (mg/d)	Vit E (mg/d)[b]	Thiamin (mg/d)	Riboflavin (mg/d)	Niacin (mg/d)[c]	Vit B_6 (mg/d)
Infants									
7–12 mo		1.0							
Children									
1–3 y	100	0.87	210	13	5	0.4	0.4	5	0.4
4–8 y	100	0.76	275	22	6	0.5	0.5	6	0.5
Males									
9–13 y	100	0.76	445	39	9	0.7	0.8	9	0.8
14–18 y	100	0.73	630	63	12	1.0	1.1	12	1.1
19–30 y	100	0.66	625	75	12	1.0	1.1	12	1.1
31–50 y	100	0.66	625	75	12	1.0	1.1	12	1.1
51–70 y	100	0.66	625	75	12	1.0	1.1	12	1.4
> 70 y	100	0.66	625	75	12	1.0	1.1	12	1.4
Females									
9–13 y	100	0.76	420	39	9	0.7	0.8	9	0.8
14–18 y	100	0.71	485	56	12	0.9	0.9	11	1.0
19–30 y	100	0.66	500	60	12	0.9	0.9	11	1.1
31–50 y	100	0.66	500	60	12	0.9	0.9	11	1.1
51–70 y	100	0.66	500	60	12	0.9	0.9	11	1.3
> 70 y	100	0.66	500	60	12	0.9	0.9	11	1.3
Pregnancy									
14–18 y	135	0.88	530	66	12	1.2	1.2	14	1.6
19–30 y	135	0.88	550	70	12	1.2	1.2	14	1.6
31–50 y	135	0.88	550	70	12	1.2	1.2	14	1.6
Lactation									
14–18 y	160	1.05	885	96	16	1.2	1.3	13	1.7
19–30 y	160	1.05	900	100	16	1.2	1.3	13	1.7
31–50 y	160	1.05	900	100	16	1.2	1.3	13	1.7

This page from the National Institutes of Health Office of Dietary Supplements provides links to DRI reports and tables: Nutrient Recommendations: Dietary Reference Intakes (DRI)

Let's look at how each of these DRI values is determined, what they mean, and how they are used.

Estimated Average Requirement

The Estimated Average Requirement (EAR) is the amount of a nutrient that meets the requirements of 50 percent of people within a group of the same life stage and sex. The requirements of half of the group will fall below the EAR, and the requirements of the other half will be above it. To understand the EAR, it's important to recognize that individuals have different nutrient requirements, depending on many factors beyond our life stage and sex (differences in genetics, metabolism, body weight, and physical activity, for example), and the EAR is like the midpoint in the range of different individual requirements.

To develop the EAR, a committee of scientists evaluates the research on that nutrient and chooses a specific bodily function as a criterion on which to base it. For example, the EAR for calcium is set using a criterion of maximizing bone health, because this is quantitatively one of the most important functions of calcium, and the effects of different levels of calcium intake on bone health can be measured. Thus, the EAR for calcium is set at a point that will meet the needs, with respect to bone health, of half of the population.

The EAR for a given nutrient is shown in the graph below, with the individual requirement on the x-axis. Imagine this graph is depicting individual calcium requirements. The people on the left side of the graph have lower calcium requirements, and the people on the right side of the graph have higher calcium requirements. If everyone was eating the EAR for calcium, half would be getting enough calcium and half would not. Therefore, it wouldn't be wise to recommend that everyone only consume the EAR, because about half of the population would fall short in calcium if this was set as the recommendation. EAR values are most important because they are used to calculate the Recommended Daily Allowance (RDA) values, which are commonly used as population-wide recommendations for nutrient intake.

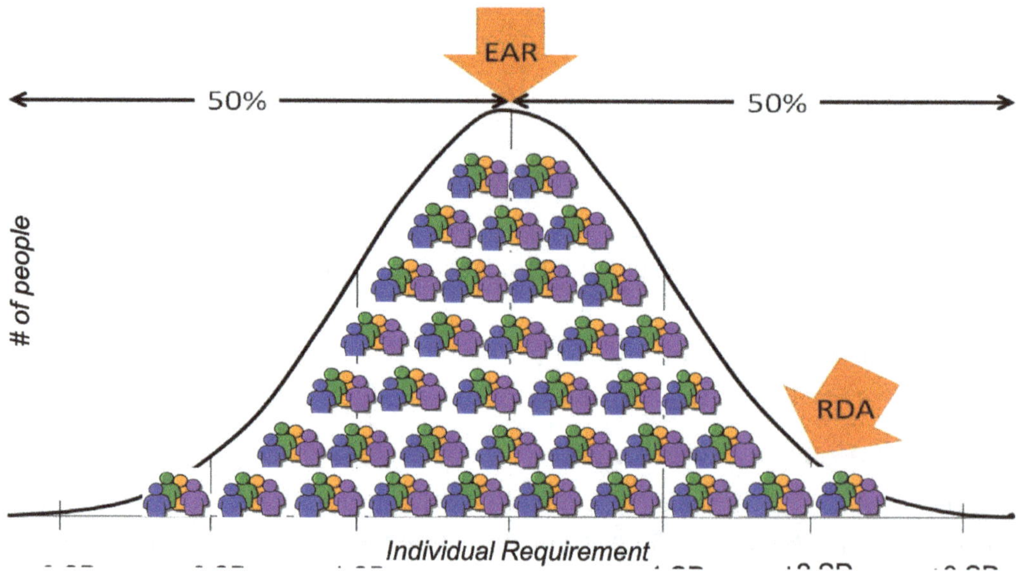

Figure 1.10. EAR and RDA relative to individual requirements for a given nutrient.

Recommended Daily Allowances

Once the EAR of a nutrient has been established, the Recommended Daily Allowances (RDA) value can be mathematically determined. While the EAR is set at a point that meets the needs of half the population, ***RDA*** values are set to meet the needs of the vast majority (97 to 98 percent) of the target healthy population. You can see this in the graph above. The RDA is a better recommendation for the population, because we can assume that if a person is consuming the RDA of a given nutrient, they are most likely meeting their nutritional needs for that nutrient.

This also explains why the RDA is not the same thing as an individual nutritional requirement. You may be consuming less than the RDA for calcium, but this does not automatically mean that your body is deficient in calcium and that you'll definitely end up with osteoporosis, because your *individual* calcium requirement may be less than the RDA. However, since you probably don't know your individual calcium requirement, the RDA is a good target amount for consumption, and the more your intake drops below the RDA, the greater your risk of later developing osteoporosis. The RDA is meant as a *recommendation*, and meeting the RDA means it is very likely that you are meeting your actual *requirement* for that nutrient.

It's interesting to compare and contrast the EER (for energy or calorie intake) and the RDA (for nutrient intake). In practice, both types of recommendations serve as a daily target for intake. However, the EER is set to meet the average caloric needs of a person, while the RDA is set to meet the needs of the vast majority of the population. Imagine if the EER was set to ensure that it met the caloric needs of the vast majority of a population. It would end up being a dramatic overestimate of caloric needs for most people. If everyone actually followed this recommendation, the majority of them would consume far more calories than they actually needed, resulting in weight gain. For nutrients, we have more flexibility in our intake, because we have ways of storing or metabolizing and excreting excess nutrients, so consuming somewhat more than our body needs is just fine.

Adequate Intake

When there is insufficient scientific evidence to set an EAR and RDA for the entire population, then the National Academies committee can decide to set an Adequate Intake (AI) level instead. The AI is based on observing healthy people and seeing how much of the nutrient in question they are consuming. An AI is less precise than an RDA, but in the absence of an RDA, the AI is our best guess of how much of a given nutrient is needed. If there is not an RDA for a nutrient, than the AI is used as the nutrient-intake goal.

For example, there has not been sufficient scientific research into the exact nutritional requirements for infants. Consequently, all of the DRI values for infants are AIs derived from nutrient values in human breast milk. For older babies and young children, AI values are derived from human milk coupled with data on adults. The AI is meant for a healthy target group and is not meant to be sufficient for certain at-risk groups, such as premature infants.

Tolerable Upper Intake Levels

Consuming inadequate amount of nutrients can cause health problems, and we use the RDA or AI values as targets to ensure that we're getting enough. However, consuming too much of many nutrients can also cause health problems. This is where the Tolerable Upper Intake Level (UL) is helpful. ULs indicate the highest level of continuous intake of a particular nutrient that may be taken without causing health problems.

It's rare to find amounts of a nutrient exceeding the UL in a balanced diet based on whole foods. However, a person who consumes dietary supplements, foods fortified with high levels of additional nutrients (protein bars, for example) or a diet based on only a few foods, might exceed the UL, and this could cause problems with nutrient toxicity. If you're selecting a supplement, be sure to choose one that does not exceed the UL for any nutrient, unless this is under specific instructions from your doctor or a registered dietitian.

When a nutrient does not have any *known* issue if taken in excessive doses, it is not assigned a UL. However, if a nutrient does not have a UL, that doesn't necessarily mean that it is safe to consume in large amounts—only that there isn't currently evidence that large amounts will cause problems. Science is an ongoing process, and the toxicity of many nutrients hasn't yet been studied.

Putting It All Together

The graph below summarizes the meaning of the 4 DRI values for nutrient intake.

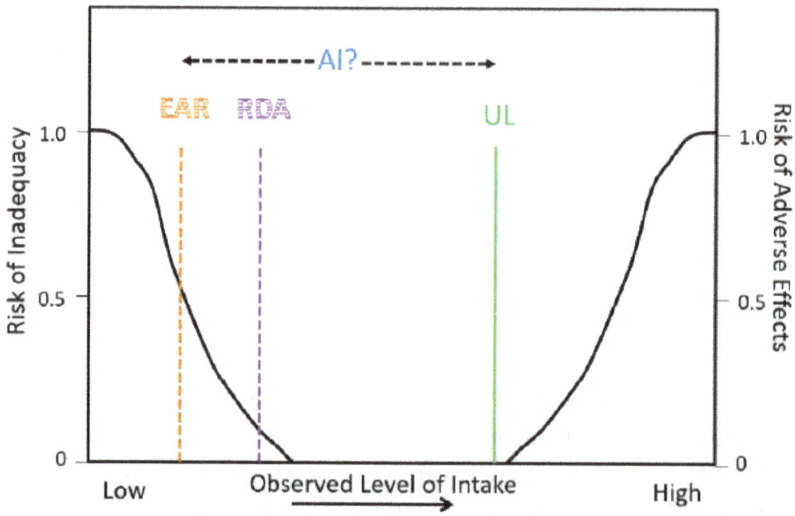

Figure 1.11. DRI values for nutrient intake. The EAR, RDA, AI, and UL are shown relative to the observed level of intake and risk of inadequacy and adverse effects.

This graph shows the risks of nutrient inadequacy and nutrient excess as we move from a low intake of a nutrient to a high intake. Starting on the left side of the graph, you can see that when you have a very low intake of a nutrient, your risk of nutrient deficiency is high. As your nutrient intake increases, the chances that you will be deficient in that nutrient decrease. The point at which 50 percent of the population meets their nutrient needs is the EAR, and the point at which 97 to 98 percent of the population meets their needs is the RDA. The UL is the highest level at which you can consume a nutrient without it being too much. As nutrient intake increases beyond the UL, the risk of health problems resulting from that nutrient increases. The AI is shown to exist somewhere between the EAR and UL, as it's an amount of the nutrient known to maintain health.

Note that there is a wide margin between the RDA and UL, showing that a person might safely eat much more than the RDA for a given nutrient without concerns of nutrient toxicity. However, be aware that the margin of safety varies depending on the nutrient. For example, fat-soluble vitamins have a smaller margin of safety between the RDA and the UL than water-soluble vitamins, meaning that it's easier to consume toxic levels of fat-soluble vitamins.

How the Dris are Used

Individuals can use the DRIs to help assess and plan their diets. Keep in mind that the values established have been devised with an ample safety margin and should be used as guidance for optimal intakes. Also, the values are meant to assess and plan the average intake over time; that is, you don't need to meet these recommendations every single day—meeting them over several days is sufficient.

The DRIs are also used by professionals, government agencies, and the food industry. Here are some examples of their applications[2]:

- Health professionals. Registered dietitians and other nutrition professionals use the DRIs to provide dietary counseling and education and to plan menus for institutions, such as hospitals, long-term care, prisons.
- Development of dietary guidelines. These include the U.S. Dietary Guidelines for Americans, MyPlate, and Canada's Food Guide. In each case, developers ensure that their advice will help people meet the DRI standards.
- Nutrition labeling. The DRIs help to inform Nutrition Facts labels on foods and Supplement Facts on supplement labels.
- Assistance programs. School meals, WIC, SNAP, Child and Adult Care, and Administration on Aging programs must ensure that their programs align with the DRI.
- Nutrition monitoring research. Data from surveys of what people in the U.S. and Canada eat are compared with the DRIs to monitor national nutritional health.

- Military. The military uses the DRIs as a reference to ensure nutrient needs are met for the armed forces, to plan meals, and to procure military rations.
- Food and supplement industries. In the development of healthy food and safe supplement products, these industries should refer to the DRI.

FOOD LABEL COMPREHENSION

Not so long ago, food choices were limited to what could be grown or raised, hunted or gathered. Today, grocery stores offer seemingly infinite choices in foods, with entire aisles dedicated to breakfast cereals and cases filled with a multitude of different yogurts. Faced with so many choices, how can we decide? Taste matters, of course. But if a healthy diet is your goal, so does nutrition. Food labels are our window into the nutritional value of a given food. Let's examine what we can learn from food labels and how reading them can help us make smart choices to contribute to a healthy diet.

The U.S. Food and Drug Administration (FDA) requires food manufacturers to accurately label foods so that consumers can be informed about their contents. There are 5 types of information required by the FDA on every food label, except for fresh produce and seafood[1,2]:

- Statement of identity *(what type of food is it?)*
- Net contents of the package *(how much is in there?)*
- Name and address of manufacturer (*where was it produced?*)
- Ingredients list *(what ingredients are included in the food?)*
- Nutrition information (*what is the amount of nutrients included in a serving of food?*)

Figure 1.12. The 5 required types of information on a food label.

The statement of identity and net contents of the package tell you what type of food you're purchasing and how much is in the package. The name and address of the manufacturer are important if there's a food recall due to an outbreak of foodborne illness or other contamination issue. Given the size of our food system and the fact that one manufacturer may make products packaged under multiple brand names, being able to trace a food's origin is critical.

The last two types of required information—the ingredients list and the nutrition information—are a bit more complex and provide valuable information to consumers, so let's look more closely at each of these parts of a food label.

Ingredients list

The ingredients list includes all ingredients, listed from most predominant to least predominant (by weight) in the product. For example, in the corn muffin mix label to the right, the most prevalent ingredient is enriched unbleached flour (with ingredients in the flour then listed in parentheses), followed by sugar, cornmeal, salt, and then a few other ingredients.

This order of ingredients comes in handy when judging the nutritional value of a product. For example, in the ingredients list for the corn muffin mix shown at right, it's interesting to note that it contains more sugar than cornmeal! The ingredients list can also help you determine whether a bread contains more whole grain flour than refined flour. Or, if you're choosing a breakfast cereal and the first ingredient is sugar, that's a red flag that it's more of a dessert than part of a nutritious breakfast.

By law, food manufacturers must also list major allergens, which include milk, egg, fish, crustacean shellfish, tree nuts, wheat, peanuts, and soybeans.[2] Allergens may be listed in a separate statement, as on the corn muffin mix label, which lists "Contains: Wheat" on the label. Alternatively, allergens can be listed in parentheses within the ingredient list, such as "lecithin (soy)." Some labels include an optional "may contain" or "made in shared equipment with..." statement that lists additional allergens that could be present, not as ingredients in the food, but in trace amounts from equipment contamination. For people with food allergies, having this information clearly and accurately displayed on food packages is vital for their safety.

The Nutrition Facts Panel

If you want to learn about the nutritional value of a food, the Nutrition Facts panel is where you'll find this information. It's very useful for comparing products and for identifying foods that will be more or less valuable in meeting your nutritional goals. For example, if you're trying to watch your intake of added sugar or saturated fat, or you're trying to incorporate more dietary sources of calcium and vitamin D, the Nutrition Facts panel is a valuable tool. There are 4 main parts of a Nutrition Facts panel, shown in the figure below. The colors are added to highlight different sections of a label; Nutrition Facts are printed in black and white.

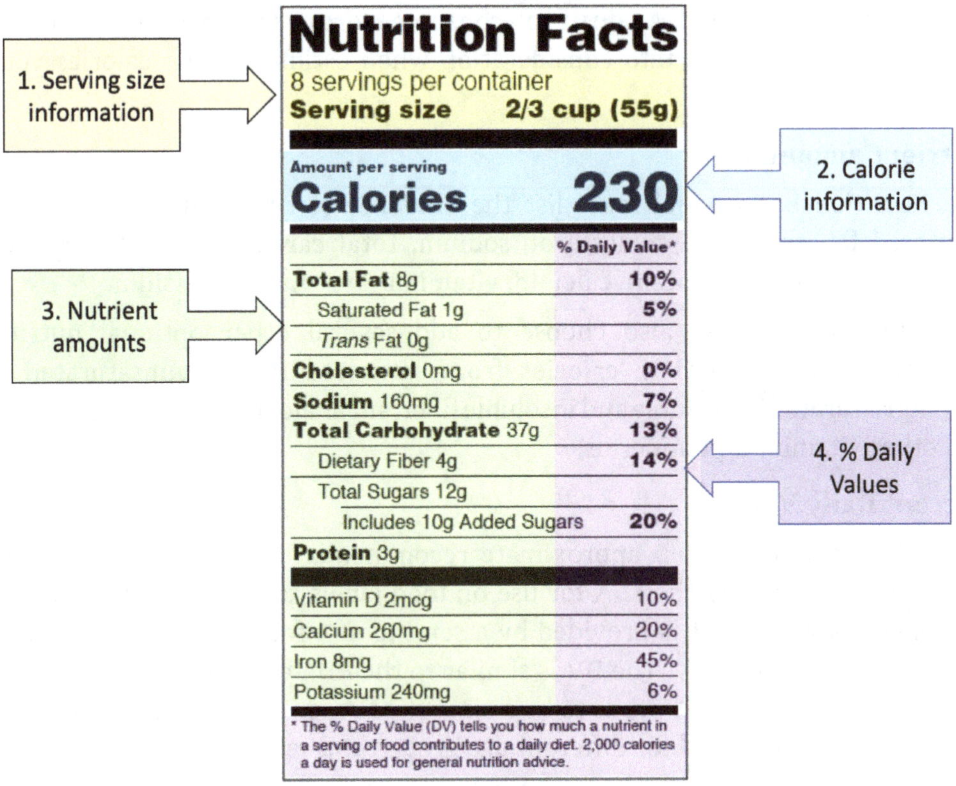

Figure 1.13. The four main sections of a Nutrition Facts label, highlighted in color.

Serving Size Information

It's fitting that serving size information is first on the Nutrition Facts panel, because all of the information that follows is based on it. The serving size of the food is the amount that is customarily eaten at one time, and all of the nutrition information on the label is based on one serving of the food. This section of the label also states the number of servings per container.[1]

It's important to note that you might not always eat one serving of a food; sometimes you might eat half of a serving, and sometimes you might eat two or more servings in one sitting. For example, if the label above is for a breakfast cereal, you might easily consume 1 ⅓ cups of cereal for breakfast. If you're interested in how many calories or nutrients you're consuming, you would need to double the nutrition values to accurately represent your breakfast, since the serving size is only ⅔ cup.

Calorie information

This section simply states the number of calories, or the amount of energy, provided in one serving of the food. Again, if you consume more or less than the serving size, you'll need to take that into consideration when estimating the calories you're consuming.

Nutrient amounts

The Nutrition Facts panel must list the amounts of these nutrients: total fat, saturated fat, trans fat, cholesterol, sodium, total carbohydrate, dietary fiber, sugars, added sugars, protein, calcium, vitamin D, iron, and potassium.[1]

Manufacturers may also choose to add several other optional nutrients or nutritional information: calories from saturated fat, polyunsaturated fat, monounsaturated fat, soluble and insoluble fiber, sugar alcohol, other carbohydrate, and other vitamins and minerals.

Percent Daily Values

The Daily Value (DV) is an approximate recommendation for daily intake for a nutrient, developed by the FDA for use on food labels so that consumers can see how much of a nutrient is provided by a serving of a food relative to about how much they need each day. The DV is similar to the RDA or AI, except that because it's used on food labels, it needs to be a simplified recommendation, with just one value rather than several for different age groups and sex, as found in the DRI.[3]

Most DVs are based on amounts for people age 4 years through adult, though there are DVs established for infants, toddlers, and pregnant and lactating women, and you'll see those used on food products specifically developed for those groups. Most of the time, the DV for a nutrient is the highest RDA or AI for the group it's intended for.

The value printed on the Nutrition Facts panel is the percent DV, which tells you how much one serving of the food contributes towards meeting the daily requirement for that nutrient.

The FDA uses the following definitions for interpreting the %DV on food labels:4

- 5%DV or less means the food is low in a nutrient.
- 10% to 19%DV means the food is a "good source" of a nutrient.
- 20%DV or greater means the food is high in a nutrient.

The DV is not as precise as the RDA, so while the %DV is useful for comparing food products or making quick judgements about the nutritional value of a food, it's better to use the RDA if you're looking for your individual nutrient requirements.

Putting the Nutrition Facts Panel to Work for you

How you use the Nutrition Facts on food labels depends on your dietary goals. If you're trying to reduce your saturated fat intake, you'll want to pay close attention to the %DV for saturated fat and try to choose foods with less than 5% DV for saturated fat. If you're watching your caloric intake, you'll want to pay attention to the calorie information. Regardless, always start by checking the serving size and comparing it to the amount you usually consume.

As an example of smart label reading, take a look at the two soup labels below. First, think about how much soup you would usually consume. There are two servings per can, but would you eat the entire can or just half of it? Many people would eat the whole can, and if that's you, you would want to double all of the calorie and nutrient information. Both soups provide 160 calories per one-cup serving, or 320 calories for the entire can.

Next, take a look at the sodium. Most Americans consume too much sodium, and this can increase the risk of developing high blood pressure. The regular soup has 680 mg or 28% of the DV for sodium. If you eat the entire can, that becomes 1360 mg or 56% DV. That's a lot of sodium. You can see how the reduced sodium soup might be the wiser choice here.

What's That Claim?

In addition to the FDA-required information on food labels, it's common for them to be peppered with claims about the nutrient content of the food and the purported health benefits of eating it. These claims are marketing tools for food manufacturers, and they're regulated by the FDA in an effort to ensure that they give the consumer accurate, science-based information about the food. Let's look at the different types of claims that you'll find on food packages.

Nutrition Facts

2 servings per container	
Serving size	**1 cup (237mL)**
Amount per serving	
Calories	**160**
	% Daily Value*
Total Fat 4g	**6%**
Saturated Fat 0.5g	**3%**
Trans Fat 0g	
Cholesterol 0mg	**0%**
Sodium 680mg	**28%**
Total Carbohydrate 24g	**8%**
Dietary Fiber 8g	**32%**
Total Sugars 5g	
Includes 0g Added Sugars	**0%**
Protein 7g	
Vitamin D 0mcg	0%
Calcium 29mg	3%
Iron 1mg	4%
Potassium 521mg	11%

* The % Daily Value (DV) tells you how much a nutrient in a serving of food contributes to a daily diet. 2,000 calories a day is used for general nutrition advice.

Nutrition Facts

2 servings per container	
Serving size	**1 cup (237mL)**
Amount per serving	
Calories	**160**
	% Daily Value*
Total Fat 4g	**6%**
Saturated Fat 0.5g	**3%**
Trans Fat 0g	
Cholesterol 0mg	**0%**
Sodium 340mg	**14%**
Total Carbohydrate 24g	**8%**
Dietary Fiber 8g	**32%**
Total Sugars 5g	
Includes 0g Added Sugars	**0%**
Protein 7g	
Vitamin D 0mcg	0%
Calcium 29mg	3%
Iron 1mg	4%
Potassium 521mg	11%

* The % Daily Value (DV) tells you how much a nutrient in a serving of food contributes to a daily diet. 2,000 calories a day is used for general nutrition advice.

Figure 1.14. Comparison of Nutrition Facts for a regular vegetable soup, and reduced sodium vegetable soup.

Nutrient Claims

Nutrient claims provide straight-forward information about the level of a nutrient or calories in the food, such as "fat-free," "low calorie," or "reduced sodium." Nutrient claims are regulated by the FDA, with very specific requirements for each one. For example, a food with a "low sodium" claim must have 140 mg of sodium or less per serving, whereas a food with a "reduced sodium" claim must have at least

25 percent less sodium than the standard product. You'll see claims that a food is "high in," "rich in," and "excellent source of" a nutrient, all of which mean that a serving of the food contains %20 DV or more. A "good source of" claim contains %19-10 DV of the nutrient.[1]

Figure 1.15. Examples of food packaging with nutrient claims. Can you spot them?

Health Claims

Health claims are statements on food packaging that link the food or a component in the food to reducing the risk of a disease. Health claims can be "authorized" or "qualified." Authorized health claims have stronger scientific evidence to back them than qualified health claims.[5]

As an example of an authorized health claim, a food that is low in sodium (per the FDA's definition of less than 140 mg per serving) can include the following claim on their packaging: "Diets low in sodium may reduce the risk of high blood pressure, a disease associated with many factors."[1]

For an authorized health claim to be approved by the FDA, the agency says "there must be significant scientific agreement (SSA) among qualified experts that the claim is supported by the totality of publicly available scientific evidence for a substance/disease relationship. The SSA standard is intended to be a strong standard that provides a high level of confidence in the validity of the substance/disease relationship."[5] In other words, the FDA requires a great deal of evidence before allowing food manufacturers to claim that their products can reduce the risk of a disease. As is evident in the low sodium claim, they also require careful language, such as "*may* reduce" (not definitely!) and "a disease associated with many factors" (as in, there are many other factors besides sodium that influence blood pressure, so a low sodium diet isn't a guaranteed way to prevent high blood pressure).

Qualified health claims have some evidence to support them, but not as much, so there's less certainty that these claims are true. The FDA reviews the evidence for a qualified claim and determines how it should be worded to convey the level of scientific certainty for it. Here's an example of a qualified health claim: "Scientific evidence suggests but does not prove that eating 1.5 ounces per day of most nuts [such as name of specific nut] as part of a diet low in saturated fat and cholesterol may reduce the risk of heart disease."

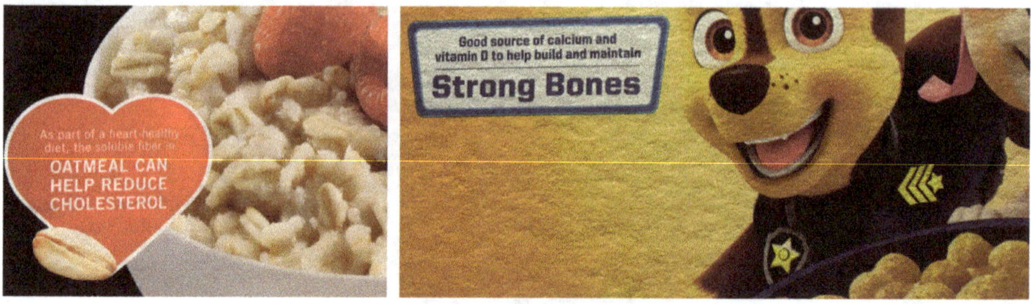

Figure 1.16. Examples of food packaging with authorized health claims. Can you spot them?

Structure-Function Claims

Health claims are very specific and precise in their language, and they convey the level of scientific certainty supporting them. In contrast, structure-function claims are intentionally vague statements about nutrients playing some role in health processes. Examples of structure-function claims are "calcium builds strong bones" and "fiber maintains bowel regularity." Note that these statements make no claims to prevent osteoporosis or treat constipation, because structure-function claims are not allowed to say that a food or nutrient will treat, cure, or prevent any disease.[6] They're allowed by the FDA, but not specifically approved or regulated, as long as their language stays within those rules.

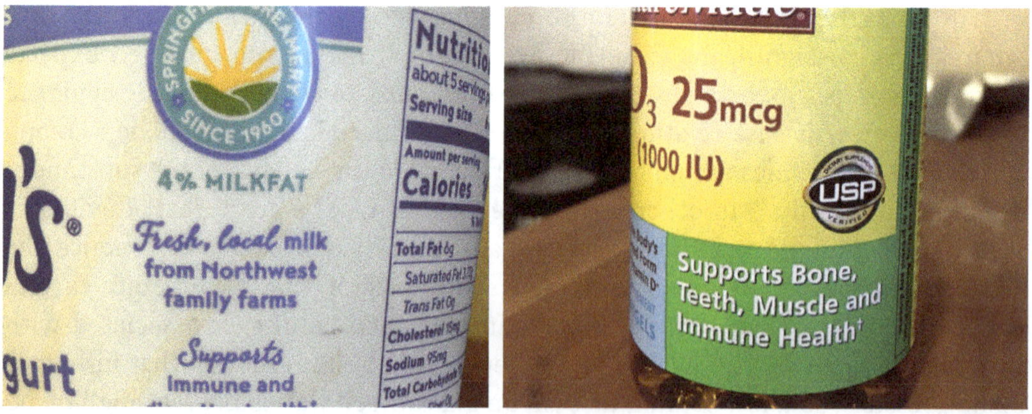

Figure 1.17. Examples of food and supplement packaging with structure-function claims. Can you spot them?

Structure-function claims were originally designed to be used on dietary supplements, but they can also be used on foods, and they're usually found on foods that are fortified with specific nutrients. They are marketing language, and because nutrients are involved in so many processes, they really don't mean much.

As you look at food labels, pay attention to what's shown on the front of the package compared with the back and side of the package. Nutrient and health claims are usually placed strategically on the front of the package, in large, colorful displays with other marketing messages, designed to sell you the product. But for consumers trying to decide which product to buy, you'll find the most useful information by turning the package around to read the Nutrition Facts panel and ingredients list. These parts of the label may appear more mundane, but if you understand how to read them, you'll find that they're rich in information.

RESOURCES FOR MAINTAINING A HEALTHY DIET

Good nutrition means eating the right foods, in the right amounts, to receive enough (but not too much) of the essential nutrients so that the body can remain free from disease, grow properly, work effectively, and feel its best. The phrase "you are what you eat" refers to the fact that the food you eat has cumulative effects on th body. And many of the nutrients obtained from food do become a part of us. For example, the protein and calcium found in milk can be used in the formation of bone. The foods we eat also impact how we feel—both today and in the future. Below we will discuss the key components of a healthy diet that will help prevent chronic disease (like heart disease and diabetes), maintain a healthy weight, and promote overall health.

Achieving A Healthy Diet

Achieving a healthy diet is a matter of balancing the quality and quantity of food that you eat to provide an appropriate combination of energy and nutrients. There are four key characteristics that make up a healthful diet:

- Adequacy
- Balance
- Moderation
- Variety

Adequacy

A diet is adequate when it provides sufficient amounts of calories and each essential nutrient, as well as fiber. Most Americans report not getting enough fruit, vegetables, whole grains or dairy, which may mean falling short in the essential vitamins and minerals found in these food groups, like Vitamin C, potassium, and calcium, as well as fiber.[1]

Balance

A balanced diet means eating a combination of foods from the different food groups, and because these food groups provide different nutrients, a balanced diet is likely to be adequate in nutrients. For example, vegetables are an important source of potassium, dietary fiber, folate, vitamin A, and vitamin C, whereas grains provide B vitamins (thiamin, riboflavin, niacin, and folate) and minerals (iron, magnesium, and selenium). No one food is more important than the other. It is the combination of all the different food groups (fruit, vegetables, grains, dairy, protein and fats/oils) that will ensure an adequate diet.

Moderation

Moderation means not eating to the extremes, neither too much nor too little of any one food or nutrient. Moderation means that small portions of higher-calorie, lower-nutrient foods like chips and candy can fit within a healthy diet. Including these types of foods can make healthy eating more enjoyable and also more sustainable. When eating becomes too extreme—where many foods are forbidden—this eating pattern is often short-lived until forbidden foods are overeaten. Too many food rules can lead to a cycle of restriction-deprivation-overeating-guilt.[2] For sustainable, long-term health benefits, it is important to give yourself permission to eat all foods.

Variety

Variety refers to consuming different foods within each of the food groups on a regular basis. Eating a varied diet helps to ensure that you consume adequate amounts of all essential nutrients required for health. One of the major drawbacks of a monotonous diet is the risk of consuming too much of some nutrients and not enough of others. Trying new foods can also be a source of pleasure—you never know what foods you might like until you try them.

Dietary Guidelines for Americans

The Dietary Guidelines are published and revised every five years jointly by the U.S. Department of Agriculture (USDA) and Health and Human Services (HHS) as a guide to healthy eating for Americans.[3]

Purpose

The purpose of the Dietary Guidelines is to give Americans evidence-based information on what to eat and drink to promote health and prevent chronic disease. Public health agencies, health care providers, and educational institutions all rely on Dietary Guidelines recommendations and strategies.[3] These agencies use the Dietary Guidelines to:

- Form the basis of federal nutrition policy and programs such as WIC and SNAP
- Help guide local, state, and national health promotion and disease prevention initiatives
- Inform various organizations and industries (for example, products developed and marketed by the food and beverage industry)

Process

Before HHS and the USDA release the new Dietary Guidelines, they assemble an Advisory Committee. This committee is composed of nationally recognized nutrition and medical researchers, academics, and practitioners. The Advisory Committee develops an Advisory Report that synthesizes current scientific and medical evidence in nutrition, which will then advise the federal government in the development of the new edition of the Dietary Guidelines.

The public also has opportunities to get involved in the development of these guidelines. The Advisory Committee holds a series of public meetings for hearing oral comments from the public, and the public also has opportunities to provide written comments to the Advisory Committee throughout the course of its work. After the Advisory Report is complete, the public has opportunities to respond with written comments and provide oral testimony at a public meeting.

2020-2025 Dietary Guidelines

The major topic areas of the Dietary Guidelines are:

- Follow a healthy dietary pattern at every life stage.
- Customize and enjoy nutrient-dense food and beverage choices to reflect personal preferences, cultural traditions, and budgetary considerations.
- Focus on meeting food group needs with nutrient-dense foods and beverages, and stay within calorie limits.
- Limit foods and beverages higher in added sugars, saturated fat, and sodium, and limit alcoholic beverages.0

Several nutrients are of special public health concern, including dietary fiber, calcium, potassium, and vitamin D. Inadequate intake of these nutrients is common among Americans and is associated with greater risk of chronic disease. People can increase their intake of these nutrients by shifting towards eating more vegetables, fruits, whole grains, dairy products, and beans. The Dietary Guidelines thus encourage the following nutrient-dense food choices:

Vegetables, including a variety of dark green, red and orange, legumes (beans and peas), starchy and other vegetables

- Fruits, especially whole fruits
- Grains, at least half of which are whole grains
- Dairy, including fat-free or low-fat milk, yogurt, and cheese, and/or fortified soy beverages and yogurt
- Protein foods, including seafood (8 or more ounces per week), lean meats and poultry, eggs, legumes (beans, peas, lentils), soy products, and nuts and seeds
- Oils, including those from plants, such as canola, corn, olive, peanut, safflower, soybean, and sunflower oils, and those present in whole foods such as nuts, seeds, seafood, olives, and avocados.

The DGA explains that most of an individual's daily caloric intake—about 85%—must be made up of nutrient-dense foods in order to meet nutrient requirements, leaving about 15% of calories available for other uses. Yet many Americans consume too much of foods with added sugars and saturated fat, in addition to excessive amounts of sodium and alcohol. Consuming too much of these dietary components is associated with development of chronic disease over time and can add calories without providing much in the way of beneficial nutrients. Therefore, the DGA recommends limiting the following:

- Added sugars – Consume less than 10 percent of calories per day from added sugars starting at age 2. (Avoid foods and beverages with added sugars for those younger than age 2.)
- Saturated fat – Consume less than 10 percent of calories per day from saturated fat starting at age 2.
- Sodium – Consume less than 2,300 milligrams (mg) per day of sodium – and even less for children younger than age 14.
- Alcoholic beverages – If alcohol is consumed, it should be consumed only in moderation (up to one drink per day for women and up to two drinks per day for men). Drinking less is better for health than drinking more.

The United States is not the only country that develops nutritional guidelines. The Food and Agriculture Organization of the United Nations has a website where you can search for dietary guidelines for different countries, such as Sweden's guidelines, illustrated below.

Figure 1.18. "Sweden's one-minute advice" by Food and Agriculture Organization of the United Nations (FAO)

The Swedish National Food Agency also has a great resource, "Find Your Way To Eat Greener, Not Too Much and Be Active" on how to put these guidelines into practice.

One way the USDA and other federal agencies implement the Dietary Guidelines is through MyPlate, which we will discuss next.

MyPlate (USDA Food Guide)

For many years, the U.S. government has been encouraging Americans to develop healthful dietary habits. For example, the food pyramid was introduced in 1992 as the symbol of healthy eating patterns for all Americans.

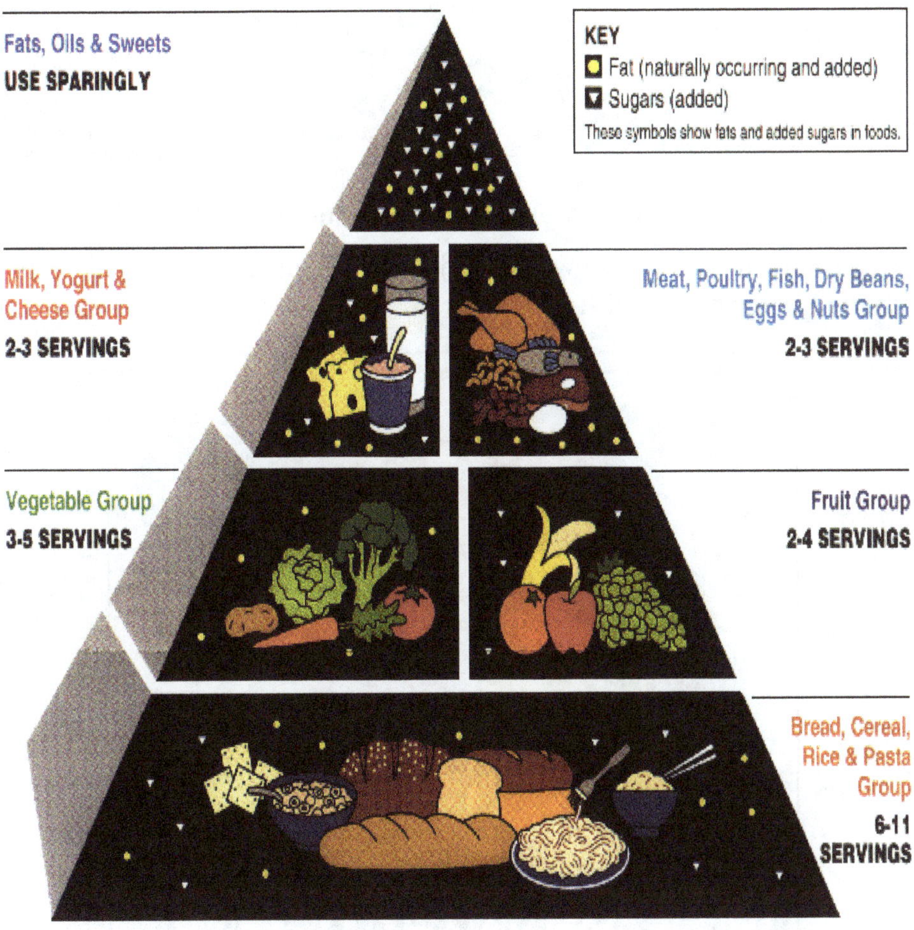

Figure 1.19. The food pyramid in 1992.

In 2005, the food pyramid was replaced with MyPyramid.

Figure 1.20. MyPyramid, introduced in 2005.

However, many felt this new pyramid was difficult to understand, so in 2011, the pyramid was replaced with MyPlate.

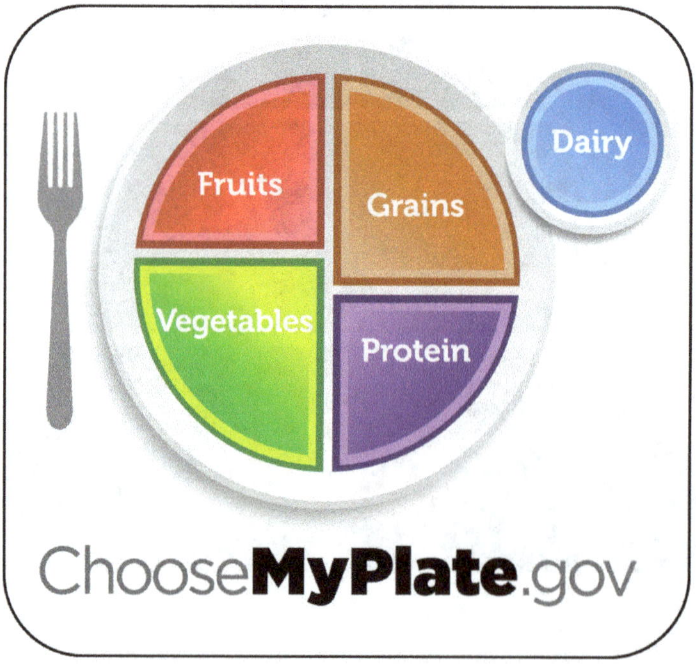

Figure 1.21. MyPlate, introduced in 2011.

MyPlate is a food guide to help Americans achieve the goals of the Dietary

Guidelines for Americans. For most people this means eating MORE:

- Whole grains
- Fruits
- Vegetables (especially dark green vegetables and red and orange vegetables)
- Legumes
- Seafood (to replace some meals of meat and poultry)
- Low-fat dairy

And LESS:

- Refined grains
- Added sugars
- Solid fats: saturated fats, trans fats, and cholesterol
- Sodium

Foods are grouped into 5 different groups based on their nutrient content. The following table summarizes the different food groups, examples of foods that fall within each group, and nutrients provided for each food group.

Table 1.4. A summary of MyPlate food groups, examples of foods that fall within each group, and nutrients provided for each food group.

Food Group	Example of Foods	Nutrients Provided
Grains	Whole grains: brown rice, oats, whole wheat bread, cereal and pasta, popcorn. Refined grains: typically tortillas, couscous, noodles, naan, pancakes (although sometimes these products can be whole grains too). For more foods and what counts as a cup check out:The Grain Group Food Gallery.	dietary fiber, several B vitamins (thiamin, riboflavin, niacin, and folate), and minerals (iron, magnesium, and selenium)
Vegetables	Dark green vegetables- broccoli, kale, and spinach. Red and orange vegetables- bell peppers, carrots and tomatoes. Starchy vegetables- corn, peas and potatoes. Beans and Peas- hummus, lentils and black beans. Other vegetables- asparagus, avocado, zucchini. For more foods and what counts as a cup check out: The Vegetable Group Food Gallery.	potassium, dietary fiber, folate, vitamin A, and vitamin C

	Fresh berries, melons, and other fruit as well as 100% fruit juice. Fruits can also be canned, frozen, or dried, and may be whole, cut-up, or pureed. For more foods and what counts as a cup check out: The Fruit Group Food Gallery.	potassium, dietary fiber, vitamin C, and folate
	Meats, poultry, seafood, beans and peas, eggs, nuts and seeds. For more foods and what counts as a cup check out: The Protein Group Food Gallery.	protein, B vitamins (niacin, thiamin, riboflavin, and B6), vitamin E, iron, zinc, and magnesium
	Milk, yogurt, cheese and calcium-fortified soy milk. Foods such as cream cheese, cream, and butter, are not part of the Dairy Group as they have little/no calcium (they count as a fat). For more foods and what counts as a cup check out: The Dairy Group Food Gallery.	calcium, potassium, vitamin D, and protein

This graphic summarizes serving sizes for each of the food groups:

Figure 1.22. Cup- and Ounce-Equivalents for Different Food Groups Within Myplate.

Planning a healthy diet using the MyPlate approach is not difficult:

- Fill half of your plate with a variety of fruits and vegetables, including red, orange, and dark green vegetables and fruits, such as kale, collard greens, tomatoes, sweet potatoes, broccoli, apples, oranges, grapes, bananas, blueberries, and strawberries in main and side dishes. Vary your choices to get the benefit of as many different vegetables and fruits as you can. One hundred percent fruit juice is also an acceptable choice as long as only half your fruit intake is replaced with juice.
- Fill a quarter of your plate with grains. Half of your daily grain intake should be whole grains such as 100 percent whole-grain cereals, breads, crackers, rice, and pasta. Read the ingredients list on food labels carefully to determine if a food is comprised of whole grains. We will discuss how to identify whole grains in more detail in later units.
- Select a variety of protein foods to improve nutrient intake and promote health benefits. Each week, be sure to include a nice array of protein sources in your diet, such as nuts, seeds, beans, legumes, poultry, soy, and seafood. The recommended consumption amount for seafood for adults is two -4ounce servings per week. When choosing meat, select lean cuts.

- If you enjoy drinking milk or eating dairy products, such as cheese and yogurt, choose low-fat or nonfat products. Low-fat and nonfat products contain the same amount of calcium and other essential nutrients as whole-milk products, but with much less fat and calories. Calcium, an important mineral for your body, is also found in lactose-free dairy products and fortified plant-based beverages, like soy milk. You can also get calcium from vegetables and other fortified foods and beverages.
- Oils are also important in your diet as they contain valuable essential fatty acids. Oils like canola oil also contain more healthful unsaturated fats compared to solid fats like butter. You can also get oils from whole foods like fish, avocados, and unsalted nuts and seeds. Although oils are essential for health, they do contain about 120 calories per tablespoon, so moderation is important.

Some people have criticized the Dietary Guidelines for Americans and MyPlate for being influenced by political and economic interests, as the meat and dairy industries and large food companies have a powerful lobbying presence that may override scientific consensus. When the 2020-2025 Dietary Guidelines were released in December of 2020, for example, they were criticized for failing to address sustainability, climate change, and the potential benefits of eating less meat and processed foods. In addition, although the 2020 Advisory Committee, made up of nutrition science experts, recommended stricter limits on added sugars and alcohol consumption, the final version of the Guidelines ignored these recommendations and stuck with the same advice about sugar and alcohol given in the 2015 version of the Guidelines.

Another guide for creating healthy, balanced meals comes from Harvard's School of Public Health. The Healthy Eating Plate (HEP) is based on the best available science and is not influenced by political or commercial pressures from food industry lobbyists.

The message of the HEP is similar to MyPlate in that the focus is on diet quality—encouraging nutrient-dense whole grains, fruits, vegetables, and beans. However, there are a few key differences between MyPlate and HEP. For example, MyPlate recommends 3 cups of dairy a day, whereas HEP recommends limiting dairy to 1-2 cups per day, and instead encourages non-dairy sources of calcium like collards, bok choy, fortified soy milk, and baked beans.

The HEP encourages protein sources from fish, poultry, beans or nuts, and it encourages consumers to limit red meat and avoid processed meat,since these foods raise the risk of heart disease, diabetes, and colon cancer. MyPlate, however, does not mention that red and processed meat should be limited.

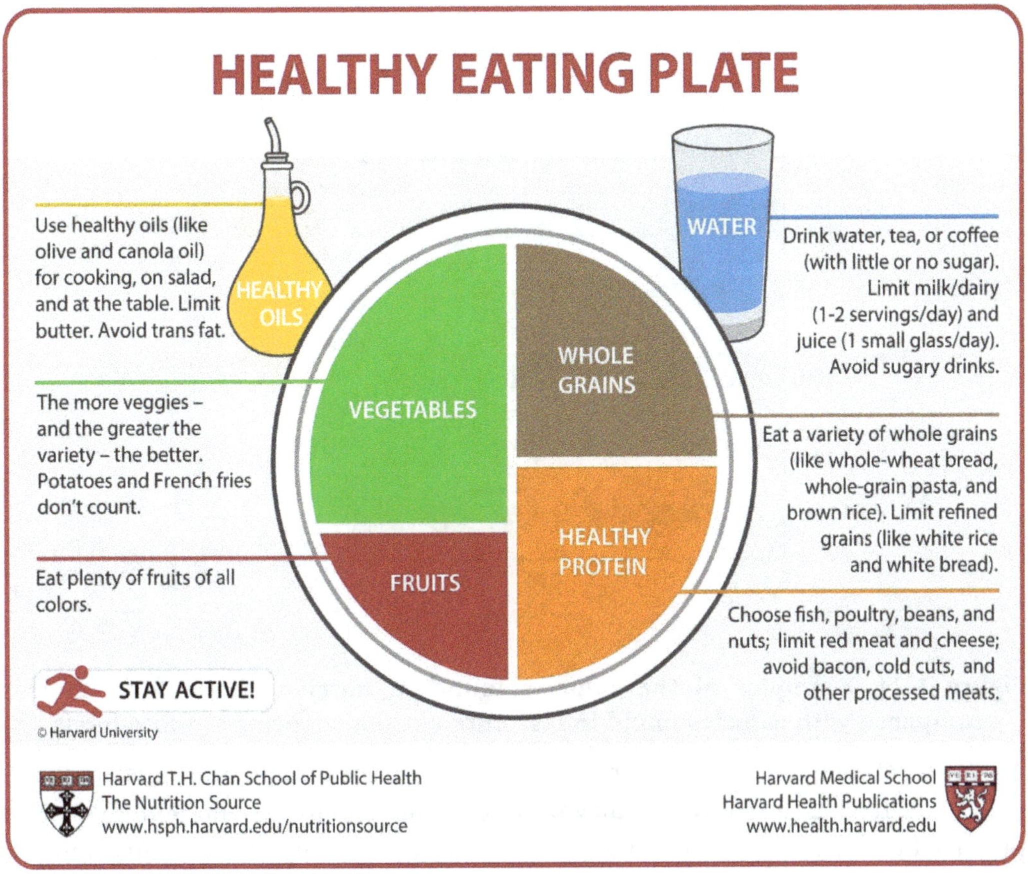

Figure 1.23. Harvard's Healthy Eating Plate.

Nutrient Density and Empty Calories

MyPlate encourages people to take a balanced approach and to eat a variety of nutrient-dense, whole foods. To help people control calories and prevent weight gain, the USDA promotes the concept of nutrient density and empty calories. Nutrient density is a measure of the nutrients that we're usually trying to consume more of—vitamins, minerals, fiber and protein—per calorie of food, coupled with little or no solid fats, added sugars, refined starches, and sodium. For example, in the screenshot below, a 90 percent lean 3-ounce ground beef patty is considered more nutrient-dense than a 75 percent lean patty. In the 90 percent lean patty, for 184 calories you get protein, iron, and other needed nutrients. On the other hand, the 75 percent lean patty has 236 calories, but the extra 52 calories add only solid fats and no other appreciable nutrients.

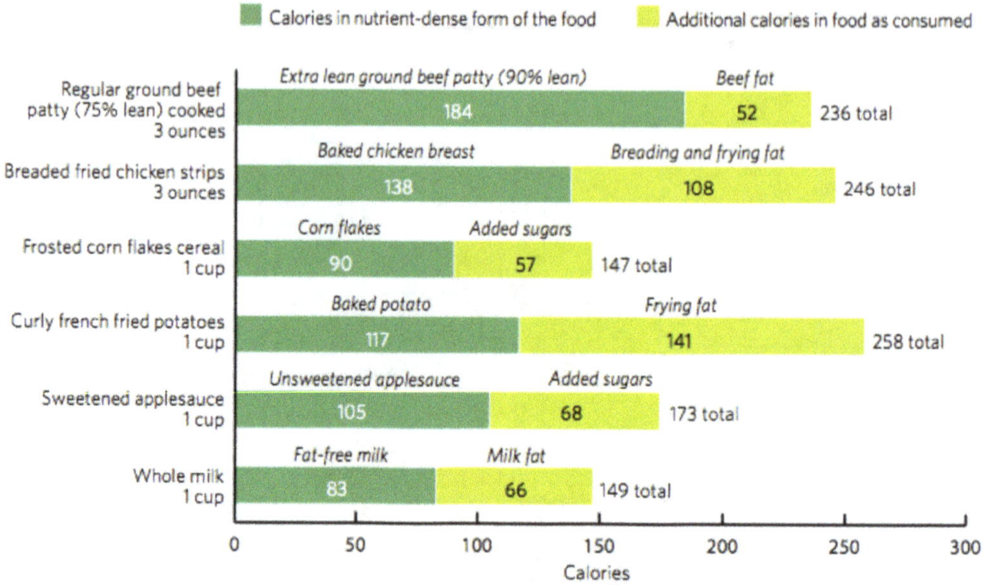

Figure 1.24. Examples of the calories found in nutrient-dense food choices compared with calories found in less nutrient-dense forms of these foods.

All vegetables, fruits, whole grains, seafood, eggs, beans and peas, unsalted nuts and seeds, fat-free and low-fat dairy products, and lean meats and poultry—when prepared with little or no added solid fats, and sugars—are nutrient-dense foods.

Foods become less nutrient dense when they contain empty calories—calories from solid fats and/or added sugars. Solid fats and added sugars add calories to a food but don't provide other nutrients. Foods with empty calories have fewer nutrients per calorie; therefore, they are less nutrient dense.

Examples of Foods High in Empty Calories

- Doughnuts, Cakes, Cookies
- Sweetened Cereals And Yogurt
- Sweetened Beverages
- High-Fat Meats
- Fried Foods
- Alcohol

Examples of nutrient-dense foods:

- Whole grains like brown rice, whole wheat bread and pasta, barley, and oatmeal

- Plain, Nonfat Milk and Yogurt
- Beans, Nuts and Seeds
- Lean Meats
- Whole, Fresh Fruits and Vegetables

You can choose more nutrient-dense foods by making small modifications to your current eating pattern. Examples include preparing foods with less fat by baking versus frying, purchasing items like cereals and fruits with less added sugar, and focusing on eating foods in their natural state versus adding a lot of extra fat, sugar and sodium.

Figure 1.25. Typical versus nutrient-dense foods.

Keep in mind that empty calories are not always a bad thing. In fact, empty calories can help promote eating more nutrient-dense foods. Adding a little fat and/or sugar to nutrient-dense foods can add flavor, making the food more enjoyable. A teaspoon of sugar in oatmeal, or a teaspoon of butter on steamed veggies is a great way to include empty calories. In these cases, the calories come packaged with other nutrients (since they are added to whole foods), whereas the empty calories in soda come with no other nutrients, only added sugar.

UNIT-2

SCIENCE AND LITERACY IN THE AREA OF NUTRITION

INTRODUCTION

If you follow nutrition science in the media for long enough, you'll start to see recurring themes. You'll see stories in the news about many of our favorite foods—like eggs, butter, coffee, and chocolate—that seem to flip-flop about whether these foods are good or bad for us. You'll notice seemingly eternal debates about whether dietary fat and carbohydrates are valuable macronutrients or villians. You'll watch as particular diets come in and out of fashion—and then back into fashion another decade or two later. And you'll see countless click-baity stories about the health benefits of eating so-called superfoods, or the dangers of eating others.

Even if you don't pay much attention to nutrition science in the news, you'll hear a ton of conflicting opinions and information just by talking to the people around you. Maybe your best friend has gone gluten-free, your dad is on a keto diet, and your coworker swears the Whole 30 diet has been life-changing. They're all trying to convince you to join them in their latest diets, but your head is swimming. They can't all be right, and you don't want to just follow the latest fad. You want to find accurate information that's based on solid scientific evidence. How can you identify it in a sea of conflicting and overwhelming information? Who can you trust?

It can be hard to filter through it all, especially when it's attached to strong opinions, emotions, and people trying to sell their product or point-of-view. And yet, we all need to make choices about what to eat, at the very least for ourselves, and often for others. You may have the responsibility of feeding family members in different stages of life, with different needs and preferences. And if you work in the health professions, you may have patients or clients who look to you as a source of reliable information about nutrition. Of course, the problem of conflicting and overwhelming information is not unique to nutrition; you'll find the same issue in many other health-related fields, and beyond.

Now, more than ever, it's essential to develop skills in ***information literacy***, including the ability to find information, evaluate whether it is accurate and useful, and apply it effectively. The purpose of this unit is to develop and hone your skills in information literacy as it applies to nutrition. You'll learn about the scientific method, because it forms the foundation of how we know what we know about nutrition. You'll learn about the different types of research studies and each of their advantages and limitations. We'll discuss various sources of information, such as scholarly and popular sources, how each of them can be useful in different ways, and how to evaluate them. We'll also discuss careers in nutrition and the different types of skills that you'll find among nutrition experts.

Unit Learning Objectives

After completing this unit, you should be able to:

- Identify the sequential steps of the scientific method, and understand the importance of reporting research results in peer-reviewed journals.
- Describe the different types of research studies used in nutrition, including the quality of evidence, advantages, and limitations of each.
- Describe some of the limitations of nutrition research, including the challenges of studying complex dietary patterns and the influence of industry funding.

- Describe differences between scholarly (peer-reviewed) and popular sources for nutrition information.
- Evaluate sources of nutrition information and distinguish between credible sources and junk science.
- Identify the qualifications of nutrition professionals and career opportunities in the field of nutrition.

THE SCIENTIFIC APPROACH

Similar to the method by which a police detective investigates a crime, nutritional scientists discover the health effects of food and nutrients by first making an observation and posing a question that they'd like to answer. Then they formulate a hypothesis, test their hypothesis through experiments, and finally interpret the results. After analyzing additional evidence from multiple sources, they may form a conclusion on whether the food suspect fits the claim. This organized process of inquiry used in forensic science, nutritional science, and every other science is called the scientific method.

The basis of what we know about nutrition is derived from research, and the scientific method underlies how research is conducted. The steps of the scientific method include:

- Observation/Question: The researcher first makes an observation and comes up with a research question to investigate.
- Hypothesis: The researcher formulates a hypothesis, or educated guess, that would explain the observation or question and that can be tested through scientific experiments.
- Experiment: The researcher designs and conducts an experiment. A good design takes into account what has been done previously. Thus, before beginning a new study, the researcher undertakes a thorough review of published research in order to ensure that their work advances the field.
- Analysis: The researcher collects and analyzes data that will either support or refute the hypothesis. If the hypothesis is not supported, researchers create a new hypothesis and conduct a new experiment. If the hypothesis is supported, researchers will design additional experiments to try to replicate the findings or to test them in different ways.
- Conclusion: After multiple experiments consistently support a hypothesis, researchers can offer a conclusion or theory.

Figure 2.1. The scientific method is a cyclical process, because it always leads to new observations and questions.

Through the scientific method, our knowledge of science builds continuously over time. No one study is enough to fully explain any one phenomenon, particularly in an area as complex as nutrition. Even experiments that go exactly as expected lead us to new questions to investigate. Science is also filled with surprises, both big and small. Experiments may not yield the results that we expect, but that can lead to new and important questions. And because scientists are human, they can make mistakes along the way or fail to acknowledge or test an important variable, which is why it's important that experiments be repeated and evaluated by other researchers along the way.

The history of nutrition is full of fascinating examples of the scientific method at work, such as the discovery that iodine is a nutrient. This story of scientific discovery began in 1811, when French chemist Bernard Courtois was isolating a substance called saltpeter, an ingredient needed to make gunpowder to be used by Napoleon's army. Part of his isolation procedure involved burning seaweed. When he did this, he observed the release of an intense violet vapor, which crystallized when he exposed it to a cold surface. He sent the violet crystals to an expert on gases, Joseph Gay-Lussac, who identified the crystal as a new element. It was named iodine, the Greek word for violet. The following scientific record is some of what took place in order to conclude that iodine is a nutrient.

Observation: Eating seaweed is a cure for goiter, an enlargement of the thyroid gland in the neck.

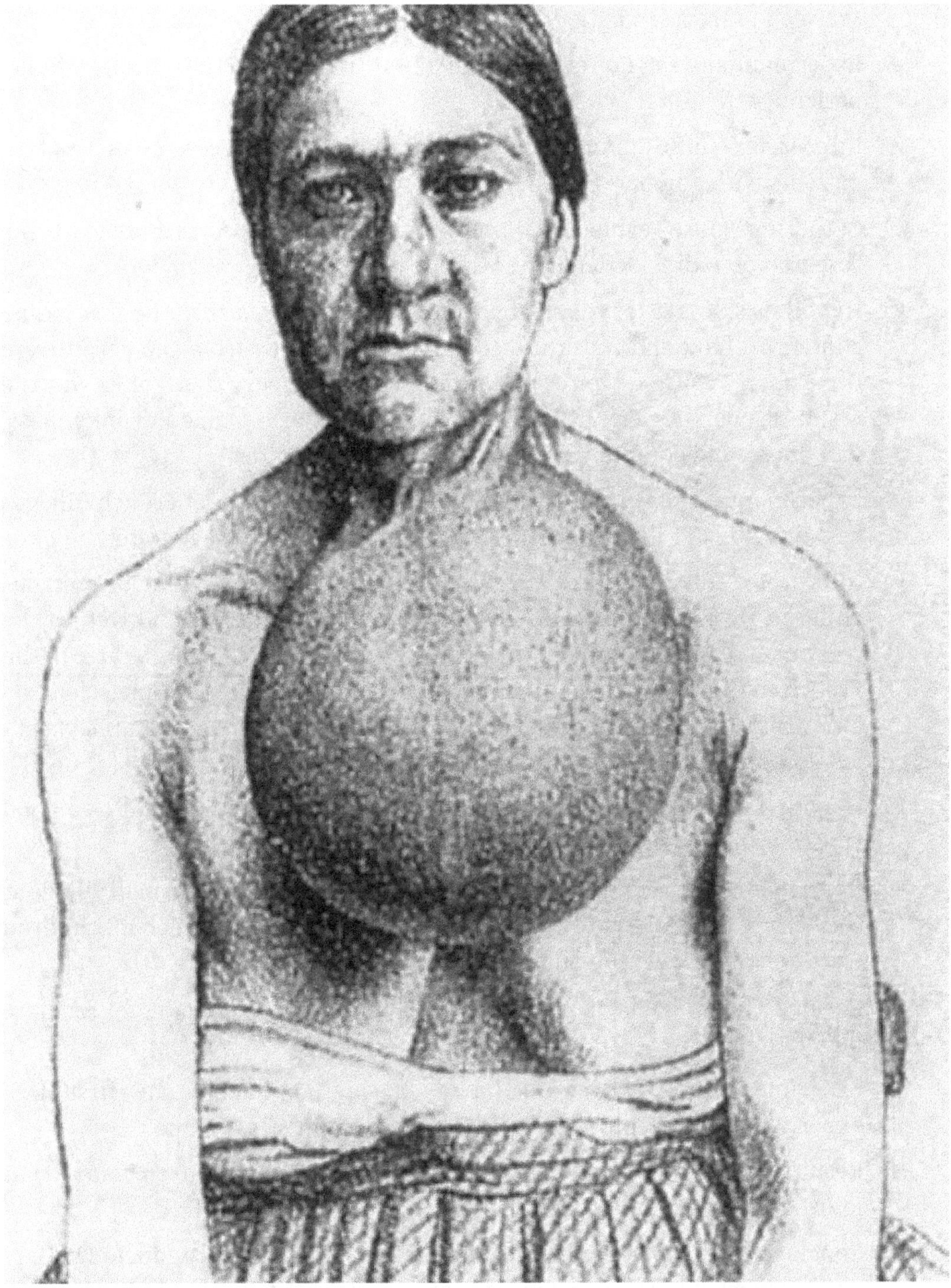

Figure 2.2. Large goiter in a woman from Bern, Switzerland.

- Hypothesis: In 1813, Swiss physician Jean-Francois Coindet hypothesized that seaweed contained iodine and that he could use iodine instead of seaweed to treat his patients.
- Experimental test: Coindet administered iodine tincture orally to his patients with goiter.
- Interpret results: Coindet's iodine treatment was successful in treating patients with goiter.
- Gathering more evidence: Many other physicians contributed to the research on iodine deficiency and goiter.
- Hypothesis: In 1851, French chemist Chatin proposed that the low iodine content in food and water of certain areas far away from the ocean were the primary cause of goiter and renounced the theory that goiter was the result of poor hygiene. (Physicians at the time also blamed drunkenness, dampness, and contaminated water as causes of goiter.)
- Experimental test: In the late 1860s, authorities in several French villages began giving out iodine tablets and salt in an effort to treat goiter.
- Interpret results: The program was effective, and 80 percent of goitrous children were cured. However, adults did not always respond well to the treatment, and because men with goiter were exempted from service in the French military, some people were opposed to treating it. Some scientists also insisted that goiter was caused by infectious disease, so iodine wasn't yet accepted as a means of preventing it.
- Hypothesis: In 1918, Swiss doctor Bayard proposed iodizing salt as a good way to treat areas endemic with goiter.
- Experimental test: Iodized salt was transported by mules to a small village at the base of the Matterhorn, where more than 75 percent of school children were goitrous. It was given to families to use for six months.
- Results: The iodized salt was beneficial in treating goiter in this remote population.
- Experimental test: Physician David Marine conducted the first U.S. experiment of treating goiter with iodized salt in Akron, Ohio.
- Results: This study conducted on over 4,000 school children found that iodized salt prevented goiter.
- Conclusions: Seven other studies similar to Marine's were conducted in Italy and Switzerland that also demonstrated the effectiveness of iodized salt in treating goiter. In 1924, U.S. public health officials initiated the program of iodizing salt and started eliminating the scourge of goiterism. Today,

more than 70 percent of American households use iodized salt, and many other countries have followed the same public health strategy to reduce the health consequences of iodine deficiency.

It took more than one hundred years from iodine's discovery as an effective treatment for goiter until public health programs recognized it as such. Although a lengthy process, the scientific method is a productive way to define essential nutrients and determine their ability to promote health and prevent disease.

Reporting scientific work

As we saw with the story of iodine research, scientists must share their findings in order for other researchers to expand and build upon their discoveries. Collaboration with other scientists when planning and conducting studies and analyzing results is important for scientific research. For this reason, communicating with peers and disseminating study results are important aspects of a scientist's work. Scientists can share results by presenting them at a scientific meeting or conference, but this approach can reach only the select few who are present. Instead, most scientists present their results in peer-reviewed manuscripts that are published in scientific journals.

Peer-reviewed manuscripts are scientific papers that are reviewed by a scientist's colleagues, or peers. These colleagues are qualified individuals, often experts in the same research area, who judge whether or not the scientist's work is suitable for publication. The process of peer review is a quality control step; its goal is to ensure that the research described in a scientific paper is original, significant, logical, and thorough. It's important to note that peer review doesn't mean a study is perfect or even good. Sometimes bad studies slip through peer review, but because they're published and other scientists read them, these are usually caught later and often retracted. Science is often messy and imperfect, but peer-review and publication of results are essential to its progress and ability to self-correct when people make mistakes.

RESEARCH STUDIES OF CERTAIN TYPES AND HOW TO INTERPRET THEM

The field of nutrition is dynamic, and our understanding and practices are always evolving. Nutrition scientists are continuously conducting new research and publishing their findings in peer-reviewed journals. This adds to scientific knowledge, but it's also of great interest to the public, so nutrition research often shows up in the news and other media sources. You might be interested in nutrition research to inform your own eating habits, or if you work in a health profession, so that you can give evidence-based advice to others. Making sense of science requires that you understand the types of research studies used and their limitations.

The hierarchy of nutrition evidence

Researchers use many different types of study designs depending on the question they are trying to answer, as well as factors such as time, funding, and ethical considerations. The study design affects how we interpret the results and the strength of the evidence as it relates to real-life nutrition decisions. It can be helpful to think about the types of studies within a pyramid representing a hierarchy of evidence, where the studies at the bottom of the pyramid usually give us the weakest evidence with the least relevance to real-life nutrition decisions, and the studies at the top offer the strongest evidence, with the most relevance to real-life nutrition decisions.

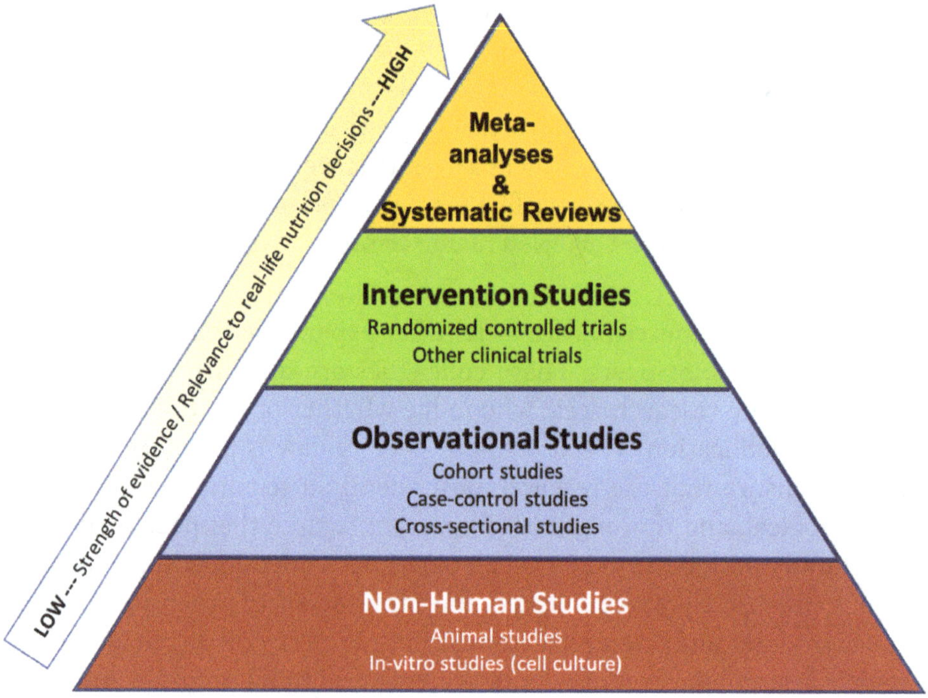

Figure 2.3. The hierarchy of evidence shows types of research studies relative to their strength of evidence and relevance to real-life nutrition decisions, with the strongest studies at the top and the weakest at the bottom.

The pyramid also represents a few other general ideas. There tend to be more studies published using the methods at the bottom of the pyramid, because they require less time, money, and other resources. When researchers want to test a new hypothesis, they often start with the study designs at the bottom of the pyramid, such as in vitro, animal, or observational studies. Intervention studies are more expensive and resource-intensive, so there are fewer of these types of studies conducted. But they also give us higher quality evidence, so they're an important next step if observational and non-human studies have shown promising results.

Meta-analyses and systematic reviews combine the results of many studies already conducted, so they help researchers summarize scientific knowledge on a topic.

Non-human studies: in vitro & animal studies

The simplest form of nutrition research is an in vitro study. In vitro means "within glass," (although plastic is used more commonly today) and these experiments are conducted within flasks, dishes, plates, and test tubes. These studies are performed on isolated cells or tissue samples, so they're less expensive and time-intensive than animal or human studies. In vitro studies are vital for zooming in on biological mechanisms, to see how things work at the cellular or molecular level. However, these studies shouldn't be used to draw conclusions about how things work in humans (or even animals), because we can't assume that the results will apply to a whole, living organism.

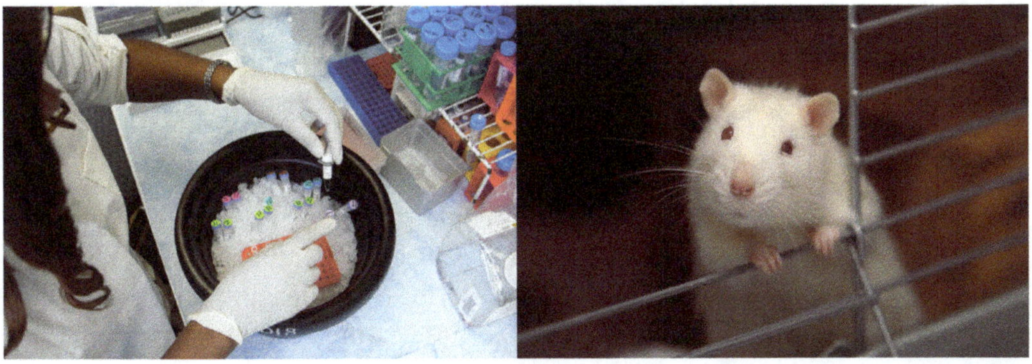

Animal studies are one form of in vivo research, which translates to "within the living." Rats and mice are the most common animals used in nutrition research. Animals are often used in research that would be unethical to conduct in humans. Another advantage of animal dietary studies is that researchers can control exactly what the animals eat. In human studies, researchers can tell subjects what to eat and even provide them with the food, but they may not stick to the planned diet. People are also not very good at estimating, recording, or reporting what they eat and in what quantities. In addition, animal studies typically do not cost as much as human studies.

There are some important limitations of animal research. First, an animal's metabolism and physiology are different from humans. Plus, animal models of disease (cancer, cardiovascular disease, etc.), although similar, are different from human diseases. Animal research is considered preliminary, and while it can be very important to the process of building scientific understanding and informing the types of studies that should be conducted in humans, animal studies shouldn't be considered relevant to real-life decisions about how people eat.

Observational Studies

Observational studies in human nutrition collect information on people's dietary patterns or nutrient intake and look for associations with health outcomes. Observational studies do not give participants a treatment or intervention; instead, they look at what they're already doing and see how it relates to their health. These types of study designs can only identify correlations (relationships) between nutrition and health; they can't show that one factor *causes* another. (For that, we need intervention studies, which we'll discuss in a moment.) Observational studies that describe factors correlated with human health are also called ***epidemiological studies.***[1]

One example of a nutrition hypothesis that has been investigated using observational studies is that eating a Mediterranean diet reduces the risk of developing cardiovascular disease. (A Mediterranean diet focuses on whole grains, fruits and vegetables, beans and other legumes, nuts, olive oil, herbs, and spices. It includes small amounts of animal protein (mostly fish), dairy, and red wine.[2]) There are three main types of observational studies, all of which could be used to test hypotheses about the Mediterranean diet:

- Cohort studies follow a group of people (a cohort) over time, measuring factors such as diet and health outcomes. A cohort study of the Mediterranean diet would ask a group of people to describe their diet, and then researchers would track them over time to see if those eating a Mediterranean diet had a lower incidence of cardiovascular disease.
- Case-control studies compare a group of cases and controls, looking for differences between the two groups that might explain their different health outcomes. For example, researchers might compare a group of people with cardiovascular disease with a group of healthy controls to see whether there were more controls or cases that followed a Mediterranean diet.
- Cross-sectional studies collect information about a population of people at one point in time. For example, a cross-sectional study might compare the dietary patterns of people from different countries to see if diet correlates with the prevalence of cardiovascular disease in the different countries.

Prospective cohort studies, which enroll a cohort and follow them into the future, are usually considered the strongest type of observational study design. Retrospective studies look at what happened in the past, and they're considered weaker because they rely on people's memory of what they ate or how they felt in the past. There are several well-known examples of prospective cohort studies that have described important correlations between diet and disease:

- Framingham Heart Study: Beginning in 1948, this study has followed the residents of Framingham, Massachusetts to identify risk factors for heart disease.
- Health Professionals Follow-Up Study: This study started in 1986 and enrolled 51,529 male health professionals (dentists, pharmacists, optometrists, osteopathic physicians, podiatrists, and veterinarians), who complete diet questionnaires every 2 years.
- Nurses Health Studies: Beginning in 1976, these studies have enrolled three large cohorts of nurses with a total of 280,000 participants. Participants have completed detailed questionnaires about diet, other lifestyle factors (smoking and exercise, for example), and health outcomes.

Observational studies have the advantage of allowing researchers to study large groups of people in the real world, looking at the frequency and pattern of health outcomes and identifying factors that correlate with them. But even very large observational studies may not apply to the population as a whole. For example, the Health Professionals Follow-Up Study and the Nurses Health Studies include people with above-average knowledge of health. In many ways, this makes them ideal study subjects, because they may be more motivated to be part of the study and to fill out detailed questionnaires for years. However, the findings of these studies may not apply to people with less baseline knowledge of health.

We've already mentioned another important limitation of observational studies—that they can only determine correlation, not causation. A prospective cohort study that finds that people eating a Mediterranean diet have a lower incidence of heart disease can only show that the Mediterranean diet is *correlated* with lowered risk of heart disease. It can't show that the Mediterranean diet directly prevents heart disease. Why? There are a huge number of factors that determine health outcomes such as heart disease, and other factors might explain a correlation found in an observational study. For example, people who eat a Mediterranean diet might also be the same kind of people who exercise more, sleep more, have higher income (fish and nuts can be expensive!), or be less stressed. These are called confounding factors; they're factors that can affect the outcome in question (i.e., heart disease) and also vary with the factor being studied (i.e., Mediterranean diet).

Intervention Studies

Intervention studies, also sometimes called experimental studies or clinical trials, include some type of treatment or change imposed by the researcher. Examples of interventions in nutrition research include asking participants to change their diet, take a supplement, or change the time of day that they eat. Unlike observational studies, intervention studies can provide evidence of cause and effect, so they are higher in the hierarchy of evidence pyramid.

The gold standard for intervention studies is the randomized controlled trial (RCT). In an RCT, study subjects are recruited to participate in the study. They are then randomly assigned into one of at least two groups, one of which is a control group (this is what makes the study controlled). In an RCT to study the effects of the Mediterranean diet on cardiovascular disease development, researchers might ask the control group to follow a low-fat diet (typically recommended for heart disease prevention) and the intervention group to eat a Mediterrean diet. The study would continue for a defined period of time (usually years to study an outcome like heart disease), at which point the researchers would analyze their data to see if more people in the control or Mediterranean diet had heart attacks or strokes. Because the treatment and control groups were randomly assigned, they should be alike in every other way except for diet, so differences in heart disease could be attributed to the diet. This eliminates the problem of confounding factors found in observational research, and it's why RCTs can provide evidence of causation, not just correlation.

Imagine for a moment what would happen if the two groups weren't randomly assigned. What if the researchers let study participants choose which diet they'd like to adopt for the study? They might, for whatever reason, end up with more overweight people who smoke and have high blood pressure in the low-fat diet group, and more people who exercised regularly and had already been eating lots of olive oil and nuts for years in the Mediterranean diet group. If they found that the Mediterranean diet group had fewer heart attacks by the end of the study, they would have no way of knowing if this was because of the diet or because of the underlying differences in the groups. In other words, without randomization, their results would be compromised by confounding factors, with many of the same limitations as observational studies.

In an RCT of a supplement, the control group would receive a placebo—***a*** "fake" treatment that contains no active ingredients, such as a sugar pill. The use of a placebo is necessary in medical research because of a phenomenon known as the placebo effect. The placebo effect results in a beneficial effect because of a subject's *belief* in the treatment, even though there is no treatment actually being administered.

For example, imagine an athlete who consumes a sports drink and then runs 100 meters in 11.0 seconds. On a different day, under the exact same conditions, the athlete is given a Super Duper Sports Drink and again runs 100 meters, this time in 10.5 seconds. But what the athlete didn't know was that the Super Duper Sports Drink was the same as the regular sports drink—it just had a bit of food coloring added. There was nothing different between the drinks, but the athlete *believed* that the Super Duper Sports Drink was going to help him run faster, so he did. This improvement is due to the placebo effect. Ironically, a study similar to this example was published in 2015, demonstrating the power of the placebo effect on athletic performance.[3]

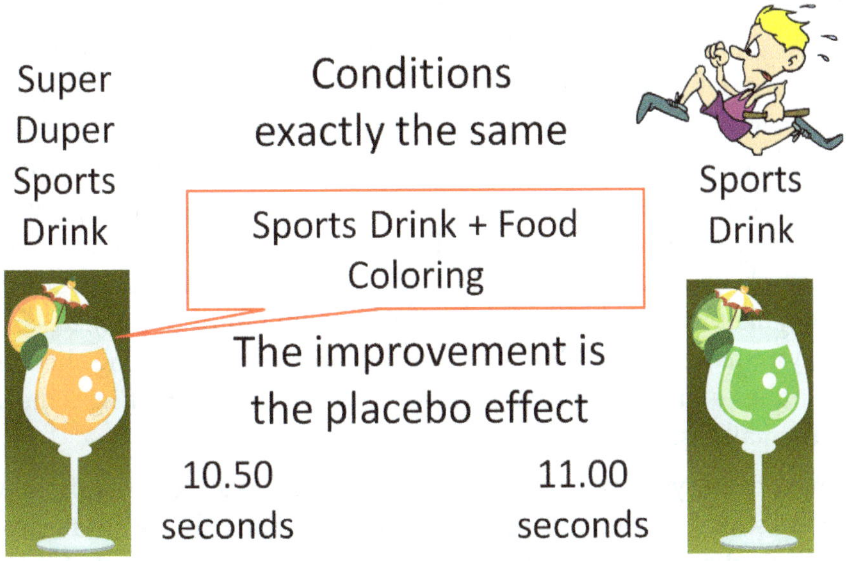

Figure 2.4. An example of the placebo effect

Blinding is a technique to prevent bias in intervention studies. In a study without blinding, the subject and the researchers both know what treatment the subject is receiving. This can lead to bias if the subject or researcher have expectations about the treatment working, so these types of trials are used less frequently. It's best if a study is double-blind, meaning that neither the researcher nor the subject know what treatment the subject is receiving. It's relatively simple to double-blind a study where subjects are receiving a placebo or treatment pill, because they could be formulated to look and taste the same. In a single-blind study, either the researcher or the subject knows what treatment they're receiving, but not both. Studies of diets—such as the Mediterranean diet example—often can't be double-blinded because the study subjects know whether or not they're eating a lot of olive oil and nuts. However, the researchers who are checking participants' blood pressure or evaluating their medical records could be blinded to their treatment group, reducing the chance of bias.

Like all studies, RCTs and other intervention studies do have some limitations. They can be difficult to carry on for long periods of time and require that participants remain compliant with the intervention. They're also costly and often have smaller sample sizes. Furthermore, it is unethical to study certain interventions. (An example of an unethical intervention would be to advise one group of pregnant mothers to drink alcohol to determine its effects on pregnancy outcomes, because we know that alcohol consumption during pregnancy damages the developing fetus.)

Meta-analyses and systematic reviews

At the top of the hierarchy of evidence pyramid are systematic reviews and meta-analyses. You can think of these as "studies of studies." They attempt to combine all of the relevant studies that have been conducted on a research question and summarize their overall conclusions. Researchers conducting a systematic review formulate a research question and then systematically and independently identify, select, evaluate, and synthesize all high-quality evidence that relates to the research question. Since systematic reviews combine the results of many studies, they help researchers produce more reliable findings. A meta-analysis is a type of systematic review that goes one step further, combining the data from multiple studies and using statistics to summarize it, as if creating a mega-study from many smaller studies.

However, even systematic reviews and meta-analyses aren't the final word on scientific questions. For one thing, they're only as good as the studies that they include. The Cochrane Collaboration is an international consortium of researchers who conduct systematic reviews in order to inform evidence-based healthcare, including nutrition, and their reviews are among the most well-regarded and rigorous in science. For the most recent Cochrane review of the Mediterranean diet and cardiovascular disease, two authors independently reviewed studies published on this question. Based on their inclusion criteria, 30 RCTs with a total of 12,461 participants were included in the final analysis. However, after evaluating and combining the data, the authors concluded that "despite the large number of included trials, there is still uncertainty regarding the effects of a Mediterranean-style diet on cardiovascular disease occurrence and risk factors in people both with and without cardiovascular disease already." Part of the reason for this uncertainty is that different trials found different results, and the quality of the studies was low to moderate. Some had problems with their randomization procedures, for example, and others were judged to have unreliable data. That doesn't make them useless, but it adds to the uncertainty about this question, and uncertainty pushes the field forward towards more and better studies. The Cochrane review authors noted that they found seven ongoing trials of the Mediterranean diet, so we can hope that they'll add more clarity to this question in the future.[5]

Science is an ongoing process. It's often a slow process, and it contains a lot of uncertainty, but it's our best method of building knowledge of how the world and human life works. Many different types of studies can contribute to scientific knowledge. None are perfect—all have limitations—and a single study is never the final word on a scientific question. Part of what advances science is that researchers are constantly checking each other's work, asking how it can be improved and what new questions it raises.

NUTRITION SCIENCE: HEALTHY SKEPTICISM

By this point in the unit, you should understand the types of studies used in nutrition research and the quality of evidence each can provide. As you sift through studies, there are a few limitations of nutrition research that are always worth keeping in mind:

- Challenges in nutrition research: There are some inherent challenges to researching what people eat and how it affects their health. This fact limits the quality of evidence and stirs controversy in nutrition research.
- Influence of food industry funding: Because government funding for nutrition research is limited, much of it is funded by the food industry. Industry's primary goal is to sell more of their products, so they have a conflict of interest that can result in bias in the way they design studies and interpret the results.

Being alert to these two things means always examining nutrition research with a healthy sense of skepticism. Let's take a closer look at these two issues in nutrition science.

Challenges in Nutrition Research

How does the food we eat affect our health? This question is exceedingly difficult to answer with certainty. We all need to eat every day, but we can choose from a huge array of possible foods in different combinations. And it's probably not what we eat on any given day that matters, but what we eat over months and years and decades—our long-term eating patterns—that matter to our long-term health.

Imagine that you're a nutrition researcher, and you've made the observation that over the last 50 years in the U.S., people have been consuming more and more processed foods (foods made with refined ingredients and industrial processes, usually with the addition of sugar, fat, and/or salt). You hypothesize that processed foods are contributing to obesity, which has also increased over the last 50 years. You might first test your hypothesis in animal studies by feeding mice and rats a buffet of potato chips, soda, and Twinkies, and measuring changes in their body weight. You might find that the animals do, in fact, gain weight on this diet. However, you know that what is true in rodents isn't always true in humans, and you'll need to study humans in order to understand the role of processed foods in the obesity epidemic.

Your next step might be to conduct an observational study, the most common type of study design in human nutrition research. For example, you might do a cross-sectional study where you compare groups of people who eat a lot of processed foods with those who eat very little. Or you might conduct a prospective cohort study in which you ask people how much processed foods they eat and then follow them over time, looking for correlations between processed food consumption and their body composition.

These types of studies have been conducted, and they've found correlations between consumption of processed foods and obesity. For example, a cross-sectional study published in 2018 compared the consumption of processed foods and the prevalence of obesity in 19 European countries and found that countries where people eat more processed foods also have a greater prevalence of obesity.[1] A prospective cohort study published in 2016 followed nearly 8,500 university graduates in Spain and found that those who ate more processed foods were more likely to be overweight or obese 9 years later.[2]

From these results, can we conclude that eating more processed food *causes* weight gain? Nope. It's a tempting conclusion, but this brings us to the first major problem with nutrition research: Observational studies can only show that two variables (eating processed foods and obesity, in this case) are ***correlated***, not that one causes the other. This distinction is especially important in nutrition because diet is intertwined with many other lifestyle and socioeconomic factors that also affect health outcomes. For example, people who eat more processed food might also eat fewer fruits and vegetables, exercise or sleep less, have more stress, or have less access to preventative healthcare. These are just a few of the confounding factors that could explain the observed correlation between processed food consumption and weight gain. Weight gain might have nothing to do with processed food and instead be driven by one or all of these factors, or others that we haven't considered.

The second major problem with observational nutrition research is that it's difficult to accurately quantify what and how much people eat, especially over long periods of time. Epidemiological studies usually rely on questionnaires that ask people to remember how much food they ate, but people are notoriously bad at remembering this type of information, and sometimes we fudge the truth. For example, you might remember that you had a cup of coffee but forget that you added cream, completely forget about a mid-morning muffin snack, or guess that you ate 2 cups of veggies when it was closer to 1.5 cups. And many diet questionnaires, called food frequency questionnaires, ask people to recollect and mentally average their food and beverage intake for the last 12 months, not just yesterday. The image below shows part of a page from a -24page National Health and Nutrition Examination Survey (NHANES) food questionnaire, a national survey often used for research on country-wide dietary patterns. As you can see, these questions are detailed, and there's plenty of room for small errors to accumulate.

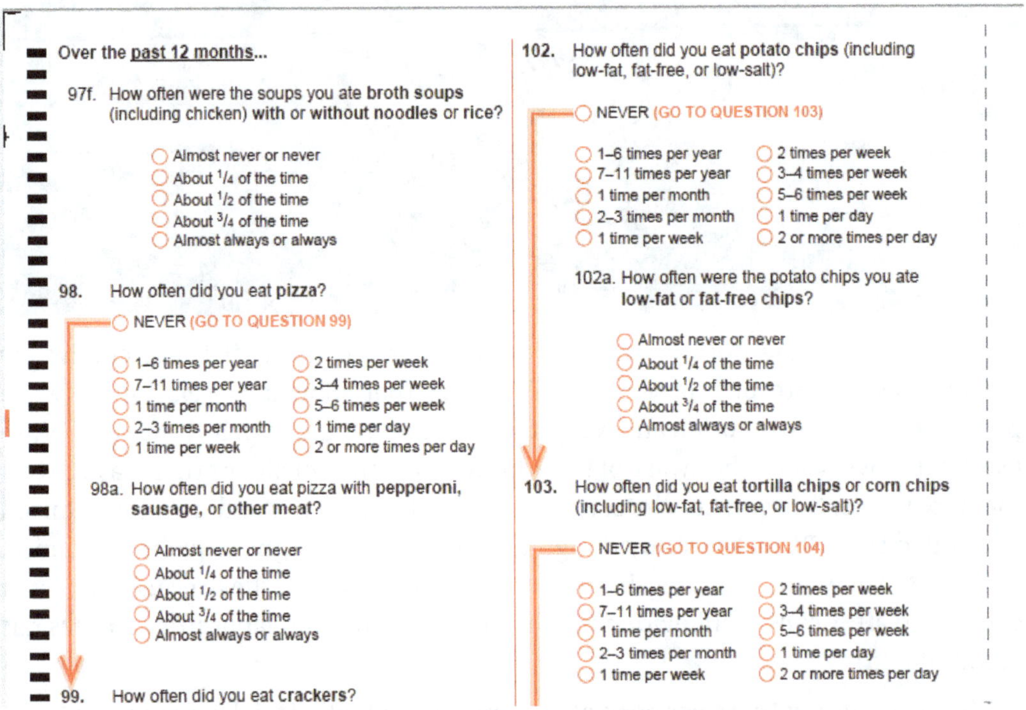
Over the past 12 months...

97f. How often were the soups you ate **broth soups** (including chicken) **with** or **without noodles** or rice?

- Almost never or never
- About 1/4 of the time
- About 1/2 of the time
- About 3/4 of the time
- Almost always or always

98. How often did you eat **pizza**?

- NEVER (GO TO QUESTION 99)
- 1–6 times per year
- 7–11 times per year
- 1 time per month
- 2–3 times per month
- 1 time per week
- 2 times per week
- 3–4 times per week
- 5–6 times per week
- 1 time per day
- 2 or more times per day

98a. How often did you eat pizza with **pepperoni, sausage**, or other meat?

- Almost never or never
- About 1/4 of the time
- About 1/2 of the time
- About 3/4 of the time
- Almost always or always

99. How often did you eat **crackers**?

102. How often did you eat **potato chips** (including low-fat, fat-free, or low-salt)?

- NEVER (GO TO QUESTION 103)
- 1–6 times per year
- 7–11 times per year
- 1 time per month
- 2–3 times per month
- 1 time per week
- 2 times per week
- 3–4 times per week
- 5–6 times per week
- 1 time per day
- 2 or more times per day

102a. How often were the potato chips you ate **low-fat or fat-free chips**?

- Almost never or never
- About 1/4 of the time
- About 1/2 of the time
- About 3/4 of the time
- Almost always or always

103. How often did you eat **tortilla chips** or **corn chips** (including low-fat, fat-free, or low-salt)?

- NEVER (GO TO QUESTION 104)
- 1–6 times per year
- 7–11 times per year
- 1 time per month
- 2–3 times per month
- 1 time per week
- 2 times per week
- 3–4 times per week
- 5–6 times per week
- 1 time per day
- 2 or more times per day

Figure 2.5. A sample page from the NHANES Food Questionnaire.

A third challenge in nutrition research is that diet is just so complex. Stanford physician and researcher John Ioannidis, a frequent critic of observational nutrition research, described the complexity of diet in a 2018 editorial published in JAMA: "Individuals consume thousands of chemicals in millions of possible daily combinations. For instance, there are more than 250,000 different foods and even more potentially edible items, with 300,000 edible plants alone." He also points out that how an individual responds to a particular dietary pattern can be influenced by genetics, age, and the way they metabolize nutrients. "Disentangling the potential influence on health outcomes of a single dietary component from these other variable is challenging, if not impossible," Ioannidis wrote.

Returning to the question of processed foods, all three of these challenges impact how we interpret observational studies that show a correlation between processed food consumption and weight gain. It doesn't mean that these studies are useless, but we want to be aware of their limitations and consider other ways to test the hypothesis. One way to overcome these challenges is to conduct a randomized controlled trial (RCT), the study design that gives us the highest quality evidence. RCTs are time and funding-intensive experiments, so they're usually only conducted after consistent evidence from observational and laboratory studies has accumulated.

It turns out that there has been a randomized controlled trial of processed foods and weight gain. It was funded by the National Institutes of Health and published in 2019 in the journal *Cell Metabolism*.[4] In this study, 20 participants lived in the NIH's Clinical Center for one month, where they consumed only processed foods for two weeks and only unprocessed foods for another two weeks, and they could eat as much or as little as they liked during each of these periods. The diets were carefully designed by dietitians so that they were matched in calories, sugar, fat, fiber, and macronutrients, and the exact amounts consumed by the participants were measured every day (solving the problem of measuring diet complexity and accurately describing what and how much people eat every day).

The study found that people ate about 500 kilocalories more and gained about a pound per week when they were eating processed foods. This study design could show causation, not just correlation, because the other nutritional factors like calories, sugar, fat, fiber, and macronutrients were held constant, and the diets were tested in the same people, so other factors such as genetics, sleep, stress, and exercise were constant between the two types of diets. (This was an example of a crossover randomized controlled trial, in which each subject serves as their own control, and they completed the processed and unprocessed phases of the trial in random order.)

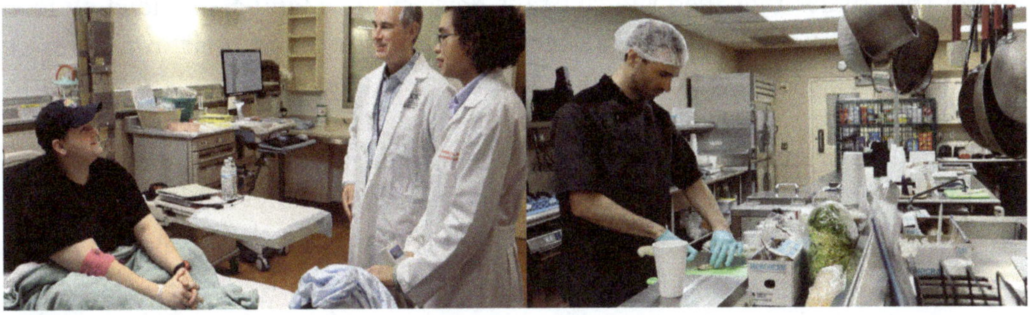

Figure 2.6. At left, researchers Kevin Hall and Stephanie Chung talk with one of the processed foods trial participants at the NIH Clinical Center, an inpatient facility where participants lived for the duration of the study. At right, an NIH worker prepares meals for participants in the center's kitchen. All meals were provided for study participants to carefully control their diet during the trial.

This study suggested something very important—that food processing causes people to eat more food and gain weight. However, even the best study design has limitations. For one thing, this study was small (just 20 participants), and it only lasted for two weeks, so we don't know if the findings apply to the general population over a lifetime of complex, ever-changing diets. The next steps will be to try to repeat the study in another group of people to see if the finding holds and to design studies to figure out why processed foods cause increased caloric intake.

As you evaluate nutrition research, especially observational studies, keep in mind the inherent challenges of nutrition research and look for randomized controlled trials that can help solve those challenges. Even for randomized controlled trials, consider their limitations, and know that one study is never enough to fully answer a question in the complex field of nutrition.

Industry Influence: Follow The Money

Understanding how diet influences health is a pressing need. By some estimates, a suboptimal diet is the single greatest changeable risk factor contributing to death and disability worldwide, and in the United States, the cost of diet-related chronic diseases are estimated to be as high as $1 trillion each year.[5] Yet, for all its importance, nutrition science has long suffered from a lack of government investment, with only about 5 percent of the National Institutes of Health (NIH) budget, or 1.8$ billion, directed towards research on how the foods we eat affect our health, according to a 2019 investigation by Politico. "In 2018, NIH funding for cancer, which affects just under 9 percent of the population, was $6.3 billion. Funding for obesity, which affects about 30 percent of the country, was about $1 billion," the article noted.[6]

With so little government funding for nutrition science, who is funding the rest of the studies that feed the constant news cycle? Many are funded by food companies and industry groups, either to conduct studies in their own research divisions or in the form of grants given to university scientists. That's problematic, because food companies and industry groups have an inherent bias or conflict of interest. Their primary goal is to promote their products and to sell more of them—not to advance knowledge of food and health—and this affects how they frame research questions, design and interpret studies, and report their results.

Marion Nestle, a retired nutrition professor at New York University, has written extensively about this problem. For a year in 2015-2016, she informally tracked industry-funded studies and found that 90 percent of the time, their conclusions benefitted the industry that funded them. In another example, a 2013 meta-analysis found that among studies that looked at whether soda consumption contributed to obesity, those funded by the soda industry were five times more likely to conclude that it doesn't contribute to obesity compared to those not funded by the industry.[7] (Consider the processed foods RCT we just discussed. It was funded by the National Institutes of Health, which doesn't have a stake in the results. Would you trust the results of a study of processed foods if it was funded by Nabisco? Or for that matter, the Broccoli Growers of America? Probably not.)

There's likely a long history of biased nutrition research influencing dietary advice. For example, in the 1960s, the sugar industry paid well-respected academic scientists to publish research concluding that it was fat—not sugar—that was detrimental to heart health.[8] (Both too much fat and too much sugar can negatively affect heart health, but it benefited the sugar industry to focus the blame on fat.)

As recently as 2015, Coca-Cola was funding research meant to promote lack of physical activity as the main cause of obesity, shifting blame away from dietary factors, such as drinking soda.[9] When food companies drive the narrative coming out of nutrition research, this can potentially impact public policy.

Media attention has made researchers and policy makers much more aware of the problems with industry funding and conflicts of interest in nutrition research, and they're working to solve them. But regardless, if you see reports of a study that shows that blueberries can block bladder infections, pistachios can prevent pancreatitis, or cinnamon can cure cancer... well, you should be skeptical, and always check the funding source. Studies on a single type of food are almost always industry-funded.

How to Find Clarity in a Complex Field

Let's review some of the key issues:

- Nutrition research is really difficult to do well. We want to know how nutrition relates to health over the long term, but it's hard to quantify how people eat over a lifetime and track them for long enough to see an impact.
- We often rely on observational studies, which can only show that two variables are *correlated*, not that one causes the other.
- Randomized controlled trials are rare, and they're often small, short-term studies that may or may not tell us what happens in the real world.
- Diet is exceptionally complex, with countless combinations of different nutrients and foods.
- One study is never enough to fully answer a question in the complex field of nutrition.
- Nutrition research is often funded by the food industry, which can be biased towards results that benefit business, not human health.

All of this can be discouraging, but you should also know that researchers are working hard to improve the quality of nutrition research and to interpret it honestly for the rest of us. As a consumer of nutrition information, use a skeptical eye when you read news of the latest nutrition research. Look for areas of consensus, where committees of experts have put their heads together to come up with the best advice they can based on the evidence we have, such as in the Dietary Guidelines for Americans. You'll find that while experts in this field are often debating the latest controversy, they also agree on a lot. As we continue on in this unit, we'll talk more about how to find accurate sources of information and who you can trust for evidence-based advice in the field of nutrition.

FINDING RELIABLE SOURCES OF NUTRITION INFORMATION

As we discussed in the previous section, science is always evolving, albeit sometimes slowly. One study is not enough to make a guideline or recommendation or to cure a disease. Science is a stepwise process that continuously builds on past evidence and develops towards a well-accepted consensus, although even that can be questioned as new evidence emerges. Unfortunately, the way scientific findings are communicated to the general public can sometimes be inaccurate or confusing. In today's world, where instant Internet access is just a click away, it's easy to be overwhelmed or misled if you don't know where to go for reliable nutrition information. Therefore, it's important to know how to find accurate sources of nutrition information and how to interpret nutrition-related stories when you see them.

Deciphering Nutrition Information

"New study shows that margarine contributes to arterial plaque."

"Asian study reveals that two cups of coffee per day can have detrimental effects on the nervous system."

How do you react when you read headlines like this? Do you boycott margarine and coffee? When reading nutrition-related claims, articles, websites, or advertisements, always remember that one study neither proves or disproves anything. Readers who may be looking for answers to complex nutrition questions can quickly misconstrue such statements and be led down a path of misinformation, especially if the information is coming from a source that isn't credible. Listed below are ways that you can develop a discerning eye when reading news highlighting nutrition science and research.

- The scientific study under discussion should be published in a peer-reviewed journal. Having gone through the peer review process, these studies have been checked by other experts in the field to ensure that their methods and analysis were rigorous and appropriate. Peer-reviewed articles also include a review of previous research findings on the topic of study and examine how their current findings relate to, support, or are in contrast to previous research. Question studies that come from less trustworthy sources (such as non peer-reviewed journals or websites) or that are not formally published.

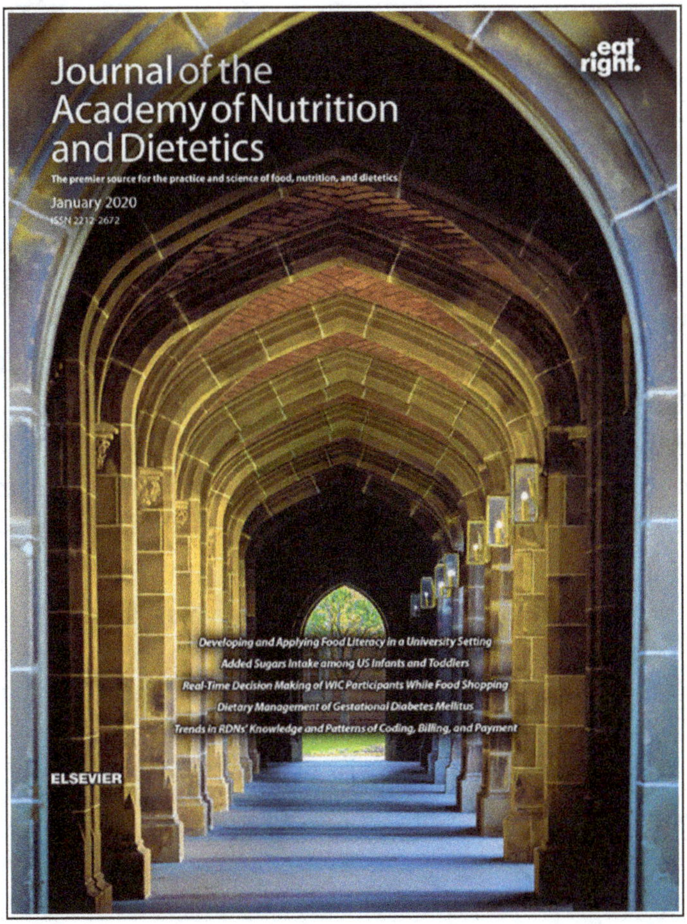

Figure 2.7. An example of a peer-reviewed journal, the Journal of the Academy of Nutrition and Dietetics, which publishes research findings of nutrition scientists. Nutrition research is also frequently published in journals like the Journal of Nutrition, American Journal of Clinical Nutrition, and Journal of Nutrition Education and Behavior, as well as medical and behavioral journals.

The report should disclose the methods used by the researcher(s).

- Identify the type of study and where it sits on the hierarchy of evidence. Keep in mind that a study in humans is likely more meaningful than one that's in vitro or in animals; an intervention study is usually more meaningful than an observational study; and systematic reviews and meta-analyses often give you the best synthesis of the science to date.
- If it's an intervention study, check for some of the attributes of high-quality research already discussed: randomization, placebo control, and blinding. If it's missing any of those, what questions does that raise for you?
- Did the study last for three weeks or three years? Depending on the research question, studies that are short may not be long enough to establish a true relationship with the issues being examined.
- Were there ten or two hundred participants? If the study was conducted on only a few participants, it's less likely that the results would be valid for a larger population.
- What did the participants actually do? It's important to know if the study included conditions that people rarely experience or if the conditions replicated real-life scenarios. For example, a study that claims to find a health benefit of drinking tea but required participants to drink 15 cups per day may have little relevance in the real world.
- Did the researcher(s) observe the results themselves, or did they rely on self reports from program participants? Self-reported data and results can be easily skewed by participants, either intentionally or by accident.
- The article should include details on the subjects (or participants) in the study. Did the study include humans or animals? If human, are any traits/characteristics noted? You may realize you have more in common with certain study participants and can use that as a basis to gauge if the study applies to you.
- Statistical significance is not the same as real-world significance. A statistically significant result is likely to have not occurred by chance, but rather to be a real difference. However, this doesn't automatically mean that the difference is relevant in the real world. For example, imagine a study reporting that a new vitamin supplement causes a statistically significant reduction in the duration of the common cold. Colds can be miserable, so that sounds great, right? But what if you look closer and see that the supplement only shortened study subjects' colds by half a day? You might decide that it isn't worth taking a supplement just to shorten a cold by half a day. In other words, it's not a real-world benefit to you.

- Credible reports should disseminate new findings in the context of previous research. A single study on its own gives you very limited information, but if a body of literature (previously published studies) supports a finding, it adds credibility to the study. A news story about a new scientific finding should also include comments from outside experts (people who work in the same field of research but weren't involved in the new study) to provide some context for what the study adds to the field, as well as its limitations.
- When reading such news, ask yourself, "Is this making sense?" Even if coffee does adversely affect the nervous system, do you drink enough of it to see any negative effects? Remember, if a headline professes a new remedy for a nutrition-related topic, it may well be a research-supported piece of news, but it could also be a sensationalized story designed to catch the attention of an unsuspecting consumer. Track down the original journal article to see if it really supports the conclusions being drawn in the news report.

The Craapp Test

While there is a wealth of information about nutrition on the internet and in books and magazines, it can be challenging to separate the accurate information from the hype and half-truths. You can use the CRAAPP Test to help you determine the validity of the resources you encounter and the information they provide. By applying the following principles, you can be confident that the information is credible. We've added several notes to the traditional CRAAPP Test to help you expand your analysis and apply it to nutrition information.

Table 2.1. The CRAAPP Test is a six-letter mnemonic device for evaluating the credibility and validity of information found through various sources, including websites and social media channels. The CRAAPP Test can be particularly useful in evaluating nutrition related news and articles.

CRAAPP Test Principle	Questions to ask
Currency	When was it written or published? Has the website been updated recently? Do you need current information, or will older sources meet your research need? Where is your topic in the information cycle? Note: In general, newer articles are more likely to provide up-to-date perspectives on nutrition science, so as a starting point, look for those published in the last 5-7 years. However, it depends on the question that you're researching. In some areas, nutrition science hasn't changed much in recent years, or you may be interested in historical background on the question. In either case, an older article would be appropriate.

Relevance	Does it meet stated requirements of your assignment? Does it meet your information needs/answer your research question? Is the information at an appropriate level or for your intended audience?
Authority	Who is the creator/author/publisher/source/sponsor? Are they reputable? What are the author's credentials and their affiliations to groups, organizations, agencies or universities? What type of authority does the creator have? For example, do they have subject expertise (scholar), social position (public office, title), or special experience? Note: The authority on nutrition information would be a registered dietitian nutritionist (RDN), a professional with advanced degree(s) in nutrition (MS or PhD), or a physician with appropriate education and expertise in nutrition. (This will be discussed in more detail on the next page). Look for sources authored or reviewed by experts with this level of authority or written by people who seek out and include their expertise in the article.
Accuracy	Is the information reliable, truthful, and correct? Does the creator cite sources for data or quotations? Who did they cite? Are they cherry-picking facts to support their argument? Is the source peer-reviewed, or reviewed by an editor? Do other sources support the information presented? Are there spelling, grammar, and typo errors that demonstrate inaccuracy? Note: Oftentimes, checking the accuracy of information in a given article or website means opening a new internet tab and doing some additional sleuthing to check the claims against other sources.
Purpose	Is the intent of the website to inform, persuade, entertain, or sell something? Does the point of view seem impartial or biased? Is the content primarily opinion? Is it balanced with other viewpoints? Who is the intended audience? Note: Particularly if you're looking at an organization's website, do some background research on the organization to see who funds it and what is the purpose of the group. That information can help you determine if their point-of-view is likely to be biased.
Process	What kind of effort was put into the creation and delivery of this information? Is it a Tweet? A blog post? A YouTube video? A press release? Was it researched, revised, or reviewed by others before published? How does this format fit your information needs or requirements of assignment?

WHO CAN YOU BELIEVE FOR INFORMATION ON NUTRITION?

Trustworthy Sources

Authoritative nutrition news is based on solid scientific evidence, supported by multiple studies, and published in peer-reviewed journals. You can obtain valid nutrition information from many reputable organizations, websites, and professionals, if you know where to look. Whatever the source of your nutrition news, remember to apply the criteria outlined previously in this unit to ensure the validity of the information presented. You can find many trustworthy sources that advocate good nutrition to promote health and prevent disease using evidence-based science.

Trusted Organizations Active in Nutrition Policy and Research

US Department of Agriculture Food and Nutrition Information Center. The USDA website has more than 2,500 links to information about diet, nutrition, disease, body weight and obesity, food safety, food labeling, packaging, dietary supplements, and consumer questions. Using this interactive site, you can find tips and resources on how to eat a healthy diet, nutritional information, and a food planner.

The Academy of Nutrition and Dietetics (AND). The AND promotes scientific, evidenced-based food and nutrition information. It is focused on informing the public about recent scientific studies, weight-loss concerns, food safety topics, nutrition issues, and disease prevention. This website also has lots of practical tips and suggestions on how to plan and prepare nutritious meals.

Department of Health and Human Services (HHS). The HHS website provides credible information about healthful lifestyles and the latest in health news. A variety of online tools are available to assist with food-planning, weight maintenance, physical activity, and dietary goals. You can also find healthful tips for all age groups, tips for preventing disease, and information on general health issues.

Centers for Disease Control and Prevention (CDC). The CDC provides up-to-date public health information and data on many nutrition-related topics, including healthful eating, cholesterol, high blood pressure, obesity, alcohol use, breastfeeding, infant and toddler nutrition, and food safety, as well as other public health issues like physical activity and tobacco usage. They also publish a monthly online newsletter called CDC Vital Signs that includes current data on the most pressing public health matters.

Many additional websites, organizations, and professionals provide valid health and nutrition information. Let's take a look at some of these other resources.

Trusted Websites and Sources

Web domains can be an indicator of the reliability of a website.

- Websites of government agencies end in .gov and are usually considered to be trustworthy sources of evidence-based health information.
- University websites typically end in .edu, indicating the source is focused primarily on providing educational resources rather than seeking financial gain.
- Many professional organizations and non-profit organizations use websites ending in .org, but this type of domain may also be used by special interest groups and biased groups promoting a specific agenda. Approach these websites with a critical eye, looking for the common signs of reliability.
- Business and company websites typically end in .com, indicating that the primary focus of the website is to promote that particular company's services and goods rather than to simply educate a consumer. News organizations also have .com websites, and while their primary mission is to inform readers, the same rules of discernment apply to make sure they're delivering news objectively. Major news organizations or those with a science or health focus usually have reporters who specialize in these areas so have more background knowledge of the field, and they're more likely to have a process for fact-checking an article.

Any of these types of web domains could contain credible information, but you must be a savvy consumer and use the knowledge gained in this unit to separate trusted sources from the more questionable options. Check out this list of websites as a starter kit for generally reliable, trusted sources for health and nutrition information.

Table 2.2. Reliable websites that provide nutrition information.

GOVERNMENT WEBSITES	
USDA Center for Nutrition Policy and Promotion	https://www.fns.usda.gov/cnpp
Food and Drug Administration	http://www.fda.gov/
Healthy People	https://www.healthypeople.gov/
Foodsafety.gov	https://www.foodsafety.gov/
Nutrition.gov	https://www.nutrition.gov/
ChooseMyPlate	https://www.choosemyplate.gov/
National Center for Complementary and Integrative Health	https://nccih.nih.gov/
National Heart, Lung, and Blood Institute	https://www.nhlbi.nih.gov/health-topics
National Institutes of Health Office of Dietary Supplements	https://ods.od.nih.gov/
INTERNATIONAL WEBSITES	
World Health Organization	https://www.who.int/
Food and Agricultural Organization of the United Nations	http://www.fao.org/
NON-GOVERNMENT WEBSITES	
Harvard School of Public Health	https://www.hsph.harvard.edu/nutrition/
Mayo Clinic	https://www.mayoclinic.org/
Linus Pauling Institute	http://lpi.oregonstate.edu/
American Society for Nutrition	http://www.nutrition.org/
American Cancer Society	https://www.cancer.org/
American Heart Association	https://www.heart.org/
American Diabetes Association	http://www.diabetes.org/
Center for Science in the Public Interest	https://cspinet.org/
Food Allergy Research & Education	https://www.foodallergy.org/
Institute of Medicine: Food and Nutrition	http://nationalacademies.org/hmd/Global/Topics/Food-Nutrition.aspx

Trusted Professionals

When looking for credible nutrition information, one of the most important aspects to consider is the expertise of the individual providing the information. Nutrition is a tricky field because the term "nutritionist" is not a legally-protected or regulated term, so it's imperative to seek experts that are formally-educated and credentialed in nutrition. Look for professionals with the following degrees or backgrounds:

- Registered dietitian nutritionist (RD or RDN)
- Professional with advanced degree(s) in nutrition (MS or PhD)
- Physician (MD) with appropriate education and expertise in nutrition

Registered dietitians or professionals with advanced degrees in the field of nutrition are the most credible sources for sound nutrition advice. Be skeptical of other official-sounding credentials, like "holistic nutrition practitioner," or even just "nutritionist." There are no standards for what these titles mean, which means that anyone can call themselves a "nutritionist," and you could be taking advice from a well-qualified individual or someone who just took an online course or got a mail-order certificate. Physicians can also be good sources for nutrition information, depending on their education and background. But be mindful that most medical schools include minimal or no education and training in nutrition so most physicians may have limited knowledge in this field unless they have sought out specific nutrition training on their own.

Careers in Nutrition

If you are considering a career in nutrition, it is important to understand the opportunities that may be available to you. Both dietitians and qualified nutritionists provide nutrition-related services to people in the private and public sectors. A dietitian is a healthcare professional who has registered credentials and can provide nutritional care in the areas of health and wellness for both individuals and groups. While registration isn't required to use the term "nutritionist," a *qualified* nutritionist will have an education similar to that of a dietitian, but most likely will not have completed an internship or passed a credentialing exam like a registered dietitian. People in both professions work to apply nutritional science, using evidence-based best practices, to help people nourish their bodies and improve their lives.

Becoming a registered dietitian requires a bachelor's or master's degree in dietetics (master's degree will be required beginning in 2024), including courses in biology, chemistry, biochemistry, microbiology, anatomy and physiology, nutrition, and food service management. Other suggested courses include economics, business, statistics, computer science, psychology, and sociology. In addition, people who pursue this path must complete a dietetic internship (including 1200 hours of supervised practice), pass a national exam, and maintain their registration through ongoing continuing education. Many states also have licensure that requires additional forms and documentation. You can learn more about the path to becoming a registered dietitian by going to cdrnet.org/certifications.

Dietitians and nutritionists plan food and nutrition programs, promote healthy eating habits, and recommend dietary modifications. But typically, to work in a clinical setting (like a hospital) or outpatient setting, the RD credential is required. For example, a dietitian might teach a patient with hypertension how to follow a lower-sodium diet. Nutrition-related careers can be extremely varied. Some individuals work in government settings, while others work in education or the private sector. Some jobs in nutrition focus on working with athletes, and others provide guidance to patients with long-term, life-threatening diseases. But no matter the circumstance or the clientele, working in the field of diet and nutrition focuses on helping people improve their dietary habits by translating nutritional science into food choices.

In the public sector, careers in nutrition span from government work to community outreach.

- Nutritionists and dietitians who work for the government may be involved with federal food programs (WIC, SNAP, school meals, etc), communication campaigns, or creating and analyzing public policy.

- On the local level, clinical careers include working in hospitals and nursing-care facilities. This requires creating meal plans and providing nutritional guidance to help patients restore their health or manage chronic conditions. Clinical dietitians consult with doctors and other health-care professionals to coordinate dietary recommendations with medical needs.
- Nutrition jobs in the community often involve working in public health clinics, cooperative extension offices, and HMOs to prevent disease and promote the health of the local community.
- Nutrition jobs in the nonprofit world involve anti-hunger organizations, public health organizations, and activist groups.

Nutritionists and dietitians can also find work in the private sector.

- Increased public awareness of food, diet, and nutrition has led to employment opportunities in advertising, marketing, and food manufacturing. Dietitians working in these areas analyze foods, prepare marketing materials, or report on issues such as the impact of vitamins and herbal supplements.
- Consultant careers can include working in wellness programs, supermarkets, physicians' offices, gyms, and weight-loss clinics.
- Consultants in private practice perform nutrition screenings for clients and use their findings to provide guidance on diet-related issues, such as weight reduction.
- Nutrition careers in the corporate world include designing wellness strategies and nutrition components for companies, working as representatives for food or supplement companies, designing marketing and educational campaigns, and becoming lobbyists.
- Others in the private sector work in food service management at health-care facilities or at company and school cafeterias.
- Sustainable agricultural practices provide interesting private sector careers on farms and in food systems.

Whether you pursue nutrition as a career or simply work to improve your own dietary choices, what you are learning in this course can provide a solid foundation for the future. Remember, your ability to think clearly, communicate, hope, dream, go to school, gain knowledge, and earn a living are impacted by your health. Good health allows you to function normally and work hard to pursue your goals. Yet, achieving optimal health is a complex process, involving multiple dimensions of wellness, along with your physical or medical reality. It's our hope that you use the knowledge gained in this class, not just to earn a good grade, but that you also apply it to make a difference in your life.

UNIT-3

PHOTOSYNTHESIS, DIGESTION, AND METABOLISM: MOLECULES OF LIFE

INTRODUCTION

Life is extraordinarily complex. That's true whether you're considering something as seemingly simple as a blade of grass or as obviously complicated as the human body. However, zooming in to look at the most basic elements of living things—from atoms to molecules to cells—we can start to see similarities and patterns that help us make sense of this complexity. And as we consider biological processes, such as photosynthesis, digestion, and metabolism, we also see how these processes fit together. Plants use photosynthesis to capture energy from the sun. Animals, humans included, eat the plants (or other animals that ate the plants) and use the processes of digestion and metabolism to utilize the energy they contain. In the process of energy metabolism, animals breathe out carbon dioxide, which is then used by plants for photosynthesis. It's truly a circle of life!

In this unit, we'll also zero in on the digestive system. While all organ systems relate to nutrition in some way, the digestive system takes center stage as the site of food processing and nutrient extraction. And while it usually functions as an efficient and coordinated system, we'll also consider some common ways that it can go awry, resulting in disorders and discomforts of the digestive tract or adverse reactions to certain foods.

Unit Learning Objectives

After completing this unit, you should be able to:

- Define and describe the levels of structural organization of the human body—from atoms and molecules to the whole organism—including the basic characteristics of cells and the organ systems.
- Describe how photosynthesis is the pathway to glucose production and describe the relationship between photosynthesis in plants and energy metabolism in the human body.
- Identify and briefly describe the functions of the organs of the gastrointestinal tract, and discuss the five fundamental activities of digestion, giving an example of each.
- Describe several common disorders and discomforts of the GI tract, including their causes, symptoms, and approaches to treatment.
- Describe food intolerances, food allergies, and celiac disease, including the different causes, symptoms, and treatments for each condition.

LIFE ORGANIZATION

Before you begin to study the different structures and functions of the human body related to nutrition, it is helpful to consider the basic architecture of the body; that is, how its smallest parts are assembled into larger structures. It is convenient to consider the structures of the body in terms of fundamental levels of organization that increase in complexity: atoms, molecules, cells, tissues, organs, organ systems, and organisms. Higher levels of organization are built from lower levels. Therefore, atoms combine to form molecules, molecules combine to form cells, cells combine to form tissues, tissues combine to form organs, organs combine to form organ systems, and organ systems combine to form organisms (Figure 3.1).

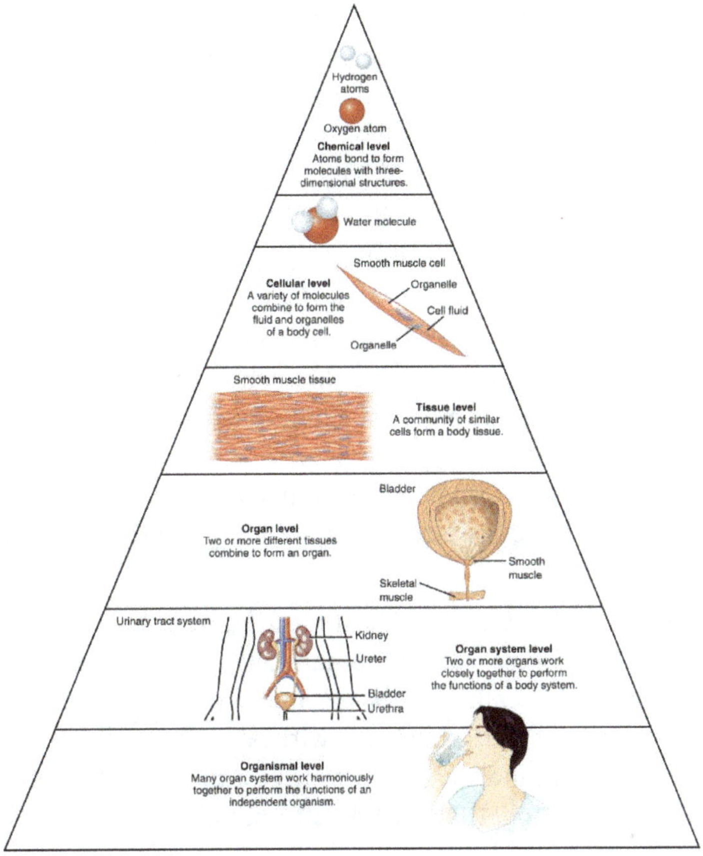

Figure 3.1. Levels of structural organization of the human body. The organization of the body often is discussed in terms of distinct levels of increasing complexity, from the smallest chemical building blocks to a unique human organism.

The levels of organization

Consider the simplest building blocks of matter: atoms and molecules. In Unit 1, you had an introduction to atoms and molecules. Remember, all matter in the universe is composed of one or more unique elements, such as hydrogen, oxygen, carbon, and nitrogen. The smallest unit of any of these elements is an atom. Atoms of individual elements combine to make molecules, and molecules bond together to make bigger macromolecules. Four macromolecules—carbohydrates, lipids, proteins, and nucleic acids (e.g., DNA, RNA)—make up all of the structural and functional units of cells.

The Basic Structural and Functional Unit of Life: The Cell

Cells are the most basic building blocks of life. All living things are composed of cells. New cells are made from preexisting cells, which divide in two. Who you are has been determined because of two cells that came together inside your mother's womb. The two cells containing all of your genetic information (DNA) fused to begin the development of a new organism. Cells divided and differentiated into other cells with specific roles that led to the formation of the body's numerous organs, systems, blood, blood vessels, bones, tissues, and skin. While all cells in an individual contain the same DNA, each cell only expresses the genetic codes that relate to that cell's specific structure and function.

As an adult, you are made up of trillions of cells. Each of your individual cells is a compact and efficient form of life—self-sufficient, yet interdependent upon the other cells within your body to supply its needs. There are hundreds of types of cells (e.g., red blood cells, nerve cells, skin cells). Each individual cell conducts all the basic processes of life. It must take in nutrients, excrete wastes, detect and respond to its environment, move, breathe, grow, and reproduce. Many cells have a short life span and have to be replaced continually. For example, enterocytes (cells that line the intestines) are replaced every 2-4 days, and skin cells are replaced every few weeks.

Although a cell is defined as the "most basic" unit of life, it is structurally and functionally complex (Figure 3.2). A human cell typically consists of a flexible outer cell membrane (also called a plasma membrane) that encloses cytoplasm, a water-based cellular fluid, together with a variety of functioning units called organelles. The organelles are like tiny organs constructed from several macromolecules bonded together. A typical animal cell contains the following organelles:

- Nucleus: houses genetic material (DNA)
- Mitochondria: often called the powerhouse of the cell, generates usable energy for the cell from energy-yielding nutrients
- Ribosomes: assemble proteins based on genetic code
- Endoplasmic reticulum: processes and packages proteins and lipids
- Golgi apparatus(golgi body): distributes macromolecules like proteins and lipids around the cell
- Lysosomes: digestive pouches which break down macromolecules and destroy foreign invaders

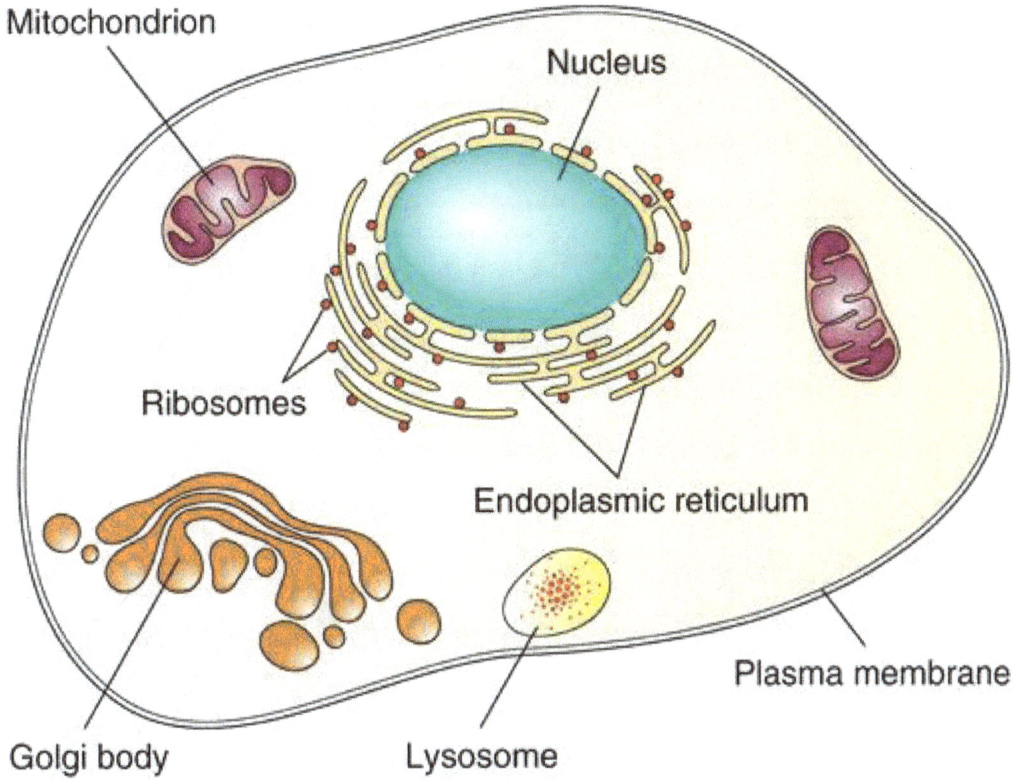

Figure 3.2. The cell structure

Tissues, Organs, Organ Systems, and Organisms

A tissue is a group of many similar cells that share a common structure and work together to perform a specific function. There are four basic types of human tissues: connective tissue, which connects tissues; epithelial tissue, which lines and protects organs; muscle, which contracts for movement and support; and nerve, which responds and reacts to signals in the environment.

An organ is a group of similar tissues arranged in a specific manner to perform a specific physiological function. Examples include the brain, liver, and heart. An organ systemis a group of two or more organs that work together to perform a specific physiological function. Examples include the digestive system and central nervous system.

There are eleven distinct organ systems in the human body (Figure 3.3). Assigning organs to organ systems can be imprecise since organs that "belong" to one system can also have functions integral to another system. In fact, many organs contribute to more than one system. And most of these organ systems are involved in nutrition-related functions within the body (Table 3.1). For example, the cardiovascular system plays a role in nutrition by transporting nutrients in the blood to the cells of the body. The endocrine system produces hormones, many of which are involved in regulating appetite, digestive processes, and nutrient levels in the blood. Even the reproductive system plays a role in providing nutrition to a developing fetus or growing baby.

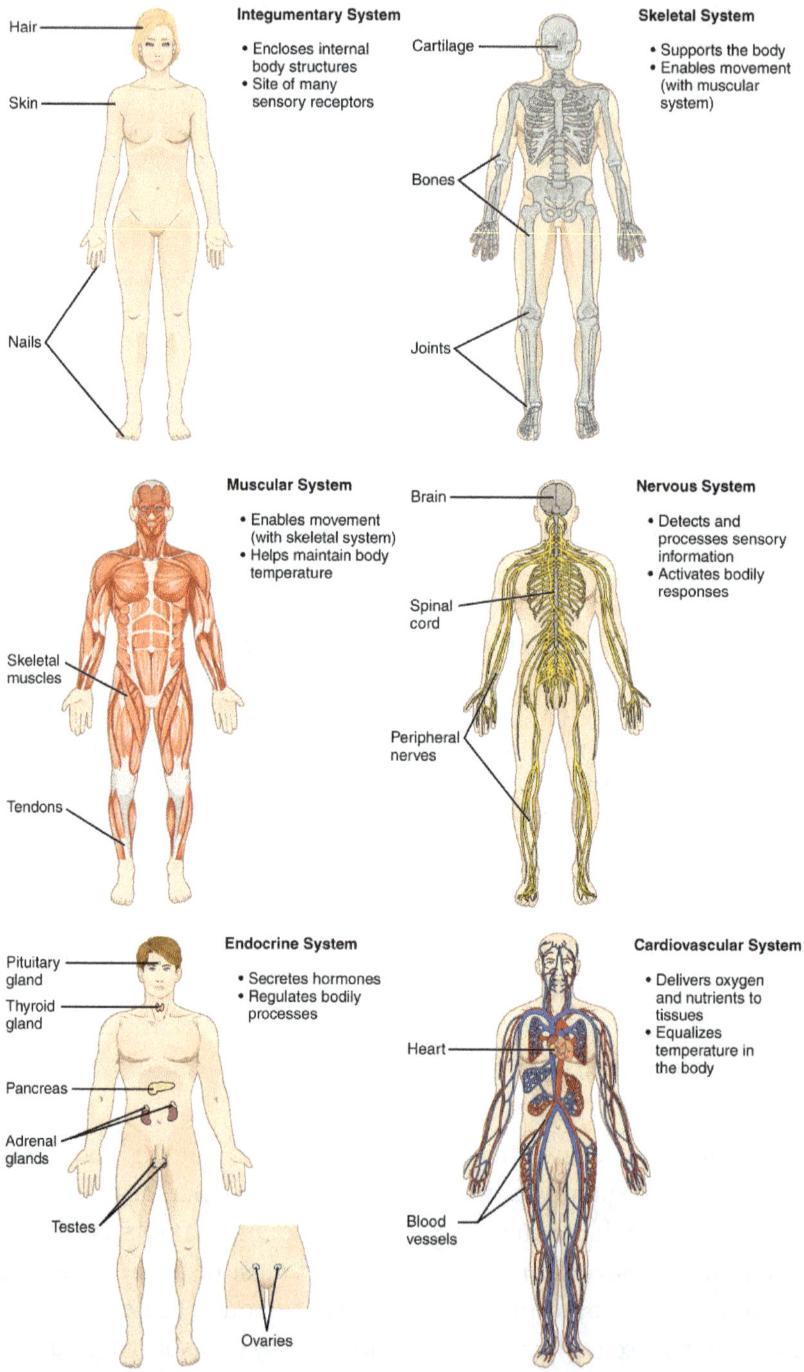

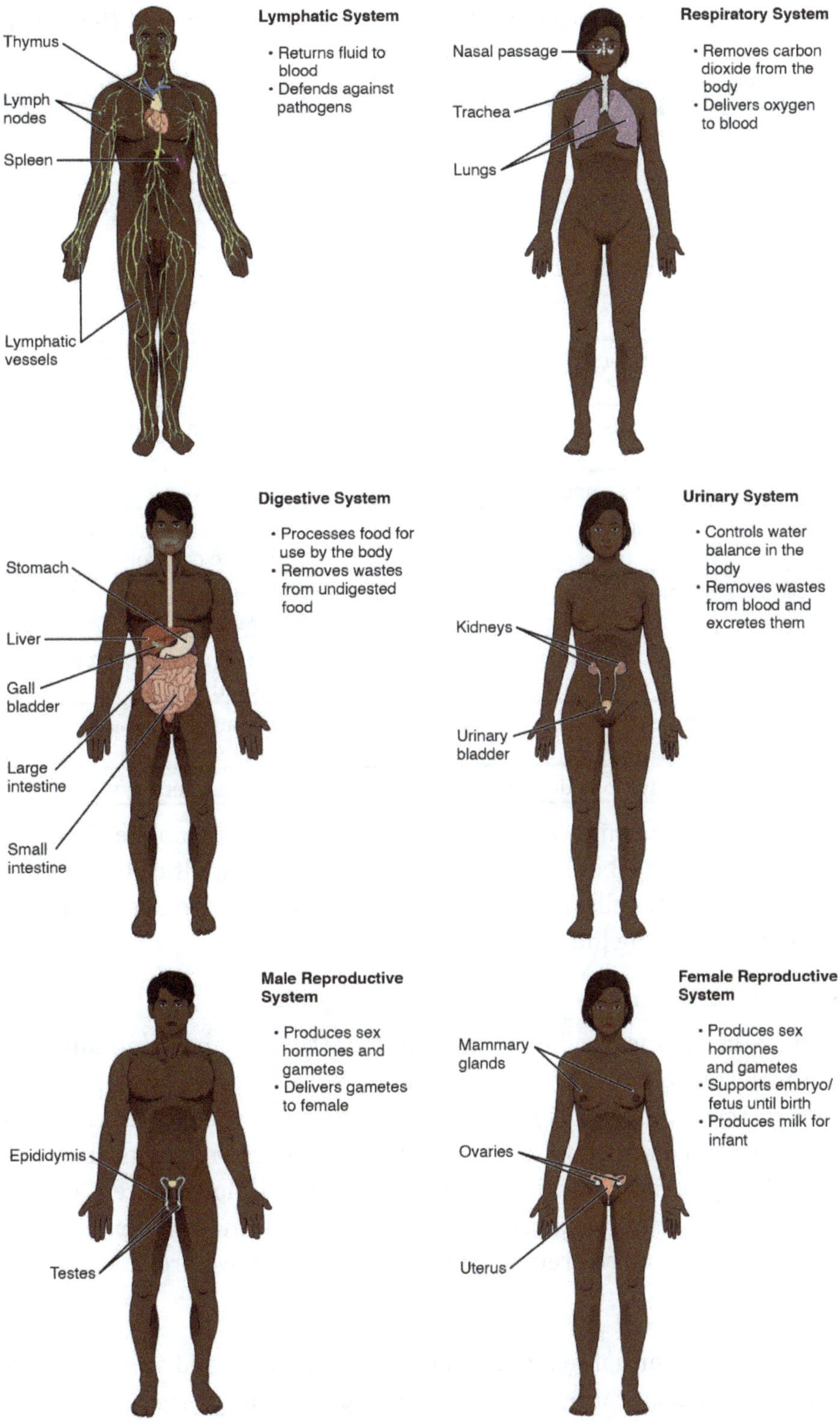

Figure 3.3. Organ systems of the human body

Table 3.1. The eleven organ systems in the human body and their major functions

Organ System	Major Organ Components	Major Functions
Cardiovascular	Heart, blood vessels, blood	Transport oxygen, nutrients, and waste products
Digestive	Mouth, esophagus, stomach, intestines, salivary glands, pancreas, liver and gallbladder	Digestion and absorption
Endocrine	Endocrine glands (e.g., thyroid, ovaries, pancreas)	Produce and release hormones, regulate nutrient levels
Lymphatic	Lymphatic vessels, lymph nodes, thymus gland, spleen	Fluid balance, defend against pathogens
Integumentary	Skin, nails, hair, sweat glands	Protection, body temperature regulation
Muscular	Skeletal, smooth, and cardiac muscle	Body movement
Nervous	Brain, spinal cord, nerves	Interpret and respond to stimuli, appetite control
Reproductive	Gonads, genitals	Reproduction and sexual characteristics
Respiratory	Lungs, nose, mouth, throat, trachea	Gas exchange (oxygen and carbon dioxide)
Skeletal	Bones, tendons, ligaments, cartilage, joints	Structure and support, calcium storage
Urinary	Kidneys, bladder, ureters	Waste excretion, water balance

An organism is the highest level of organization—a complete living system capable of conducting all of life's biological processes. In multicellular organisms, including humans, all cells, tissues, organs, and organ systems of the body work together to maintain the life and health of the organism.

PHOTOSYNTHESIS AND METABOLISM

As we just learned in the previous section, cells are the most basic building blocks of life. Cells make up your tissues, organs, and ultimately, you as a human being. And every one of those cells needs energy to perform their specific functions. Where does that energy come from? It comes from the macronutrients that we eat—carbohydrates, protein, and fat. In order to understand how the cells of the body put that energy to use, you must have a basic understanding of photosynthesis, cellular respiration, and the relationship between these two processes.

Photosynthesis

Photosynthesis is essential to all life on earth; both plants and animals depend on it. It is the only biological process that can capture energy that originates from sunlight and convert it into a chemical compound (glucose) that every organism uses to power its daily functions. Photosynthesis is also a source of oxygen necessary for many living organisms.

The importance of photosynthesis is not just that it can capture sunlight's energy. Photosynthesis is vital because it provides a way to capture the energy from solar radiation (the "photo-" part) and store that energy in the carbon-carbon bonds of glucose (the "-synthesis" part). Glucose is the main energy source that animals and humans use to power the synthesis of adenosine triphosphate (ATP). ATP is the energy-containing molecule found in the cells of all animals and humans. Energy from the foods we eat is captured in ATP and used to fuel the workload of cells.[1]

The energy stored in carbohydrate molecules from photosynthesis passes through the food chain. Consider a predator, such as a wolf, preying on a deer. The wolf is at the end of an energy path that went from atoms colliding on the surface of the sun, to visible light, to photosynthesis, to vegetation, to the deer, and finally to the wolf. The wolf, by feeding on the deer, receives a portion of the energy that originated in the photosynthetic vegetation that the deer consumed.

Our food supply is also directly linked to photosynthesis. Major grocery stores in the United States are organized into departments, such as dairy, meats, produce, bread, cereals, and so forth. Each aisle contains hundreds, if not thousands, of different products for customers to buy and consume. Although there is a large variety, each item ultimately can be linked back to photosynthesis. Meat and dairy link because the animals were fed plant-based foods. The breads, cereals, and pastas come largely from starchy grains, which are the seeds of photosynthesis-dependent plants. What about desserts and drinks? All of these products contain sugar—sucrose is a plant product, a carbohydrate molecule, which is also derived from photosynthesis. Many items are less obviously derived from plants: virtually every spice and flavoring in the spice aisle was produced by a plant as a leaf, root, bark, flower, fruit, or stem. (Salt is a notable exception.) Ultimately, photosynthesis connects to every meal and every food a person consumes.

Main Structures and Summary of Photosynthesis

Photosynthesis generally takes place in the leaves of plants. It is a multi-step process that requires sunlight, carbon dioxide (CO_2, found in the air), and water (H_2O, from the soil). After the process is complete, the plant releases oxygen into the air (O_2, essential for many living organisms) and produces the simple carbohydrate molecule of glucose, which can be used as an energy source by the plant, converted to starch and stored for a later energy source, or converted into other organic molecules such as fats, proteins and vitamins. This glucose contains the energy that all living organisms need to survive.

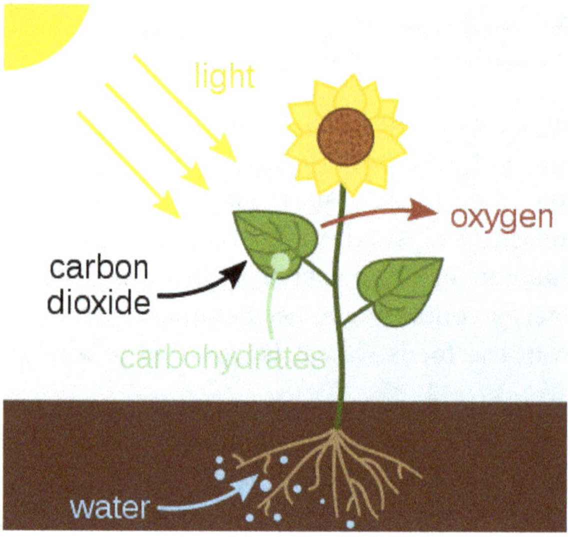

Figure 3.5. Depiction of photosynthesis in plants. The carbohydrates produced are stored in or used by the plant.

The basic formula for photosynthesis is as follows:

$$6CO_2 + 6H_2O + \text{sun's energy} = C_6H_{12}O_6 + 6O_2$$

Another way of saying this:

Photosynthesis uses:	6 molecules of carbon dioxide ($6CO_2$) 6 molecules of water ($6H_2O$) the sun's energy

Photosynthesis produces:	1 molecule of glucose ($C_6H_{12}O_6$) 6 molecules of oxygen ($6O_2$)

Starch is the storage form of glucose in plants, stored in seeds, roots, and tubers for later use as an energy source for the plant to reproduce. When a seed is buried deep in the soil, this starch can be broken down into glucose to be used for energy for the seed to sprout. As the seed sprouts, and shoots go above the ground and leaves start to form, the new plant can then photosynthesize glucose for an energy source. When we eat foods that contain starch, we must digest that starch down into single sugars (glucose) in order for the glucose to be absorbed into the intestinal cells, where it will enter the bloodstream to be carried to all cells of the body to use as an energy source. The basic process of digestion of these foods will be covered in the next section.

Cellular Respiration

All living things require energy to survive. For humans, and many other organisms, that energy is generated by the complex interaction of photosynthesis and cellular respiration. Cellular respiration is a key pathway in energy metabolism (the process of converting food into energy) of all aerobic organisms. Respiration refers to breathing: taking in oxygen and removing carbon dioxide. But ultimately, the reason we need to breathe is to provide the oxygen needed to carry out cellular respiration in our cells and to remove the carbon dioxide that is produced as a byproduct.

In the process of cellular respiration, energy that is stored in the food we eat is converted to the body's energy currency, ATP, while a small amount is lost as heat. During cellular respiration, glucose is broken down to carbon dioxide and water; in the process, ATP is released. Cellular respiration occurs in part in the mitochondria of cells and is an aerobic process, which means that oxygen is required. It is a series of reactions that can be summarized as follows:

$$\text{glucose} + 6O_2 \rightarrow 6CO_2 + 6H_2O + \text{energy (ATP and heat)}$$

Another way of saying this:

Cellular respiration uses:	1 molecule of glucose ($C_6H_{12}O_6$) 6 molecules of oxygen ($6O_2$)
Cellular respiration produces:	6 molecules of carbon dioxide ($6CO_2$) 6 molecules of water ($6H_2O$) Energy

Even though glucose is the starting substance used in cellular respiration, we do not consume only glucose as an energy source. Instead, many different kinds of food molecules are broken down into smaller molecules, metabolized, and then enter the cellular respiration pathway. For example, complex carbohydrates like starch are readily converted to glucose. Fats and proteins can also be used in cellular respiration, but they must be modified before they can feed into the process.

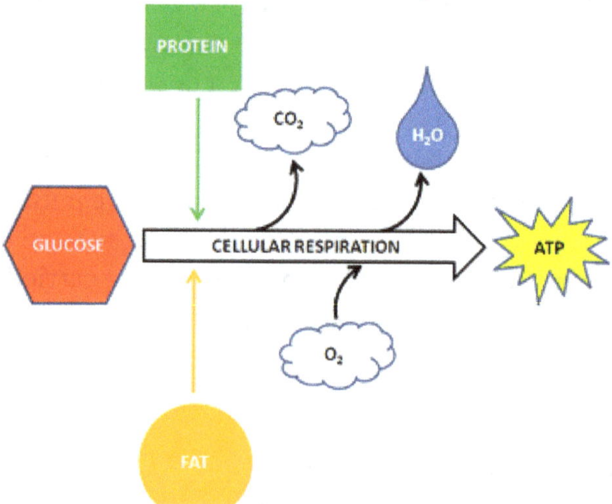

Figure 3.6. Nutrients fuel cellular respiration. Other carbohydrates, like starch and sugars, are converted to glucose before entering cellular respiration.

Photosynthesis-cellular respiration cycle

If you compare the summary reactions of photosynthesis and cellular respiration, you can see that cellular respiration is the opposite of photosynthesis. Because each process starts where the other ends, they form a cycle. What one reaction uses, the other reaction produces, and what one produces the other uses.

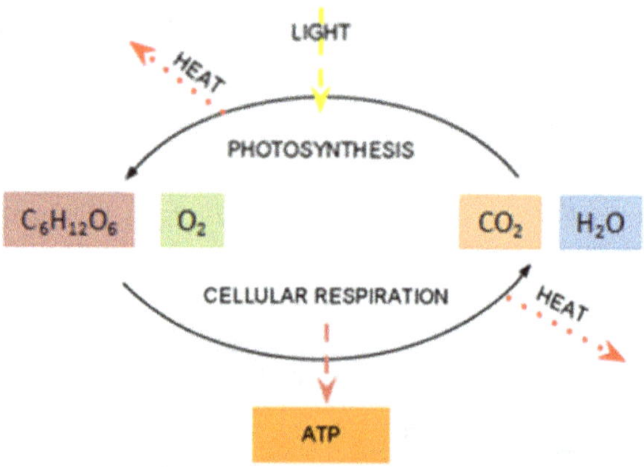

Figure 3.7. The relationship between the reactions of photosynthesis and cellular respiration.

The cycling that occurs between photosynthesis and cellular respiration is vital to the health of planet Earth. If there was no way for the carbon dioxide produced through cellular respiration to be utilized, breathing organisms (like humans and animals) would soon suffocate. Additionally, photosynthetic organisms are at the base of almost every food chain on the planet, so without these organisms, mass starvation would result. Luckily, this planet is full of organisms capable of photosynthesis (e.g., trees and grass on land and algae and bacteria in the ocean). Without this vital connection between photosynthesis and cellular respiration, life as we know it would cease to exist.

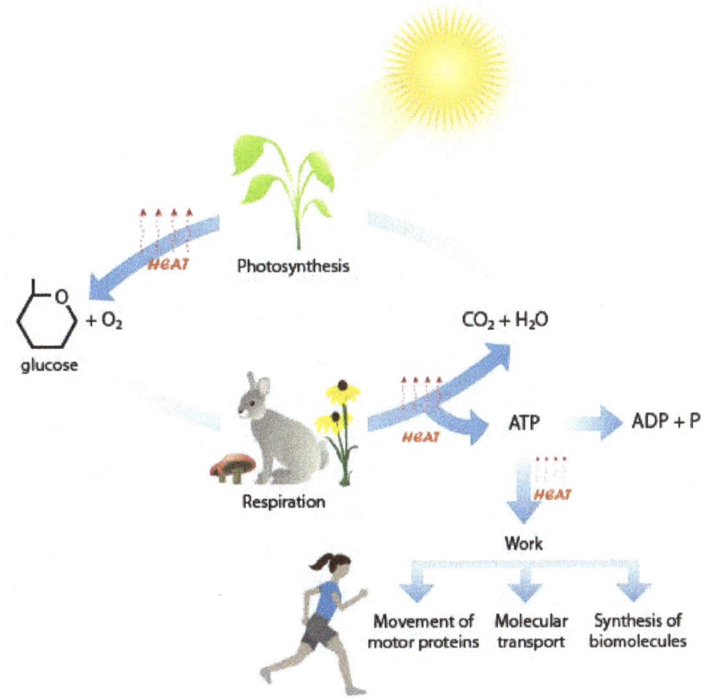

Figure 3.8. The photosynthesis-cellular respiration cycle. The two processes are intimately linked.

THE DIGESTIVE SYSTEM

We just learned that our body is composed of billions of cells. To function, these cells need essential nutrients—carbohydrates, proteins, fats, vitamins, and minerals—which we obtain from foods. However, before our cells can access these nutrients, foods need to be broken down or digested into their simplest units, so that the nutrients can be absorbed and enter the bloodstream. Digestion is a complex process that involves many organs and chemicals, as we'll explore on this page.

An overview of the organs involved in digestion

The function of the digestive system is to break down the foods you eat, release their nutrients, and absorb those nutrients into the body. Although the small intestine is the workhorse of the system where the majority of digestion and absorption occurs, each of the digestive system organs makes a vital contribution to this process.

The easiest way to understand the digestive system is to divide its organs into two main categories: the gastrointestinal tract (GI tract) and the accessory organs.

- The GI tract is a one-way tube about 25 feet in length, beginning at the mouth and ending at the anus. Between these two points, the GI tract also contains the pharynx, esophagus, stomach, small and large intestines, and the rectum. The small intestine is comprised of three parts: the duodenum, the jejunum, and the ileum. The large intestine, also called the colon, is similarly divided into three sections: the ascending colon, transverse colon, and descending colon. Both the mouth and anus are open to the external environment; thus, food and wastes within the GI tract are technically considered to be outside the body. Only through the process of absorption do the nutrients in food enter into and nourish the body's "inner space."
- Accessory organs, despite their name, are critical to the function of the digestive system. They are considered accessory organs since they are not actually part of the intestinal tract itself, but have ducts that deliver digestive juices into the tract to help aid in digestion. There are four accessory organs: the salivary glands, liver, gallbladder, and pancreas. All of these organs secrete fluids containing a variety of chemicals such as enzymes and acids that aid in digestion.

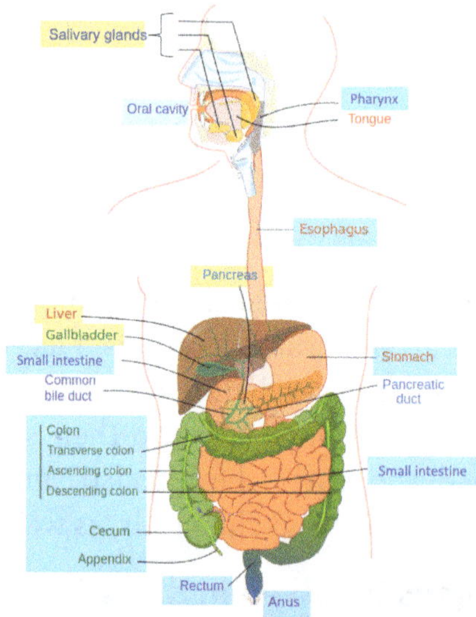

Figure 3.9. An overview of the organs involved in digestion. The parts of the GI tract are highlighted in blue, and the accessory organs are highlighted in yellow.

An Overview of the Digestive Process

The process of digestion includes five main activities: ingestion, mechanical digestion, chemical digestion, absorption, and excretion.

The first of these processes, ingestion, refers to the entry of food into the GI tract through the mouth. There, the food is chewed and mixed with saliva, which contains enzymes that begin breaking down the carbohydrates and lipids in food. Mastication (chewing) increases the surface area of the food and allows for food to be broken into small enough pieces to be swallowed safely.

Food (now called a bolus since it has been chewed and moistened) leaves the mouth when the tongue and pharyngeal muscles propel the bolus into the esophagus. The bolus will travel down the esophagus through an involuntary process called peristalsis. Peristalsis consists of sequential, alternating waves of contraction and relaxation of the smooth muscles in the GI tract, which act to propel food along (Figure 3.10). These waves also play a role in mixing food with digestive juices. Peristalsis is so powerful that foods and liquids you swallow enter your stomach even if you are standing on your head.

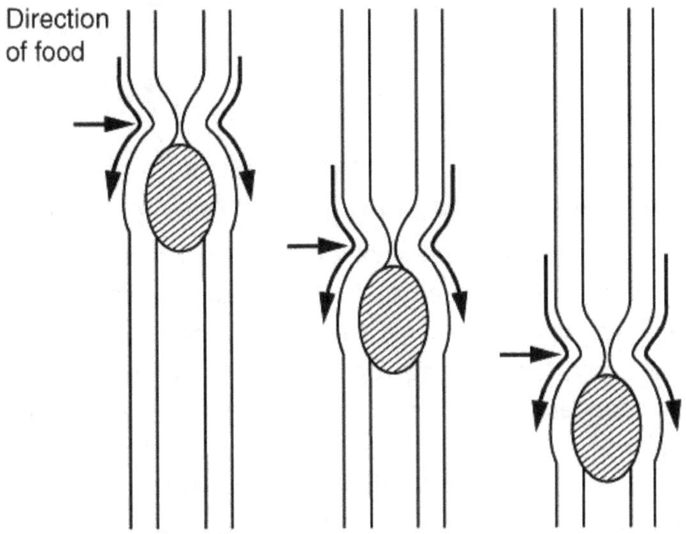

Figure 3.10. Peristalsis moves food through the digestive tract with alternating waves of muscle contraction and relaxation.

Digestion includes both mechanical and chemical processes. Mechanical digestion is a purely physical process of making food particles smaller to increase both surface area and mobility. Mechanical digestion does not change the chemical nature of the food. It includes mastication, tongue movements that help break food into smaller bits and mix it with saliva, mixing and churning of the stomach to further break food apart and expose more of its surface area to digestive juices, and peristalsis to help move food along the intestinal tract. Segmentation is also an example of mechanical digestion. Segmentation, which occurs mainly in the small intestine, consists of localized contractions of circular muscle of the GI tract. These contractions isolate small sections of the intestine, moving their contents back and forth while continuously subdividing, breaking up, and mixing the contents. By moving food back and forth in the intestinal tract, segmentation mixes food with digestive juices and facilitates absorption.

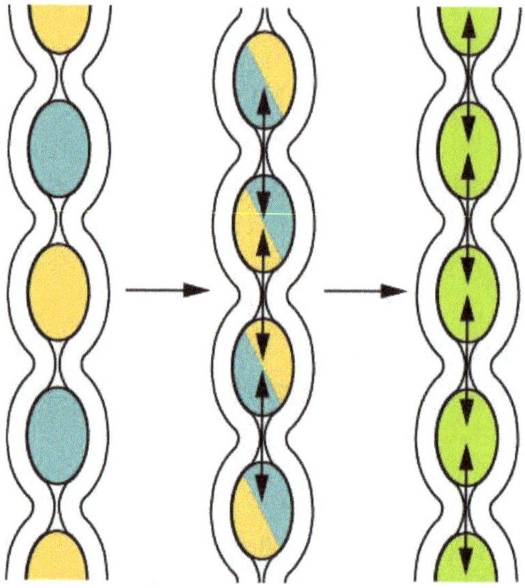

Figure 3.11. Segmentation separates chyme and then pushes it back together, mixing it and providing time for digestion and absorption.

In chemical digestion, digestive secretions that contain enzymes start to break down the macronutrients into their chemical building blocks (for example, starch into glucose). Enzymes are chemicals that help speed up or facilitate chemical reactions in the body. They bring together two compounds to react, without undergoing any changes themselves. For example, the main chemical reaction in digestion is hydrolysis.Hydrolysis is the splitting of one molecule into two with the addition of water. For example, the sugar sucrose (a double sugar) needs to be broken down to its building blocks, glucose and fructose (both single sugars), before it can be absorbed. This breakdown happens through hydrolysis, and the enzyme, sucrase, brings together the sucrose molecule and the water molecule to react. This process is illustrated in the following animation.

Nutrients are of little to no value to the body unless they enter the bloodstream. This occurs through the process of absorption, which takes place primarily within the small intestine. There, most nutrients are absorbed from the lumen (or inside space) of the GI tract into the bloodstream. Larger lipids are absorbed into lymph but eventually enter the bloodstream as well.

In excretion, the final step of digestion, undigested materials are removed from the body as feces. The feces is stored in the rectum until it leaves the body through the anus.

Functions of the Digestive Organs

Now that you have an overview of the digestive organs and the digestive process, let's discuss in more detail what types of mechanical and chemical digestion take place in each of the organs of the GI tract. Let's imagine eating a peanut butter and jelly sandwich that contains carbohydrates, proteins, fats, vitamins, and minerals. How does each organ participate in breaking this sandwich down into units that can be absorbed and utilized by cells throughout the body?

Mouth

Ingestion of the peanut butter and jelly sandwich happens in the mouth or oral cavity. This is where mechanical and chemical digestion also begin. Teeth physically crush and grind the sandwich into smaller particles and mix the food particles with saliva. Salivary amylase (a digestive enzyme) is secreted by salivary glands (salivary glands produce saliva which is a mixture of water, enzymes, and other chemicals) and begins the chemical breakdown of carbohydrates in the bread, while lingual lipase (another digestive enzyme) starts the chemical breakdown of triglycerides (the main form of fat in food) in the peanut butter.

Esophagus

The esophagus is a muscular tube that transports food from the mouth to the stomach. No chemical digestion occurs while the bolus is mechanically propelled through this tube by peristalsis.

Stomach

The stomach is an expansion of the GI tract that links the esophagus to the first part of the small intestine (the duodenum). The empty stomach is only about the size of your fist but can stretch to hold as much as 4 liters of food and fluid—more than 75 times its empty volume—and then return to its resting size when empty. An important function of the stomach is to serve as a temporary holding chamber. You can ingest a meal far more quickly than it can be digested and absorbed by the small intestine. Thus, the stomach holds food and secretes only small amounts into the small intestine at a time. (The length of time food spends in the stomach varies by the macronutrient composition of the meal. A high-fat or high-protein meal takes longer to break down than one rich in carbohydrates. It usually takes a few hours after a meal to empty the stomach contents completely into the small intestine.)

When the peanut butter and jelly sandwich enters the stomach, a highly muscular organ, powerful peristaltic contractions help mash, pulverize, and churn it into chyme. Chyme is a semiliquid mass of partially digested food along with gastric juices secreted by cells in the stomach. These gastric juices contain hydrochloric acid, which lowers the pH of the chyme in the stomach. This acidic environment kills many bacteria or other germs that may have been present in the food, and it causes the three-dimensional structure of dietary proteins to unfold. Gastric juices also contain the enzyme **pepsin**, which begins the chemical breakdown of proteins in the peanut butter and bread. Gastric lipase continues the breakdown of fat from the peanut butter.

Small Intestine

Chyme released from the stomach enters the small intestine, where most digestion and absorption occurs. The small intestine is divided into three parts, all part of one continuous tube: the duodenum, the jejunum, and the ileum.

Once the chyme enters the duodenum (the first segment of the small intestine), the pancreas and gallbladder are stimulated to release juices that aid in digestion. The pancreas (located behind the stomach) produces and secretes pancreatic juices which consist mostly of water, but also contain bicarbonate that neutralizes the acidity of the stomach-derived chyme and enzymes that further break down proteins, carbohydrates, and lipids. The small intestine's absorptive cells also synthesize digestive enzymes that aid in the breakdown of sugars and proteins.

The gallbladder (a small sac located behind the liver) stores, concentrates, and secretes a fluid called bile that helps to digest fats. Bile is made in the liver and stored in the gallbladder. Bile is an emulsifier; it acts similar to a detergent (that would remove grease from a frying pan) by breaking large fat droplets into smaller fat droplets so they can mix with the watery digestive juices. Peristalsis and segmentation control the movement and mixing of chyme through the small intestine. As in the esophagus and stomach, peristalsis consists of circular waves of smooth muscle contractions that propel food forward. Segmentation helps to mix food with digestive juices and facilitates absorption.

Nutrient absorption takes place mainly in the latter part of the small intestine, the ileum. The small intestine is perfectly structured for maximizing nutrient absorption. Its surface area is greater than 200 square meters—about the size of a tennis court! The large surface area is due to the multiple levels of folding, villi, and microvilli that cover the internal tissue of the small intestine. Villi are tiny finger-like projections that are covered with enterocytes or absorptive cells. The absorptive cell membrane is made of even smaller projections, called microvilli (Figure 3.12). These microvilli are referred to collectively as the brush border since their appearance resembles the bristles on a brush.

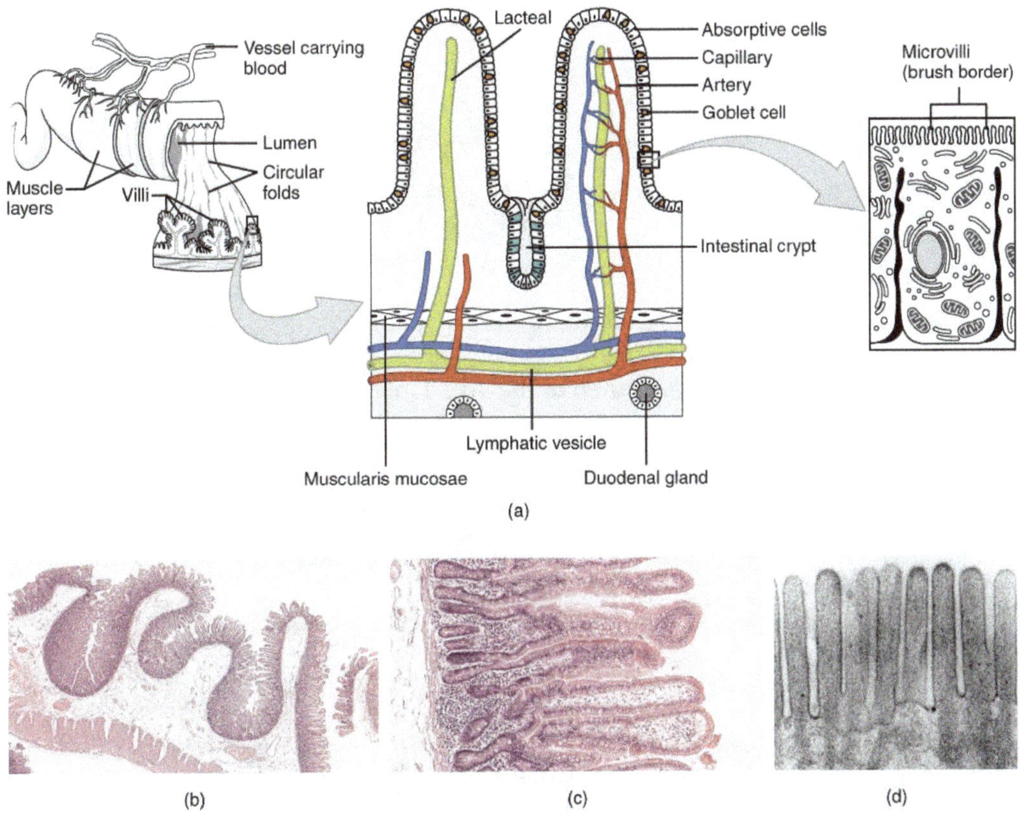

Figure 3.12. Histology of the small intestine. (a) The absorptive surface of the small intestine is vastly enlarged by the presence of circular folds, villi, and microvilli. (b) Micrograph of the circular folds. (c) Micrograph of the villi. (d) Electron micrograph of the microvilli.

Digested nutrients are absorbed into either capillaries or lymphatic vessels contained within each villus. Amino acids (from protein digestion), small fatty acids (from triglyceride digestion), sugars (from carbohydrate digestion), water-soluble vitamins, and minerals are transported from the intestinal cells into the bloodstream through capillaries. The larger fatty acids, fat-soluble vitamins, and other lipids (that are packaged in lipid transport particles) are transported first through lymphatic vessels and then eventually meet up with the blood. Water-soluble nutrients that enter the bloodstream are transported directly to the liver where the liver processes, stores, or releases these nutrients to other body cells.

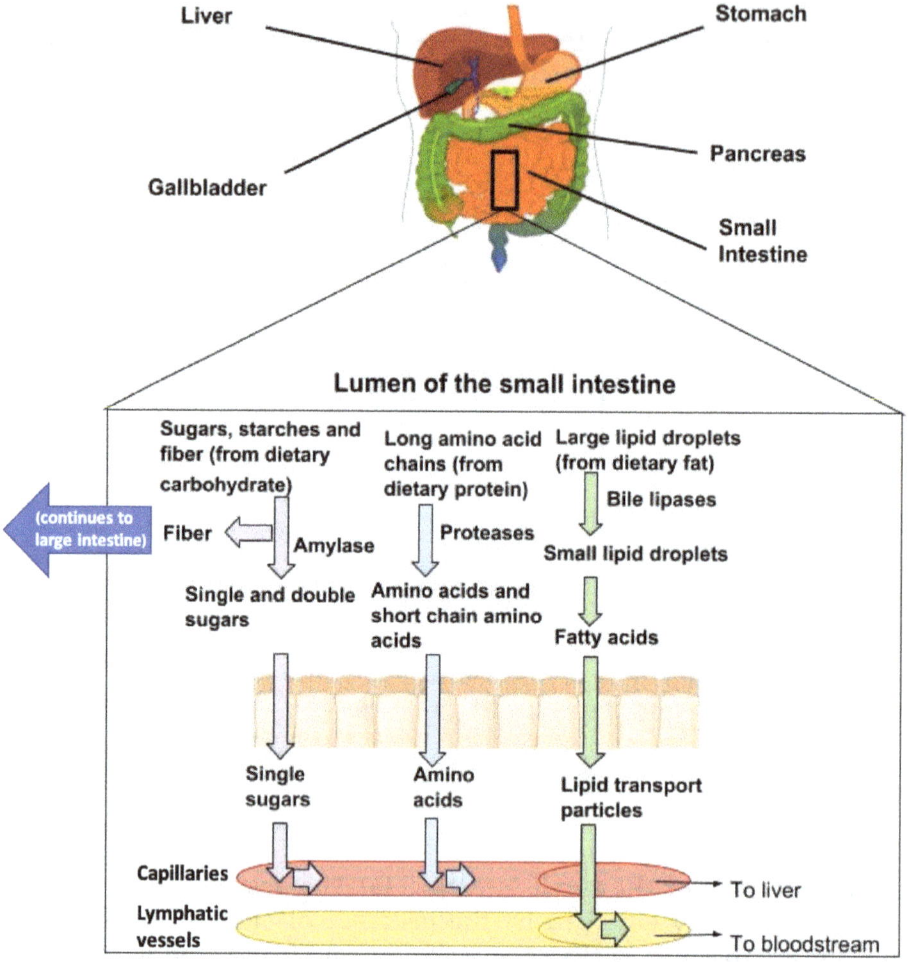

Figure 3.13. The digestion and absorption of nutrients in the small intestine.

Large Intestine

Most of the nutrients from the peanut butter and jelly sandwich have now been digested and absorbed. Any components that still remain (usually less than ten percent of food consumed) and the indigestible fiber move from the small intestine to the large intestine (colon). A main task of the large intestine is to absorb much of the remaining water. Water is present not only from the solid foods and beverages consumed, but also the digestive juices released by the stomach and pancreas. As water is reabsorbed, liquid chyme becomes a semisolid, referred to as feces. **Feces** is composed of undigested food residues, unabsorbed digested substances, millions of bacteria, old cells from the lining of the GI tract, inorganic salts, and enough water to let it pass smoothly out of the body.

Feces is stored in the rectum (a temporary holding area) until it is expelled through the anus via defecation. No further chemical breakdown of food takes place in the large intestine except that accomplished by the bacteria that inhabit this portion of the GI tract. There are trillions of bacteria residing in the large intestine (referred to as the bacterial flora), exceeding the total number of cells in the human body. This may seem rather unpleasant, but the great majority of bacteria in the large intestine are harmless and many are even beneficial—facilitating chemical digestion and absorption, improving immune function, and synthesizing vitamins such as biotin, pantothenic acid, and vitamin K.

The figure below summarizes the functions of the digestive organs.

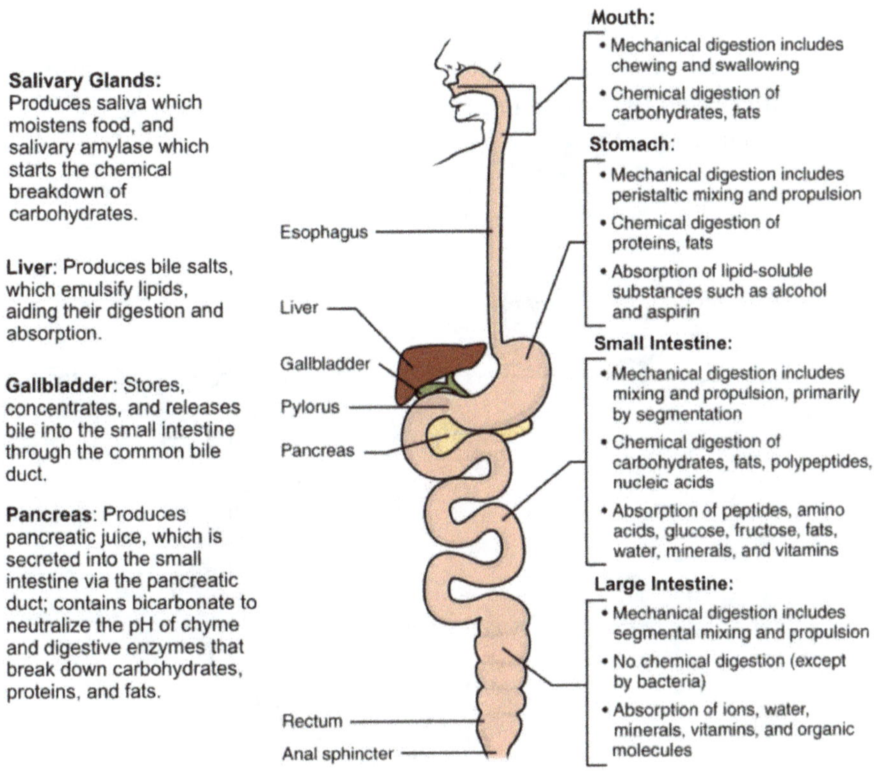

Figure 3.14. Summary of digestion and absorption. Digestion begins in the mouth and continues as food travels through the small intestine. Most absorption occurs in the small intestine.

THE GI TRACT'S DISORDERS

Now that we've covered the structures and functions of the digestive system, it should be clear that the digestion of food requires the coordinated action of multiple organs. If any of these aren't working well, it can wreak havoc on the function of the entire system and interfere with health and quality of life. Let's look at some common discomforts and disorders of the GI tract.

Heartburn and Gastroesophageal Reflux

Heartburn is a burning, often painful, sensation in the chest (behind the breastbone) or throat. Heartburn is caused by gastroesophageal reflux (GER), when the acidic chyme in the stomach escapes back into the esophagus and even into the mouth. Normally, this reflux is prevented by the lower esophageal sphincter (a tight ring of muscle) that sits between the esophagus and stomach. The muscles of the sphincter contract to keep it closed, only relaxing to allow food boluses and liquid to pass from the esophagus into the stomach and then quickly contracting again to keep the contents of the stomach separate from the esophagus. The lower esophageal sphincter can be weakened because of increased pressure on the abdomen from obesity or pregnancy, exposure to tobacco smoke, and some medications, so the risk of GER is increased in these scenarios.

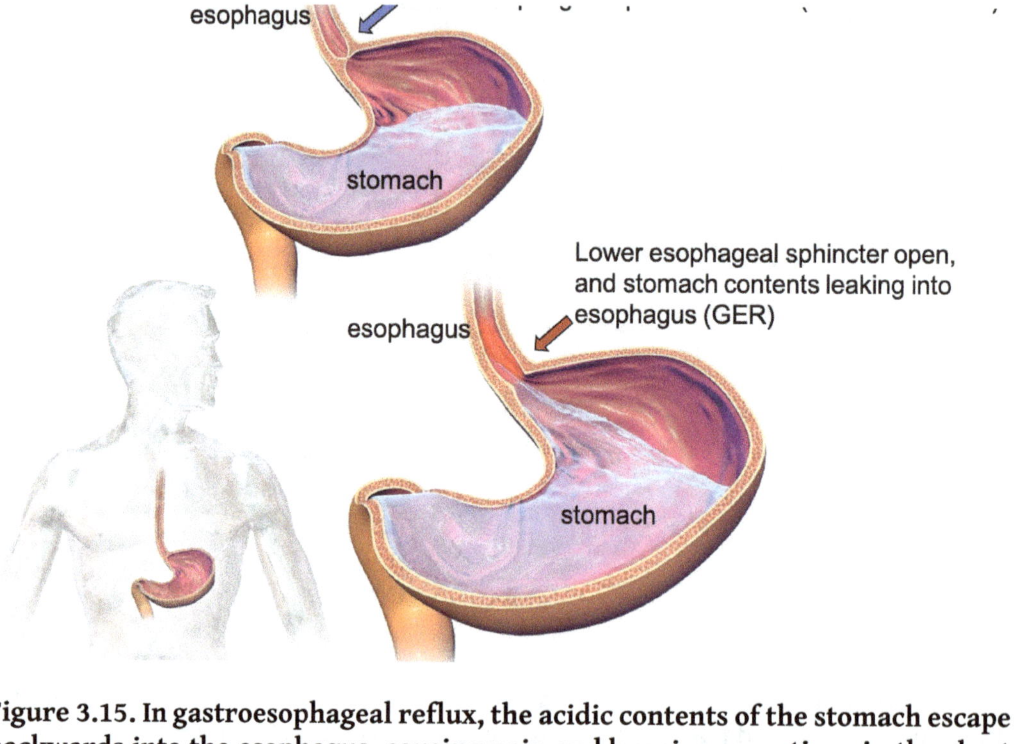

Figure 3.15. In gastroesophageal reflux, the acidic contents of the stomach escape backwards into the esophagus, causing pain and burning sensations in the chest and/or throat.

Occasional heartburn is a common complaint, especially after eating large greasy or spicy meals. However, if it occurs more than twice per week, it may be diagnosed as gastroesophageal reflux disease (GERD), which should be treated not only to relieve the discomfort that it causes but also to prevent damage to the tissues of the esophagus, which can increase the risk of cancer. In addition to heartburn, GER and GERD can cause difficult and painful swallowing, a persistent sore throat or cough, a sense that there's a lump in your throat, and nausea and vomiting.

The symptoms of GER and GERD can often be addressed through diet and lifestyle changes, including the following:

- Avoid foods that seem to trigger symptoms. Common culprits are greasy or spicy foods, chocolate, coffee, peppermint, alcohol, and acidic foods such as tomatoes or citrus.
- Eat smaller, more frequent meals instead of large meals.
- Avoid eating 3 hours before bedtime, and stay upright for 3 hours after eating.
- Wear clothing that is loose around the abdomen.
- Raise the head of your bed by 6 to 8 inches by placing blocks under the bedposts. (Extra pillows will not help.)
- Quit smoking and avoid secondhand smoke, if needed.
- Lose weight, if needed.

Over-the-counter medications like antacids (Maalox, Mylanta, Rolaids) can also help with occasional heartburn. If symptoms are persistent and frequent, it's wise to see a doctor to be checked out for GERD and to discuss other treatment and medication options.[1,2]

Peptic Ulcers

Peptic ulcers are sores on the tissues lining the esophagus, stomach, or duodenum (the first section of the small intestine). They occur when the mucous coating the GI tissues is damaged, exposing the tissue to pepsin and hydrochloric acid. This further erodes away the tissues, causing pain if it damages a nerve and bleeding if it damages a blood vessel. Ulcers are most serious if they perforate the wall of the GI tract, which can cause a serious infection. Peptic ulcers cause stomach pain, often when the stomach is empty, and may go away when you eat or take antacids. In the most serious cases, blood may be seen in vomit and/or the stool, and the patient may have very sharp and persistent stomach pain.

Doctors used to believe that stress and excessive stomach acid caused peptic ulcers, so they would recommend a bland diet, stress reduction, and acid-suppressing medications as treatment. However, these treatments often didn't work, because they weren't actually treating the root cause of the ulcers. We now know that there are two main causes of peptic ulcers. The first and most common cause is infection with a bacteria called Helicobacter pylori (H. pylori). *H. pylori* infection is very common, with about %50 of the population worldwide harboring the bacteria, most of them living without any symptoms. It's not certain how people are infected with *H. pylori* or why they cause ulcers in some people and not in others. However, understanding the link between *H. pylori* and ulcers was an important discovery, because it led to effective treatments. The *H. pylori* bacteria are able to survive the acidic environment of the stomach, and they damage the mucous coating of the GI tract, leaving it vulnerable to further damage from acid and pepsin. Ulcers caused by *H. pylori* infection are treated with antibiotics to kill the bacteria, with the addition of an acid-suppressing medication to allow the tissue to heal.

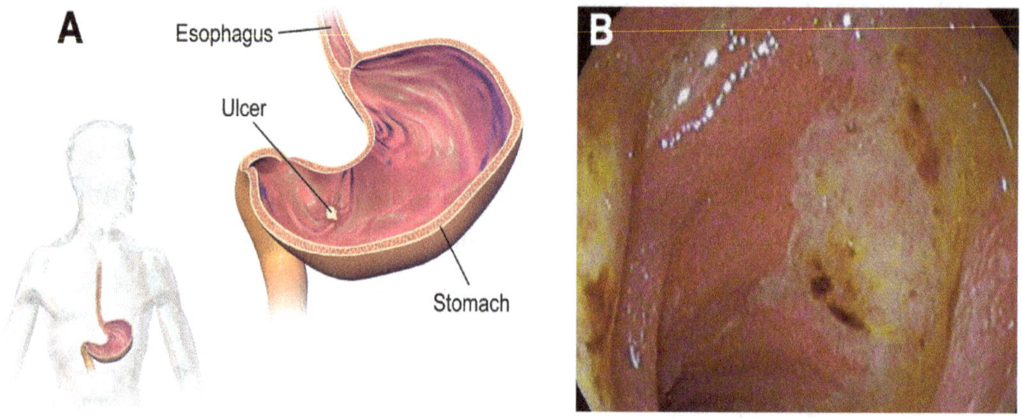

Figure 3.16. Peptic ulcers. (A) The location of a peptic ulcer in the stomach. (B) A photo from an endoscopy of a patient with an ulcer in the duodenum.

The second major cause of peptic ulcers is long-term use of nonsteroidal anti-inflammatory drugs (NSAIDs), such as aspirin or ibuprofen. One of the side effects of NSAIDS is that they block the production of an enzyme that protects the stomach lining, so using these medications frequently and chronically can increase the risk of developing a stomach ulcer. To treat an ulcer caused by NSAIDs, doctors recommend stopping or reducing the use of NSAIDs. They may also prescribe an acid-reducing medication to allow the tissue to heal.[3]

Diarrhea and Constipation

Both diarrhea and constipation can occur if the normal function and rhythm of the GI tract is disrupted. If waste matter moves too quickly through the large intestine, not enough water is absorbed, resulting in the loose, watery stools characteristic of diarrhea. This is most commonly caused by ingesting food or water contaminated with bacteria (e.g., *E. coli, Salmonella*), viruses (e.g., norovirus, rotavirus), or parasites (e.g., *Cryptosporidium enteritis, Giardia lamblia*). Dietary allergies and intolerances can also cause diarrhea, as we'll discuss on the next page.

Complications of diarrhea include dehydration and malabsorption of nutrients. On the other end of the spectrum is constipation, characterized by infrequent bowel movements (less than 3 times per week) with stools that are hard, dry, or lumpy, and often painful to pass. Sometimes, constipation is caused by holding stool and delaying defecation. That gives the colon and rectum additional time to absorb water, making the feces too hard and dry. Delaying defecation is common in children or others who may fear that it will hurt to pass a stool, but of course, holding it only worsens the problem. Constipation can also occur due to other disruptions in daily rhythms, such as changing what or how much you eat, travel, or medication changes. Constipation is common in pregnancy due to hormonal changes. It also becomes more common with age, which may be due to decreased physical activity, medication use, or weakness in the smooth muscle of the intestine. Constipation can be a sign of another medical problem, so chronic constipation should be checked out by a doctor.

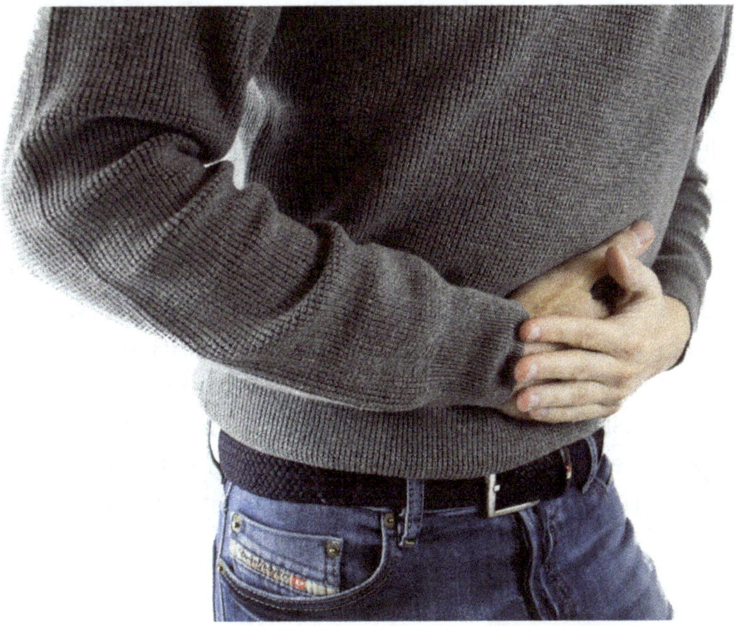

Constipation can often be addressed by dietary changes, including eating more high-fiber foods (whole grains, legumes, fruits, vegetables, nuts, etc.) and drinking more water. It can also be helpful to attempt a bowel movement after meals, when the intestine is more active, and to make that a habit to try to establish more regularity in bowel movements. A caffeinated beverage with breakfast can help, as can increasing physical activity. Fiber supplements such as Metamucil, Citrucel, or Benefiber can be helpful for increasing fiber intake and addressing constipation, at least in the short term. However, it's preferable to transition to dietary sources of fiber, as they come packaged with many other valuable nutrients.

Laxatives may also be helpful to address constipation in the short-term but are usually not a good long-term solution. It's possible to become dependent on some types of laxatives for bowel movements, meaning that the colon doesn't contract normally on its own. In these cases, a doctor can help make a plan to gradually reduce laxative use and find other ways to improve bowel regularity.

Irritable Bowel Syndrome (IBS)

Irritable bowel syndrome (IBS) is a type of functional GI disorder, meaning that it's caused by a disruption in the signals between the brain and gut. People suffering from IBS often experience abdominal pain, bloating, the feeling that they can't finish a bowel movement, as well as diarrhea or constipation or both, often in cycles. IBS is common; about 12% of people in the U.S. are thought to have it. It's more common in women, seems to run in families, and is often associated with stress, history of trauma, or severe GI infections. IBS isn't well understood. It's not clear what causes it, and it may have different causes in different people.

Of course, since the cause of IBS isn't understood, that lack of understanding makes it difficult to treat. Some people find that eating more fiber-rich foods and increasing physical activity improve their symptoms, so these are good first steps (and good for health regardless of their effect on IBS). Others find that following a diet that is low in carbohydrates called FODMAPs helps their symptoms. FODMAPs are fermentable carbohydrates found in many foods and can usually be eaten without issue by most people. Foods that are high in FODMAPs include fruits, vegetables, legumes, dairy products, wheat, and honey, so this is a *very* restrictive diet and should only be attempted with the guidance of a dietitian. Without careful planning, a low-FODMAP diet can be deficient in fiber, vitamins, and minerals. It is usually followed for just a few weeks, and if it helps with symptoms, foods are gradually added back to see what can be tolerated. In addition to dietary strategies, physicians sometimes prescribe medications to treat the symptoms of diarrhea or constipation associated with IBS.

Inflammatory Bowel Disease (IBD)

Inflammatory bowel disease (IBD) includes two types of disorders: ulcerative colitis and Crohn's disease. Ulcerative colitis is specific to the large intestine (colon) and rectum, whereas Crohn's disease can affect any part of the GI tract. Both are chronic inflammatory conditions in which symptoms may periodically flare and become more severe. IBD is often confused with IBS, because of the similarities in their names and some symptoms. However, they are different disorders with different causes. IBD is generally more severe and long-lasting, and it causes damage to the GI tract that can be seen on endoscopy (when a camera is inserted into the GI tract to visualize the interior). It's important to get an accurate diagnosis of IBD in order to treat the disorder appropriately.[9]

Common symptoms of IBD are diarrhea, cramping and abdominal pain, feeling tired, and weight loss. IBD may be caused by autoimmune reactions (in which the immune system attacks the body's own cells, in this case the cells of the GI tract) or certain genes, and other causes are being investigated. IBD often develops in people during adolescence or in their 20s. It may be treated with medications to reduce inflammation or modulate the immune system, or sometimes surgery.[10,11]

FOOD INTOLERANCES, ALLERGIES, AND CELIAC DISEASE

Food is a source of nutrients for our bodies, and as we've learned, the GI tract functions to extract those nutrients from food and absorb them into the body. But sometimes, specific foods can cause problems for the GI tract and the body, including food intolerances, food allergies, and celiac disease. These conditions are often confused for one another, but they have different causes, symptoms, and approaches to treatment.

FOOD INTOLERANCES

A food intolerance occurs when a person has difficulty digesting a specific food or nutrient, causing unpleasant GI symptoms such as gas, bloating, flatulence, cramping, and diarrhea. Food intolerances are commonly caused by the body not producing enough of a particular digestive enzyme, so the symptoms generally involve the digestive system, and the severity of symptoms usually correlates with how much of the food was eaten. Unlike food allergies, the immune system does not play a role in food intolerance, and while the symptoms are unpleasant, they are generally not dangerous and will subside once the food passes out of the GI tract. People with food intolerances can also often consume small amounts of the offending food without symptoms.[1]

Lactose intolerance is a common food intolerance. People with lactose intolerance do not produce enough of the enzyme lactase, which is responsible for digesting the milk sugar lactose into single sugar molecules that can be absorbed in the small intestine. Undigested lactose can't be absorbed, so it continues on to the large intestine. There, it draws more water into the large intestine, and bacteria metabolize the lactose, resulting in gas and acid production. These conditions cause the uncomfortable symptoms of gas, bloating, and diarrhea within about 30 minutes to two hours of consuming dairy foods.

As with most food intolerances, people with lactose intolerance can often consume some amount of lactose without discomfort, although this varies from person to person. Aged hard cheese, buttermilk, and yogurt (as long as it doesn't include added milk solids) are often well-tolerated because they are low in lactose, which is consumed by bacteria during fermentation and aging. In addition, lactose-free milk and lactase enzyme supplements are available. Dairy products are some of the main dietary sources of calcium and vitamin D, so people who avoid dairy need to take special care to include other sources of these nutrients in their diets.[2]

Figure 3.17. Taking a lactase enzyme supplement allows many people with lactose intolerance to eat dairy products without suffering symptoms.

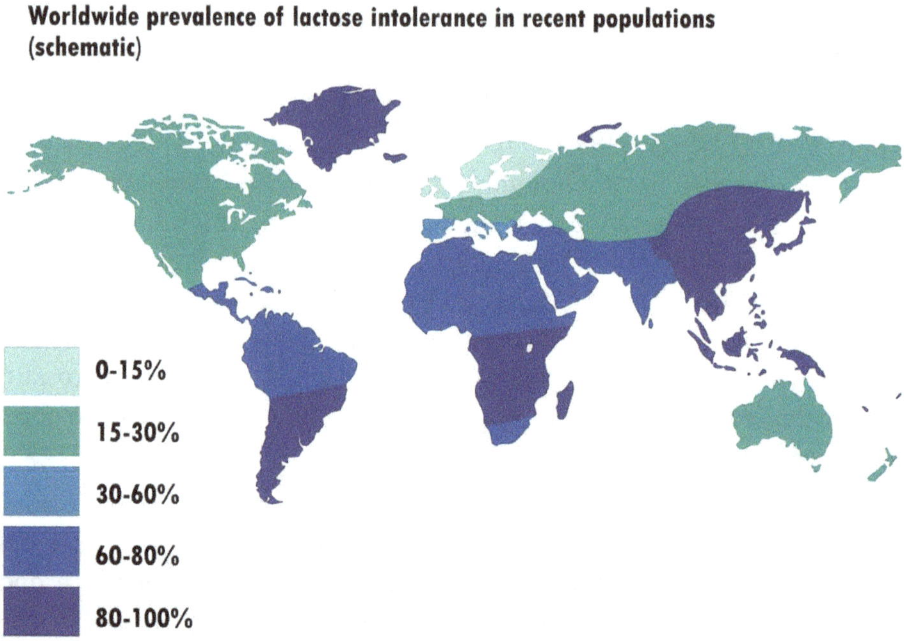

Figure 3.18. The prevalence of lactose intolerance worldwide.

The vast majority of humans are born with the ability to digest lactose. All mammalian milk, including human milk, contains lactose, so historically, infants with lactose intolerance wouldn't have survived. (Today, infants with lactose intolerance can consume soy-based infant formula.) Beyond infancy, lactose intolerance depends on your genes. In much of the world, it's common for the activity of the lactase gene to decline with age, resulting in less lactase production and more lactose intolerance. Worldwide, 65% of the human population has some degree of lactose intolerance in adulthood. On the other hand, lactose tolerance is common in cultures where early domestication of dairy animals provided an important source of nutrition. In the U.S., adults of European descent can often tolerate lactose, whereas lactose intolerance is common among Asian Americans, African Americans, Mexican Americans, and Native Americans.

FOOD ALLERGIES

In addition to its role in digestion, the GI tract serves an important immune function. Intestinal cells form the barrier between the interior of the body and the lumen, or tube, of the GI tract, which is technically outside of the body and teeming with potential pathogens. Immune tissue in the GI tract and other parts of the body produce immune cells that target foreign invaders, in part through the production of antibodies, protective proteins that bind to foreign substances. However, this function requires the immune system to accurately distinguish between normal food proteins and invading pathogens. A food allergy is what happens if the immune system mistakenly identifies a food protein as an invasive threat.

The most common type of food allergy involves immunoglobulin E (IgE), a type of antibody produced by the immune system in response to a specific substance, orallergen. Symptoms of an allergic reaction usually occur immediately after consuming the food (i.e., within seconds to minutes), although reactions can sometimes be delayed by two hours or more. Because an allergic reaction is caused by the immune system, it can lead to symptoms all over the body, including skin rashes; swollen lips, face, or throat; wheezing and difficulty breathing; nausea and vomiting; cramping; diarrhea; and rarely, a dangerous drop in blood pressure. A severe allergic reaction involving more than one organ system—a rash coupled with difficulty breathing, for example—is called anaphylaxis. Anaphylaxis can be life-threatening and should be treated immediately with epinephrine, commonly administered by injection with a device such as an EpiPen, and then person should seek immediate medical attention.

The most common food allergies in the U.S. are caused by proteins in peanuts, tree nuts, milk, shellfish, eggs, fish, wheat, soy and sesame.[4,5] A 2019 study reported that %19 of adults in the U.S. believe they're allergic to at least one food. After asking people about their symptoms, the researchers estimated that the true incidence of food allergies is closer to %11, while the remaining %8 of people likely have a food intolerance.

Figure 3.19. Common food allergens: peanuts, dairy, wheat, eggs, and shrimp. At right, an EpiPen, containing injectable epinephrine, is pictured.

Food allergies are common in children, affecting about 8% of U.S. children, although allergies can also develop later in life.[5] It's common for young children to outgrow allergies to egg, dairy, wheat, or soy, but peanut, tree nut, and shellfish allergies are often lifelong. Recent research has found that letting babies eat common food allergens, particularly peanut products, can prevent the development of allergies, perhaps by allowing the immune system an early opportunity to learn to differentiate between food proteins and invading pathogens.[6]

If you think you may have a food allergy, it's important to see an allergist to ensure you have an accurate diagnosis. Food allergies are diagnosed based on symptoms after consuming a food, specific IgE blood tests, and/or skin prick tests, where a tiny amount of food protein is scratched onto the skin to test for a reaction. Blood and skin tests determine whether a person is sensitized to an allergen, meaning that they're producing IgE antibodies to the food. However, the presence of IgE antibodies doesn't definitively mean a person has a food allergy; it's common to have a positive IgE test but still be able to eat the food without symptoms. The gold standard test for diagnosing is an oral food challenge—consuming a small amount of the food and watching for signs of a reaction—although due to time, cost, and risk, these are not always conducted. If you are diagnosed with a food allergy, you will be counseled to strictly avoid the food and carry injectable epinephrine in case of accidental consumption. Unlike with food intolerances, consuming even a small amount of food can cause a serious reaction in those with food allergies.[1]

There are some promising new therapies for treating food allergies that involve exposing a person to small amounts of the allergen to try to teach the body to tolerate it. These don't cure the allergy completely but may reduce the risk of a severe allergic reaction. The first of these therapies was approved by the U.S. Food and Drug Administration in January 2020.[7]

There are blood tests available that claim to screen for as many as 90 to 100 food allergies from one blood sample. These measure a different type of antibody called IgG, but the presence of IgG does not indicate a food allergy. Therefore, these tests are not recommended by allergy experts, because they may cause a person to unnecessarily fear and avoid a long list of foods to which they are not allergic.[8]

Celiac Disease

Celiac disease is an autoimmune disorder affecting between 0.5 and %1.0 of people in the U.S., or one in every 100 to 200 people.[9] In autoimmune diseases, the immune system produces antibodies that attack and damage the body's own tissues. In the case of celiac disease, the body has an abnormal immune reaction to gluten (a group of proteins found in wheat, rye, and barley), causing antibodies to attack the cells lining the small intestine. This results in damage to the villi, decreasing the surface area for nutrient absorption. There is no cure for celiac disease, but it's very effectively treated by eliminating gluten from the diet.[10]

Symptoms of celiac disease can range from mild to severe and can include pale, fatty, loose stools, gastrointestinal upset, constipation, abdominal pain, and skin conditions. Nutrient malabsorption can lead to weight loss and in children, a failure to grow and thrive. Symptoms can appear in infancy or much later in life, as late as age seventy. Celiac disease is not always diagnosed, because the symptoms may be mild. Even without symptoms, the disease can still damage the small intestine and impair nutrient absorption. Nutrient deficiencies can cause health problems over time, particularly in children and the elderly. For example, poor absorption of iron and folic acid can cause anemia, which impairs oxygen transport to cells in the body. Calcium and vitamin D deficiencies can lead to osteoporosis, a disease in which bones become brittle.

Diagnosis of celiac disease begins with a blood test for specific antibodies that are elevated in those with the disease. If the blood test is positive, the diagnosis is confirmed with a biopsy of the small intestine, a procedure in which a small amount of tissue is removed for examination. These tests may not accurately detect celiac disease in people already consuming a gluten-free diet, because without gluten, antibody levels may be low and damage to the small intestine may not be visible. This is why it's best to be tested for celiac before eliminating gluten from the diet.

It isn't clear what causes celiac disease; genetics play a role, but other factors also seem to influence its development. Celiac disease is most common in people of European descent and is rare in people of African American, Japanese, and Chinese descent. It is more prevalent in women and in people with type 1 diabetes, autoimmune thyroid disease, and Down and Turner syndromes.

Celiac disease is treated by completely avoiding gluten, as consuming even small amounts can cause intestinal damage. People with celiac can consume grains that don't contain gluten, including rice, corn, millet, buckwheat, and quinoa.

Oats can be consumed, although they are often contaminated with gluten from neighboring fields or shared processing equipment, so it's best to buy oats labeled gluten-free. There are also an increasing number of gluten-free products available in stores. After eliminating gluten from the diet, the tissues of the small intestine usually heal within six months.

Figure 3.20. Celiac disease is caused by an autoimmune response to gluten, found in wheat, rye, and barley. At right is a section of the small intestine from a biopsy, visualized under a microscope, from a celiac patient. The villi, which would normally be finger-like projections, are blunted and flattened by damage caused by the disease.

Celiac disease is different from a wheat allergy in both cause and symptoms. Because celiac disease is an autoimmune condition, the cells of the small intestine come under attack, and damage causes chronic symptoms. A wheat allergy, on the other hand, is caused by antibodies attacking an allergen in the wheat itself, and the symptoms are usually immediate and acute.

Sometimes, people test negative for celiac disease but still believe that consuming gluten is causing symptoms, usually gastrointestinal in nature. They may eliminate gluten from their diet and find that they feel better. This is often called non-celiac gluten sensitivity (NCGS). It's not clear what causes NCGS or why a gluten-free diet is helpful. In some cases, it may be a placebo effect. In others, it seems that it's not gluten causing the problem but other dietary components, such as FODMAPs, that happen to also be low in a gluten-free diet. And in some cases, gluten does seem to cause symptoms in people without celiac disease, although researchers don't understand why.

Except in the case of celiac disease, wheat allergy, or confirmed NCGS, gluten-free foods or a gluten-free diet are not inherently more healthful. In fact, packaged gluten-free foods are often more highly processed, with more added sugar, salt, and fat, compared to foods containing wheat. A gluten-free diet can also be lower in fiber, so those following the diet should be sure to include naturally gluten-free whole grains, legumes, nuts, fruits, and vegetables to provide adequate fiber.

UNIT-4

CARBOHYDRATES

CARBOHYDRATES: AN INTRODUCTION

If someone says to you, "I love carbohydrates, and I eat them all day long!" what would you assume they're eating?

Do You Picture this?

Figure 4.1. Examples of carbohydrate-rich snack foods.

And This?

Figure 4.2. Examples of grain-based foods.

When we ask this question in class, most students describe foods like the ones above. However, carbohydrates are found not just in grains, or in sweets and processed foods, but in **every** food group.

In fact, carbohydrates are the most abundant nutrient (except water) in the diets of most humans around the world. Since the dawn of agriculture, human cultures have relied on staple grains, such as corn, rice, and wheat, as the foundation of their diets, and these foods are rich in carbohydrates. But fruits and vegetables, dairy products, legumes, and nuts also have naturally-occurring carbohydrates. And of course, carbohydrates are a key ingredient in desserts, sugar-sweetened beverages like sodas, and many of the packaged snack foods that are readily available and—let's face it—can be hard to stop eating.

In other words, if someone says they eat a high carbohydrate diet, that could mean many different things. They very well could be talking about a balanced diet focused on whole foods, like this:

Figure 4.3. Examples of whole foods containing carbohydrates, including fresh fruit, legumes and grains, and cheese.

The diet industry likes to sell us simple messages about "good" and "bad" foods, and these days, we tend to hear that carbohydrates are in the "bad" group. But given that carbohydrates are in so many different types of foods, that's obviously an oversimplified message—and it's not fair to all of the awesome sources of carbohydrates in the world of food. Not all carbohydrate-rich foods are the same. In this unit, you'll learn to appreciate the nutrient-dense carbohydrate foods, identify which don't offer as valuable a nutritional package, and understand how a balanced diet can include all of them.

Unit Learning Objectives

After Completing this Unit, You Should be Able to:

- Classify the different types of carbohydrates, identify their food sources, and discuss how these carbohydrates are digested and absorbed in the body.
- Define the guidelines for total carbohydrate, fiber, and added sugar intake.
- Explain how glucose is regulated and utilized in the body and describe how the body adapts to a low carbohydrate intake.
- List the causes, complications, and treatment for different types of diabetes.
- Describe the health benefits, types of, and food sources of dietary fiber.

- Differentiate between whole and refined grains in foods by examining food labels.
- Distinguish between added and natural-occurring sugars in foods, and discuss health implications of too much added sugar.
- Identify sugar substitutes in foods, and describe potential benefits and drawbacks of sugar substitutes.

DIFFERENT TYPES OF CARBOHYDRATES

On this page, we'll get acquainted with the chemical structure of different types of carbohydrates and learn where we find them in foods.

First, all carbohydrates are made up of the same chemical elements:

- Carbon (that's the "carbo-" part)
- Hydrogen and oxygen, in about a two-to-one proportion, just like in H_2O (that's the "-hydrate" part)

For this reason, you may see carbohydrates abbreviated as "CHO" in our class.

Carbohydrates can be divided into two main types: simple and complex. Simple carbohydrates are made up of just one or two sugar units, whereas complex carbohydrates are made up of many sugar units. We'll look at each of these in turn. This figure gives you an overview of the types of carbohydrates that we'll cover.

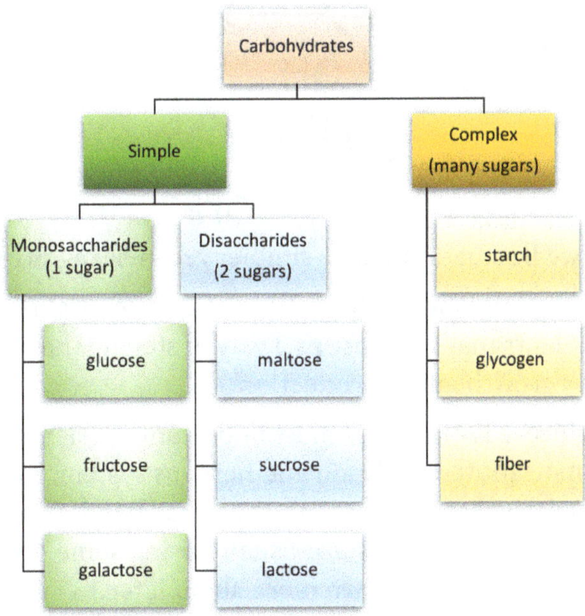

Figure 4.4. Carbohydrates can be divided into two main types: simple (including monosaccharides and disaccharides) and complex.

Simple Carbohydrates

Simple carbohydrates are sometimes called "sugars" or "simple sugars." There are 2 types of simple carbohydrates: monosaccharides and disaccharides.

Monosaccharides contain just one sugar unit, so they're the smallest of the carbohydrates. (The prefix "mono-" means "one.") The small size of monosaccharides gives them a special role in digestion and metabolism. Food carbohydrates have to be broken down to monosaccharides before they can be absorbed in the gastrointestinal tract, and they also circulate in blood in monosaccharide form.

There are 3 monosaccharides

- Glucose
- Fructose
- Galactose

Note that all three have the same chemical formula ($C_6H_{12}O_6$); the atoms are just arranged a bit differently.

- 1 - Glucose

Here's the chemical structure of glucose:

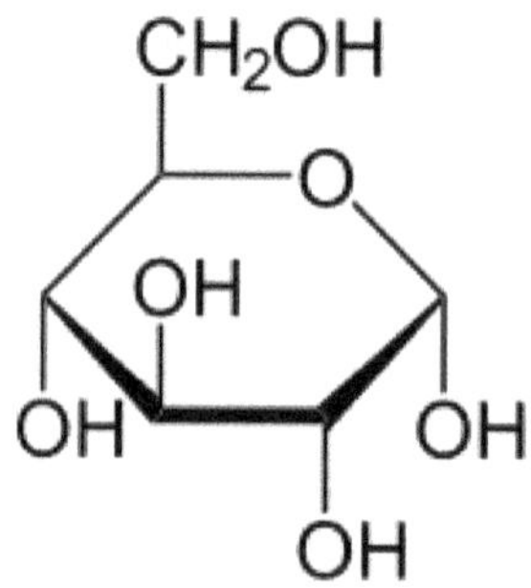

In this class, we'll sometimes use a simpler green hexagon to represent glucose:

You're already familiar with glucose, because it's the main product of photosynthesis. Plants make glucose as a way of storing the sun's energy in a form that it can use for growth and reproduction.

In humans, glucose is one of the most important nutrients for fueling the body. It's especially important for the brain and nervous system, which aren't very good at using other fuel sources. Muscles, on the other hand, can use fat as an energy source. (In practice, your muscles are usually using some combination of fat and glucose for energy, which we'll learn more about later.)

Food sources of glucose: Glucose is found in fruits and vegetables, as well as honey, corn syrup, and high fructose corn syrup. (All plants make glucose, but much of the glucose is used to make starch, fiber, and other nutrients. The foods listed here have glucose in its monosaccharide form.)

2 – Fructose

Here's the chemical structure of fructose:

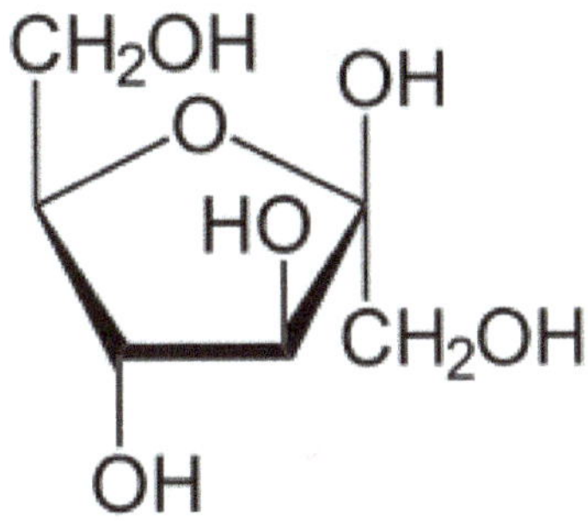

In this class, we'll sometimes use a simpler purple pentagon to represent fructose:

Fructose is special because it is the sweetest carbohydrate. Plants make a lot of fructose as a way of attracting insects and animals, which help plants to reproduce. For example, plants make nectar, which is high in fructose and very sweet, to attract insects that will pollinate it. Plants also put fructose into fruit to make it tastier. Animals eat the fruit, wander away, and later poop out the seeds from the fruit, thereby sowing the seeds of the next generation. Animal gets a meal, and the plant gets to reproduce: win-win!

Figure 4.5. Fructose in nature: A bee collects sweet nectar from a flower, in the process spreading pollen from flower to flower and helping plants to reproduce. Bees use nectar to make honey, which humans harvest for use as a sweetener. (Honey contains a mix of sucrose, fructose, and glucose). A kiwi is sweetened in part by fructose. Animals enjoy the sweet fruit and then later poop out the seeds, sowing them for a new generation of kiwi trees.

Food Sources of Fructose: Fruits, Vegetables, Honey, High Fructose Corn Syrup

3 – Galactose

Here is the chemical structure of galactose:

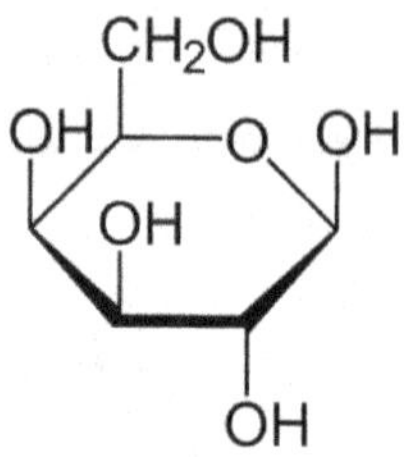

In this class, we'll sometimes use a blue hexagon to represent galactose:

Food sources of galactose: Galactose is found in milk (and dairy products made from milk), but it's almost always linked to glucose to form a disaccharide (more on that in a minute). We rarely find it in our food supply in monosaccharide form.

The second type of simple carbohydrates is disaccharides. They contain two sugar units bonded together.

There are 3 disaccharides:

- Maltose (glucose + glucose)
- Sucrose (glucose + fructose)
- Lactose (glucose + galactose)

Maltose

Here is the chemical structure of maltose:

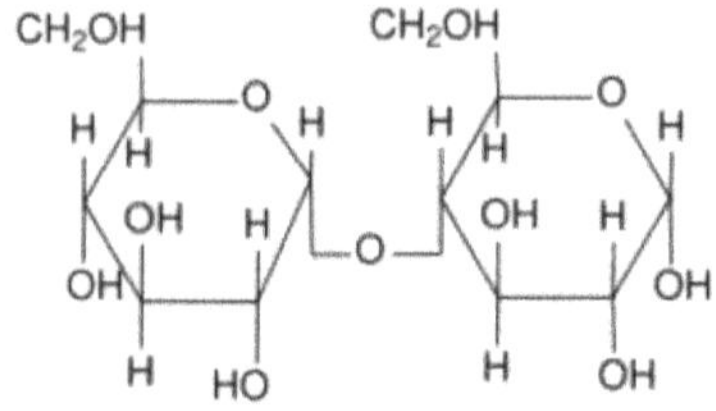

Maltose is made of two glucose molecules bonded together. It doesn't occur naturally in any appreciable amount in foods, with one exception: sprouted grains. Grains contain a lot of starch, which is made of long chains of glucose (more on this in a minute), and when the seed of a grain starts to sprout, it begins to break down that starch, creating maltose. If bread is made from those sprouted grains, that bread will have some maltose. Sprouted grain bread is usually a little heavier and sweeter than bread made from regular flour.

Maltose also plays a role in the production of beer and liquor, because this process involves the fermentation of grains or other carbohydrate sources. Maltose is formed during the breakdown of those carbohydrates, but there is very little remaining once the fermentation process is complete.

You can taste the sweetness of maltose if you hold a starchy food in your mouth for a minute or so. Try this with a simple food like a soda cracker. Starch is not sweet, but as the starch in the cracker begins to break down with the action of salivary amylase, maltose will form, and you'll taste the sweetness!

2 – Sucrose

Here is the chemical structure of sucrose:

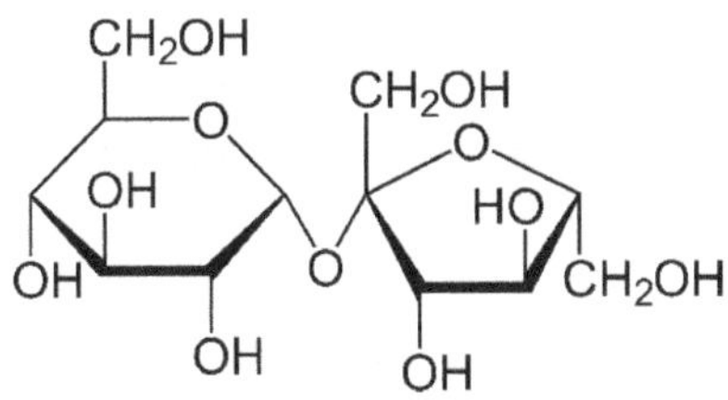

Sucrose is made of a glucose molecule bonded to a fructose molecule. It's made by plants for the same reason as fructose — to attract animals to eat it and thereby spread the seeds.

Sucrose is naturally-occurring in fruits and vegetables. (Most fruits and vegetables contain a mixture of glucose, fructose, and sucrose.) But humans have also figured out how to concentrate the sucrose in plants (usually sugar cane or sugar beets) to make refined table sugar. We also find sucrose in maple syrup and honey.

The sucrose found in a sweet potato is chemically identical to the sucrose found in table sugar. Likewise, the fructose found in a fig is chemically identical to the fructose found in high fructose corn syrup. As we'll discuss more later, what's different is the package the sugars come in. When you eat a sweet potato or a fig, you also get lots of fiber, vitamins, and minerals in that package, whereas sugar and high fructose corn syrup only provide sugar, nothing else. It's not a bad thing to eat sugar. After all, it's a vital fuel for our brain and nervous system. But paying attention to the package it comes in can help us make good overall choices for health.

3 – Lactose

Here is the chemical structure of lactose:

Lactose is made of a glucose molecule bonded to a galactose molecule. It is sometimes called "milk sugar" as it is found in dairy products like milk, yogurt, and cheese. These are the only animal foods that have significant amounts of carbohydrate. Most of our carbohydrates come from plant foods.

COMPLEX CARBOHYDRATES

Complex carbohydrates are also called polysaccharides, because they contain many sugars. (The prefix "poly-" means "many.") There are 3 main polysaccharides:

- Starch
- Glycogen
- Fiber

All three of these polysaccharides are made up of many glucose molecules bonded together, but they differ in their structure and the type of bonds.

1 – Starch

Starch is made up of long chains of glucose. If these chains are straight, they're called amylose; if they're branched, they're called amylopectin.

Here is an amylose segment containing 3 glucose units.

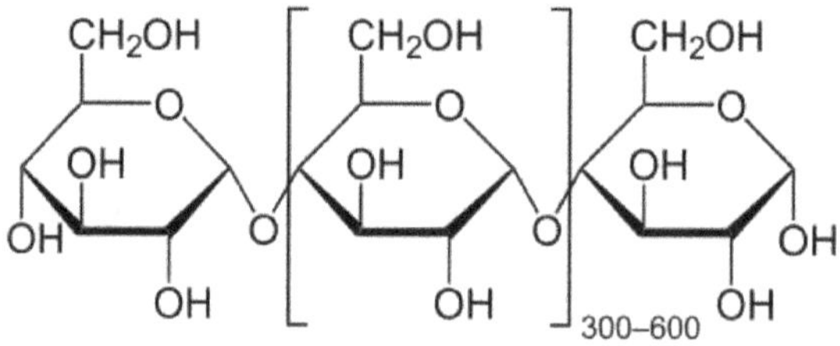

The next figure shows an amylopectin segment containing 4 glucose units. The chemical structure is represented differently, but can you spot the place where it branches?

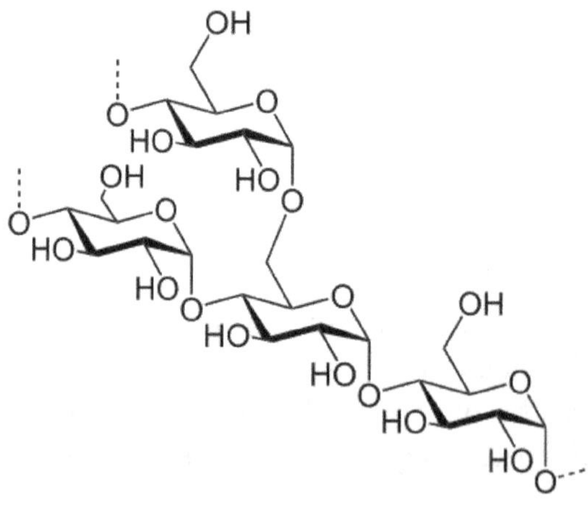

Using our green hexagon to represent glucose, you can picture starch as something like this:

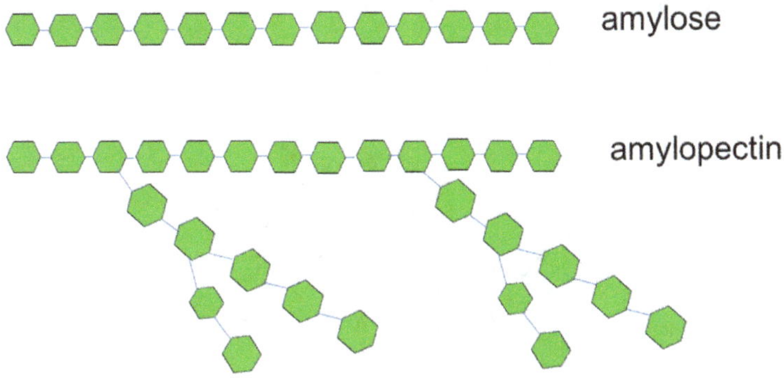

Humans have digestive enzymes to break down both types of starch, which we'll discuss on the next page.

Starch is the storage form of carbohydrate in plants. Plants make starch in order to store glucose. For example, starch is in seeds to give the seedling energy to sprout, and we eat those seeds in the form of grains, legumes (soybeans, lentils, pinto and kidney beans, for example), nuts, and seeds. Starch is also stored in roots and tubers to provide stored energy for the plant to grow and reproduce, and we eat these in the form of potatoes, sweet potatoes, carrots, beets, and turnips.

When we eat plant foods with starch, we can break it down into glucose to provide fuel for our body's cells. In addition, starch from whole plant foods comes packaged with other valuable nutrients. We also find refined starch—such as corn starch—as an ingredient in many processed foods, because it serves as a good thickener.

2 – Glycogen

Glycogen is structurally similar to amylopectin, but it's the storage form of carbohydrate in animals, humans included. It's made up of highly branched chains of glucose, and it's stored in the liver and skeletal muscle. The branched structure of glycogen makes it easier to break down quickly to release glucose to serve as fuel when needed on short notice.

Liver glycogen is broken down to glucose, which is released into the bloodstream and can be used by cells around the body. Muscle glycogen provides energy only for muscle, to fuel activity. That can come in handy if you're being chased by a lion, or sprinting to make your bus! Both liver and muscle glycogen serve as relatively short-term forms of energy storage; together, they can only provide enough glucose to last for about 24 hours in a person fasting or eating a very low carbohydrate diet.

Even though glycogen is stored in the liver and muscles of animals, we don't find it in meat, because it's broken down soon after slaughter. Thus, glycogen is not found in our food. Instead, we have to make it in our liver and muscle from glucose.

Here's a Beautiful Depiction of Glycogen.

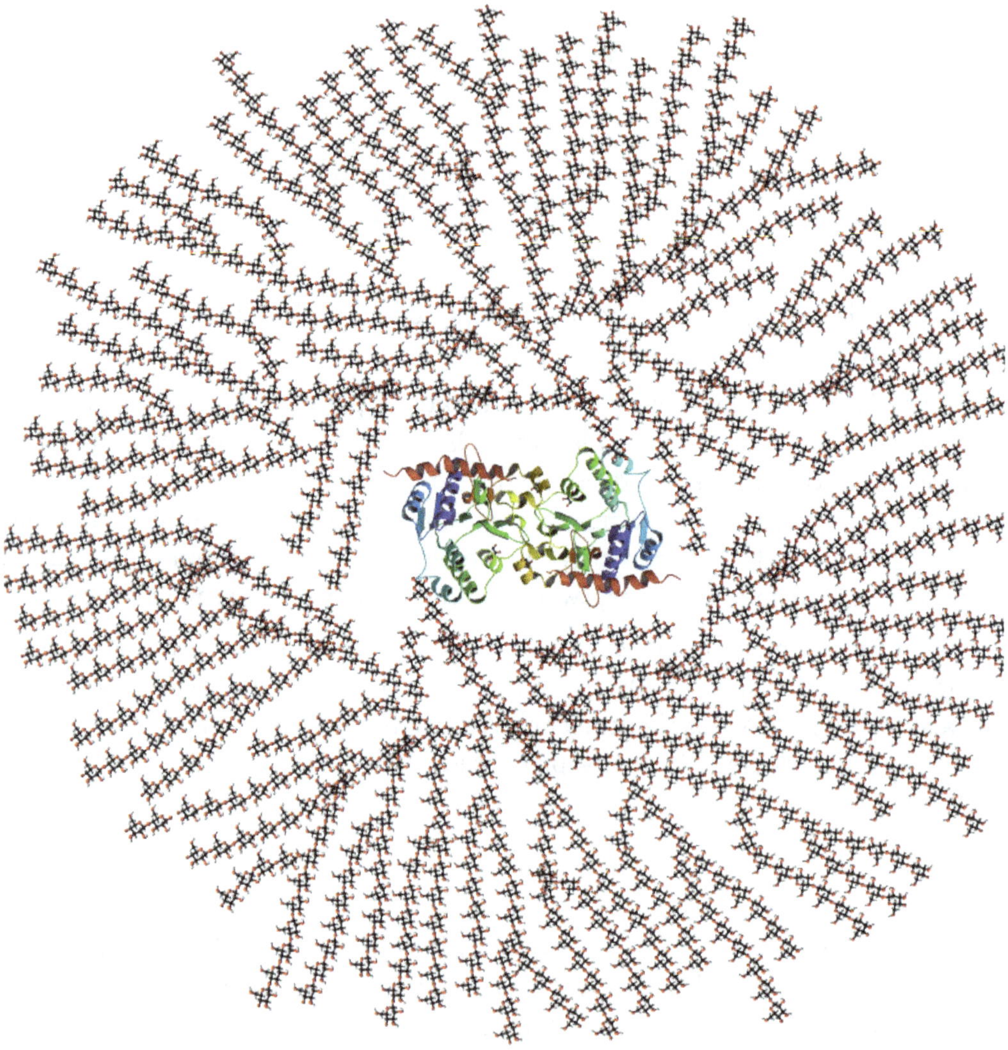

Figure 4.6. Glycogen is made from long, branching chains of glucose, radiating around a central protein.

3 – Fiber

Fiber includes carbohydrates and other structural substances in plants that are indigestible to human enzymes. Fiber is made by plants to provide protection and structural support. Think about thick stems that help a plant stand upright, tough seed husks, and fruit skin that protect what's growing inside. These are full of fiber.

Figure 4.7. Examples of food plants high in fiber, including wheat, broccoli, and apples.

In our food, we find fiber in whole plant foods like whole grains, seeds, nuts, fruits, vegetables, and legumes.

One of the most common types of fiber is cellulose, the main component in plant cell walls. The chemical structure of cellulose is shown in the figure below, with our simplified depiction next to it. You can see that cellulose has long chains of glucose, similar to starch, but they're stacked up, and there are hydrogen bonds linking the stacks.

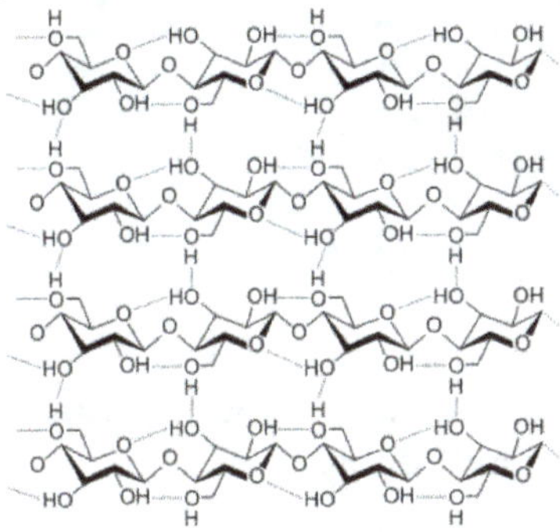

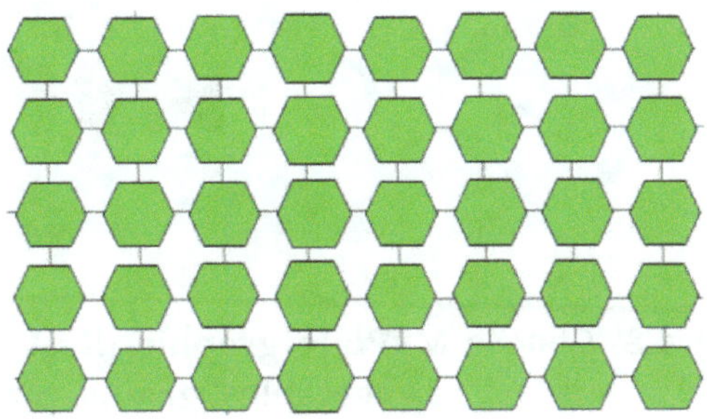

When we eat fiber, it passes through the small intestine intact, because we don't have digestive enzymes to break it down. Then, in the large intestine, our friendly microbiota—the bacteria that live in our colons—go to work on the fiber. Some fiber can be fermented by those bacteria. We'll discuss fiber more later in the unit.

CARBOHYDRATE FOOD SOURCES AND INTAKE RECOMMENDATIONS

Carbohydrates in Foods

Looking at the food groups represented in MyPlate below, which food groups do you think contain carbohydrates? If you answered, all of them, you're correct! This section will review which food groups contain the different types of carbohydrates. One of the goals of this course is to learn more about the different nutrients in foods and to understand the importance of eating a wide variety of foods from the different food groups.

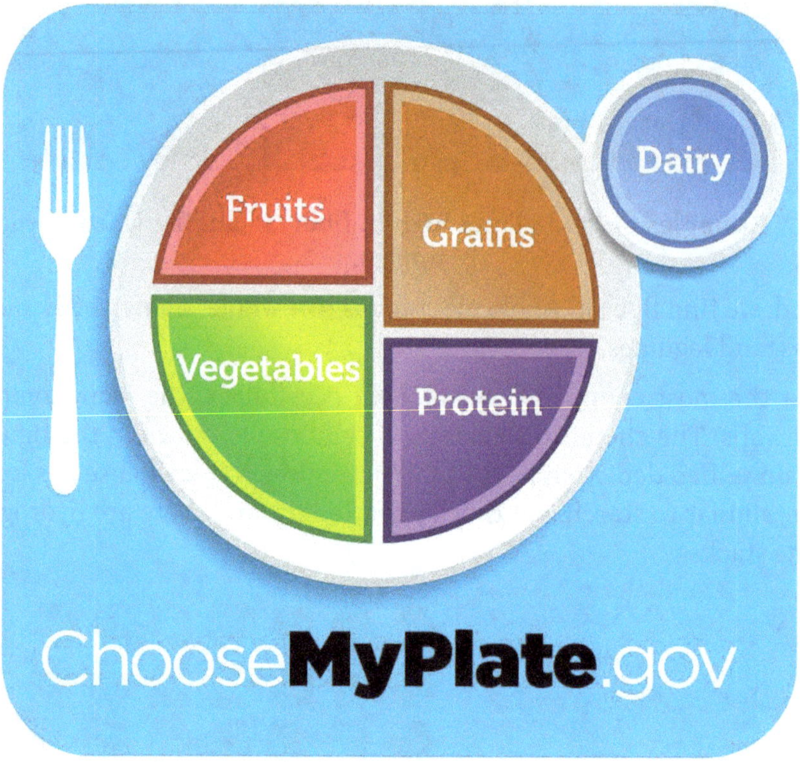

Figure 4.8. Choose MyPlate graphic illustrating the USDA food groups: fruits, vegetables, grains, protein and dairy.

- **Fruits**– Fruits are sweet, so we know they must contain sugar. Fruits contain sucrose, glucose, and fructose. This sugar is naturally-occurring and comes packaged with other great nutrients, like Vitamin C and potassium. Whole fruit also contains fiber, since fiber is found in all whole plant foods. Juice has little to no fiber, even high pulp orange juice.
- **Vegetables**– Some vegetables are sweet and also contain sugar, although much less than fruit. Similar to fruits, some vegetables (like carrots and green beans) contain small amounts of sucrose, glucose, and fructose. Starchy vegetables (corn, peas, and potatoes, for example) primarily contain starch but some are also sweet and contain sucrose, glucose, and fructose (sweet potatoes and sweet corn, for example). Just like whole fruit, any whole vegetable also contains fiber.
- **Grains**– Grains naturally contain starch and fiber. Sprouted grains also contain maltose. If grains are sweetened (sugar is added), they might contain sucrose (white cane sugar) or fructose and glucose (honey and/or HFCS).

- **Dairy**-This is the one animal food that contains carbohydrate. Milk, cheese, and yogurt contain naturally-occurring lactose. If dairy (like yogurt) is sweetened, then it will also contain added sugar like sucrose (white cane sugar) or fructose and glucose (honey and/or HFCS).
- **Protein**– Meats do not contain carbohydrate, but many plant foods that fall into the protein group, like beans and nuts, contain starch and fiber.
- **Fats**– Concentrated fats like butter and oil do not contain carbohydrate.

This information is summarized in the table below:

Table 4.1. USDA food groups with examples of foods and type of carbohydrate present within each food group.

Food Group	Example of Food	Type of Carbohydrate Present
Fruits	Apple, orange, banana	Sucrose, glucose, fructose, and fiber
	Orange juice	Sucrose, glucose, fructose
Vegetables	Non-starchy veggies	Sucrose, glucose, fructose, and fiber
	Starchy veggies (corn, potatoes, sweet potatoes, peas)	Starch and fiber, with varying amount of sucrose, glucose, and fructose
Dairy	Milk, plain yogurt, cheese	Lactose
Grains	Wheat, rice, oatmeal, barley	Starch and fiber
	Sprouted grains	Starch, fiber, and maltose
Protein	Meat	None
	Beans and nuts	Starch and fiber
Fats	Oils, Butter	None

Looking at all the foods that contain carbohydrates, you might be able to guess why eliminating carbohydrates from the diet can lead to weight loss. It drastically reduces the variety of choices one has, leaving you primarily with low carbohydrate veggies and meats. Not surprisingly, people usually consume less calories with this way of eating. However, for most people, this is not a sustainable or enjoyable way of eating, and it can also be hard to consume a nutritionally balanced diet with so many foods off-limits.

CARBOHYDRATE GUIDELINES FOR INTAKE

Total Carbohydrate Intake

The recommended dietary allowance (RDA) for total carbohydrate intake is 130 grams. This is the minimum amount of glucose utilized by the brain, so if you consume less than this, you will probably go into ketosis. In order to meet the body's high energy demand for glucose, the acceptable macronutrient distribution range (AMDR) for an adult is 45%-65% of total calories. This is about 225 grams to

325 grams of carbohydrate per day if eating a 2,000 Calorie diet. (*REMEMBER: 1 gram of carbohydrate contains 4 calories.*)

Fiber Intake

The Adequate Intake (AI) for fiber is 14 grams of fiber for every 1,000 calories consumed. This is about 28 grams for an adult female (19-30 years old) and 38 grams for an adult male (19-30 years old). Most people in the United States only get half the amount of fiber they need in a day—about 12 to 18 grams.

Added Sugar Intake

The 2020 Dietary Guidelines recommend that less than 10% of total calories come from added sugars because of its link to obesity and chronic disease. This means that someone eating a 2,000 calorie diet would want to limit their added sugar intake to about 12 teaspoons per day. To put that in perspective, a 12 oz can of soda has about 10 teaspoons of sugar. We will discuss added sugar in more detail later in the unit.

Below is a chart summarizing the above recommendations.

Table 4.2 Dietary Recommendations for Carbohydrates

	Recommendations
RDA for Total Carbohydrate	130 grams
AMDR for Total Carbohydrate	45% – 65% of total calories
AI for Fiber	14 grams for every 1,000 calories consumed
Dietary guidelines for added sugar	Less than 10% of total calories

DIGESTION AND ABSORPTION OF CARBOHYDRATES

Imagine taking a bite of pizza. It tastes amazing, but it's also full of fuel for your body, much of it in the form of carbohydrates.

What types of carbohydrates would you find in that bite?

- Lactose from the cheese
- Sucrose, glucose, and fructose from the naturally-occurring sugars in the tomatoes, as well as sugar that may have been added to the sauce
- Starch in the flour used to make the crust
- Fiber in the flour, tomatoes, and basil

In order to use these food carbohydrates in your body, you first need to digest them. Last unit, we explored the gastrointestinal system and the basic process of digestion. Now that you know about the different types of carbohydrates, we'll take a closer look at how these molecules are digested as they travel through the GI system.

Carbohydrate Digestion

In the image below, follow the numbers to see what happens to carbohydrates at each site of digestion.

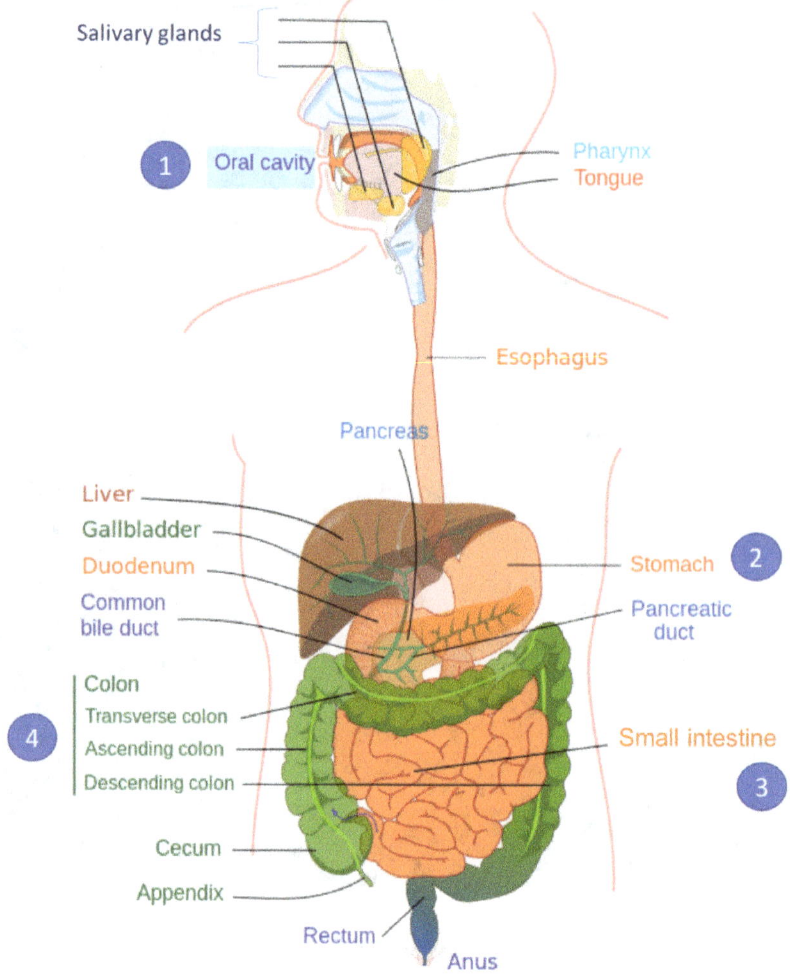

Figure 4.9. The digestive system

1 – Mouth or Oral Cavity

As you chew your bite of pizza, you're using mechanical digestion to begin to break it into smaller pieces and mix it with saliva, produced by several salivary glands in the oral cavity.

Some enzymatic digestion of starch occurs in the mouth, due to the action of the enzyme salivary amylase. This enzyme starts to break the long glucose chains of starch into shorter chains, some as small as maltose. (The other carbohydrates in the bread don't undergo any enzymatic digestion in the mouth.)

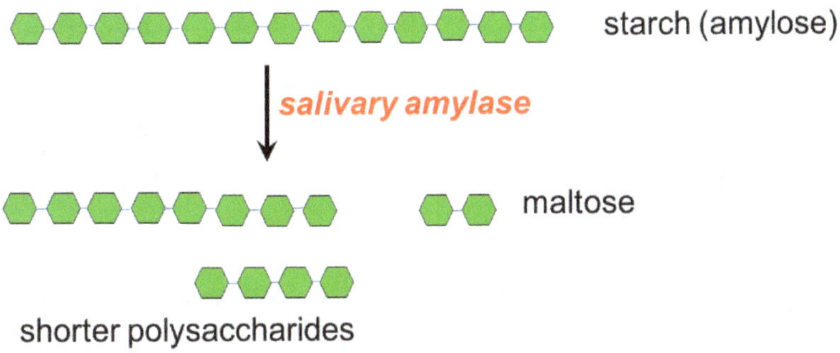

Fig. 4.10. The enzyme salivary amylase breaks starch into smaller polysaccharides and maltose.

2 – Stomach

The low pH in the stomach inactivates salivary amylase, so it no longer works once it arrives at the stomach. Although there's more mechanical digestion in the stomach, there's little chemical digestion of carbohydrates here.

3 – Small intestine

Most carbohydrate digestion occurs in the small intestine, thanks to a suite of enzymes. Pancreatic amylase is secreted from the pancreas into the small intestine, and like salivary amylase, it breaks starch down to small oligosaccharides (containing 3 to 10 glucose molecules) and maltose.

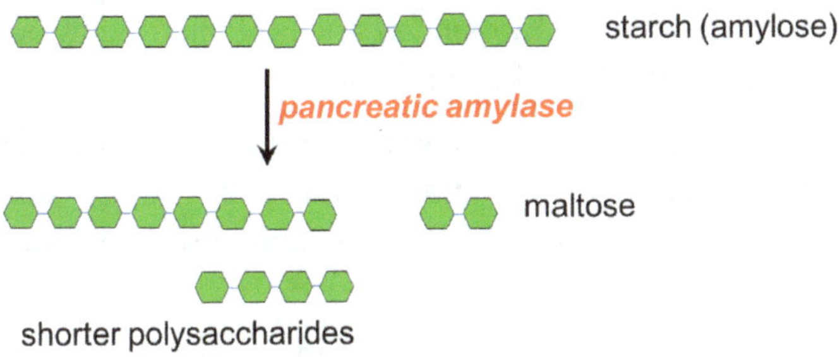

Figure 4.11. The enzyme pancreatic amylase breaks starch into smaller polysaccharides and maltose.

The rest of the work of carbohydrate digestion is done by enzymes produced by the enterocytes, the cells lining the small intestine. When it comes to digesting your slice of pizza, these enzymes will break down the maltose formed in the process of starch digestion, the lactose from the cheese, and the sucrose present in the sauce.

Maltose is digested by maltase, forming 2 glucose molecules.

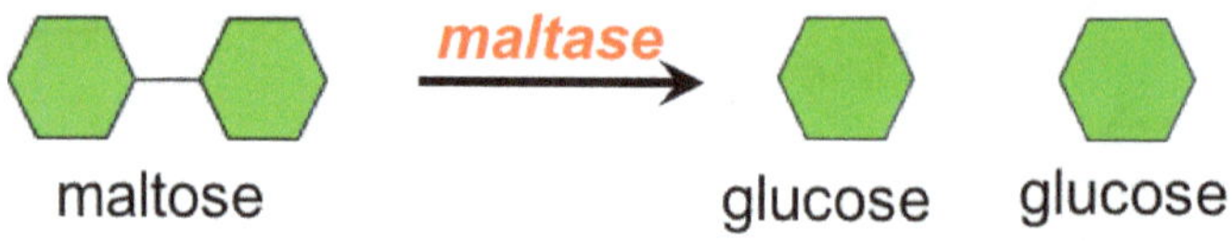

Lactose is digested by lactase, forming glucose and galactose.

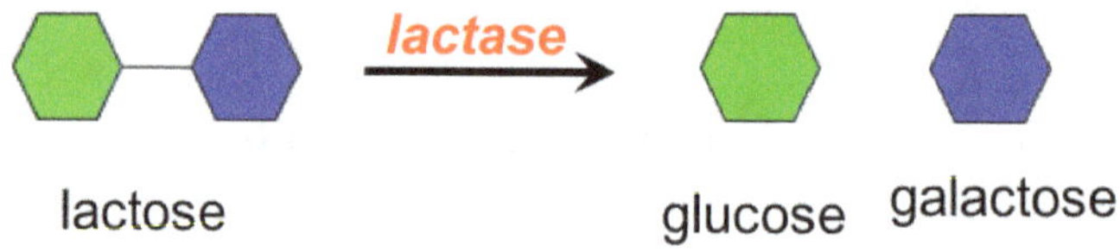

Sucrose is digested by sucrase, forming glucose and fructose.

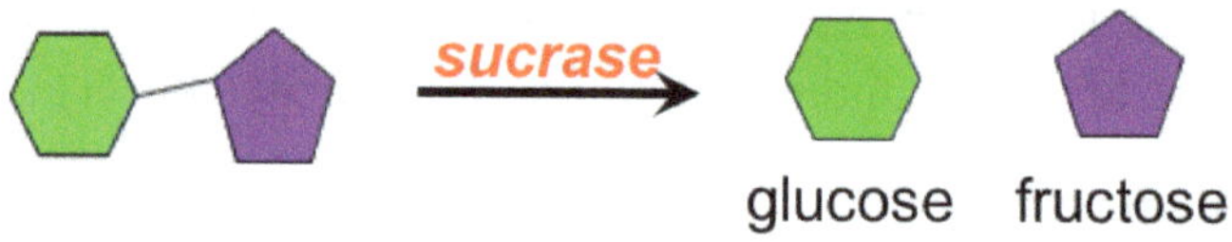

Figure 4.12. Action of the enzymes maltase, lactase, and sucrase.

(Recall that if a person is lactose intolerant, they don't make enough lactase enzyme to digest lactose adequately. Therefore, lactose passes to the large intestine. There it draws water in by osmosis and is fermented by bacteria, causing symptoms such as flatulence, bloating, and diarrhea.)

By the end of this process of enzymatic digestion, we're left with three monosaccharides: glucose, fructose, and galactose. These can now be absorbed across the enterocytes of the small intestine and into the bloodstream to be transported to the liver.

Digestion and absorption of carbohydrates in the small intestine are depicted in a very simplified schematic below. (*Remember that the inner wall of the small intestine is actually composed of large circular folds, lined with many villi, the surface of which are made up of microvilli. All of this gives the small intestine a huge surface area for absorption.*)

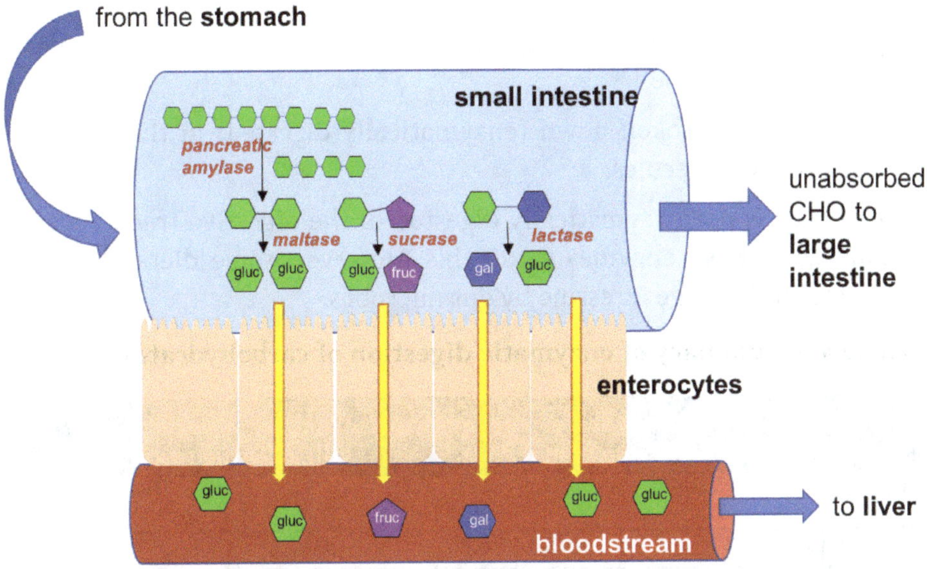

Figure 4.13. Digestion and absorption of carbohydrates in the small intestine.

Fructose and galactose are converted to glucose in the liver. Once absorbed carbohydrates pass through the liver, glucose is the main form of carbohydrate circulating in the bloodstream.

4 – Large Intestine or Colon

Any carbohydrates that weren't digested in the small intestine—mainly fiber—pass into the large intestine, but there's no enzymatic digestion of these carbohydrates here. Instead, bacteria living in the large intestine, sometimes called our gut microbiota, ferment these carbohydrates to feed themselves. Fermentation causes gas production, and that's why we may experience bloating and flatulence after a particularly fibrous meal. Fermentation also produces short-chain fatty acids, which our large intestine cells can use as an energy source. Over the last decade or so, more and more research has shown that our gut microbiota are incredibly important to our health, playing important roles in the function of our immune response, nutrition, and risk of disease. A diet high in whole food sources of fiber helps to maintain a population of healthy gut microbes.

Carbohydrate digestion:Summary

The primary goal of carbohydrate digestion is to break polysaccharides and disaccharides into monosaccharides, which can be absorbed into the bloodstream.

- After eating, nothing needs to happen in the digestive tract to the monosaccharides in a food like grapes, because they are already small enough to be absorbed *as is*.

- Disaccharides in that grape or in a food like milk are broken down (enzymatically digested) in the digestive tract to monosaccharides (glucose, galactose, and fructose).
- Starch in food is broken down (enzymatically digested) in the digestive tract to glucose molecules.
- Fiber in food is not enzymatically digested in the digestive tract, because humans don't have enzymes to do this. However, some dietary fiber is fermented in the large intestine by gut microbes.

Table 4.3. Summary of enzymatic digestion of carbohydrates

Carbohydrates in food	Is this carbohydrate enzymatically digested? (enzyme name)	What is absorbed into the villi after digestion?
Monosaccharides		
Glucose	No	Glucose
Fructose	No	Fructose. It is then transported to the liver where it is converted to glucose.
Galactose	No	Galactose. It is then transported to the liver where it is converted to glucose.
Disaccharides		
Maltose	Yes **(maltase)**	Glucose
Sucrose	Yes **(sucrase)**	Glucose, Fructose
Lactose	Yes **(lactase)**	Glucose, Galactose
Polysaccharides		
Starch	Yes **(amylase, maltase)**	Glucose
Fiber	No **(Humans don't have the digestive enzymes to break down fiber, but some is fermented by gut microbes in the large intestine.)**	N/A

GLUCOSE USE AND REGULATION IN THE BODY

On the last page, we traced the process of digesting the carbohydrates in a slice of pizza through the gastrointestinal tract, ending up with the absorption of monosaccharides across the cells of the small intestine and into the bloodstream. From there, they travel to the liver, where fructose and galactose are converted to glucose. After any meal containing carbohydrates, you experience a rise in blood glucose that can serve as fuel for cells around the body. But during the periods between meals, including while you're sleeping and exercising, your body needs fuel, too. To ensure that you have enough glucose in your blood at any given time, your body has a finely-tuned system to regulate your blood glucose concentration. This system allows you to store glucose when you have excess available (when your blood glucose is high) and to pull glucose out from your stores when needed (when your blood supply gets low).

Your body's ability to maintain equilibrium or a steady state in your blood glucose concentration is called homeostasis. It's a critical part of normal physiology, because if your blood glucose gets too low (called hypoglycemia, cellular function starts to fail, especially in the brain. If blood glucose gets too high (called hyperglycemia, it can cause damage to cells.

Hormones involved in blood glucose regulation

Central to maintaining blood glucose homeostasis are two hormones, insulin and glucagon, both produced by the pancreas and released into the bloodstream in response to changes in blood glucose.

- Insulin is made by the beta-cells of the pancreas and released when blood glucose is high. It causes cells around the body to take up glucose from the blood, resulting in lowering blood glucose concentrations.
- Glucagon is made by the alpha-cells of the pancreas and released when blood glucose is low. It causes glycogen in the liver to break down, releasing glucose into the blood, resulting in raising blood glucose concentrations. (Remember that glycogen is the storage form of glucose in animals.)

The image below depicts a mouse islet of Langerhans, a cluster of endocrine cells in the pancreas. The beta-cells of the islet produce insulin, and the alpha-cells produce glucagon.

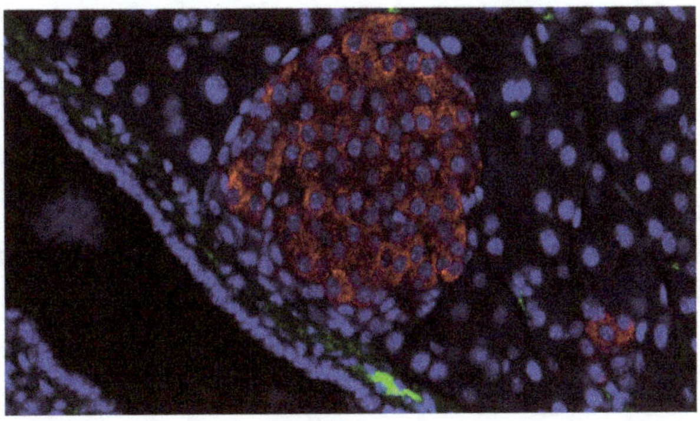

Figure 4.14. A mouse islet of Langerhans, visualized with immunofluorescent microscopy. In this image, cell nuclei are stained blue, insulin is stained red, and blood vessels are stained green. You can see that this islet is packed with insulin and sits right next to a blood vessel, so that it can secrete the two hor-

mones, insulin and glucagon, into the blood. Glucagon is not stained in this image, but it's there!

In the figure below, you can see blood glucose and insulin throughout a 24-hour period, including three meals. You can see that when glucose rises, it is followed immediately by a rise in insulin, and glucose soon drops again. The figure also shows the difference between consuming a sucrose-rich food and a starch-rich food. The sucrose-rich food results in a greater spike in both glucose and insulin. Because more insulin is required to handle that spike, it also causes a more precipitous decline in blood glucose. This is why eating a lot of sugar all at once may increase energy in the short-term, but soon after may make you feel like taking a nap!

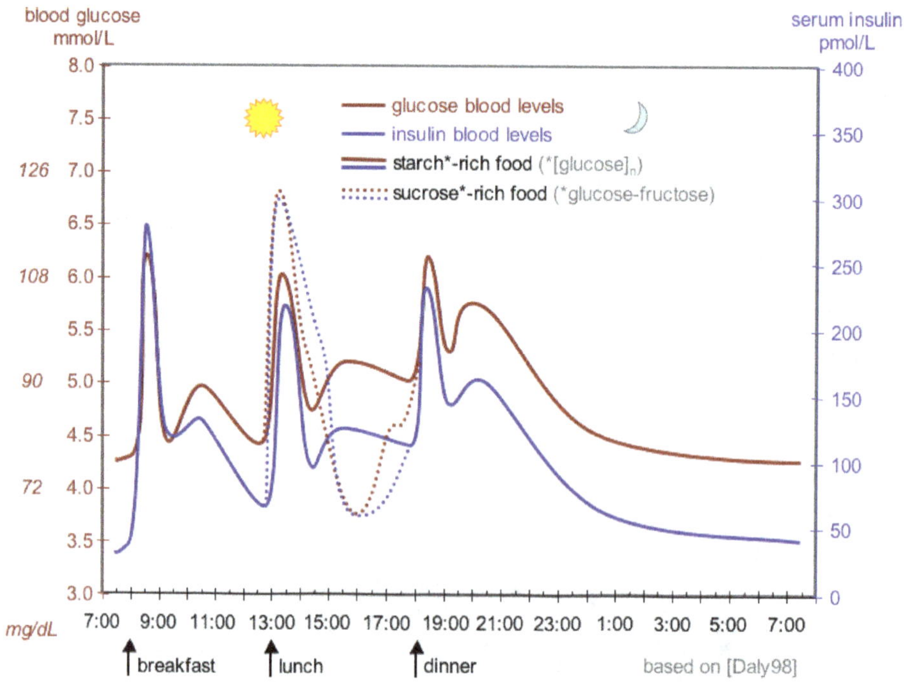

Figure 4.15. Typical pattern of blood glucose and insulin during a 24-hour period, showing peaks for each of 3 meals and highlighting the effects of consumption of sugar-rich foods.

Let's look a little closer at how insulin works, illustrated in the figure below. Insulin is released by the pancreas into the bloodstream. Cells around the body have receptors for insulin on their cell membranes. Insulin fits into its receptors (labeled as step 1 in Figure 4.16), kind of like a key in a lock, and through a series of reactions (step 2), triggers glucose transporters to open on the surface of the cell (step 3). Now glucose can enter the cell, making it available for the cell to use and at the same time lowering the concentration of glucose in the blood.

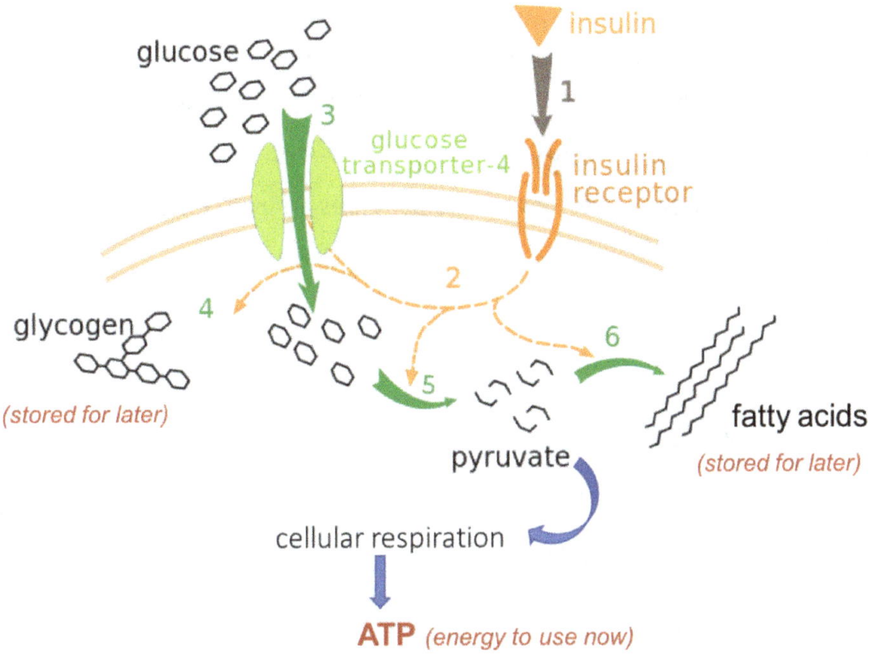

Figure 4.16. Insulin binds to its receptors on the cell membrane, triggering GLUT-4 glucose transporters to open on the membrane. This allows glucose to enter the cell, where it can be used in several ways.

The figure also shows several different ways glucose can be used once it enters the cell.

- If the cell needs energy right away, it can metabolize glucose through cellular respiration, producing ATP (step 5).
- If the cell doesn't need energy right away, glucose can be converted to other forms for storage. If it's a liver or muscle cell, it can be converted to glycogen (step 4). Alternatively, it can be converted to fat and stored in that form (step 6).

In addition to its role in glucose uptake into cells, insulin also stimulates glycogen and fat synthesis as described above. It also stimulates protein synthesis. You can think of its role as signaling to the body that there's lots of energy around, and it's time to use it and store it in other forms.

On the other hand, when blood glucose falls, several things happen to restore homeostasis.

- You receive messages from your brain and nervous system that you should eat.
- Glucagon is released from the pancreas into the bloodstream. In liver cells, it stimulates the breakdown of glycogen, releasing glucose into the blood.

- In addition, glucagon stimulates a process called gluconeogenesis, in which new glucose is made from amino acids (building blocks of protein) in the liver and kidneys, also contributing to raising blood glucose.

How glucose provides energy

Now let's zoom in on how exactly glucose provides energy to the cell. We can trace this process in the figure below.

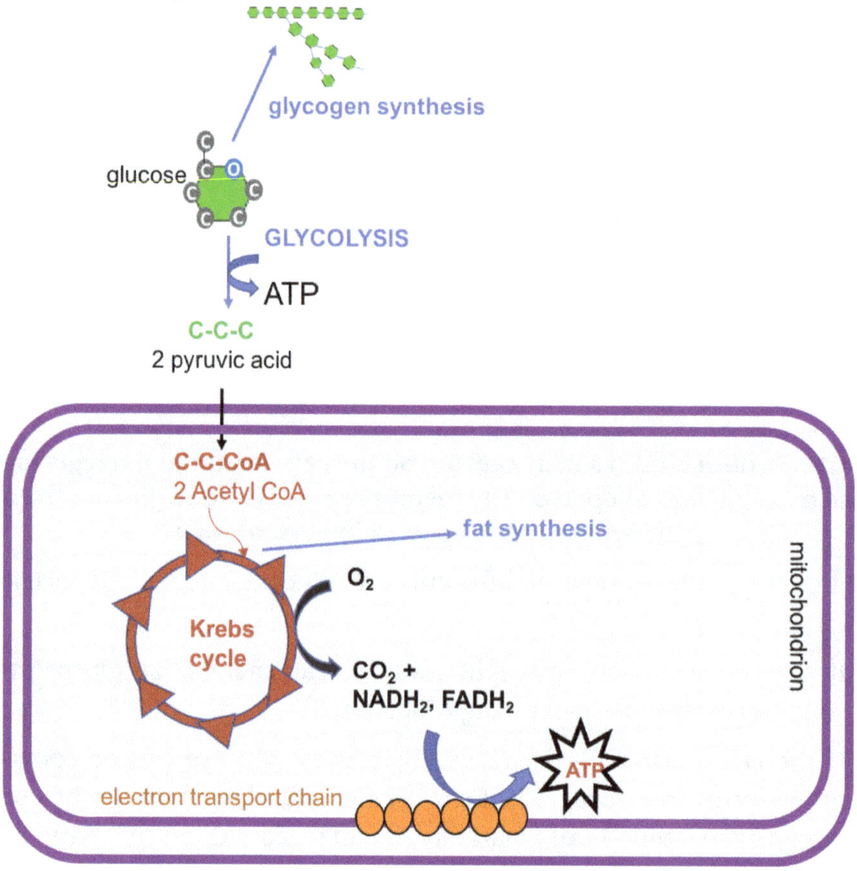

- Figure 4.17. Overview of glucose metabolism in the fed state, when there is adequate glucose available. Glucose can be used to generate ATP for energy, or it can be stored in the form of glycogen or converted to fat for storage in adipose tissue.
- Glucose, a 6-carbon molecule, is broken down to two 3-carbon molecules called pyruvate through a process called glycolysis.
- Pyruvate enters a mitochondrion of the cell, where it is converted to a molecule called acetyl CoA.

- Acetyl CoA goes through a series of reactions called the Krebs cycle. This cycle requires oxygen and produces carbon dioxide. It also produces several important high energy electron carriers called $NADH_2$ and $FADH_2$.
- These high energy electron carriers go through the electron transport chain to produce ATP—energy for the cell!
- Note that the figure also shows that glucose can be used to synthesize glycogen or fat, if the cell already has enough energy.

What happens when there isn't enough glucose?

We've already talked about what happens when blood glucose falls: glucagon is released, and that stimulates the breakdown of glycogen as well as the process of gluconeogenesis from amino acids. These are important mechanisms for maintaining blood glucose levels to fuel the brain when carbohydrate is limited. Hypoglycemia (low blood glucose) can cause you to feel confused, shaky, and irritable, because your brain doesn't have enough glucose. If it persists, it can cause seizures and eventually coma, so it's good we have these normal mechanisms to maintain blood glucose homeostasis!

What happens if your carbohydrate supply is limited for a long time? This might happen if a person is fasting, starving, or consuming a very low carbohydrate diet. In this case, your glycogen supplies will become depleted within about 24 hours. How will you get enough glucose (especially for the brain) and energy? You'll have to use the other two macronutrients in the following ways:

- **Protein**: You'll continue to use some amino acids to make glucose through gluconeogenesis and others as a source of energy through acetyl CoA. However, if a person is starving, they also won't have extra dietary protein. Therefore, they start breaking down body proteins, which will cause muscle wasting.
- **Fat:** You can break down fat as a source of energy, but you can't use it to make glucose. Fatty acids can be broken down to acetyl CoA in the liver, but acetyl CoA can't be converted to pyruvate and go through gluconeogenesis. It can go through the Krebs cycle to produce ATP, but if carbohydrate is limited, the Krebs cycle gets overwhelmed. In this case, acetyl CoA is converted to compounds called ketones or ketone bodies. These can then be exported to other cells in the body, especially brain and muscle cells.

These Pathways are Shown in the Figure Below

Ketone production is important, because ketones can be used by tissues of the body as a source of energy during starvation or a low carbohydrate diet. Even the brain can adapt to using ketones as a source of fuel after about three days of starvation or very low-carbohydrate diet. This also helps to preserve the protein in the muscle.

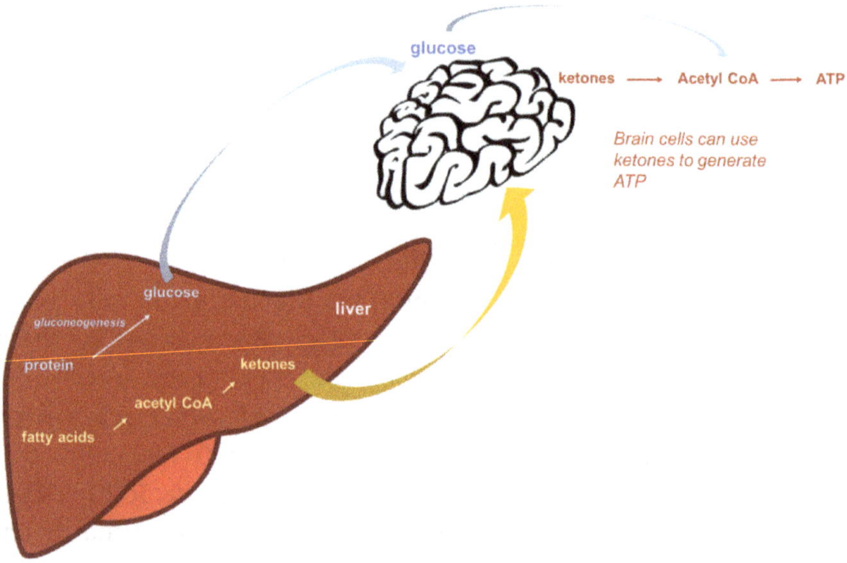

Figure 4.18. During starvation or when consuming a low-carbohydrate diet, protein (amino acids) can be used to make glucose by gluconeogenesis, and fats can be used to make ketones in the liver. The brain can adapt to using ketones as an energy source in order to conserve protein and prevent muscle wasting.

Ketones can be excreted in urine, but if ketone production is very high, they begin to accumulate in the blood, a condition called ketosis. Symptoms of ketosis include sweet-smelling breath, dry mouth, and reduced appetite. People consuming a very low carbohydrate diet may be in ketosis, and in fact, this is a goal of the currently popular ketogenic diet. (Ketones are acidic, so severe ketosis can cause the blood to become too acidic, a condition called ketoacidosis. This mainly happens with uncontrolled diabetes.)

Is following a ketogenic diet an effective way to lose weight? It can be, but the same can be said of any diet that severely restricts the types of foods that you're allowed to eat. Following a ketogenic diet means eating a high fat diet with very little carbohydrate and moderate protein. This means eating lots of meat, fish, eggs, cheese, butter, oils, and low carbohydrate vegetables, and eliminating grain products, beans, and even fruit. With so many fewer choices, you're likely to spend more time planning meals and less time mindlessly snacking. Being in ketosis also seems to reduce appetite, and it causes you to lose a lot of water weight initially. However, studies show that being in ketosis doesn't seem to increase fat-burning or metabolic rate. There are also concerns that the high levels of saturated fat in most ketogenic diets could increase risk of heart disease in the long term. Finally, it's a very difficult diet to maintain for most people, and reverting back to your previous dietary patterns usually means the weight will come back. The ketogenic diet is also very similar to the Atkins diet that was all the rage in the 1990's, and we tend to be skeptical of such fad diets, preferring to focus instead on balance, moderation, and enjoyment of a wide variety of foods.

Diabetes

Diabetes is a chronic disease in which your normal system of regulating blood glucose doesn't work. There are three main types of diabetes: type 1, type 2, and gestational diabetes.

Type 1 Diabetes:

This is an autoimmune disease in which the beta-cells of the pancreas are destroyed by your own immune system. Without the beta-cells, you can't make enough insulin, so in type 1 diabetes, you simply don't have enough insulin to regulate your blood glucose levels. Remember how we said insulin is like the key that lets glucose into the body's cells? In type 1 diabetes, you're missing the key, so glucose stays in the blood and can't get into cells.

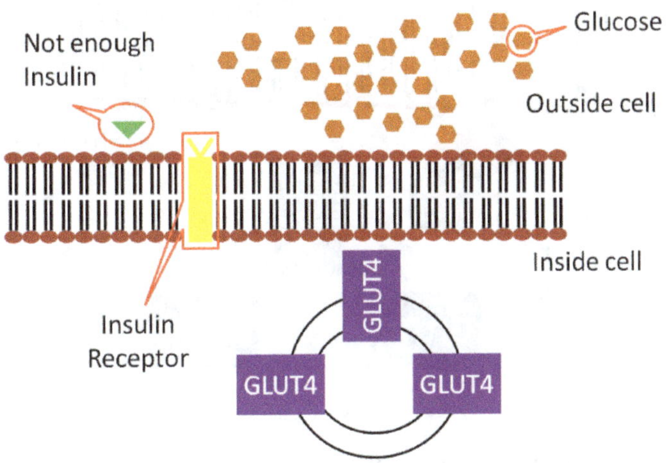

Figure 4.19. In type 1 diabetes, the pancreas does not make enough insulin, so glucose transporters (GLUT-4) do not open on the cell membrane, and glucose is stuck outside the cell.

Common symptoms include weight loss and fatigue, because the body's cells are starved of glucose. Excess glucose from the blood is also excreted in the urine, increasing urination and thirst.

Once diagnosed, type 1 diabetics have to take insulin in order to regulate their blood glucose. Traditionally, this has required insulin injections timed with meals. New devices like continuous glucose monitors and automatic insulin pumps can track glucose levels and provide the right amount of insulin, making managing type 1 diabetes a little easier. Figuring out the right amount of insulin is important, because chronically elevated blood glucose levels can cause damage to tissues around the body. However, too much insulin will cause hypoglycemia, which can be very dangerous.

Type 1 diabetes is most commonly diagnosed in childhood, but it has been known to develop at any age. It's much less common than type 2 diabetes, accounting for 5-10% of cases of diabetes.

Type 2 Diabetes:

Development of type 2 diabetes begins with a condition called insulin resistance. At least initially, the pancreas is producing enough insulin, but the body's cells don't respond appropriately. It's as if you still have the insulin key but can't find the keyhole to unlock the doors and let the glucose in.

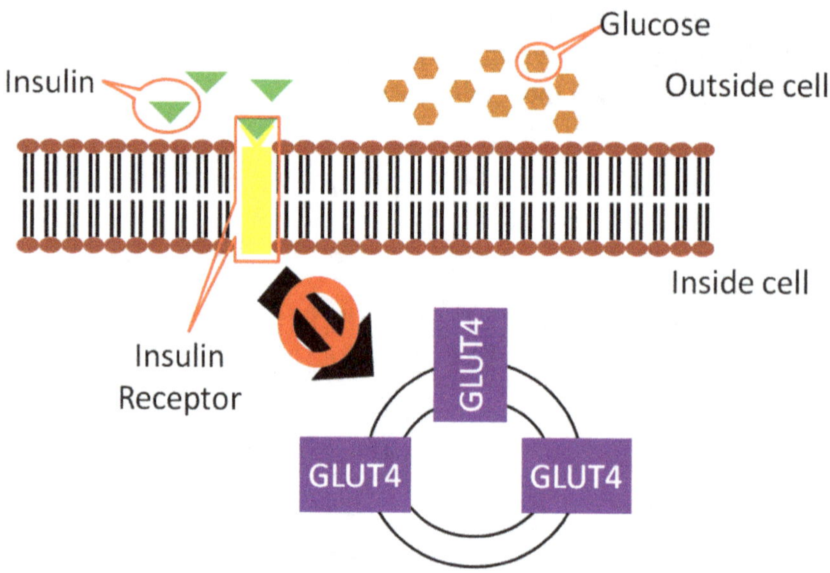

Figure 4.20. In type 2 diabetes, the cell does not respond appropriately to insulin, so glucose is stuck outside the cell.

The result is the same: high blood glucose. At this point, you may be diagnosed with a condition called prediabetes. The pancreas tries to compensate by making more insulin, but over time, it becomes exhausted and eventually produces less insulin, leading to full-blown type 2 diabetes. According to the CDC, 100 million Americans are living with diabetes (30.3 million) or prediabetes (84.1 million).

Although people of all shapes and sizes can get type 2 diabetes, it is strongly associated with abdominal obesity. In the past, it was mainly diagnosed in older adults, but it is becoming more and more common in children and adolescents as well, as obesity has increased in all age groups. In the maps below, you can see that as obesity has increased in states around the country, so has diabetes.

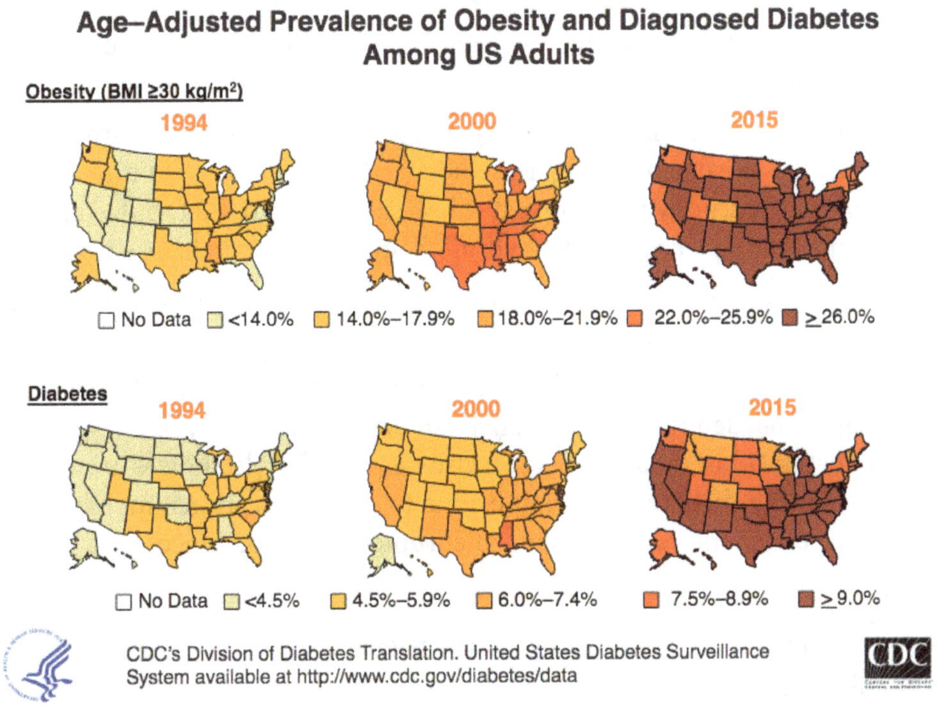

Figure 4.21. Data from the CDC show the increasingly prevalence of both obesity and type 2 diabetes between 1994 and 2015.

The complications of type 2 diabetes result from long-term exposure to high blood glucose, or hyperglycemia. This causes damage to the heart, blood vessels, kidneys, eyes, and nerves, increasing the risk of heart disease and stroke, kidney failure, blindness, and nerve dysfunction. People with uncontrolled Type 2 diabetes can also end up with foot infections and ulcers because of impaired nerve function and wound healing. If left untreated, this results in amputation.

Gestational Diabetes

Gestational diabetes is diabetes that develops during pregnancy in women that did not previously have diabetes. It affects approximately 6 percent of pregnancies in the U.S. It can cause pregnancy complications, mostly associated with excess fetal growth because of high blood glucose. Although it usually goes away once the baby is born, women who have gestational diabetes are more likely to develop type 2 diabetes later in life, so it is a warning sign for them.[6]

Diabetes Management:

All of the following have been shown to help manage diabetes and reduce complications. Diabetes management, as well as prevention (particularly if you've been diagnosed with prediabetes), starts with lifestyle choices.

- Exercise helps to improve your body's insulin response and can also help maintain a healthy weight.
- Eating well with diabetes doesn't require a special diet but instead regular, balanced meals following the Dietary Guidelines. It isn't necessary to eliminate carbohydrates or eat a low-carbohydrate diet, but emphasizing whole food sources of carbohydrate helps with blood glucose regulation.
- Managing stress levels and getting enough sleep can also help with blood glucose regulation.
- Medications may be needed. Insulin is needed for type 1 diabetes and may be needed for more advanced or severe cases of type 2 or gestational diabetes. Other medications can also help. If lifestyle choices aren't enough to manage diabetes, it is important to use medications appropriately to help reduce the complications of chronic high blood glucose.

REFERENCES:

Salway, J. G. (2004). Metabolism at a Glance (3rd ed.). Malden, Mass.: Blackwell Publishing.

Smolin, L., & Grosvenor, M. (2016). Nutrition Science and Applications. Danvers, Mass.: John Wiley & Sons, Ltd.

Gibson, A. A., Seimon, R. V., Lee, C. M. Y., Ayre, J., Franklin, J., Markovic, T. P., ... Sainsbury, A. (2015). Do ketogenic diets really suppress appetite? A systematic review and meta-analysis. Obesity Reviews: An Official Journal of the International Association for the Study of Obesity, 16(1), 64–76. https://doi.org/10.1111/obr.12230

Hall, K. D., Chen, K. Y., Guo, J., Lam, Y. Y., Leibel, R. L., Mayer, L. E., ... Ravussin, E. (2016). Energy expenditure and body composition changes after an isocaloric ketogenic diet in overweight and obese men. The American Journal of Clinical Nutrition, 104(2), 324–333. https://doi.org/10.3945/ajcn.116.133561

Abbasi, J. (2018). Interest in the Ketogenic Diet Grows for Weight Loss and Type 2 Diabetes. JAMA, 319(3), 215–217. https://doi.org/10.1001/jama.2017.20639

Belluz, J. (2018, February 21). The keto diet, explained. Retrieved November 15, 2019, from Vox website: https://www.vox.com/science-and-health/2018/2/21/16965122/keto-diet-reset

Centers for Disease Control and Prevention. (2019, June 11). Diabetes Basics. Retrieved November 15, 2019, from https://www.cdc.gov/diabetes/basics/index.html

Deputy, N. P., Kim, S. Y., Conrey, E. J. & Bullard, K. M. Prevalence and Changes in Preexisting Diabetes and Gestational Diabetes Among Women Who Had a Live Birth — United States, 2012–2016. MMWR Morb Mortal Wkly Rep 67, (2018). https://www.cdc.gov/mmwr/volumes/67/wr/mm6743a2.htm

WHOLE VERSUS REFINED GRAINS, FOOD SOURCES, HEALTH BENEFITS, AND TYPES OF FIBER

Dietary fiber is defined by the Institute of Medicine's Food and Nutrition Board as "nondigestible carbohydrates and lignin that are intrinsic and intact in plants." Fiber plays an important role in giving plants structure and protection, and it also plays an important role in the human diet.

Cellulose is one type of fiber. The chemical structure of cellulose is shown in the figure below, with our simplified depiction next to it. You can see that cellulose has long chains of glucose, similar to starch, but they're stacked up, and there are hydrogen bonds linking the stacks. The special bonds between these glucose units in fiber are not enzymatically digested in the digestive tract, and therefore, fiber passes undigested to the colon or large intestine.

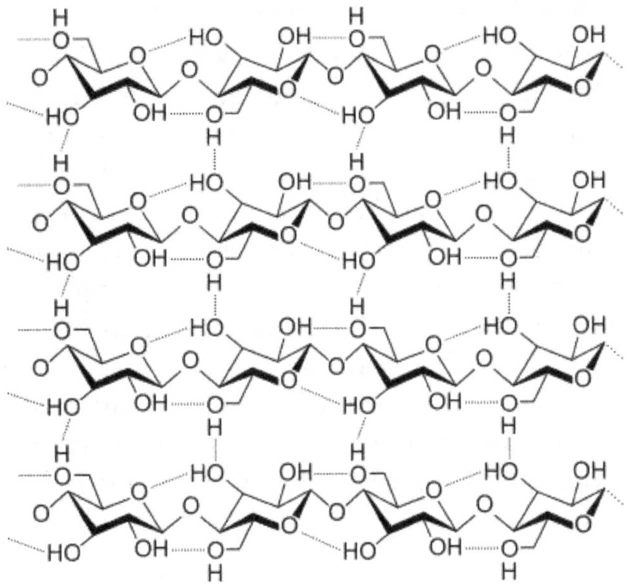

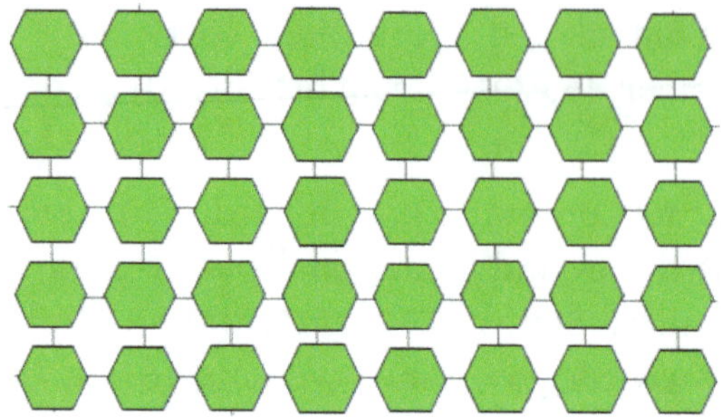

Figure 4.22. The chemical structure of cellulose, and a simplified illustration of cellulose.

You might be wondering how fiber has any benefit to us if we can't digest it. However, it doesn't just pass through the digestive tract as a waste product. Instead, it serves many functions on its journey, and these contribute to our health. Let's explore the different types of fiber, where we find them in foods, and what benefits they provide!

TYPES OF FIBER

Whole plant foods contain many different types of molecules that fit within the definition of fiber. One of the ways that types of fiber are classified is by their solubility in water. Whole plant foods contain a mix of both soluble and insoluble fiber, but some are better sources of one than the other.

- Soluble Fiber – These fibers dissolve in water, forming a viscous gel in the GI tract, which helps to slow digestion and the absorption of glucose. This means that including soluble fiber in a meal helps to prevent sharp blood sugar spikes, instead making for a more gradual rise in blood glucose. Consuming a diet high in soluble fiber can also help to lower blood cholesterol levels, because soluble fiber binds cholesterol and bile acids (which contain cholesterol) in the GI tract. Soluble fiber is also highly fermentable, so it is easily digested by bacteria in the large intestine. Pectins and gums are common types of soluble fibers, and good food sources include oat bran, barley, nuts, seeds, beans, lentils, peas, and some fruits and vegetables. (Psyllium fiber supplements like Metamucil are composed mainly of soluble fiber, so if you've ever stirred a spoonful of this into a glass of water, you've seen the viscous consistency characteristic of soluble fiber.)
- Insoluble Fiber – These fibers typically do not dissolve in water and are nonviscous. Some are fermentable by bacteria in the large intestine but to a lesser degree than soluble fibers. Insoluble fibers help prevent constipation, as they create a softer, bulkier stool that is easier to eliminate. Lignin,

cellulose, and hemicellulose are common types of insoluble fibers, and food sources include wheat bran, vegetables, fruits, and whole grains.

FOOD sources of dietary fiber

Since fiber provides structure to plants, fiber can be found in all whole plant foods, including whole grains (like oatmeal, barley, rice and wheat), beans, nuts, seeds, and whole fruits and vegetables.

Figure 4.23. A bowl of oatmeal topped with blueberries and sunflower seeds.

This meal is packed with fiber from the oatmeal, blueberries, and sunflower seeds.

When foods are refined, parts of the plant are removed, and during this process, fiber and other nutrients are lost. For example, fiber is lost when going from a whole fresh orange to orange juice. A whole orange contains about 3 grams of fiber, whereas a glass of orange juice has little to no fiber. Fiber is also lost when grains are refined. We will discuss this more a little later. Take a look at the list of foods below to see the variety of foods which provide dietary fiber.

Table 4.4. Common foods listed with standard portion size, and calories and fiber in a standard portion.

Food	Standard Portion Size	Calories in Standard Portion	Dietary Fiber in Standard Portion (g)
Shredded wheat ready-to-eat cereal (various)	1-1 ¼cup	155-220	5.0-9.0
Wheat bran flakes ready-to-eat cereal (various)	¾ cup	90-98	4.9-5.5

Food	Standard Portion Size	Calories in Standard Portion	Dietary Fiber in Standard Portion (g)
Lentils, cooked	½ cup	115	7.8
Black beans, cooked	½ cup	114	7.5
Refried beans, canned	½ cup	107	4.4
Avocado	½ cup	120	5.0
Pear, raw	1 medium	101	5.5
Pear, dried	¼ cup	118	3.4
Apple, with skin	1 medium	95	4.4
Raspberries	½ cup	32	4.0
Mixed vegetables, cooked from frozen	½ cup	59	4.0
Potato, baked, with skin	1 medium	163	3.6
Pumpkin seeds, whole, roasted	1 ounce	126	5.2
Chia seeds, dried	1 Tbsp	58	4.1
Sunflower seed kernels, dry roasted	1 ounce	165	3.1
Almonds	1 ounce	164	3.5
Plain rye wafer crackers	2 wafers	73	5.0
Bulgur, cooked	½ cup	76	4.1
Popcorn, air-popped	3 cups	93	3.5
Whole wheat spaghetti, cooked	½ cup	87	3.2
Quinoa, cooked	½ cup	111	2.6

Although you can get fiber from supplements, whole foods are are a better source, because the fiber comes packaged with other essential nutrients and phytonutrients.

HEALTH BENEFITS OF DIETARY FIBER

A High-Fiber Diet Has Many Benefits, Which Include:

- Helps prevent constipation. Many fibers (but mostly insoluble fibers) help provide a softer, bulkier stool which is then easier to eliminate.
- Helps maintain digestive and bowel health. Dietary fiber promotes digestive health through its role in supporting elimination and fermentation, and it's positive impact on gut microbiota. Since fiber provides a bulkier stool, this helps keeps the digestive tract muscles toned and strong which can help prevent hemorrhoids and diverticula.
- Lowers risk of cardiovascular disease. Higher fiber intake has been shown to improve blood lipids by reducing total cholesterol, triglycerides, and low density cholesterol ("bad cholesterol," associated with a higher risk

of cardiovascular disease), and increasing high density cholesterol ("good cholesterol," associated with lower risk of cardiovascular disease). Higher fiber intake has also been associated with lower blood pressure and reduced inflammation.

- Lowers risk of type 2 Diabetes. Higher fiber intake (especially viscous, or soluble fibers) has been shown to slow down glucose digestion and absorption, benefiting glucose metabolism. A higher fiber diet may also decrease diabetes risk by reducing inflammation.
- Lowers risk of colorectal cancer. More evidence is supporting the idea that higher fiber intake lowers the risk of colorectal cancer, although researchers aren't sure why. One hypothesis is that dietary fiber decreases transit time (the time it takes for food to move through the digestive tract), thereby exposing the cells of the gastrointestinal tract to carcinogens from food for a shorter time.
- Helps maintain a healthy body weight. Research has shown a relationship between higher dietary fiber intake and lower body weight. The mechanisms for this are unclear, but perhaps high-fiber foods are more filling and therefore keep people satisfied longer with fewer calories. High-fiber foods also tend to be more nutrient-dense compared to many processed foods, which are more energy-dense.

Whole *vs.* Refined Grains

Figure 4.24. Wheat growing in a field

Before they are harvested, all grains are whole grains. They contain the entire seed (or kernel) of the plant. This seed is made up of three edible parts: the bran, the germ, and the endosperm. The seed is also covered by an inedible husk that protects the seed.

- The bran is the outer skin of the seed. It contains antioxidants, B vitamins and fiber.
- The endosperm is by far the largest part of the seed and provides energy in the form of starch to support reproduction. It also contains protein and small amounts of vitamins and minerals.
- The germ is the embryo of the seed—the part that can sprout into a new plant. It contains B vitamins, protein, minerals like zinc and magnesium, and healthy fats.

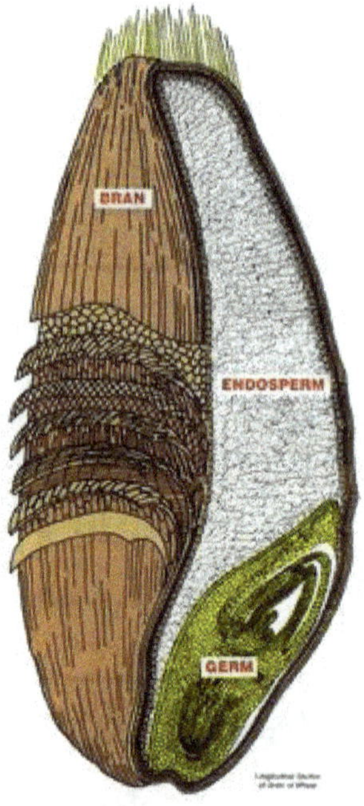

Figure 4.25. The anatomy of a wheat kernel which includes the bran, endosperm, and germ.

The Dietary Guidelines for Americans define whole grains and refined grains in the following way:

Whole Grains—Grains and grain products made from the entire grain seed, usually called the kernel, which consists of the bran, germ, and endosperm. If the kernel has been cracked, crushed, or flaked, it must retain the same relative proportions of bran, germ, and endosperm as the original grain in order to be called whole grain. Many, but not all, whole grains are also sources of dietary fiber."

Whole grains include foods like barley, corn (whole cornmeal and popcorn), oats (including oatmeal), rye, and wheat. (For a more complete list of whole grains, check out the Whole Grain Council.)

Refined Grains—Grains and grain products with the bran and germ removed; any grain product that is not a whole-grain product. Many refined grains are low in fiber but enriched with thiamin, riboflavin, niacin, and iron, and fortified with folic acid."

Refined grains include foods like white rice and white flour. According to the Whole Grain Council, "Refining a grain removes about a quarter of the protein in a grain, and half to two thirds or more of a score of nutrients, leaving the grain a mere shadow of its original self."

Refined grains are often enriched with vitamins and minerals, meaning that some of the nutrients lost during the refining process are added back in after processing. However, many vitamins and minerals are not added back, and neither are the fiber, protein, and healthy fats found in whole grains. In the chart below you can see the differences in essential nutrients between whole wheat flour, refined wheat flour, and enriched wheat flour.

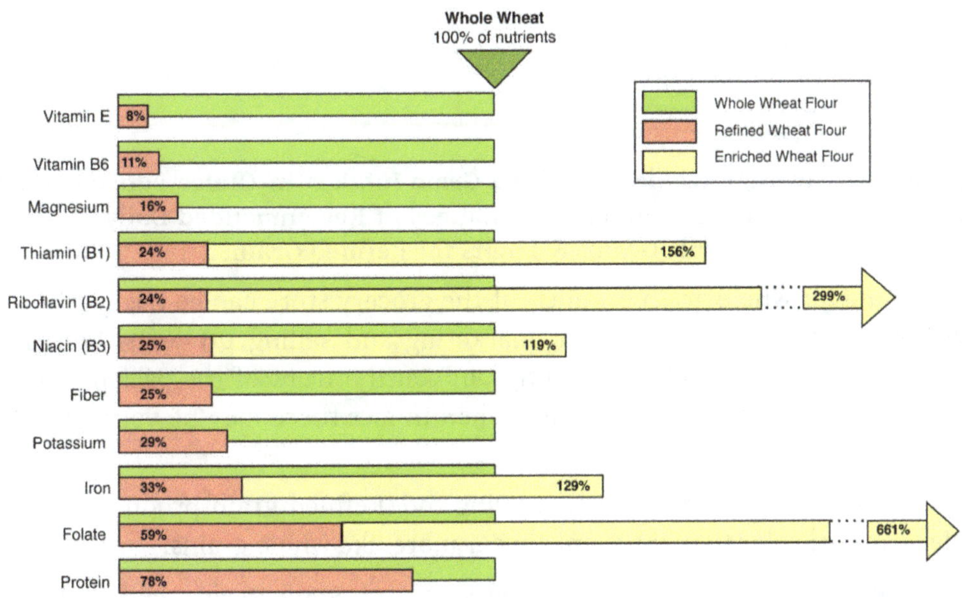

Figure 4.26. The nutrient content of refined wheat and enriched wheat as compared to whole wheat flour.

Because whole grains offer greater nutrient density, MyPlate and the Dietary Guidelines recommend that at least half of our grains are whole grains. Yet current data show that while most Americans are eating enough grains overall, they're eating too many refined grains and not enough whole grains, as shown in this graphic from the Dietary Guidelines:

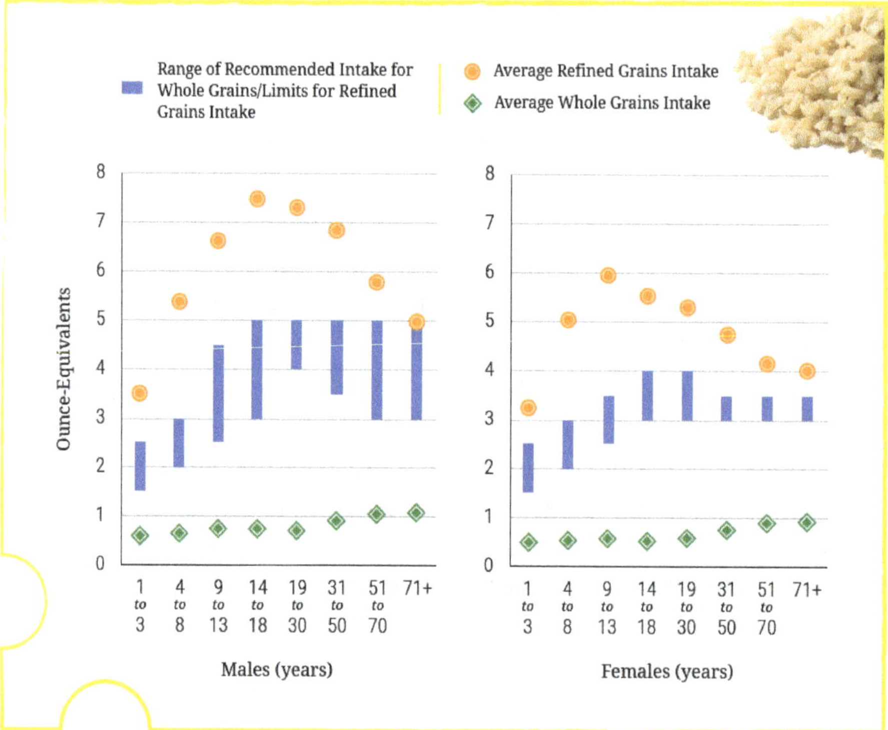

Figure 4.27. Average Whole & Refined Grain Intakes in Ounce-Equivalents per Day by Age-Sex Groups, Compared to Ranges of Recommended Daily Intake for Whole Grains & Limits for Refined Grains.

Looking for whole grain products at the grocery store can be tricky, since the front-of-package labeling is about marketing and selling products. Words like "made with whole grain" and "multigrain" on the front of the package make it appear like a product is whole grain, when in fact there may be very few whole grains present.

The color of a bread can be deceiving too. Refined grain products can have added caramel color to make them appear more like whole grains.

To determine if a product is a good source of whole grain, the best place to look is the ingredient list on the Nutrition Facts panel. The ingredients should list a whole grain as the first ingredient (e.g., 100% whole wheat), and it should not be followed by a bunch of refined grains such as enriched wheat flour.

Getting familiar with the name of whole grains will help you identify them. Common varieties include wheat, barley, brown rice, buckwheat, corn, rye, oats, and wild rice. Less known varieties include teff, amaranth, millet, quinoa, black rice, black barley, and spelt.

Most, but not all, whole grains are a good source of fiber, and that is one of the benefits of choosing whole grains. Keep in mind that some products add extra fiber as a separate ingredient, like wheat bran, inulin, or cellulose. These boost the grams of fiber on the Nutrition Facts label and may make the product a good source of fiber, but it doesn't mean it's a good source of whole grains. In fact, it may be a product made mostly of refined grains, so it would still be missing the other nutrients that come packaged in whole grains and may not have the same health benefits. Therefore, just looking at fiber on the Nutrition Facts label is not a good indicator of whether or not the product is made with whole grains.

Also, some products that are 100% whole wheat but do not appear to be a good source of fiber, because the serving size is small. The bread label below is an example of this. The first ingredient is "stone ground whole wheat flour" with no refined flours listed, but it still has only 2g of fiber and 9% DV. But of course, that still contributes to your fiber intake for the day, and if you made yourself a sandwich with two slices of bread, that would provide 18% of the DV.

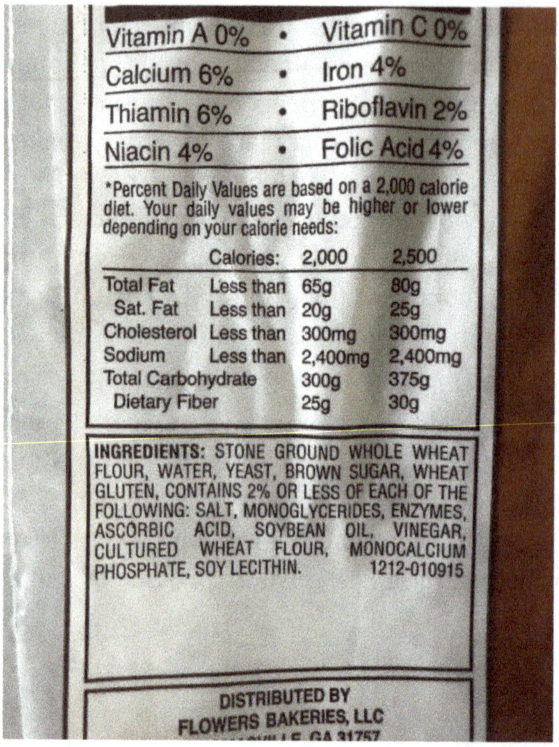

Figure 4.28. Example of 100% whole wheat bread with Nutrition Facts and ingredient list.

REFERENCES

Institute of Medicine, Food and Nutrition Board. (2005). Dietary Reference Intakes for Energy, Carbohydrate, Fiber, Fat, Fatty Acids, Cholesterol, Protein, and Amino Acids (Macronutrients). Retrieved from https://www.nap.edu/read/10490/chapter/1

U.S Department of Agriculture, Agricultural Research Service, Nutrient Data Laboratory. (2014). USDA National Nutrient Database for Standard Reference, Release 27. Available at: http://www.ars.usda.gov/nutrientdata.

Academy of Nutrition and Dietetics. (2015). Position of the Academy of Nutrition and Dietetics: Health Implications of Dietary Fiber. Journal of the Academy of Nutrition and Dietetics, 115(11), 1861–1870. doi: 10.1016/j.jand.2015.09.003

Whole Grain Council. Definition of a Whole Grain. Retrieved from https://wholegrainscouncil.org/definition-whole-grain

Mayo Clinic Staff. (2018). Dietary Fiber: Essential for a healthy diet. Retrieved from https://www.mayoclinic.org/healthy-lifestyle/nutrition-and-healthy-eating/in-depth/fiber/art-20043983

SUGAR: LABEL READING, FOOD SOURCES, AND HEALTH IMPLICATIONS

Most of us enjoy the taste of sweetness, but you've also probably heard that you shouldn't eat too much sugar. Maybe you've even heard that sugar is *toxic.* The truth about sugar is more complex. It's true that most Americans eat more added sugar than recommended and would benefit from cutting back. It's also true that added sugars are hidden in many foods, and it can take savvy label-reading to find them. But let's also remember the big picture when we think about sugar. Some sugar is naturally-occurring in whole foods, packaged with other valuable nutrients. There's also room in the diet for some added sugar, and it can be valuable for making nutrient-dense foods more palatable or just for the pleasure of enjoying a treat. Let's look closer at the role of sugar in the diet.

Food Sources of Naturally-Occurring and Added Sugars

As we've already discussed, sugars are naturally found in fruits, veggies, and dairy. These are nutrient dense foods that come packaged with other essential nutrients too.

Figure 4.29. Examples of food that contain naturally occurring sugars: fruit, vegetables, and dairy.

Fresh fruits and veggies contain naturally-occurring sugars like glucose, fructose, and sucrose, but also come packaged with fiber, potassium, and Vitamin C. Dairy foods like unsweetened yogurt, milk, and cheese contain naturally-occurring lactose but also come packaged with calcium, potassium, phosphorus, and riboflavin.

Another food that contains natural sugar in the form of maltose is sprouted grain bread. In the example below, the only ingredients are sprouted organic rye kernels and water, yet there are 7 grams of sugar per slice. This sugar must be naturally-occurring maltose, and as you can imagine, it comes packaged with nutrients like fiber, protein, and iron.

Nutrition Facts

Serving Size: 2oz (56g / 3/4" slice)
Servings Per Container: about 7

Amount Per Serving	
Calories 150	Calories from Fat 0
	% Daily Value*
Total Fat 0 g*	0%
Saturated Fat 0 g	0%
Trans Fat 0 g	
Cholesterol 0 mg	0%
Sodium 10 mg	0%
Total Carbohydrate 32 g	10%
Dietary Fiber 5 g	20%
Sugars 7 g	
Protein 6 g	

Vitamin A 0% • Vitamin C 0%
Calcium 2% • Iron 10%

Figure 4.30. An example of a sprouted wheat bread that contains naturally occuring maltose from sprouted rye kernels.

In contrast, added sugars are concentrated sweeteners that are added as ingredients to foods to make them sweeter. They add calories to a food but contribute little to no essential nutrients, so they decrease the nutrient density of foods. Among the most common sources of added sugar are table sugar (sucrose) and high fructose corn syrup, but they come in many different forms with different names. For example, honey, maple syrup, agave nectar, and brown rice syrup may all sound more wholesome and natural, but they're still added sugars, because they are concentrated sweeteners that contribute little to no other nutrients. Other names for added sugar you might not recognize as sweeteners at all, like barley malt or treacle. Here's a list of 61 different names for added sugars:

61 Names for Added Sugar

- Agave nectar
- Barbados sugar
- Barley malt
- Barley malt syrup
- Beet sugar
- Brown sugar
- Buttered syrup
- Cane juice
- Cane juice crystals
- Cane sugar
- Caramel
- Carob syrup
- Castor sugar
- Coconut palm sugar
- Coconut sugar
- Confectioner's sugar
- Corn sweetener
- Corn syrup
- Corn syrup solids
- Date sugar
- Dehydrated cane juice
- Demerara sugar
- Dextrin
- Dextrose
- Evaporated cane juice
- Free-flowing brown sugars
- Fructose
- Fruit juice
- Fruit juice concentrate
- Glucose
- Glucose solids
- Golden sugar
- Golden syrup
- Grape sugar
- High-Fructose Corn Syrup (HFCS)
- Honey
- Icing sugar
- Invert sugar
- Malt syrup
- Maltodextrin
- Maltol
- Maltose
- Mannose
- Maple syrup
- Molasses
- Muscovado
- Palm sugar
- Panocha
- Powdered sugar
- Raw sugar
- Refiner's syrup
- Rice syrup
- Saccharose
- Sorghum Syrup
- Sucrose
- Sugar (granulated)
- Sweet Sorghum
- Syrup
- Treacle
- Turbinado sugar
- Yellow sugar

Figure 4.31. Names of sugar commonly added to food.

We find added sugars in some expected places, like cookies, ice cream, and soda, but there can also be a surprising amount of added sugar in yogurt, breakfast cereals, energy bars, and plant-based milk alternatives, like soy milk. We also find added sugars hiding in unexpected places, like ketchup, salad dressings, bread, and pasta sauce. In fact, nearly 75% of packaged products in the U.S. food supply are now sweetened.

In general, most people don't need to worry much about how much naturally-occurring sugar they consume. This goes back to the fact that naturally-occurring sugars are packaged with other nutrients. For example, a large apple contains about 23 grams of sugar, more than half of it in the form of fructose.[1] However, it also has more than 5 grams of fiber, plus a significant amount of vitamin C and potassium. The fiber slows down the digestion and absorption of the sugar into your bloodstream, giving your body more time to metabolize it and giving you a greater feeling of fullness.

A single can of soda, on the other hand, contains about 33 grams of sugar.[1] It's in a similar chemical form as the sugar in the apple—a mix of fructose and glucose—but it's not accompanied by any fiber to slow down digestion. Therefore, it's rapidly absorbed into your bloodstream, and your body has to quickly metabolize the fructose to glucose and increase insulin secretion to process the spike in sugar. Plus, although the soda contains 150 calories and the apple has just 116, the apple is probably going to leave you feeling more satisfied and less hungry compared with the soda.

For all of these reasons, it's the added sugars that we worry about, not the naturally-occurring ones. That said, there is room for some added sugar in a balanced diet, and you can use it to make nutrient-dense food tastier. For example, you can drizzle honey into plain yogurt or sprinkle some brown sugar on roasted winter squash. You get far more nutritional "bang for your buck" using added sugars in this way than consuming them in something like a soda. (And of course, there's also room in a balanced diet for occasional treats!)

HOW MUCH ADDED SUGAR ARE WE EATING?

On average, Americans consume 22 to 30 teaspoons of added sugar daily, up to 17% of calories, well in excess of the recommendation to limit added sugar intake to 10% of calories or less. This is shown in the image below from the Dietary Guidelines.

Where are all of these added sugars coming from? Nearly half of them come from soda, juices, and other sugary drinks, as illustrated below. Therefore, the Dietary Guidelines recommend that people drink more water and less sugary drinks.

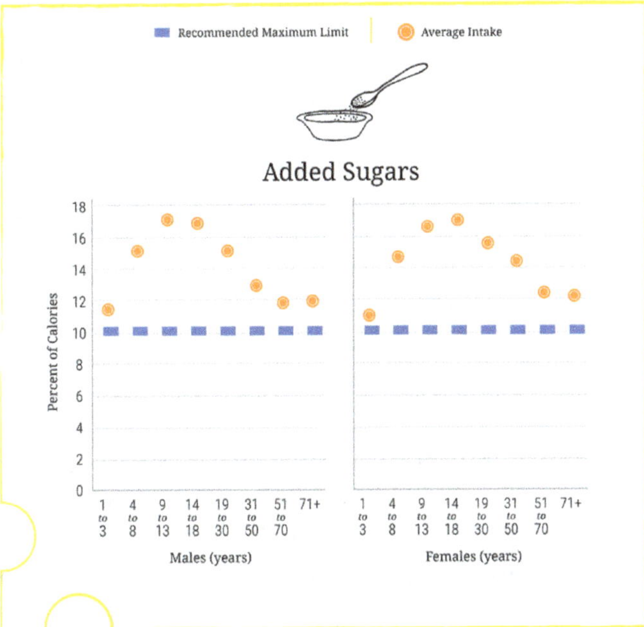

Figure 4.32. Average intakes of added sugars as a percent of calories per day by age-sex group, in comparison to the Dietary Guidelines' maximum limit of less than 10 percent of calories.

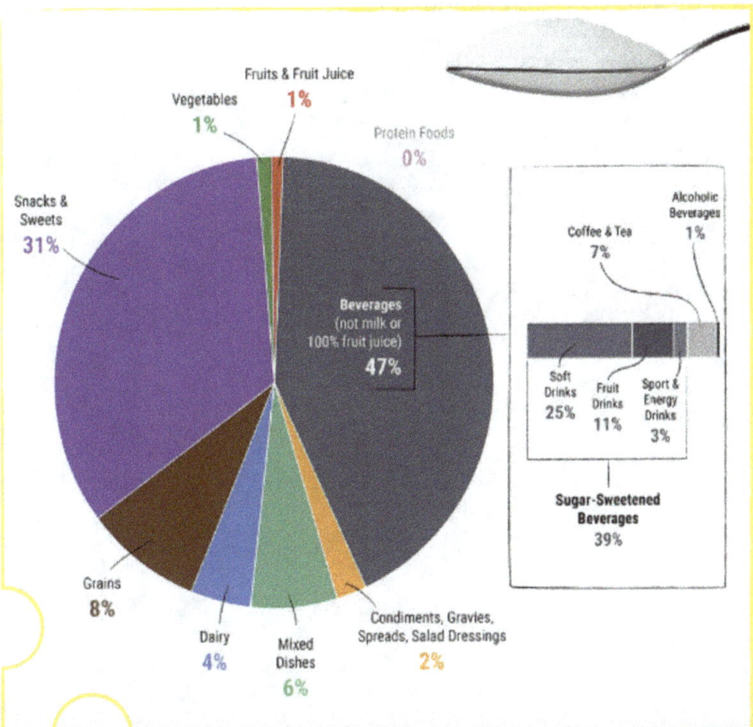

Figure 4.33. Food category sources of added sugars in the U.S. population Ages 2 years and older.

On the Nutrition Facts panel, sugar is expressed in grams, but most of us don't think in grams. Therefore, it can be helpful to convert gram amounts to teaspoons, which are easier to visualize. Use the conversions shown in the graphic below to make these calculations.

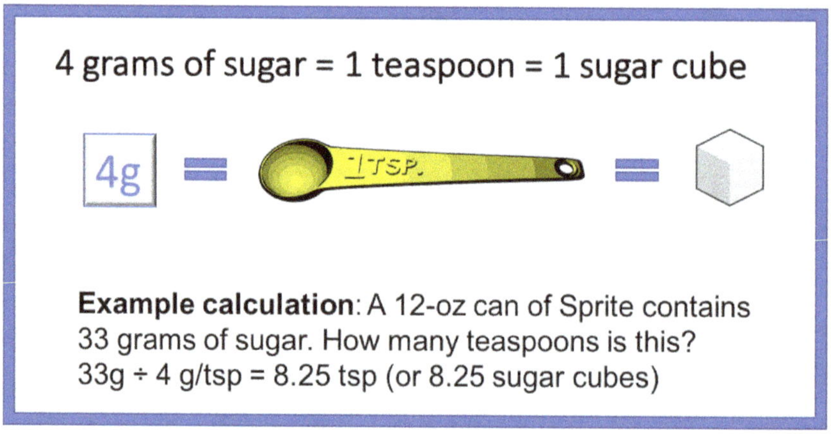

Figure 4.34. One teaspoon is equal to 4 grams of sugar or a sugar cube.

The sugar in soda adds up fast, especially with our super sized portions. For example, a 64-ounce soda has 186 grams of sugar, or about 46 teaspoons. (186 grams divided by 4g/tsp = 46 teaspoons.)

Figure 4.35. Forty six sugar cubes stacked next to a big gulp to illustrate the 46 teaspoons of sugar that the soda contains.

It can be eye-opening to track your added sugar intake for a few days, and this may give you an idea of sources of added sugar that you can live without and replace with something else. However, tracking added sugar intake can be tedious. In the big picture, it's most important to focus on eating whole foods that are minimally processed and to consume added sugars in moderation.

BENEFITS OF EATING LESS ADDED SUGAR

Research shows that adopting an eating pattern that is relatively low in added sugars has many benefits, including a lower risk of:

- Cardiovascular disease
- Obesity
- Type 2 diabetes
- Some cancers
- Dental cavities

Why does too much added sugar cause health problems? The reasons are complex, and this is an ongoing area of research and controversy. One possible explanation is that a diet high in added sugar means the pancreas has to work hard to make enough insulin, and over time, it can begin to fail and the body's cells start to become insulin resistant. The liver also has to work hard to metabolize fructose, and too much fructose increases fat synthesis, which can raise blood lipid and cholesterol levels, increasing risk of heart disease.

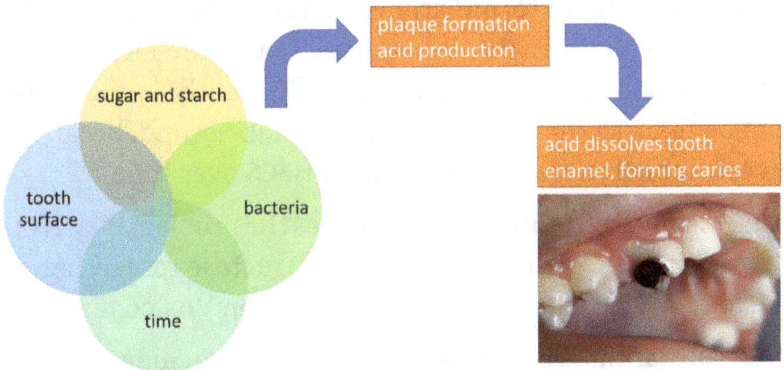

Fig. 4.36. Dental caries are formed because of a combination of factors: the presence of oral bacteria; a supply of sugar and/or starch for them to eat; tooth surface where they can form colonies, or plaque; and time.

Both dietary sucrose and starch are associated with tooth decay. Bacteria living in the mouth can utilize the carbohydrates passing through the oral cavity for their own benefit. Those bacteria happily metabolize carbohydrates, especially sucrose, but also starchy foods, which stick to teeth and linger there. Acid is formed in the process, and this can dissolve your tooth enamel, eventually causing cavities, also known as dental caries. Reducing sugar intake, limiting between-meal snacks, and brushing after meals to remove lingering carbohydrates can help reduce the risk of dental caries. The use of fluoride and regular dental care also help.

ARE SOME ADDED SUGARS BETTER THAN OTHERS?

Students often ask which sugar is healthiest: high fructose corn syrup, honey, agave syrup, or sugar? In general, as far as the body is concerned, sugar is sugar. These are all concentrated sweeteners that contain calories with very few/no other nutrients, so all should be used only in moderation.

High fructose corn syrup (HFCS) has gotten a lot of attention in the last several decades and has been blamed for the obesity epidemic and many other poor health outcomes. This is in part because it's widely used to sweeten soda and so has become a large part of the American diet. It's true that fructose is more work for the body to process, because it has to be converted to glucose. Here's what the website Sugar Science, written by researchers and scientists from the University of California, San Francisco, has to say about the difference between table sugar and high fructose corn syrup:

"Table sugar (sucrose), derived from sugar cane and beets, is made up of equal portions of two types of sugars. It's half (50%) glucose and half (50%) fructose. High-fructose corn syrup (HFCS) is derived from corn syrups that have undergone enzymatic processing to convert some of their glucose into fructose to produce a desired sweetness. HFCS comes in different formulations, depending on the manufacturer. More common formulations contain 42% fructose or 55% percent, but some contain as much as 90%. Why should we care? First, because there is significant evidence that fructose is processed differently in the body than other sugars and can be toxic to the liver, just like alcohol. Second, because as a nation, we have been consuming more of our sugars in HFCS over time."

But focusing too much attention on fructose as the problem may risk missing the forest for the trees. Here's what Dr. Luc Tappy, a fructose researcher at University of Lausanne, had to say about the issue in an article on Vox.com:

"Given the substantial consumption of fructose in our diet, mainly from sweetened beverages, sweet snacks, and cereal products with added sugar, and the fact that fructose is an entirely dispensable nutrient, it appears sound to limit consumption of sugar as part of any weight loss program and in individuals at high risk of developing metabolic diseases. There is no evidence, however, that fructose is the sole, or even the main factor in the development of these diseases, nor that it is deleterious to everybody, and public health initiatives should therefore broadly focus on the promotion of healthy lifestyles generally, with restriction of both sugar and saturated fat intakes, and consumption of whole grains, fresh fruits and vegetables rather than focusing exclusively on reduction of sugar intake."

Are sweeteners such as honey, maple syrup, and molasses any better than more refined and processed sweeteners? Maybe. These sweeteners do contain minerals and antioxidants, so they may offer a slight edge in terms of nutrition. However, keep in mind that minerals and antioxidants are abundant in whole foods such as whole grains, vegetables, and fruits, and these obviously offer many other benefits. These sweeteners are still considered sources of added sugar and should be used in moderation. That said, each of them offers different delicious flavors, and honey has the added benefit that it can be purchased locally, so there are good reasons to turn towards these products when you want to add some sweetness to your food.

LABEL-READING TO IDENTIFY SUGAR

If you're trying to figure out if a food is high in added sugar and what ingredients are contributing the added sugar, there are two places you should look on the label. First, check the Nutrition Facts panel to see how many grams of added sugar are in one serving. Be aware that the "total sugars" on the label includes both added and naturally-occurring sugars. The ingredients list will identify the sources of added sugar, which may be listed as any of the 61 different names in the graphic high on this page.

Let's take a look a some labels to practice identifying naturally occurring and added sugars in foods. Below are labels from a 6-ounce serving of plain yogurt. There are 6 grams of total sugar and 0 grams of added sugar listed on the label. What ingredients contain naturally-occurring sugar?

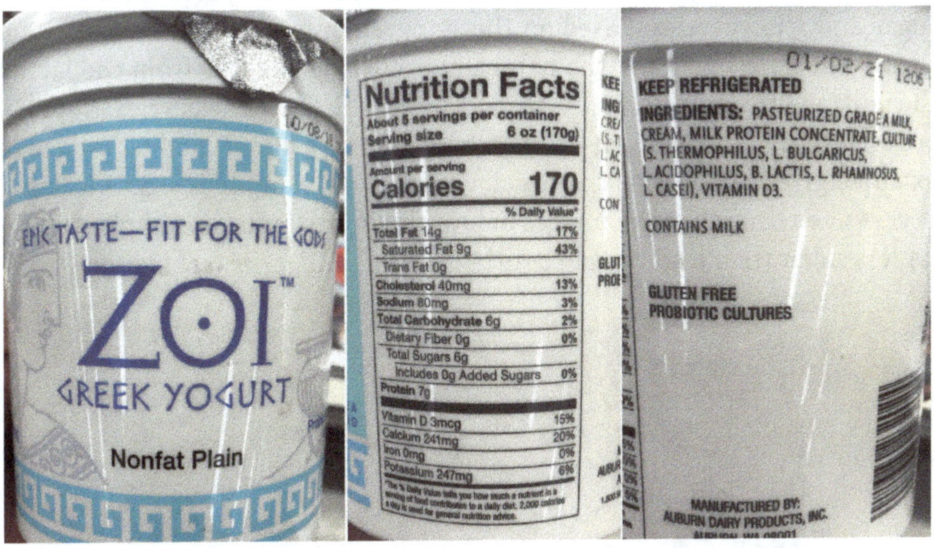

Figure 4.37. Plain yogurt with Nutrition Facts and ingredient list.

The ingredients include nonfat milk, maltodextrin (a food additive that is a polysaccharide), milk protein concentrate, vitamins, and bacteria. There are no sources of added sugar in the ingredient list and zero grams of added sugar shown in the Nutrition Facts, so the 6 grams of total sugars are from naturally-occurring lactose in the milk. Next, look at the label below for a 6-ounce serving of sweetened strawberry yogurt. There are 28 grams of total sugar and 21 grams of added sugar listed on the label. What ingredients are contributing naturally-occurring and added sugar in this product?

Figure 4.38. Strawberry yogurt with Nutrition Facts and ingredient list.

To answer this question, we again have to look at the ingredients list. Like the plain yogurt, the first ingredient is milk, but this strawberry yogurt also contains cane sugar and strawberries. Based on these ingredients, the added sugar comes from sucrose in cane sugar,and the naturally-occurring sugar is from the lactose from the milk and the glucose, fructose, and sucrose in the strawberries. Not all yogurts are created equal, and many of them have less ingredients and less sugar than the example given above with the Greek yogurt. One example is siggi's Icelandic style skyr. As you can see in the images below, the ingredients are simple, and there is a lot less sugar than traditional yogurts.

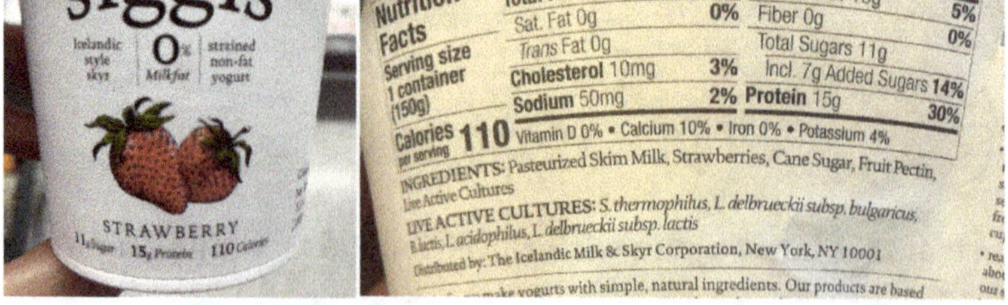

Figure 4.39. Siggi's strawberry yogurt and Nutrition Facts.

In the siggi's strawberry yogurt, a 5.3 oz (150g) serving has 11 grams of total sugar and 7 grams of added sugar. This is a combination of naturally-occurring sugar from the milk and strawberries and added sugar from the cane sugar. However, this product only has 7 grams of added sugar (just under 2 teaspoons), which is a lot less than the strawberry Greek yogurt with 21 grams of added sugar (just over 5 teaspoons) shown above. It is important to pay attention to labels when shopping for nutrient-dense foods.

REFERENCES:

US Department of Health and Human Services and U.S. Department of Agriculture. (2015). Dietary Guidelines for Americans. Retrieved from https://health.gov/dietaryguidelines/2015/

Institute of Medicine, Food and Nutrition Board. (2005). Dietary Reference Intakes for Energy, Carbohydrate, Fiber, Fat, Fatty Acids, Cholesterol, Protein, and Amino Acids (Macronutrients). Retrieved from https://www.nap.edu/read/10490/chapter/1

Ng, S.W., Slining, M.M., & Popkin, B.M. (2012). Use of caloric and noncaloric sweeteners in US consumer packaged foods, 2005-2009. *Journal of the Academy of Nutrition and Dietetics.* 112(11), 1828-1834. e1-6. doi: 10.1016/j.jand.2012.07.009

[1]U.S Department of Agriculture, Agricultural Research Service, Nutrient Data Laboratory. (2014). USDA National Nutrient Database for Standard Reference, Release 27. Available at: http://www.ars.usda.gov/nutrientdata.

Belluz and Zarracina. (2017). Sugar, explained. *Vox.* Retrieved from https://www.vox.com/science-and-health/2017/1/13/14219606/sugar-intake-dietary-nutrition-science

Sugar Science, "Too Much Can Make Us Sick," University of California, San Franciso, http://sugarscience.ucsf.edu/too-much-can-make-us-sick/#.W5lUKy-ZP-Y, accessed September 12, 2018.

Whitaker, E.M., "The Sweet Science of Honey," Sugar Science, UCSF, http://sugarscience.ucsf.edu/the-sweet-science-behind-honey.html#.W5qtp1InYdU, accessed September 13, 2018.

Phillips, K.M., et al. (2009). Total antioxidant content of alternatives to refined sugar. *J Am Diet Assoc.* 109(1), 64-71. Retrieved from https://www.ncbi.nlm.nih.gov/pubmed/19103324

SUBSTITUTES FOR SUGAR

You should now understand the problems with consuming too much added sugar, but what if you've sworn off regular soda and switched to diet versions? What if you're choosing "sugar-free" products, sweetened not with sugar but with sugar substitutes like aspartame, saccharin, or stevia? Are these a better choice?

Figure 4.40. Examples of products containing high-intensity sweeteners: diet soda, sugar-free chocolate, and bulk containers of Splenda and stevia extract.

Diet sodas are the biggest source of sugar substitutes in the American diet, but these ingredients are found in a range of foods, including ice cream, yogurt, cereals, iced tea, energy drinks, candy, cookies, granola bars, salad dressings, frozen dinners, and energy bars.[1] Products containing sugar substitutes are often labeled as sugar-free or "lite," but some don't have any front-of-package labeling with this information, and you may not even realize that you're consuming them. With more consumers watching their sugar intake, the use of sugar substitutes is growing, and the food industry is working hard to market them as a healthier choice. We can expect to see them in more and more products, so it's important to understand what these substances are and what they may mean for our health.

WHAT ARE SUGAR SUBSTITUTES?

You may find sugar substitutes called lots of different things, including artificial, non-nutritive, high-intensity, or low-calorie sweeteners. Regardless of the name, these are substances that have a sweet taste but few or no calories.[2] In fact, they are much sweeter than sucrose, so a tiny amount can add a lot of sweetness to food. (Sweetener packets like Splenda and Equal contain a small amount of sweetener and a lot of filler ingredients.)

Table 4.5. Sugar substitutes approved by the FDA for use in the United States with their brand names and sweetness relative to sucrose.

Sweetener	Brand Names	Sweetness (relative to sucrose)
Acesulfame Potassium (Ace-K)	Sweet One® Sunett®	200x
Advantame		20,000x
Aspartame	Nutrasweet® Equal® Sugar Twin®	200x
Neotame	Newtame®	7,000-13,000
Saccharin	Sweet'N Low® Sweet and Low® Sweet Twin® Necta Sweet®	200-700x
Sucralose	Splenda®	600x
Luo Han Guo or monk fruit extracts	Nectresse® Monk Fruit in the Raw® PureLo®	100-250 x
Stevia	Truvia® PureVia® Enliten®	200-400 x

Unlike regular sugar, the sweeteners listed in the table above are not associated with dental caries, and they generally don't raise blood glucose.

Sugar alcohols are another type of sugar substitute. They include sorbitol, mannitol, lactitol, erythritol, and xylitol. They are chemically similar to monosaccharides but different enough that they aren't processed in the body to the same extent. However, they are at least partially metabolized and contain about 2 kcal/gram (compared with 4 kcal/gram for sucrose). (An exception is erythritol, which contains just 0.2 kcal/g.) Unlike the sweeteners listed in the table above, they are not "high-intensity" but instead are generally less sweet than sucrose. Because they are not fully digested, consuming large amounts of them can cause bloating, gas, and diarrhea.1

Sugar alcohols are often used in sugar-free chewing gums and breath mints and can carry a health claim that they don't promote tooth decay, because mouth bacteria can't easily metabolize them. Xylitol in particular has been studied for its ability to decrease the incidence of tooth decay. However, these studies generally use large doses. For example, a person might have to chew xylitol gum five times per day to see a benefit. The American Academy of Pediatric Dentistry supports the use of xylitol but says the evidence for benefit is not clear and that amounts required may not be practical in real life.[4]

Can Sugar Substitutes Help With Weight Loss?

When people choose diet soda or a sugar-free dessert, they're probably assuming that it's a healthier choice and perhaps that it could help them lose weight. However, studies show this isn't necessarily the case.

In the short-term, if someone who drinks a lot of sugar-sweetened beverages switches to diet versions, studies show that this can result in weight loss. That makes sense, because you're removing a lot of empty calories from the diet.[5]

However, in the long-term, studies show there isn't a clear benefit to consuming sugar substitutes. A recent systematic review and meta-analysis combined the results of studies that lasted at least 6 months.[6] Among the randomized controlled trials, they found no difference in body mass index (BMI—a measure of the ratio of body weight to height) between people who consumed sugar and those who consumed sugar substitutes. Observational studies that tracked large groups over years found that people who consumed sugar substitutes tended to have a higher BMI, greater weight and waist circumference, and a higher incidence of obesity, hypertension, metabolic syndrome, type 2 diabetes, and cardiovascular events.[7] Because these are observational studies, we can't conclude that the sugar substitutes *cause* these health outcomes, but we can conclude that their use is not associated with better health.

When it comes to weight management, the goal is to adopt eating habits that support a sustainable healthy body weight. Sugar substitutes might help in the short-term with decreasing calorie intake and perhaps gradually moving away from sweetened beverages, but better long-term goals for health would be to shift to water and other unsweetened beverages. If you're looking for a way to sweeten your oatmeal or yogurt, you might try adding fresh fruit rather than sugar or an artificial sweetener packet. (Or go ahead and add a bit of brown sugar or a drizzle of honey, keeping in mind the overall goal of moderation.)

Are High-Intensity Sweeteners Safe?

Over the years, there have been a number of concerns about non-nutritive sweeteners. For example, in the 1970s, studies showed that saccharin was linked to bladder cancer in lab rats, so it was labeled as a potential carcinogen, although its use as a sweetener continued. In 2000, after many studies showed no link between cancer and saccharin, the warning labels were no longer required. Some studies have also raised concerns about a link between aspartame and sucralose and cancer, but the FDA has reviewed this evidence and concluded: "Based on the available scientific evidence, the agency has concluded that the high-intensity sweeteners approved by FDA are safe for the general population under certain conditions of use." The National Cancer Institute also says there is no clear evidence that high intensity sweeteners cause cancer.[8]

Figure 4.41. A "Saccharin Notice" sign warns consumers that a grocery store shelf contains products with saccharin, which has been shown to cause cancer in laboratory animals. Between 1977 and 2000, products containing saccharin had to include a cancer warning label. This requirement was removed after the U.S. Department of Health and Human Services determined it was not a concern in humans at doses typically consumed.

There are other emerging safety concerns about sugar substitutes, though. Small studies on both mice and humans show that consuming artificial sweeteners can change our gut bacteria and cause glucose intolerance.[9-11] Glucose intolerance means that blood glucose is abnormally elevated, showing that glucose metabolism is not working properly, and it is a precursor to the development of diabetes. Other researchers worry that having the taste of sweetness signaled to the brain without accompanying calories could derail our normal pathways for sensing hunger and satiety and for regulating glucose metabolism.[12] This research is alarming but still preliminary. However, it is an active area of study, and we can expect more information to emerge in the years to come.

ARE NATURAL SWEETENERS BETTER THAN ARTIFICIAL ONES?

Sweeteners made from the stevia plant and from monk fruit extracts are both derived from plants and so are considered more natural than the other choices. However, it's important to not confuse natural with safe. Remember that many things in nature are dangerous, even deadly. (Consider cyanide, poisonous mushrooms, and botulinum toxin, for example.) Stevia sweeteners, which are growing in popularity and are often marketed as a more natural alternative, are made through a highly industrial extraction process, and some are produced by genetically-modified yeast. None of that makes them inherently less safe, but it does highlight that they aren't exactly natural.

Figure 4.42. A box of Sweetleaf sweetener, marketed as "Natural Stevia Sweetener."

What's important is how well these products are tested and studied for their safety. The Center for Science in the Public Interest, a consumer advocacy nonprofit organization, has criticized the FDA for not requiring more testing of stevia and monk fruit extracts, although they recommend stevia as one of the safer options for sugar substitutes based on existing data. However, recent research has shown that, like artificial sweeteners, stevia also affects the growth of gut bacteria.

What's The Bottom Line?

Sugar substitutes can add sweetness to a food without the calories, and they aren't associated with tooth decay. Despite concerns over the years, they probably don't cause cancer. However, they may not help with weight loss or maintenance in the long-term, and recent research shows that they may alter the gut microbiota and metabolic health.

UNIT-5

LIPIDS

LIPIDS: AN INTRODUCTION

When you think of foods important to the Pacific Northwest, salmon may first come to mind. But to many indigenous people who have long made their home along the Pacific coast, a smaller, humbler fish is considered even more vital: the eulachon smelt.

Traditionally, one of the first signs of spring in the region was the migration of the eulachon smelt into the region's rivers. These thin, blue and silver fish spend most of their lives in the cold Pacific Ocean, but when it's time for them to lay their eggs in early spring, they swim up into the rivers of Oregon, Washington, Canada, and Alaska to sow the seeds of their next generation.

For indigenous people living in this region, who historically subsisted through long winters on stored and preserved foods, the arrival of the eulachon smelt would have been a welcome infusion of calories and flavor. Legends describe the small oily fish saving entire villages from starvation; it's also known as "*halimotkw*," translated as "savior fish" or "salvation fish."

Figure 5.1. Fresh-caught eulachon smelt from the Kuskokwim River, Alaska, 2008.

Beyond the timing of its late winter arrival, what makes the eulachon so valuable is its high lipid content. It's so oily that dried eulachon will ignite and burn like a candle, and nutritionally, it's a dense source of calories. (Remember: Fat contains 9 kilocalories per gram compared to just 4 for carbohydrate and protein.) It's also a good source of fat-soluble vitamins, especially vitamin A, and high in omega-3 fatty acids. William Clark (of the Lewis and Clark expedition), after tasting eulachon from the Columbia River in 1805, wrote: "They are so fat they require no additional sauce, and I think them superior to any fish I ever taste[d], even more delicate and luscious than the white fish of the lakes which have heretofore formed my standard of excellence among the fishes."

Clark may have been the first person of European descent to document the eulachon, but it had long been valued by indigenous people, including members of the Tsimshian, Tlingit, Haida, Nisga's, and Bella Coola tribes. The eulachon run was an annual community event, and people camped for several weeks at the mouths of rivers to net and process the fish. They smoked eulachon to preserve it, but even more importantly, they fermented it in large batches and then cooked it to extract its oils. Once cooled, the fat was solid at room temperature—similar to lard or butter—and could be used for fat and flavor in cooking for the year ahead. It was so valuable that it was traded hundreds of miles inland, forming the great "grease trials" of the Northwest.

Figure 5.2 (left): A page from William Clark's journal of 1805, describing his observations of the euchalon smelt, including its "delicate and luscious" taste. Figure 5.3 (right): Mural panel by Will S. Taylor, entitled "A Tsimshian Family Making Eulachon Butter," circa 1825, with this description: "The glow of the ember fire is on the girls face as she waits for stones to heat. In the box at the right, fish are being boiled by means of the heated stones: the oil thus removed from the fish forms "butter." The residue is being strained by the woman at the left."

These traditional uses of the eulachon continue on a smaller scale, but since the 1990s, the eulachon population has collapsed. Researchers say its biggest threat is climate change, and eulachon have been classified as a threatened species under the Endangered Species Act since 2010.

In today's world—when we can obtain a day's worth of calories and more than enough fat with just a quick trip through a fast food drive-through—it's easy to forget the biological and cultural importance of lipids. But the historical significance of the eulachon remind us of how vital these molecules are to our survival.

REFERENCES

Wolf, E. C., Woody, E., & Zuckerman, S. (2011). *Salmon Nation: People, Fish, and Our Common Home* (2nd ed.). Corvallis, Oregon: Oregon State University Press.

MacKinnon, J. B. (2015, July 7). 'Salvation Fish' That Sustained Native People Now Needs Saving. Retrieved September 30, 2019, from National Geographic News website: https://www.nationalgeographic.com/news/2015/07/150707-salvation-fish-canada-first-nations-animals-conservation-world/

National Marine Fisheries Service. September 2017. Recovery Plan for the Southern Distinct Population Segment of Eulachon (*Thaleichthys pacificus*). National Marine Fisheries Service, West Coast Region, Protected Resources Division, Portland, OR, 97232.

MAIN FUNCTIONS OF FATS

Fats serve useful functions in both the body and the diet. In the body, fat functions as an important depot for energy storage, offers insulation and protection, and plays important roles in regulating and signaling. Large amounts of dietary fat are not required to meet these functions, because most fat molecules can be synthesized by the body from other organic molecules like carbohydrate and protein (with the exception of two essential fatty acids). However, fat also plays unique roles in the diet, including increasing the absorption of fat-soluble vitamins and contributing to the flavor and satisfaction of food. Let's take a closer look at each of these functions of fats in the body and in the diet.

THE FUNCTIONS OF FATS IN THE BODY

Storing Energy

The excess energy from the food we eat is incorporated into adipose tissue, or fatty tissue. Most of the energy required by the human body is provided by carbohydrates and lipids. As discussed in the Carbohydrates unit, glucose is stored in the body as glycogen. While glycogen provides a ready source of energy, it is quite bulky with heavy water content, so the body cannot store much of it for long. Fats, on the other hand, can serve as a larger and more long-term energy reserve. Fats pack together tightly without water and store far greater amounts of energy in a reduced space. A fat gram is densely concentrated with energy, containing more than double the amount of energy as a gram of carbohydrate.

We draw on the energy stored in fat to help meet our basic energy needs when we're at rest and to fuel our muscles for movement throughout the day, from walking to class, playing with our kids, dancing through dinner prep, or powering through a shift at work. Historically, when humans relied on hunting and gathering wild foods or on the success of agricultural crops, having the ability to store energy as fat was vital to survival through lean times. Hunger remains a problem for people around the world, and being able to store energy when times are good can help them endure a period of food insecurity. In other cases, the energy stored in adipose tissue might allow a person to weather a long illness.

Unlike other body cells that can store fat in limited supplies, fat cells are specialized for fat storage and are able to expand almost indefinitely in size. An overabundance of adipose tissue can be detrimental to your health not only from mechanical stress on the body due to excess weight, but also from hormonal and metabolic changes. Obesity can increase the risk for many diseases, including type 2 diabetes, heart disease, stroke, kidney disease, and certain types of cancer. It can also interfere with reproduction, cognitive function, and mood. Thus, while some body fat is critical to our survival and good health, in large quantities it can be a deterrent to maintaining good health.

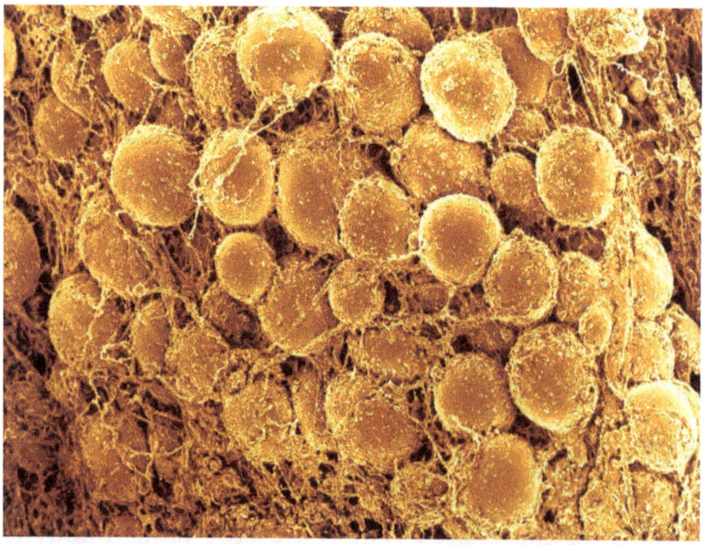

Figure 5.3. Scanning electron micrograph of adipose tissue, showing adipocytes. Computer-coloured orange.

Insulating and Protecting

The average body fat for a man is 18 to 24 percent and for a woman is 25 to 31 percent[1], but adipose tissue can comprise a much larger percentage of body weight depending on the degree of obesity of the individual. Some of this fat is stored within the abdominal cavity, called visceral fat, and some is stored just underneath the skin, called subcutaneous fat. Visceral fat protects vital organs—such as the heart, kidneys, and liver. The blanket layer of subcutaneous fat insulates the body from extreme temperatures and helps keep the internal climate under control. It pads our hands and buttocks and prevents friction, as these areas frequently come in contact with hard surfaces. It also gives the body the extra padding required when engaging in physically demanding activities such as ice skating, horseback riding, or snowboarding.

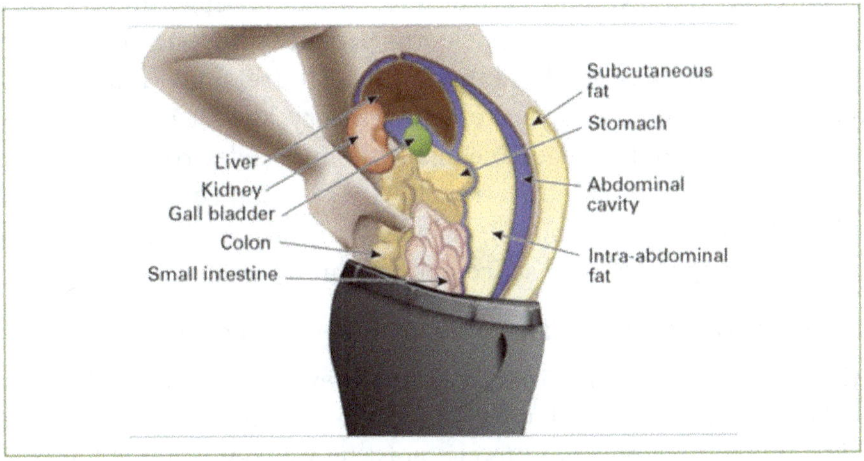

Figure 5.4. There are two types of fat stored as adipose tissue: subcutaneous fat and visceral fat.

Regulating and Signaling

Fats help the body to produce and regulate hormones. For example, adipose tissue secretes the hormone leptin, which signals the body's energy status and helps to regulate appetite. Fat is also required for reproductive health; a woman who lacks adequate amounts may stop menstruating and be unable to conceive until her body can store more energy as fat. Omega-3 and omega-6 essential fatty acids help regulate cholesterol and blood clotting and control inflammation in the joints, tissues, and bloodstream. Fats also play important functional roles in sustaining nerve impulse transmission, memory storage, and tissue structure. Lipids are especially focal to brain activity in structure and in function, helping to form nerve cell membranes, insulate neurons, and facilitate the signaling of electrical impulses throughout the brain.

The FunctionoOf Fats in the Diet

Aiding Absorption and Increasing Bioavailability

The dietary fats in the foods we eat aid in the transport of fat-soluble vitamins, carrying them through the digestive process and improving their intestinal absorption. This improved absorption is known as increased bioavailability. Dietary fats can also increase the bioavailability of compounds known as phytochemicals—non-essential plant compounds considered beneficial to human health. Many phytochemicals are fat-soluble, such as lycopene found in tomatoes and beta-carotene found in carrots, so dietary fat improves the absorption of these molecules in the digestive tract.

In addition to improving bioavailability of fat-soluble vitamins, some of the best dietary sources of these vitamins are also foods that are high in fat. For example, good sources of vitamin E are nuts (including peanut butter and other nut butters), seeds, and plant oils such as those found in salad dressings, and it's difficult to consume enough vitamin E if you're eating a very low-fat diet. (Although fried foods are usually cooked in vegetable oils, vitamin E is destroyed by high heat, so you won't find a lot of vitamin E in french fries or onion rings. Your best bets are minimally-processed, whole foods.) Vegetable oils also provide some vitamin K, and fatty fish and eggs are good sources of vitamins A and D.

Contributing to the Smell, Taste, and Satiety of Foods

Fats satisfy appetite (the desire to eat) because they add flavor to foods. Fat contains dissolved compounds that contribute to mouth-watering aromas and flavors. Fat also adds texture, making baked foods moist and flakey, fried foods crispy, and adding creaminess to foods like ice cream and cream cheese. Consider fat-free cream cheese; when fat is removed from the cream, much of the flavor is also lost. As a result, it is grainy and flavorless—nothing like its full-fat counterpart—and many additives are used in an attempt to replace the lost flavor.

Fats satisfy hunger (the need to eat) because they're slower to be digested and absorbed than other macronutrients. Dietary fat thus contributes to satiety—the feeling of being satisfied or full. When fatty foods are swallowed, the body responds by enabling the processes controlling digestion to slow the movement of food along the digestive tract, giving fats more time to be digested and absorbed and promoting an overall sense of fullness. Sometimes, before the feeling of fullness arrives, people overindulge in fat-rich foods, finding the delectable taste irresistible. Slowing down to appreciate the taste and texture of foods can give your body time to send signals of satiety to your brain, so you can eat enough to be satisfied without feeling overly full.

Providing Essential Fatty Acids

Most lipid molecules can be synthesized in the body from other organic molecules, so they don't specifically need to be provided in the diet. However, there are two that are considered essential and must be included in the diet: linoleic acid and alpha-linolenic acid. We'll discuss these two fatty acids in detail later in the unit.

REFERENCES

ACE (2009) *What are the guidelines for percentage of body fat loss?* American Council on Exercise (ACE). Ask the Expert Blog. December 2009 ,2.

Obesity Prevention Source. Harvard T.H. Chan School of Public Health. Health Risks: Weight Problems Take a Hefty Toll on Body and Mind. Retrieved from: https://www.hsph.harvard.edu/obesity-prevention-source/obesity-consequences/health-effects/

Images

- Figure 5.3. "Adipose tissue, close-up showing adipocytes, SEM" by David Gregory & Debbie Marshall is licensed under CC BY 4.0
- Figure 5.4. "Pandemic of Lifestyle Disease" by Sandra Cohen-Rose and Colin Rose is licensed under CC-BY-2.0
- "French Fries in Paris" by Jim Larrison is licensed under CC BY 2.0
- "Eating a Burger Jonny B'z Dog and More Lourdie Dinner March 24, 20116" by Steven Depolo is licensed under CC-BY-2.0

LIPID: TYPES AND STRUCTURES

Lipids are a family of organic compounds that are mostly insoluble in water, meaning they do not mix well with water. There are three main types of lipids: triglycerides, phospholipids, and sterols. On this page, we'll learn about the structures of these three types of lipids, as well as their functions in the body and where you can find them in foods.

Triglycerides

Triglycerides are the main form of lipids in the body and in foods. More than 95 percent of lipids in the diet are in the form of triglycerides, some having a visible presence and some hidden in foods. Concentrated fats (butter and vegetable oil, for example) and marbling of fat in meat are obviously visible. But fat can also be hidden in foods, as in baked goods, dairy products like milk and cheese, and fried foods. Naturally occurring triglycerides are found in many foods, including avocados, olives, corn, and nuts. We commonly call the triglycerides in our food "fats" and "oils." Fats are lipids that are solid at room temperature, whereas oils are liquid. The terms fats, oils, and triglycerides are often used interchangeably. In this unit, when we use the word fat, we are referring to triglycerides.

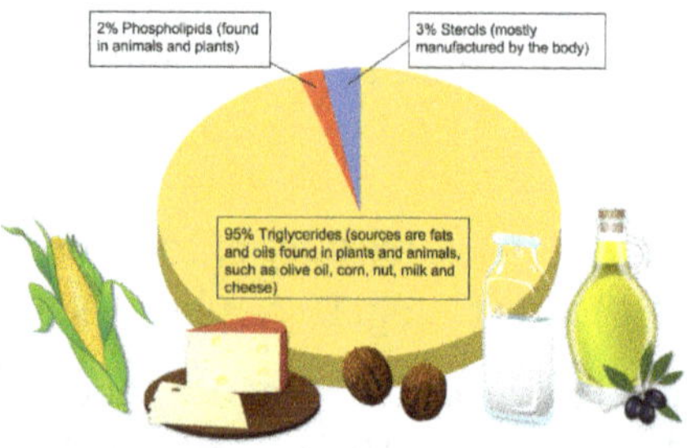

Figure 5.5. 95% of fats in the diet are in the form of triglycerides. Sterols (like cholesterol) make up about 3% of dietary fat intake and phospholipids make up roughly 2% of dietary fat intake.

The structure of a triglyceride is made up of glycerol and three fatty acids. Glycerol is the three-carbon backbone of triglycerides, while fatty acids are longer chains of carbon molecules attached to the glycerol backbone. The "glyceride" in the word "triglyceride" refers to this glycerol backbone, while the "tri" refers to the fact that there are three fatty acids attached. Fatty acids are called acids because they have an acid group (-COOH) on one end of a carbon chain. A monoglyceride contains glycerol with one fatty acid attached, and a diglyceride contains glycerol with two fatty acids attached.

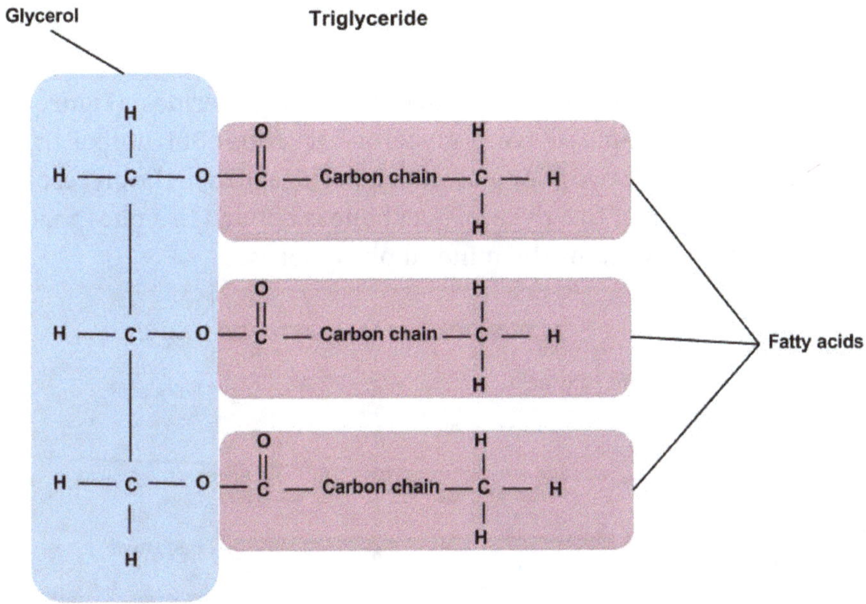

Figure 5.6. The chemical structure of a triglyceride, showing the glycerol backbone and three attached fatty acids.

Figure 5.7. The structure of a triglyceride is often depicted as a simplified drawing of the glycerol backbone and three fatty acids.

There are different types of fatty acids, and triglycerides can contain a mixture of them. Fatty acids are classified by their carbon chain length and degree of saturation. Foods contain different proportions of fatty acid types, and this influences disease risks associated with dietary patterns. We will take a closer look at these differences, along with food sources, in the next section.

Phospholipids

Phospholipids are found in both plants and animals but make up only about 2 percent of dietary lipids. However, they play many important roles in the body and in foods. Phospholipids can also be synthesized by the body, so they don't have to be consumed in the diet.

Phospholipids are similar in structure to triglycerides (Figure 5.8). Like triglycerides, phospholipids have a glycerol backbone. But unlike triglycerides, phospholipids only have two fatty acid molecules attached to the glycerol backbone, while the third carbon of the glycerol backbone is bonded to a phosphate group—a chemical group that contains the mineral phosphorus.

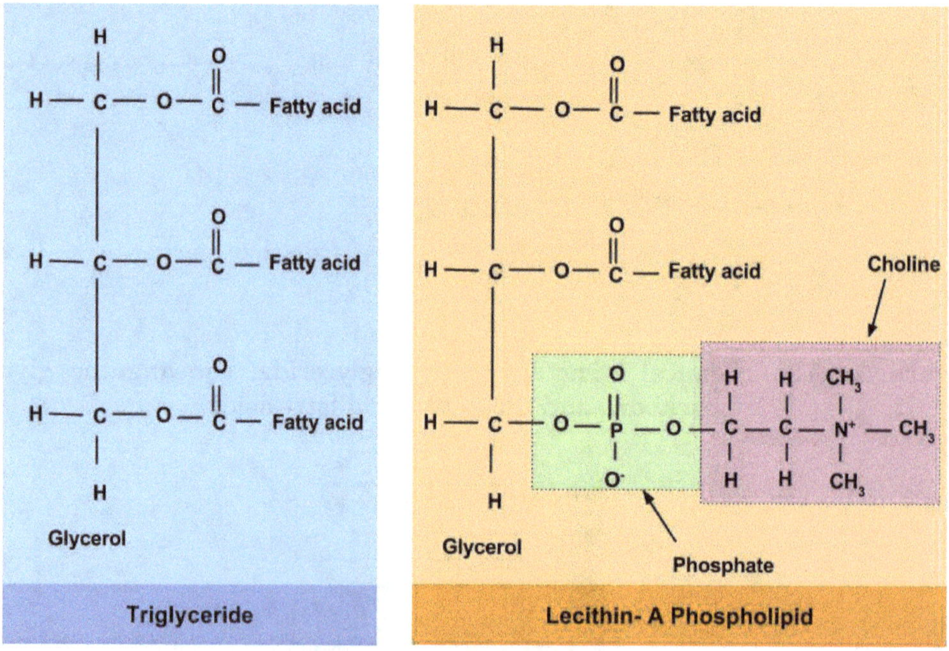

Figure. 5.8. The structural difference between a triglyceride (on the left) and a phospholipid (on the right) is in the third carbon position, where the phospholipid contains a phosphate group instead of a fatty acid.

The unique structure of phospholipids makes them both fat- and water-soluble, or amphiphilic. The fatty-acids are hydrophobic (dislike water), and the phosphate group and glycerol are hydrophilic (attracted to water).

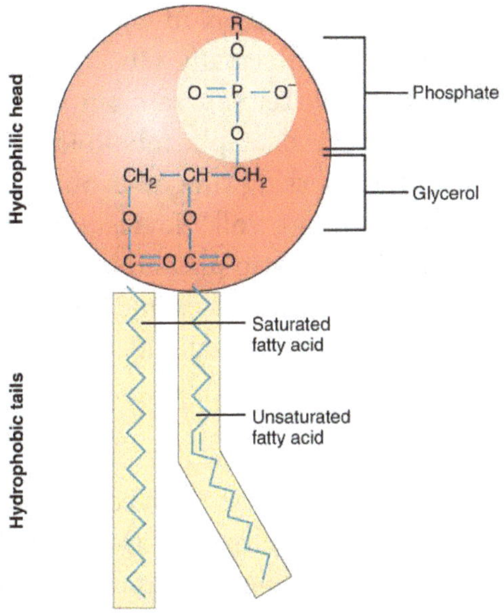

Figure 5.9. A phospholipid molecule consists of a polar phosphate "head," which is hydrophilic, and a non-polar lipid "tail," which is hydrophobic.

The amphiphilic nature of phospholipids makes them very useful for several functions in the body. Every cell in the body is encased in a membrane composed primarily of a double layer of phospholipids (also known as the phospholipid bilayer), which protects the inside of the cell from the outside environment while at the same time allowing for transport of fat and water through the membrane. Phospholipids also play a role in transporting fats in the blood, as we'll learn later in this unit.

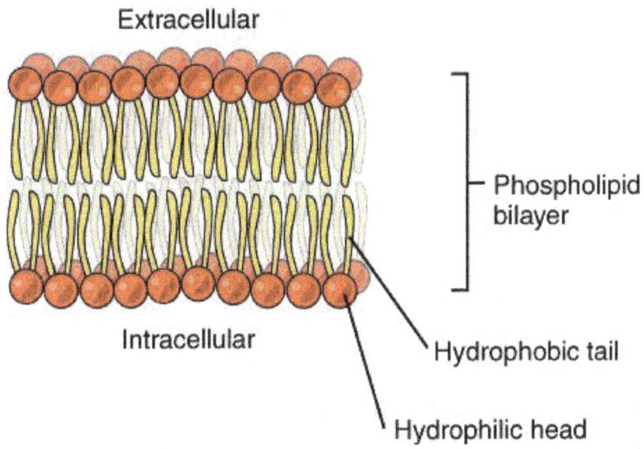

Figure 5.10. The phospholipid bilayer consists of two adjacent sheets of phospholipids, arranged tail to tail. The hydrophobic tails associate with one another, forming the interior of the membrane. The polar heads contact the fluid inside and outside of the cell.

Another important role of phospholipids is to act as emulsifiers. Emulsions are mixtures of two liquids that do not normally mix (oil and water, for example). Without an emulsifier, the oil and water separate out into two layers. Because of their ability to mix with both water and fat, phospholipids are ideal emulsifiers that can keep oil and water mixed, dispersing tiny oil droplets throughout the water. Lecithin—a phospholipid found in egg yolk, soybean, and wheat germ—is often used as a food emulsifier. Emulsifiers also play an important role in making food appetizing; their inclusion in foods like sauces and creams makes for a smoother texture and prevents the oil and water ingredients from separating out. They also can extend shelf life.

Sterols

Sterols have a very different structure from triglycerides and phospholipids. Most sterols do not contain any fatty acids but rather are multi-ring structures, similar to chicken wire. They are complex molecules that contain interlinking rings of carbon atoms, with side chains of carbon, hydrogen, and oxygen attached.

Cholesterol is the best-known sterol because of its role in heart disease. It forms a large part of the fatty plaques that narrow arteries and obstruct blood flow in atherosclerosis. However, cholesterol also has many essential functions in the body. Like phospholipids, cholesterol is present in all body cells as it is an important substance in cell membrane structure. Cholesterol is also used in the body as a precursor in the synthesis of a number of important substances, including vitamin D, bile, and sex hormones such as progesterone, testosterone, and estrogens.

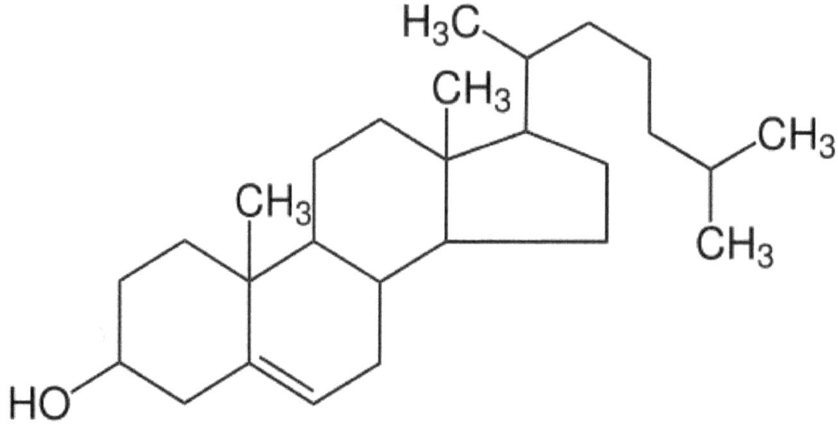

Figure 5.11. Cholesterol is made up of multiple carbon rings bonded together.

Cholesterol is not an essential nutrient; it does not need to be consumed in the diet, because it is manufactured in the liver. Only foods that come from animal sources contain cholesterol. Cholesterol is found in foods like meat, poultry, fish, egg yolks, butter, and dairy products made from whole milk.

Plant foods do not contain cholesterol, but sterols found in plants resemble cholesterol in structure. Plant sterols inhibit cholesterol absorption in the human body, which can contribute to lower cholesterol levels, particularly lower LDL ("bad") cholesterol levels. Plant sterols occur naturally in vegetable oils, nuts, seeds, and whole grains. In addition, some foods like margarines and dressings are fortified with plant sterols.

TYPES OF FATTY ACIDS AND FOOD SOURCES

On the previous page, we learned that triglycerides—the main form of fat in the body and in food—are made up of a glycerol backbone with three fatty acids attached. As mentioned, fatty acids can differ from one another in both carbon chain length and degree of saturation. These characteristics influence the resulting fat in many ways.

Chain length of fatty acids

Fatty acids have different chain lengths, typically between four and 24 carbons, and most contain an even number of carbon atoms. When the carbon chain length is shorter, the melting point of the fatty acid becomes lower (such as fats found in dairy products) and the fatty acid becomes more liquid. Longer chain lengths tend to result in more solid fats, although melting point is also influenced by the degree of saturation.

Degrees of Saturation of Fatty Acids

Fatty acid chains are composed primarily of carbon and hydrogen atoms that are bonded to each other. The term "saturation" refers to whether the carbon atom in a fatty acid chain is filled (or "saturated") to capacity with hydrogen atoms. In a saturated fatty acid, each carbon is bonded to two hydrogen atoms, with single bonds between the carbons.

Alternatively, fatty acids can have points where hydrogen atoms are missing, because there is a double bond between carbons (C=C). This is referred to as a point of unsaturation, because the carbon is only bonded to one hydrogen atom instead of two. Unsaturated fatty acids have one or more points of unsaturation, or double bonds between the carbons. A monounsaturated fatty acid is a fatty acid with one double bond, and a polyunsaturated fatty acid is a fatty acid with two or more double bonds.

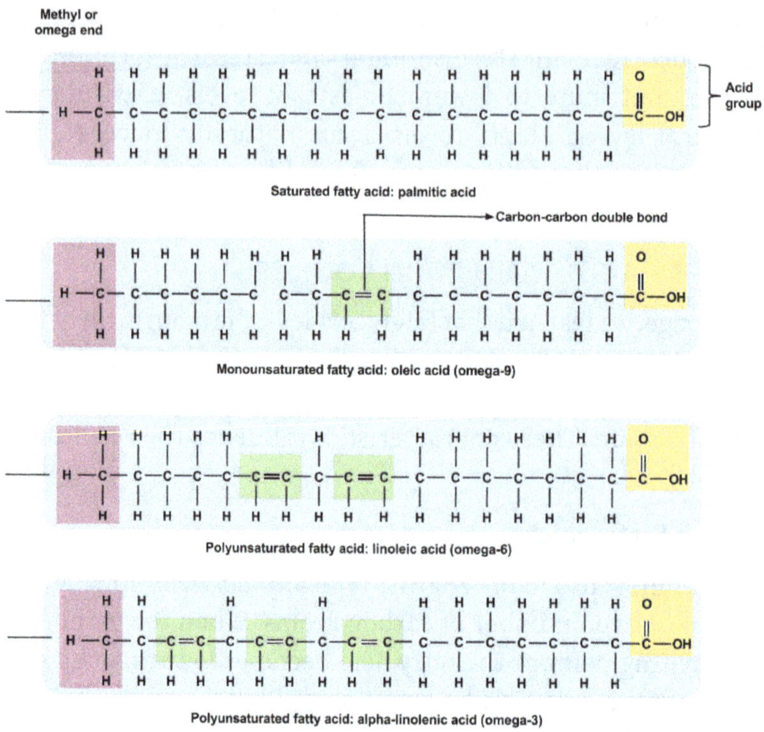

Figure 5.12. The structures of a saturated, monounsaturated, and polyunsaturated fat. Note the differences in points of unsaturation (C=C double bonds) on some of the fatty acids.

Triglycerides in food contain a mixture of saturated, monounsaturated, and polyunsaturated fatty acids, but some foods are better sources of these types of fatty acids than others (Figure 5.13). For example, coconut oil is very high in saturated fat, but it still contains some monounsaturated and polyunsaturated fatty acids. Peanut oil is often thought of as a good source of monounsaturated fat, because that is the predominant fatty acid in the oil, but peanut oil also contains a fair amount of polyunsaturated fatty acids and even some saturated fatty acids.

Saturated Fatty Acids

Fat sources with a high percentage of saturated fatty acids tend to be solid at room temperature. This is because the lack of double bonds in the carbon chains of saturated fatty acids makes them very straight, so they pack together well (like a box of toothpicks). Fats that have mostly saturated fatty acids, like butter and coconut oil, are solid at room temperature, as are the visible fat layers in a strip of bacon or a cut of beef. Consuming a diet high in saturated fats is associated with an increased risk of heart disease, because such a diet increases blood cholesterol, specifically the LDL ("bad") cholesterol level. (More on this later.) Food sources of predominately saturated fatty acids include most animal fats (with the exception of poultry and eggs, which contain more unsaturated fatty acids), dairy products, tropical oils (like coconut and palm oil), cocoa butter, and partially or fully hydrogenated oils.

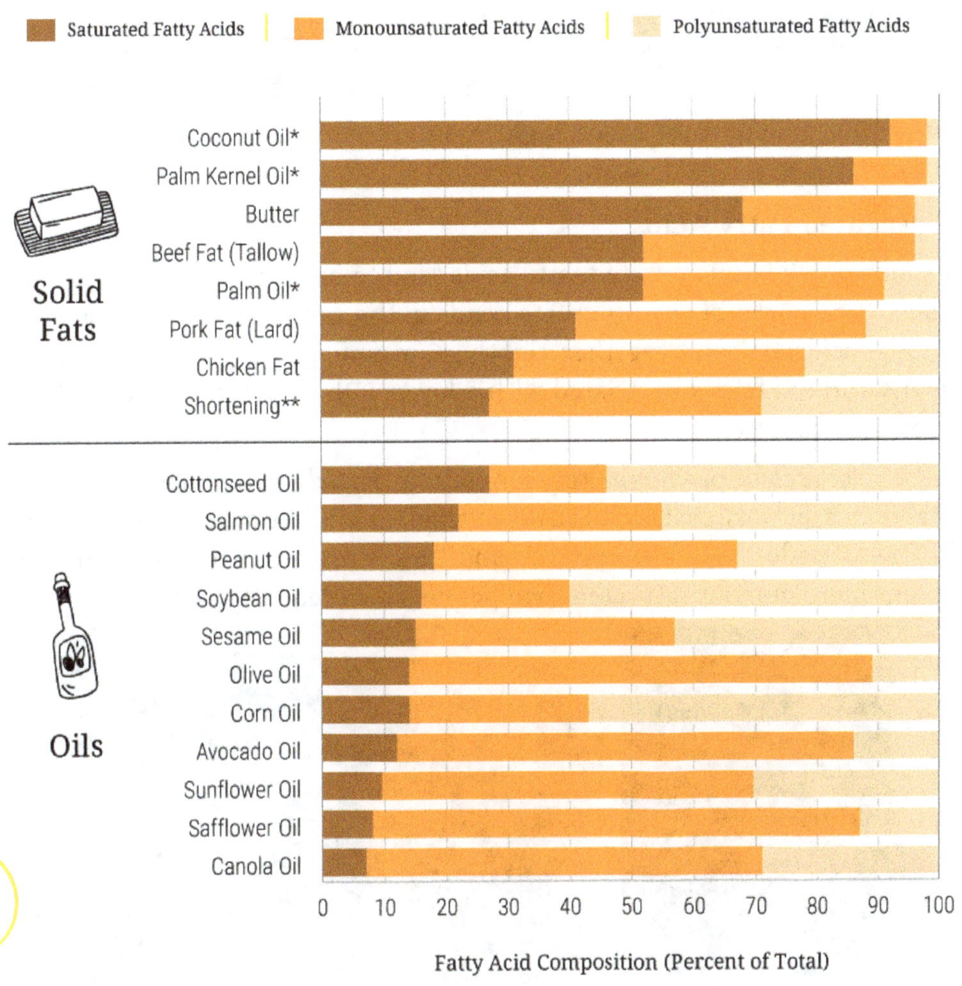

Figure 5.13. Dietary fats contain a mixture of saturated, monounsaturated, and polyunsaturated fatty acids. Foods are often categorized by the predominant type of fatty acids they contain, even though foods contain all three types.

Figure 5.14. Examples of foods high in saturated fat, such as meat and dairy products

Unsaturated Fatty Acids

Fat sources rich in unsaturated fatty acids tend to be liquid at room temperature, because the C=C double bonds create bends in the carbon chain, making it harder for fatty acids to pack together tightly. Consuming a diet rich in mono- and polyunsaturated fats is associated with a lower LDL cholesterol level and a lower risk of heart disease.

Food sources of predominately monounsaturated fats include nuts and seeds like almonds, pecans, cashews, and peanuts; plant oils like canola, olive, and peanut oils; and avocados. The fat in poultry and eggs is predominantly unsaturated and contains more monounsaturated than polyunsaturated fatty acids.

Figure 5.15. Example s of foods high in monounsaturated fat, such as olive oil, avocado, nut butters, and seeds

Food sources of predominately polyunsaturated fats include plant oils (soybean, corn); fish; flaxseed; and some nuts like walnuts and pecans.

Figure 5.16. Examples of foods high in polyunsaturated fats, like fish and nuts

Omega-3, Omega-6, and Essential Fatty Acids

In addition to the length of the carbon chain and the number of double bonds, unsaturated fatty acids are also classified by the position of the first double bond relative to the methyl (-CH3) or "omega" end of the carbon chain (the end furthest from the glycerol backbone in a triglyceride). Fatty acids with the first double bond at the third carbon from the omega end are called omega3- fatty acids. Those with the first double bond at the sixth carbon from the omega end are called omega-6 fatty acids. (There are also omega-9 fatty acids.)

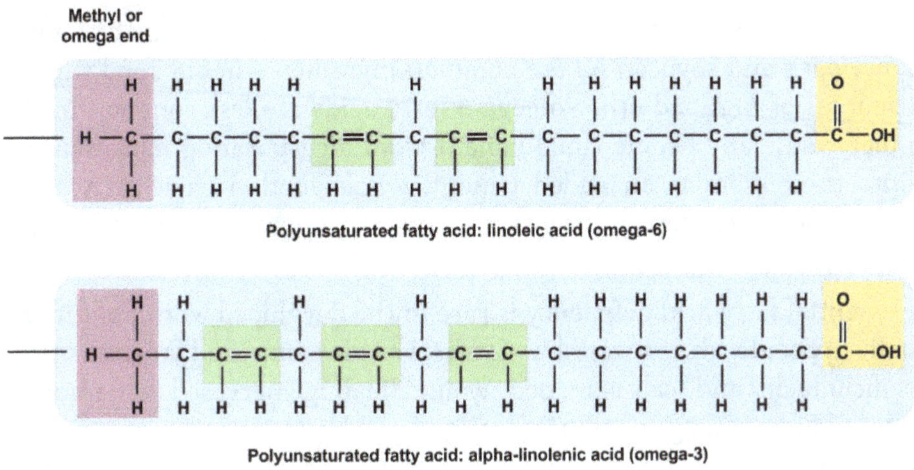

Figure 5.17. The position of the first C=C double bond determines whether an unsaturated fatty acid is classified as omega-3 or omega-6. The two essential fatty acids, linoleic acid (an omega-6) and alpha-linolenic acid (an omega-3) are shown here.

Fatty acids are vital for the normal operation of all body systems, but the body is capable of synthesizing most of the fatty acids it needs. However, there are two fatty acids that the body cannot synthesize: linoleic acid (an omega6-) and alpha-linolenic acid (ALA, an omega3-). These are called essential fatty acids because they must be consumed in the diet. Other fatty acids are called nonessential fatty acids, but that doesn't mean they're unimportant; the classification is based solely on the ability of the body to synthesize the fatty acid. Excellent food sources of linoleic fatty acid include plant oils such as corn oil and soybean oil, often found in salad dressings and margarine. Rich food sources of alpha-linolenic acid (ALA) include nuts, flaxseed, whole grains, legumes, and dark green leafy vegetables.

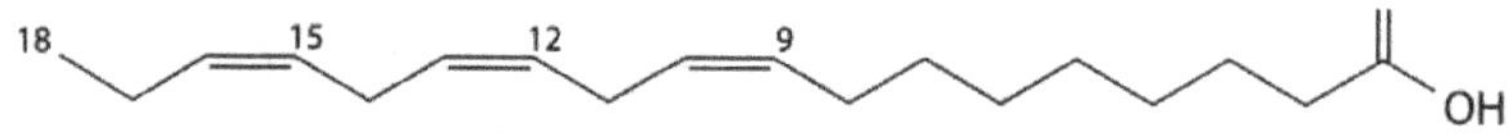

Alpha-linolenic acid (ALA). Note the first double bond occurs at the third carbon from the omega end of the fatty acid, thus making it an omega-3 fatty acid.

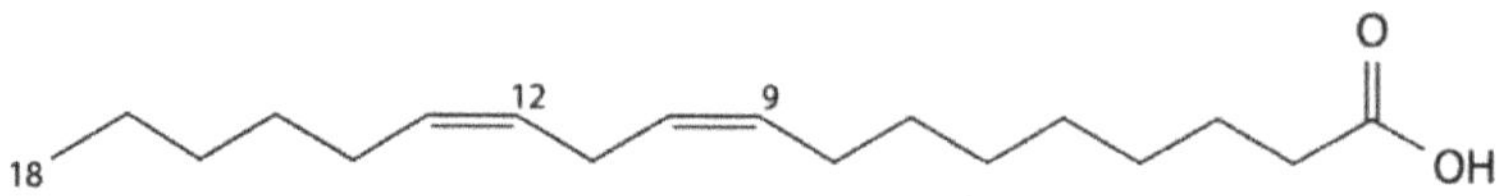

Linoleic acid. Note the first double bond occurs at the sixth carbon from the omega end of the fatty acid, thus making it an omega-6 fatty acid.

Figure 5.18. The chemical structure of the essential fatty acids shown in shorthand, without individual carbon and hydrogen atoms marked.

Most Americans easily consume enough linoleic acid and other omega-6 fatty acids, because corn and soybean oil are common ingredients in our food supply. However, sources of ALA and other omega-3 fatty acids are less common in the American diet, and many people could benefit from incorporating more sources of these into their diet. As an added benefit, whole foods rich in ALA come packaged with other healthful nutrients, like fiber, protein, vitamins, minerals, and phytochemicals.

A true essential fatty acid deficiency is rare in the developed world, but it can occur, usually in people who eat very low-fat diets or have impaired fat absorption. Symptoms include dry and scaly skin, poor wound healing, increased vulnerability to infections, and impaired growth in infants and children.[1]

Omega-3 and omega-6 fatty acids are precursors to a large family of important signaling molecules called eicosanoids (prostaglandins are one type of eicosanoid). Among the many functions of eicosanoids in the body, one of the most important is to regulate inflammation. Without these hormone-like molecules, the body would not be able to heal wounds or fight off infections each time a foreign germ presented itself. In addition to their role in the body's immune and inflammatory processes, eicosanoids also help to regulate circulation, respiration, and muscle movement.

Eicosanoids derived from omega-6 fatty acids tend to increase blood pressure, blood clotting, immune response, and inflammation. These are necessary functions, but they can be associated with disease when chronically elevated. In contrast, eicosanoids derived from omega-3 fatty acids tend to lower blood pressure, inflammation, and blood clotting, functions that can benefit heart health. Omega-3 and omega-6 fatty acids compete for the same enzymatic pathways in the formation of different eicosanoids, so increasing omega-3 fatty acids in the diet may have anti-inflammatory effects.

Two additional omega-3 fatty acids with important health benefits are eicosapentaenoic acid (EPA) and docosahexaenoic acid (DHA). These long-chain polyunsaturated fatty acids have been shown to help lower blood triglycerides and blood pressure, reduce inflammation, and prevent blood clot formation. They also promote normal growth and development in infants, especially in the development of the brain and eyes. Both of these important omega-3 fatty acids can be synthesized in the body from ALA, so they are not considered essential fatty acids. However, the rate of conversion of ALA to these omega-3s is limited, so it is beneficial to consume them regularly in the diet. Fish, shellfish, fish oils, seaweed, and algae are all good sources of EPA and DHA. DHA is also found in human breast milk in quantities dependent on the mother's own intake of DHA sources.

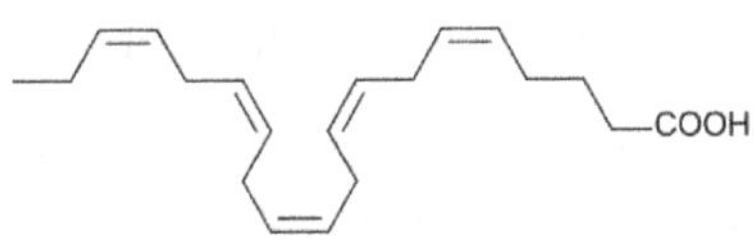

Eicosapentaenoic acid (EPA)

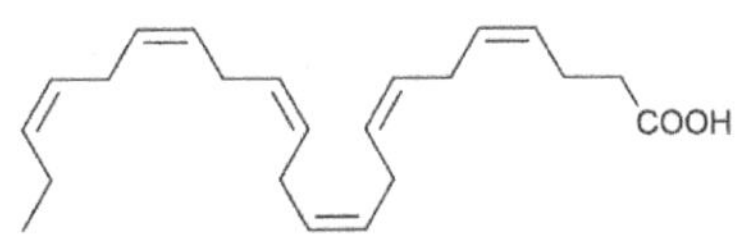

Docosahexaenoic acid (DHA)

Figure 5.19. EPA and DHA are important but non-essential omega-3 fatty acids that can be made in the body from ALA.

Fish oil and omega-3 supplements are among the most commonly used dietary supplements in the United States. Researchers have hypothesized that these supplements might decrease risk of cardiovascular disease, be helpful for those with rheumatoid arthritis, and improve infant brain development when taken in pregnancy or in infancy. Some studies have found such benefits of the supplements, but others haven't. One reason for these inconsistent results may be that studies often don't measure participants' baseline omega-3 levels and intake from foods, and those already consuming adequate omega-3s are less likely to benefit from a supplement. The Dietary Guidelines for Americans recommends consuming 8 ounces of a variety of seafood each week, and in general, people who meet this recommendation likely consume enough omega-3 fatty acids already (along with the other healthful nutrients found in fish) and are unlikely to see an added benefit of taking a fish oil supplement. Some doctors may recommend that people at risk of cardiovascular disease take a fish oil or omega-3 supplement, especially if they don't eat fish regularly.[2]

A Word About *Trans* Fats

The carbon-carbon double bond in an unsaturated fatty acid chain can result in different shapes depending on whether the fatty acid is in a *cis* or *trans* configuration. When the hydrogen atoms are bonded to the same side of the carbon chain, it is called a *cis* fatty acid. Because the hydrogen atoms are on the same side (and repelling one another), the carbon chain has a bent structure. Naturally-occurring fatty acids usually have a *cis* configuration.

In a trans fatty acid, the hydrogen atoms are bonded on opposite sides of the carbon chain, resulting in a more linear structure. Unlike *cis* fatty acids, most *trans* fatty acids are not found naturally in foods, but instead are a result of an industrial process called hydrogenation. Hydrogenation is the process of adding hydrogen to the carbon-carbon double bonds, thus making the fatty acid saturated (or less unsaturated, in the case of partial hydrogenation).

Hydrogenation creates both saturated and trans fatty acids. Trans fatty acids are actually unsaturated fatty acids, but they have the linear shape of saturated fatty acids. (The carbon chains are not bent like naturally-occurring unsaturated fats.) The trans fatty acids formed through partial hydrogenation have an unusual shape, which makes their properties and actions in the body similar to saturated fatty acids.

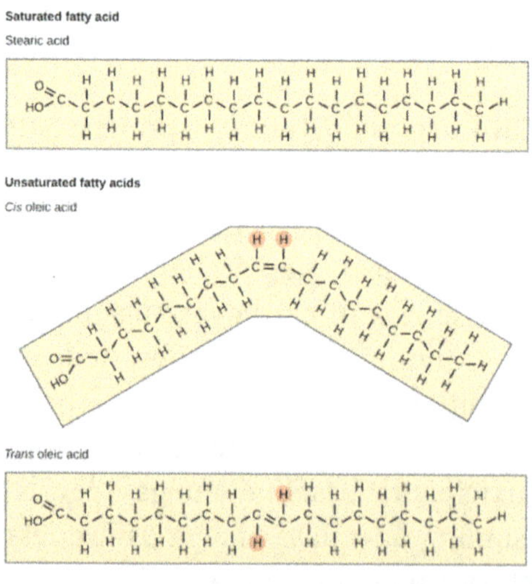

Figure 5.20. Comparison of a saturated fatty acid to both the cis and trans forms of an unsaturated fatty acid.

Hydrogenation was developed in order to make oils semi-solid at room temperature, enabling production of spreadable margarine and shortening from inexpensive ingredients like corn oil. Hydrogenation also makes oils more stable and less likely to go rancid, so partially hydrogenated oils were favored by fast food restaurants for frying, and manufacturers of processed baked goods like cookies and chips found they gave their products a longer shelf life. And because trans fats are unsaturated, nutrition scientists and the medical community believed that they were a healthier alternative to saturated fats.

But around the 1990s, evidence that trans fats were not healthy—far worse than saturated fats, in fact—began to accumulate. Like saturated fat, trans fats increase LDL ("bad") cholesterol, but they also have the effect of decreasing HDL ("good") cholesterol and of increasing inflammatory processes in the body. Researchers found that consuming trans fats, even at low levels (1 to 3 percent of total energy intake), was associated with an increased risk of coronary heart disease. They estimated that eliminating industrial trans fats from the food supply might prevent as many as 19 percent of heart attacks in the U.S. at the time, coming to 228,000 heart attacks averted.[3]

In 2006, the U.S. Food and Drug Administration (FDA) began requiring food companies to list trans fat information on the Nutrition Facts panel of food labels to keep consumers informed of their intake of these fats. That prompted the food industry to mostly eliminate partially hydrogenated oils from their products, often substituting palm oil and coconut oil in their place (both of which are high in saturated fat and may promote heart disease). In 2013, the FDA determined that trans fats were no longer considered safe in the food supply, and in 2015, the agency issued a ruling requiring that manufactured trans fats no longer be included in the U.S. food supply. A one-year extension was granted in 2018, and foods produced prior to that date were given time to work through the food supply. The final ruling requires all manufactured trans fats to be eliminated from the U.S. food supply by 2021.[4]

REFERENCES:

Oregon State University. (2014, April 28). Essential Fatty Acids. Retrieved October 17, 2019, from Linus Pauling Institute website: https://lpi.oregonstate.edu/mic/other-nutrients/essential-fatty-acids

National Institutes of Health. Office of Dietary Supplements—Omega-3 Fatty Acids. Retrieved October 17, 2019, from https://ods.od.nih.gov/factsheets/Omega3FattyAcids-HealthProfessional/

Mozaffarian, D., Katan, M. B., Ascherio, A., Stampfer, M. J., & Willett, W. C. (2006). Trans fatty acids and cardiovascular disease. The New England Journal of Medicine, 354(15), 1601–1613. https://doi.org/10.1056/NEJMra054035

Center for Food Safety and Applied Nutrition. (2018, May 18). Final Determination Regarding Partially Hydrogenated Oils. Retrieved from https://www.fda.gov/food/food-additives-petitions/final-determination-regarding-partially-hydrogenated-oils-removing-trans-fat

Figure 5.14. "Assorted Raw Meats" by Benjamin Ashton is in the Public Domain, CC0; "Butter" by oatsy40 is licensed under CC BY 2.0; "Steak" by Taryn is licensed under CC BY-SA 2.0

Figure 5.15. "Clear Glass Cruet Bottle" by Roberta Sorge is in the Public Domain, CC0; "Fruit + Veggie Toast" by ella.o is licensed under CC BY 2.0

Figure 5.16."raw fish meat on brown chopping board" by Caroline Atwood is in the Public Domain, CC0; "Nuts" by Felix Mittermeier is in the Public Domain, CC0

Figure 5.17 "Structures of a Saturated, Monounsaturated, and Polyunsaturated Fat" by Allison Calabrese is licensed under CC BY 4.0

Figure5.18. "Alpha-linolenic Acid" is in the Public Domain; "Linoleic Acid" is in the Public Domain

Figure 5.19. "Eicosapentaenoic Acid (EPA)" by Minutemen is in the Public Domain; "Docosahexaenoic Acid (DHA)" by Minutemen is in the Public Domain

Figure 5.20. "Structures of Saturated, Unsaturated, Cis and Trans fatty Acids" by Allison Calabrese is licensed under CC BY 4.0

LIPIDS' DIGESTION AND ABSORPTION

Lipid digestion and absorption pose some special challenges. Triglycerides are large molecules, and unlike carbohydrates and proteins, they're not water-soluble. Because of this, they like to cluster together in large droplets when they're in a watery environment like the digestive tract. The digestive process has to break those large droplets of fat into smaller droplets and then enzymatically digest lipid molecules using enzymes called lipases. The mouth and stomach play a small role in this process, but most enzymatic digestion of lipids happens in the small intestine. From there, the products of lipid digestion are absorbed into circulation and transported around the body, which again requires some special handling since lipids are not water-soluble and do not mix with the watery blood.

Let's start at the beginning to learn more about the path of lipids through the digestive tract.

1. Lipid digestion in the mouth

A few things happen in the mouth that start the process of lipid digestion. Chewing mechanically breaks food into smaller particles and mixes them with saliva. An enzyme called lingual lipase is produced by cells on the tongue ("lingual" means relating to the tongue) and begins some enzymatic digestion of triglycerides, cleaving individual fatty acids from the glycerol backbone.

2. Lipid digestion in the stomach

In the stomach, mixing and churning helps to disperse food particles and fat molecules. Cells in the stomach produce another lipase, called gastric lipase ("gastric" means relating to the stomach) that also contributes to enzymatic digestion of triglycerides. Lingual lipase swallowed with food and saliva also remains active in the stomach. But together, these two lipases play only a minor role in fat digestion (except in the case of infants, as explained below), and most enzymatic digestion happens in the small intestine.

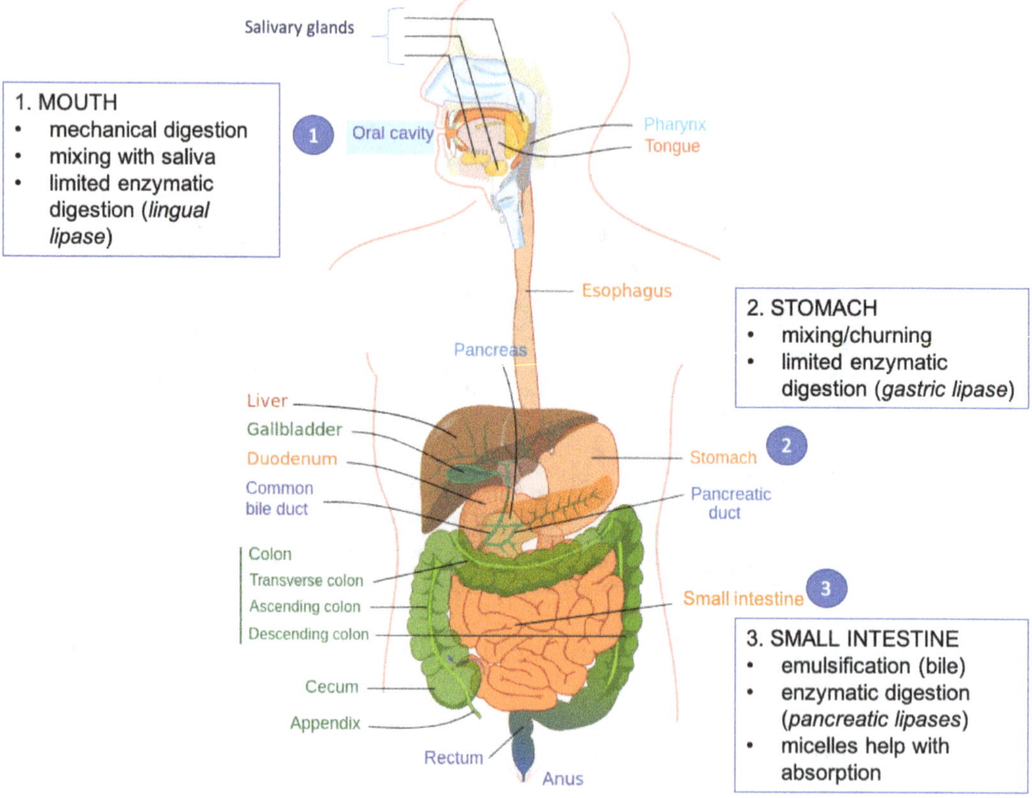

Figure 5.21. Overview of lipid digestion in the human gastrointestinal tract.

3. lipid digestion in the small intestine

As the stomach contents enter the small intestine, most of the dietary lipids are undigested and clustered in large droplets. Bile, which is made in the liver and stored in the gallbladder, is released into the duodenum, the first section of the small intestine. Bile salts have both a hydrophobic and a hydrophilic side, so they are attracted to both fats and water. This makes them effective emulsifiers, meaning that they break large fat globules into smaller droplets. Emulsification makes lipids more accessible to digestive enzymes by increasing the surface area for them to act (see Fig. *5.22* below).

The pancreas secretes pancreatic lipases into the small intestine to enzymatically digest triglycerides. Triglycerides are broken down to fatty acids, monoglycerides (glycerol backbone with one fatty acid still attached), and some free glycerol. Cholesterol and fat-soluble vitamins do not need to be enzymatically digested (see Fig. *5.22* below).

Lipid Absorption from the Small Intestine

Next, those products of fat digestion (fatty acids, monoglycerides, glycerol, cholesterol, and fat-soluble vitamins) need to enter into the circulation so that they can be used by cells around the body. Again, bile helps with this process. Bile salts cluster around the products of fat digestion to form structures called micelles, which help the fats get close enough to the microvilli of intestinal cells so that they can be absorbed. The products of fat digestion diffuse across the membrane of the intestinal cells, and bile salts are recycled back to do more work emulsifying fat and forming micelles.

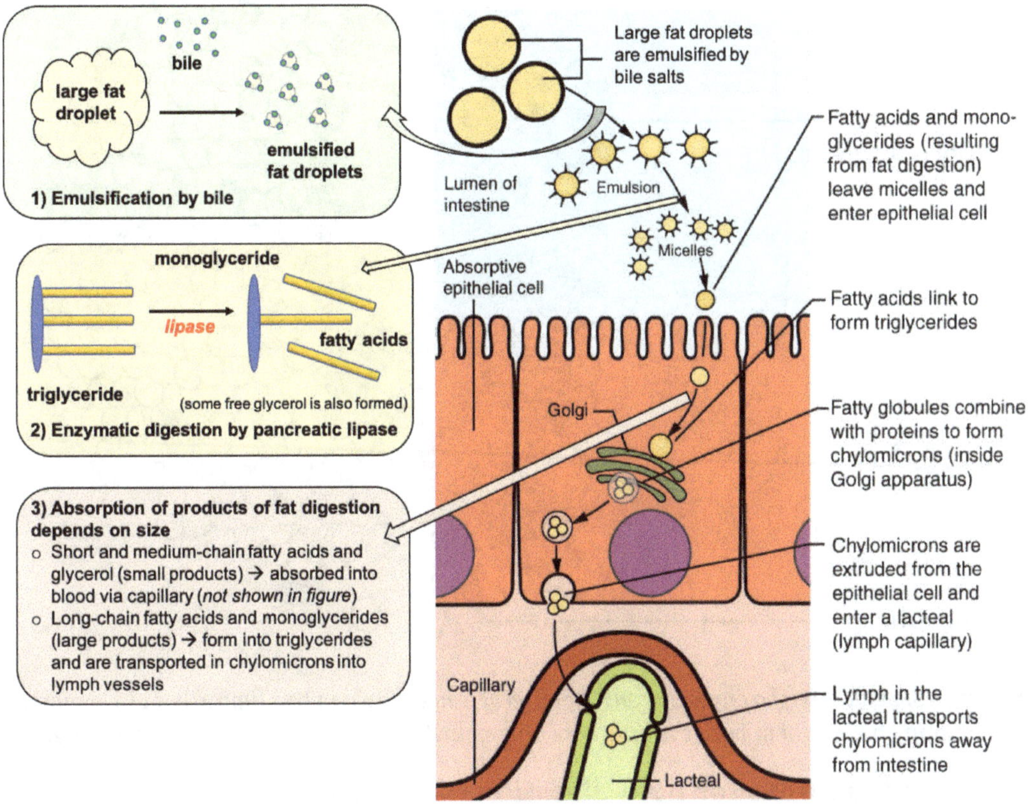

Figure 5.22. Lipid digestion and absorption in the small intestine.

Once inside the intestinal cell, short- and medium-chain fatty acids and glycerol can be directly absorbed into the bloodstream, but larger lipids such as long-chain fatty acids, monoglycerides, fat-soluble vitamins, and cholesterol need help with absorption and transport to the bloodstream. Long-chain fatty acids and monoglycerides reassemble into triglycerides within the intestinal cell, and along with cholesterol and fat-soluble vitamins, are then incorporated into transport vehicles called chylomicrons. Chylomicrons are large structures with a core of triglycerides and cholesterol and an outer membrane made up of phospholipids, interspersed with proteins (called apolipoproteins) and cholesterol. This outer membrane makes them water-soluble so that they can travel in the aqueous environment of the body. Chylomicrons from the small intestine travel first into lymph vessels, which then deliver them to the bloodstream.

Chylomicrons are one type of lipoprotein—transport vehicles for lipids in blood and lymph. We'll learn more about other types of lipoproteins on the next page.

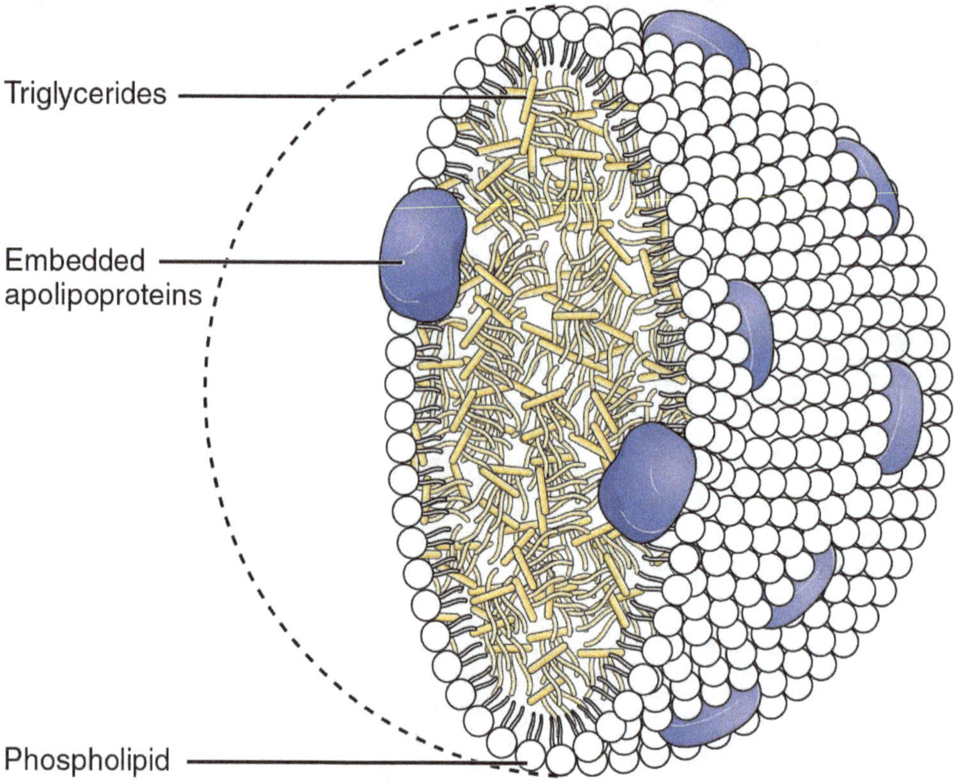

Figure 5.23. Structure of a chylomicron. Cholesterol is not shown in this figure, but chylomicrons contain cholesterol in both the lipid core and embedded on the surface of the structure.

Special adaptations for lipid digestion in infants

Lipids are an important part of an infant's diet. Breast milk contains about 4 percent fat, similar to whole cow's milk. Whether breastfed or formula-fed, fat provides about half of an infant's calories, and it serves an important role in brain development. Yet, infants are born with low levels of bile and pancreatic enzyme secretion, which are essential contributors to lipid digestion in older children and adults. So, how do babies digest all of the fat in their diet?

Infants have a few special adaptations that allow them to digest fat effectively. First, they have plenty of lingual and gastric lipases right from birth. These enzymes play a much more important role in infants than they do in adults.

Second, breast milk actually contains lipase enzymes that are activated in the baby's small intestine. In other words, the mother makes lipases and sends them in breast milk to help her baby digest the milk fats. Amazing, right? Between increased activity of lingual and gastric lipases and the lipases contained in breast milk, young infants can efficiently digest fat and reap its nutritional value for growth and brain development. Studies show that fat digestion is more efficient in premature infants fed breast milk compared with those fed formula. Even pasteurized breast milk, as is used when breast milk is donated for feeding babies in the hospital, is a little harder to digest, because heat denatures the lipases. (Infants can still digest pasteurized breast milk and formula; they're just less efficient at doing so and absorb less of the products of triglyceride digestion.)[1]

REFERENCES

- Lindshield, B. L. Kansas State University Human Nutrition (FNDH 400) Flexbook. goo.gl/vOAnR
- OpenStax, Anatomy and Physiology. OpenStax CNX. Aug 28, 2019 http://cnx.org/contents/14fb4ad7-39a1-4eee-ab6e-3ef2482e3e22@16.7.
- University of Hawai'i at Mānoa Food Science and Human Nutrition Program, "Digestion and Absorption of Lipids," CC BY-NC 4.0
- 1American Academy of Pediatrics Committee on Nutrition, 2014. Chapter 2: Development of Gastrointestinal Function. In: Kleinman RE, Greer FR, eds. Pediatric Nutrition. 7th ed. Elk Grove Village, IL: American Academy of Pediatrics.

TRANSPORT, STORAGE, AND USE OF LIPID

Once dietary lipids are digested in the gastrointestinal tract and absorbed from the small intestine, they need to be transported around the body so they can be utilized by cells or stored for later use. Once again, the fact that lipids aren't water-soluble means that they need some help getting around the watery environment of the body. Let's take a look at how this works.

lipoproteins transport lipids around the body

Lipoproteins are transport vehicles for moving water-insoluble lipids around the body. There are different types of lipoproteins that do different jobs. However, all are made up of the same four basic components: cholesterol, triglycerides, phospholipids, and proteins.

The interior of a lipoprotein—called the lipid core—carries the triglycerides and cholesterol esters, both of which are insoluble in water. Cholesterol esters are cholesterol molecules with a fatty acid attached. The exterior of lipoproteins—called the surface coat—is made up of components that are at least partially soluble in water: proteins (called apolipoproteins), phospholipids, and unesterified cholesterol. The phospholipids are oriented so that their water-soluble heads are pointed to the exterior, and their fat-soluble tails are pointed towards the interior of the lipoprotein. Apoliproteins are similarly amphipathic (soluble in both fat and water), a property that makes them useful for aiding in the transport of lipids in the blood.

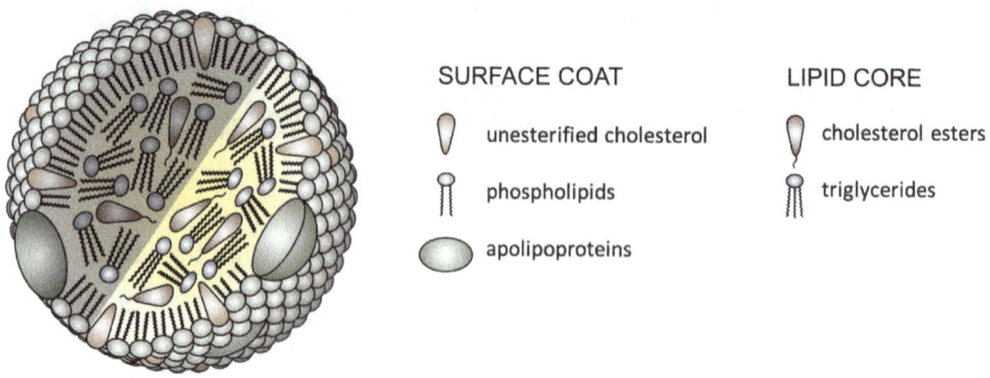

Figure 5.24. Basic structure of all lipoproteins. Note the orientation of phospholipids on the surface coat.

While all lipoproteins have this same basic structure and contain the same four components, different types of lipoproteins vary in the relative amounts of the four components, in their overall size, and in their functions. These are summarized in the graph and table below, and the following sections give more details on the role of each type of lipoprotein.

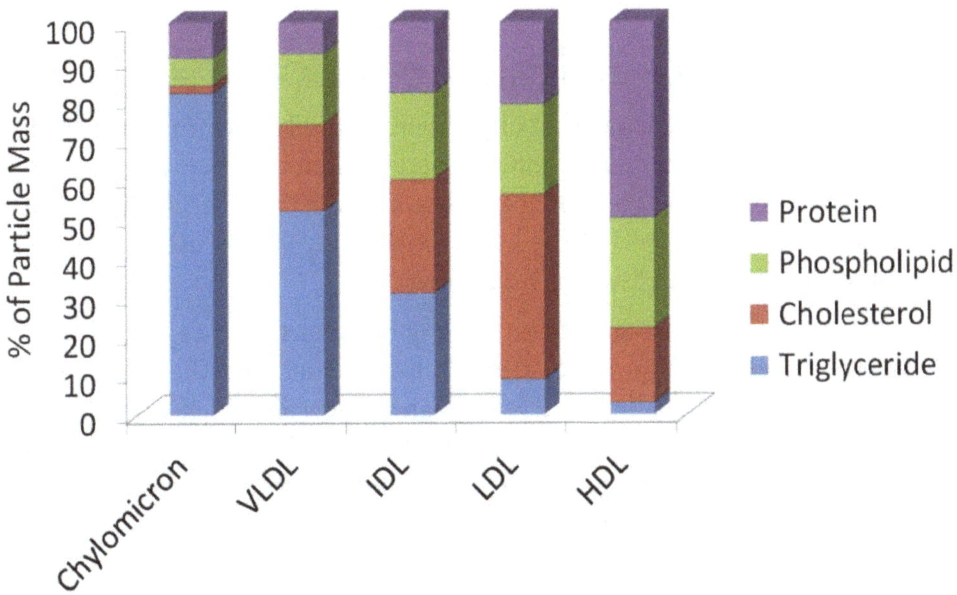

Figure 5.25. Comparison of composition of lipoproteins.

Table 5.1. Comparison of composition, size, density, and function of lipoproteins. (TG = triglycerides)

	Chylomicrons	Very-low-density lipoproteins (VLDL)	Intermediate-density lipoproteins (IDL)	Low-density lipoproteins (LDL)	High-density lipoproteins (HDL)
Diameter (nm)	75-1200 (largest)	30-80	25-35	18-25	5-12 (smallest)
Density (g/dL)	0.95 (least dense)	0.95-1.006	1.006-1.019	1.019-1.063	1.063 (most dense)
Function	Transports lipids from the small intestine, delivers TG to the body's cells	Transports lipids from the liver, delivers TG to body's cells	Formed as VLDL become depleted in TG; either returned to liver or made into LDL	Deliver cholesterol to cells	Pick up cholesterol in the body and return to the liver for disposal

Except for chylomicrons, the names of the lipoproteins refer to their density. Of the four components of lipoproteins, protein is the most dense and triglyceride is the least dense. (This is why one pound of muscle is much more compact in size than one pound of adipose or fat tissue.) High-density lipoproteins are the most dense of the lipoproteins, because they contain more protein and less triglyceride. Chylomicrons are the least dense, because they contain so much triglyceride and relatively little protein.

Chylomicrons Deliver Lipids to Cells for Utilization and Storage

On the previous page, we learned that chylomicrons are formed in the cells of the small intestine, absorbed into the lymph vessels, and then eventually delivered into the bloodstream. The job of chylomicrons is to deliver triglycerides (originating from digested food) to the cells of the body, where they can be used as an energy source or stored in adipose tissue for future use.

How do the triglycerides get from the chylomicrons into cells? An enzyme called lipoprotein lipase sits on the surface of cells that line the blood vessels. It breaks down triglycerides into fatty acids and glycerol, which can then enter nearby cells. If those cells need energy right away, they'll oxidize the fatty acids to generate ATP. If they don't need energy right away, they'll reassemble the fatty acids and glycerol into triglycerides and store them for later use.

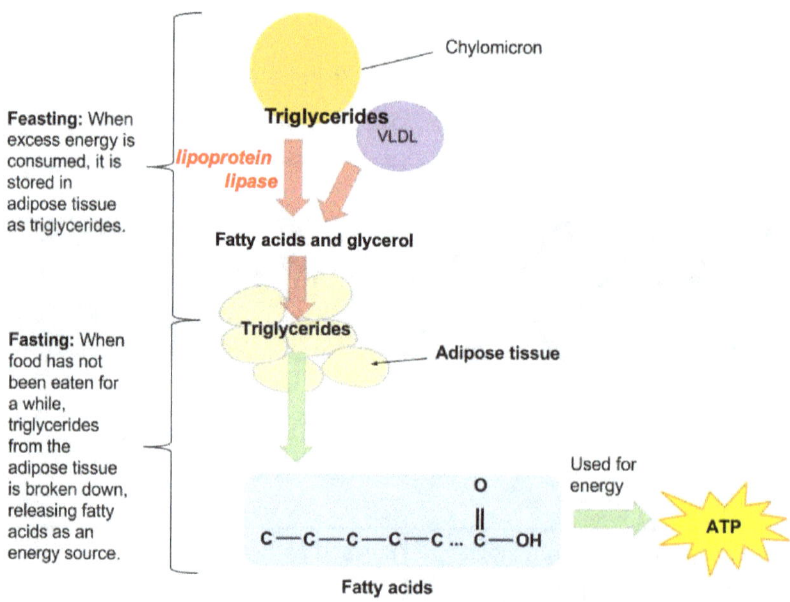

Figure 5.26. Triglycerides in chylomicrons and VLDL are broken down by lipoprotein lipase so that fatty acids and glycerol can be used for energy—or stored for later—in cells.

As triglycerides are removed from the chylomicrons, they become smaller. These chylomicron remnants travel to the liver, where they're disassembled.

Lipid Transport from the Liver

The contents of chylomicron remnants, as well as other lipids in the liver, are incorporated into another type of lipoprotein called very-low-density lipoprotein (VLDL). Similar to chylomicrons, the main job of VLDL is delivering triglycerides to the body's cells, and lipoprotein lipase again helps to break down the triglycerides so that they can enter cells (Figure 5.27).

As triglycerides are removed from VLDL, they get smaller and more dense, because they now contain relatively more protein compared to triglycerides. They become intermediate-density lipoproteins (IDL) and eventually low-density lipoproteins (LDL). The main job of LDL is to deliver cholesterol to the body's cells. Cholesterol has many roles around the body, so this is an important job. However, too much LDL can increase a person's risk of cardiovascular disease, as we'll discuss below.

High-density lipoproteins (HDL) are made in the liver and gastrointestinal tract. They're mostly made up of protein, so they're very dense. Their job is to pick up cholesterol from the body's cells and return it to the liver for disposal.

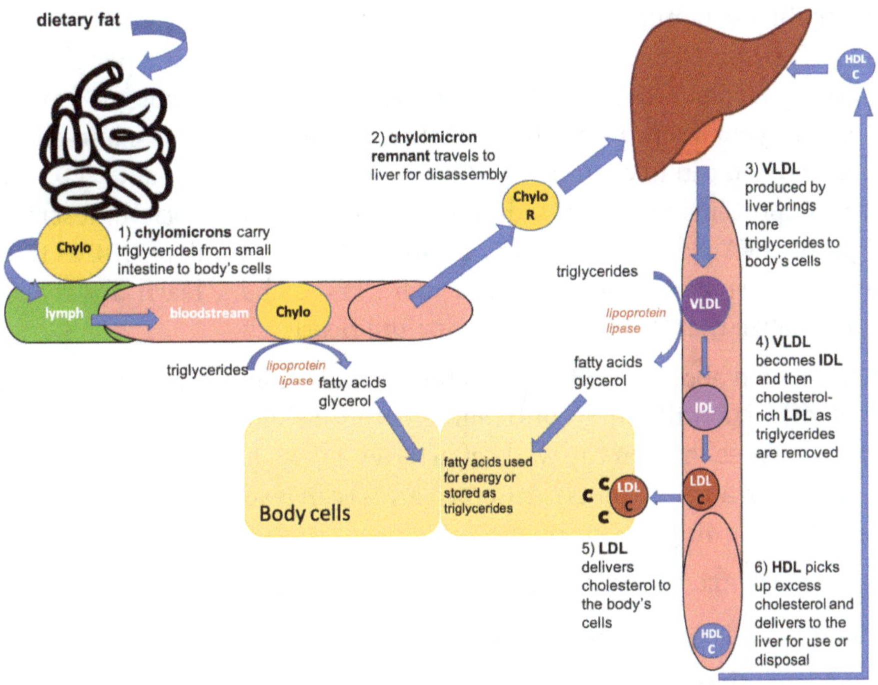

Figure 5.27. Overview of lipoprotein functions in the body.

Does Eating a Higher Fat Diet Mean You Will Store More Fat?

No. How much fat a person stores depends on how many calories they consume relative to how many calories they need to fuel their body. If they consume more calories than needed to meet their body's daily needs—whether those calories come from dietary fat, carbohydrate, or protein—then they'll store most of the excess calories in the form of fat in adipose tissue. If they consume a high-fat diet but not excess calories, then they'll utilize that fat to generate ATP for energy. That said, remember that fat is more calorically dense (9 kilocalories per gram) than protein or carbohydrates (both 4 kcal/g), so if you eat a high-fat diet, you may need to eat smaller portions. And, as we'll discuss later in this unit, there are good reasons to watch the *type* of fats that you eat, because of the relationship between dietary fat intake and risk of developing cardiovascular disease.

Understanding Blood Cholesterol Numbers

A person's blood cholesterol numbers can be one indicator of their risk of developing cardiovascular disease. This is a standard blood test, also called a lipid panel, that reports total cholesterol, LDL, HDL, and triglycerides. When doctors assess a person's risk of cardiovascular disease, they consider these numbers—along with other risk factors like family history, smoking, diabetes, and high blood pressure—in determining their recommendations for lifestyle changes (such as improving diet and getting more exercise) or prescribing medications.

You might be familiar with LDL and HDL as "bad cholesterol" and "good cholesterol," respectively. This is an oversimplification to help people interpret their blood lipid values, because cholesterol is cholesterol; it's not good or bad. The cholesterol in your food or synthesized in your body is all the same cholesterol molecule, and you can't consume good or bad cholesterol. In reality, LDL and HDL are both lipoproteins that carry cholesterol. A more appropriate descriptor for LDL might be the "bad cholesterol transporter." We can think of HDL as the "good cholesterol transporter," although the more researchers learn about HDL, the more they realize that this is also an oversimplification.

What's so bad about LDL? If there's too much LDL in the blood, it can become lodged in arterial walls and contribute to the development of ***atherosclerosis***, when fatty plaques thicken the walls of arteries and reduce the flow of blood (and therefore oxygen and nutrients). Atherosclerosis can lead to a number of problems, including the following:

- Coronary Artery Disease (can lead to angina and heart attack)
- Carotid Artery Disease (increases risk of stroke)
- Peripheral Artery Disease
- Chronic Kidney Disease

If a broken piece of plaque or a blood clot completely blocks an artery supplying the brain or the heart, it can cause a stroke or a heart attack, respectively. If you have high LDL cholesterol, then making changes like exercising more, eating less saturated fat, and stopping smoking (if applicable) can help lower it. Sometimes medications are also necessary to keep LDL in check.

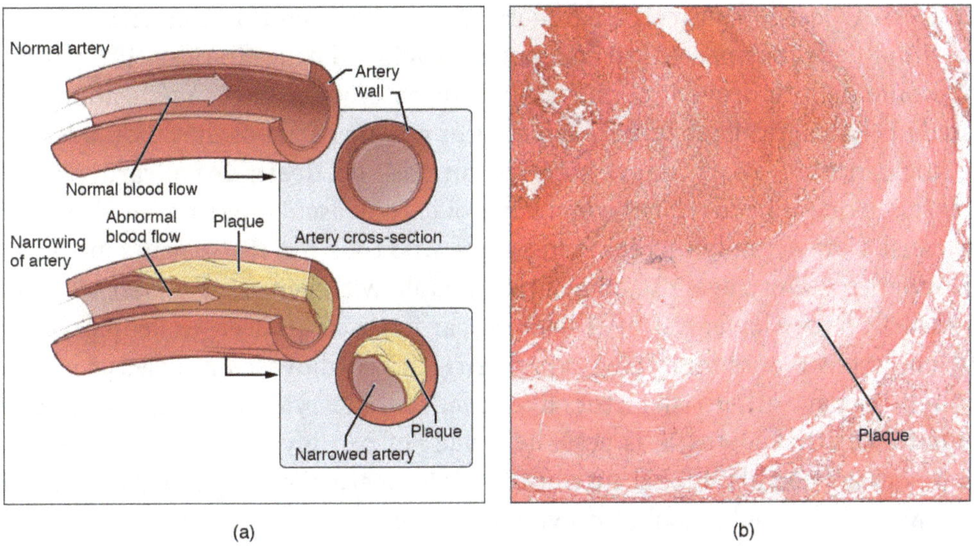

Fig. 5.28. Atherosclerosis. (a) Atherosclerosis can result from plaques formed by the buildup of fatty, calcified deposits in an artery. (b) Plaques can also take other forms, as shown in this micrograph of a coronary artery that has a buildup of connective tissue within the artery wall. LM × 40. (Micrograph provided by the Regents of University of Michigan Medical School © 2012)

Physicians sometimes run additional blood tests to measure the LDL particle size and the number of LDL particles. The standard LDL test measures the total *amount* of cholesterol that is carried by LDL, but the reality is that LDL comes in a range of sizes, and small LDL particles are more strongly associated with the risk of atherosclerosis and cardiovascular disease than large LDL particles. For two people with the same total LDL cholesterol measurement, a person with more small particles will have a greater *number* of LDL particles circulating and a higher risk of developing heart disease. A person with more large LDL particles will have fewer particles overall and a lower risk of developing heart disease. Measuring particle size is not recommended for all patients because of the cost of the test and the fact that it rarely changes treatment course or improves outcomes. However, it can be useful in patients with diabetes or insulin resistance, as they tend to have more small LDL particles, which may call for using medications sooner or in higher doses.

HDL has been considered the "good cholesterol" or "good cholesterol transporter" because it scavenges cholesterol, including LDL lodged in the arterial walls, and helps to remove it from the body. Previously, it was thought that high HDL could prevent atherosclerosis and protect people from cardiovascular disease. But over the last few years, researchers have discovered that this view of HDL is oversimplified. Pharmaceutical companies developed drugs to raise HDL, thinking this would help to prevent cardiovascular disease. When these medications were tested in clinical trials, they were effective at raising HDL, but they didn't decrease the incidence of heart attack, stroke, angina, or death from cardiovascular disease.[1] In one clinical trial, the incidence of cardiovascular events and death from any cause were actually increased in people who took the HDL-raising medication (2). Genetic studies have also shown that people with genes for higher HDL don't necessarily have a lower risk of developing cardiovascular disease.[3] People with low HDL cholesterol do seem to have a higher risk of cardiovascular disease, but they also tend to have other risk factors like sedentary lifestyle, smoking, and diabetes. It's not clear that low HDL is actually a *cause* of cardiovascular disease; we only know that it's correlated. Because of these discoveries, raising HDL cholesterol is no longer considered a goal of prevention of heart disease. Rather, lowering LDL cholesterol is the primary target.[4,5]

REFERENCES:

Rosenson, R. S. (2018). Measurement of blood lipids and lipoproteins—UpToDate. Retrieved October 7, 2019, from UpToDate website: https://www.uptodate.com/contents/measurement-of-blood-lipids-and-lipoproteins?search=lipid%20particle%20size&source=search_result&selectedTitle=1~150&usage_type=default&display_rank=1

Rosenson, R. S., & Durrington, P. (2017). HDL cholesterol: Clinical aspects of abnormal values—UpToDate. Retrieved October 7, 2019, from UpToDate website: https://www.uptodate.com/contents/hdl-cholesterol-clinical-aspects-of-abnormal-values?search=dyslipidemia%20guidelines&source=search_result&selectedTitle=7~150&usage_type=default&display_rank=7

Lincoff, A. M. *et al.* Evacetrapib and Cardiovascular Outcomes in High-Risk Vascular Disease. *New England Journal of Medicine* 376, 1933–1942 (2017).

Barter, P. J. et al. Effects of Torcetrapib in Patients at High Risk for Coronary Events. *New England Journal of Medicine* 357, 2109–2122 (2007).

Voight, B. F. *et al.* Plasma HDL cholesterol and risk of myocardial infarction: a mendelian randomisation study. *Lancet* 380, 572–580 (2012).

Jacobson, T. A., Ito, M. K., Maki, K. C., Orringer, C. E., Bays, H. E., Jones, P. H., ... Brown, W. V. (2015). National lipid association recommendations for patient-centered management of dyslipidemia: Part 1–full report. *Journal of Clinical Lipidology*, *9*(2), 129–169. https://doi.org/10.1016/j.jacl.2015.02.003

Catapano, A. L., Graham, I., De Backer, G., Wiklund, O., Chapman, M. J., Drexel, H., ... Wald, D. (2016). 2016 ESC/EAS Guidelines for the Management of Dyslipidaemias. *European Heart Journal*, *37*(39), 2999–3058. https://doi.org/10.1093/eurheartj/ehw272

RECOMMENDATIONS FOR LIPID AND HEART HEALTH

"In past decades, dietary guidance has almost universally advocated reducing the intake of total and saturated fat, with the emphasis shifting more recently from total fat to the replacement of saturated fat with polyunsaturated fats and the elimination of trans fat. These recommendations and the link between fat consumption and the risk of cardiovascular disease have been among the most vexed issues in public health: are dietary fats "villains," are they benign, or are they even "heroes" that could help us consume better overall diets and promote health?"

This quote from a 2018 analysis on diet and heart health in the medical journal, *The BMJ*, sums up the challenge of making recommendations for dietary fat intake. You may reason that if some fats are healthier than other fats, why not eliminate all "bad" fats and consume as much healthy fat as desired? Remember, everything in moderation. As we review the established guidelines for daily fat intake, we'll explain the importance of balancing fat consumption with proper fat sources.

Lipids and disease

Heart disease is the leading cause of death in the United States, claiming the lives of nearly 650,000 Americans in 2017. Deaths from stroke, which shares many of the same risk factors, accounts for an additional 150,000 fatalities each year. The burden of cardiovascular disease makes it critical to address dietary and lifestyle choices that can decrease risk factors for these diseases. According to the American Heart Association, the following risk factors for heart disease are modifiable, meaning they can be managed or changed by the individual:

- high blood pressure
- high blood cholesterol
- tobacco use
- diabetes
- poor diet
- physical inactivity
- overweight or obesity

In light of these risk factors, there are many dietary strategies related to fat intake that can reduce the risk of heart disease. The amount and type of fat that a person eats can have a profound effect on the way their body metabolizes fat and cholesterol.

If left unchecked, improper dietary fat consumption can lead down a path to severe health problems. An elevated level of lipids—including triglycerides and cholesterol—in the blood is called hyperlipidemia. Many diseases and health conditions are associated with high blood lipid levels, including:

- Cardiovascular disease
- Heart attack
- Stroke
- Congestive heart failure
- Obesity

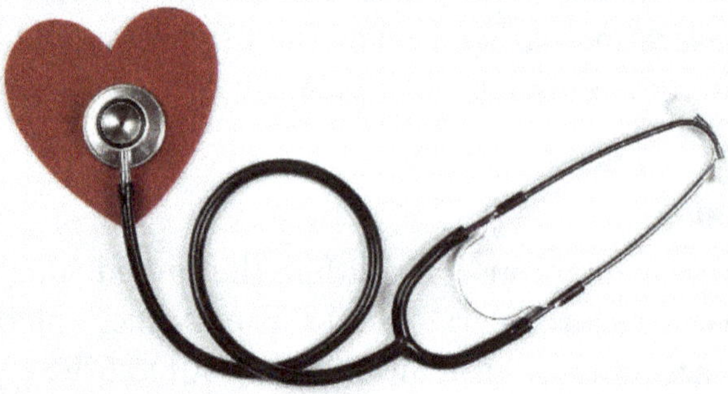

Recommended Fat Intake

Because of the association between dietary fat consumption and hyperlipidemia, health experts have developed recommendations for fat intake to help guide food choices. The acceptable macronutrient distribution range (AMDR) from the Dietary Reference Intake Committee and recommendations for adult fat consumption from the Dietary Guidelines for Americans are as follows:

- Fat calories should be limited to 20-35 percent of total calories, with most fats coming from polyunsaturated and monounsaturated fats, such as those found in fish, nuts, and vegetable oils.
- Consume fewer than 10 percent of calories from saturated fats. Some studies suggest that lowering the saturated fat content to less than 7 percent can

further reduce the risk of heart disease, but other studies contradict this recommendation.

- Keep the consumption of trans fats (any food label that reads partially hydrogenated oil) as low as possible. (As manufactured trans fats are being phased out of the U.S. food supply, Americans don't have to worry as much about identifying and avoiding trans fats. However, they're still used in other parts of the world.)
- Choose more lean and low-fat products when selecting meat, poultry, milk, and dairy products.

The Complicated Relationship of Saturated Fat with Heart Health

Being conscious of the need to reduce blood cholesterol levels to improve health often means limiting the consumption of saturated fats. Multiple studies have shown an association between the saturated fats found in some meat, whole-fat dairy products, butter, and tropical oils (coconut and palm oils) and higher LDL cholesterol levels. However, while saturated fats have a historical reputation for being unhealthy, more recent research has suggested that the link between saturated fat and heart disease may not be as clear cut as we once thought.

The BMJ's analysis on dietary fat and heart health explains the complicated relationship as such:

"Synergism and interactions between different components of foods together with the degree of processing and preparation or cooking methods lead to a 'food matrix' effect which is not captured by considering single nutrients. Different types of food that are high in saturated fats are likely to have different effects on health. For example, dairy products and processed meats, both high in saturated fats, are differentially associated with many health outcomes in prospective epidemiological studies, often in opposite directions. One explanation for this divergence is that despite their similar fat content, other components of these two food groups are associated with different health effects. For example, dairy products contain minerals such as calcium and magnesium and have probiotic features if fermented, whereas processed red meat has a high salt and preservative content.

To produce public health guidelines on which foods to eat or avoid to reduce the risk of chronic disease is complicated because dietary fats are typically mixtures of different types of fatty acids. Animal fats, for instance, are the main sources of saturated fats in many modern diets, but some animal fats are higher in monounsaturated fats than saturated fats, and polyunsaturated fats in vegetable oils will typically contain both omega-3 and omega-6 fatty acids in different concentrations. Hence, conclusions about the health effects of saturated and polyunsaturated fatty acids are unlikely to consistently translate to the health effects of the fats, oils, and foods in which those fatty acids are present."

In an article from Tufts University Health & Nutrition Letter, *Links Between Saturated Fat, Blood Cholesterol & Heart Disease Prove Complex*, researcher Ronald M. Krauss, MD, said, "there is insufficient evidence from prospective epidemiological studies to conclude that dietary saturated fat is associated with an increased risk of cardiovascular disease (CVD)." Researchers state that reducing saturated fat intake may not clearly decrease CVD risk in part because people tend to replace saturated fat with processed carbohydrates. Recent studies have attempted to tease out different approaches to lowering saturated fat in the diet, showing that the types of foods used to replace saturated fat affect the risk of heart disease.

- If saturated fat in the diet is replaced with unsaturated fats, risk for heart disease decreases.
- If saturated fat in the diet is replaced with refined carbohydrates, risk for heart disease increases.

The 2015 DGA report described the evidence as follows:

"Strong and consistent evidence shows that replacing saturated fats with unsaturated fats, especially polyunsaturated fats, is associated with reduced blood levels of total cholesterol and of low-density lipoprotein-cholesterol (LDL-cholesterol). Additionally, strong and consistent evidence shows that replacing saturated fats with polyunsaturated fats is associated with a reduced risk of CVD events (heart attacks) and CVD-related deaths.

Some evidence has shown that replacing saturated fats with plant sources of monounsaturated fats, such as olive oil and nuts, may be associated with a reduced risk of CVD. However, the evidence base for monounsaturated fats is not as strong as the evidence base for replacement with polyunsaturated fats. Evidence has also shown that replacing saturated fats with carbohydrates reduces blood levels of total and LDL-cholesterol, but increases blood levels of triglycerides and reduces high-density lipoprotein-cholesterol (HDL-cholesterol). Replacing total fat or saturated fats with carbohydrates is not associated with reduced risk of CVD. Additional research is needed to determine whether this relationship is consistent across categories of carbohydrates (e.g., whole versus refined grains; intrinsic versus added sugars), as they may have different associations with various health outcomes. Therefore, saturated fats in the diet should be replaced with polyunsaturated and monounsaturated fats."

Dr. Krauss summed up the recommendations as such: "An overall eating pattern that emphasizes whole grains rather than refined carbs such as white flour, along with foods high in polyunsaturated fats, such as fish, seeds, nuts and vegetable oils, is of more value for reducing coronary heart disease risk than simply aiming to further reduce saturated fat."

Figure 5.29. Examples of heart healthy meals, including fish, nuts, seeds, whole grains, and unsaturated fats

What About Dietary Cholesterol?

Dietary cholesterol also has a small impact on overall blood cholesterol levels, but not as much as some people may think. For most people, decreasing dietary cholesterol intake has little impact on blood cholesterol, because their bodies respond by reducing synthesis of cholesterol in favor of using the cholesterol obtained from food. Genetic factors may also influence the way a person's body responds to changes in cholesterol intake. The 2015 U.S. Dietary Guidelines for Americans (DGA) suggest limiting saturated fats, thereby indirectly limiting dietary cholesterol since foods that are high in cholesterol tend to also be high in saturated fats. (Eggs and fish are notable exceptions; both are high in cholesterol but low in saturated fats.)

The 2015 DGA dropped the previous recommendation (from the 2010 DGA) to limit the consumption of dietary cholesterol to 300 mg per day, but with a caveat:

"This change does not suggest that dietary cholesterol is no longer important to consider when building healthy eating patterns. As recommended by the Institute of Medicine, individuals should eat as little dietary cholesterol as possible while consuming a healthy eating pattern."

Practical Tips for a Heart Healthy Diet

- Focus on whole food sources of dietary fat, because they come packaged with vitamins, minerals, phytochemicals, and sometimes fiber and protein.
- Remember that saturated fats are found in large amounts in foods of animal origin. They should be limited within the diet.
- Some highly processed foods, such as stick margarines, cookies, pastries, crackers, fried foods, and snack foods are sources of saturated fat (and historically, trans fats) that can elevate your cholesterol levels, so use them sparingly.
- Fatty fish like salmon, tuna, and halibut are heart-healthy due to their high levels of omega-3 fatty acids, which can reduce inflammation and lower cholesterol levels. The American Heart Association recommends consuming fish—especially fatty fish—twice per week.
- Nuts and seeds contain high levels of unsaturated fatty acids that aid in lowering LDL when they replace saturated fat in the diet.
- Most plant-based oils (except tropical oils like palm and coconut oils) are good sources of polyunsaturated and monounsaturated fats and may help to lower LDL. Substitute oils for solid fats when possible.
- Choose whole-grain and high-fiber foods. Reduced risk for cardiovascular disease has been associated with diets that are high in whole grains and fiber. Fiber also slows down cholesterol absorption. The AHA recommends

that at least half of daily grain intake come from whole grains.

- Soluble fiber reduces cholesterol absorption in the bloodstream. Plant-based foods rich in soluble fiber include oatmeal, oat bran, legumes, apples, pears, citrus fruits, barley, and prunes.
- Check out additional information on What is the Best Oil for Cooking?

In addition to dietary changes, don't forget the value of physical activity. Physical *inactivity* increases the risk of heart disease. Physical exercise can help manage or prevent high blood pressure and high blood cholesterol. Regular activity raises HDL while also decreasing triglycerides and plaque buildup in the arteries. And activity consistently burns calories, which can help with weight loss or maintaining a healthy weight.

REFERENCES

Forouhi, N. G., Krauss, R. M., Taubes, G., & Willett, W. (2018). Dietary fat and cardiometabolic health: evidence, controversies, and consensus for guidance. *Bmj*, *361*, k2139.

Links Between Saturated Fat, Blood Cholesterol & Heart Disease Prove Complex – Tufts University Health & Nutrition Letter Article. (2010, May). Retrieved from https://www.nutritionletter.tufts.edu/issues/6_5/current-articles/Links-Between-Saturated-Fat-Blood-Cholesterol-Heart-Disease-Prove-Complex_592-1.html.

Dietary Guidelines for Americans 2015–2020 8th Edition. (n.d.). Retrieved from https://health.gov/dietaryguidelines/2015/guidelines/

UNIT-6

PROTEIN

PROTEIN: AN INTRODUCTION

Protein makes up approximately 20 percent of the human body and is present in every single cell. The word protein is a Greek word, meaning "of utmost importance." Proteins are called the workhorses of life as they provide the body with structure and perform a vast array of functions. You can stand, walk, run, skate, swim, and more because of your protein-rich muscles. Protein is necessary for proper immune system function, digestion, and hair and nail growth, and is involved in numerous other body functions. In fact, it is estimated that more than one hundred thousand different proteins exist within the human body. In this lesson you will learn about the structure of protein, the important roles that protein serves within the body, how the body uses protein, the risks and consequences associated with too much or too little protein, and where to find healthy sources of it in your diet.

STRUCTURE OF PROTEIN

What is protein?

Proteins are macromolecules composed of amino acids. For this reason, amino acids are commonly called the building blocks of protein. There are 20 different amino acids, and we require all of them to make the many different proteins found throughout the body. Proteins are crucial for the nourishment, renewal, and continuance of life.

Just like carbohydrates and fats, proteins contain the elements carbon, hydrogen, and oxygen, but proteins are the only macronutrient that also contain nitrogen as part of their core structure. In each amino acid, the elements are arranged into a specific conformation, consisting of a central carbon bound to the following four components:

- A hydrogen
- A nitrogen-containing *amino* group
- A carboxylic *acid* group (hence the name "amino acid")
- A side chain

The first three of those components are the same for all amino acids. The side chain—represented by an "R" in the diagram below—is what makes each amino acid unique.

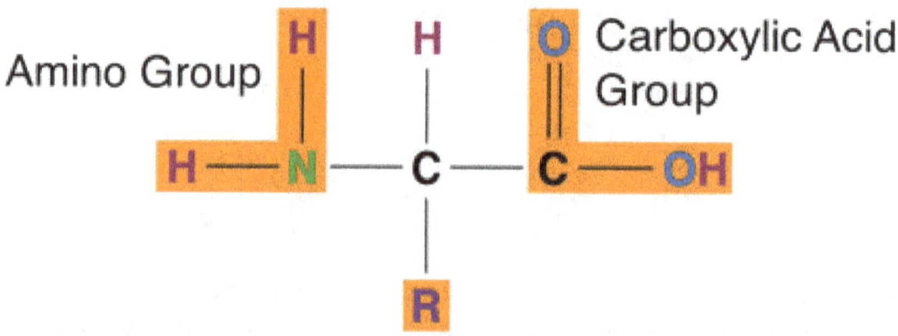

Figure 6.1. Amino Acid Structure.

Amino acid side chains vary tremendously in their size and can be as simple as one hydrogen (as in glycine, shown in Figure 6.1) or as complex as multiple carbon rings (as in tryptophan). They also differ in their chemical properties, thus impacting the way amino acids act in their environment and with other molecules. Because of their side chains, some amino acids are polar, making them hydrophilic and water-soluble, whereas others are nonpolar, making them hydrophobic or water-repelling. Some amino acids carry a negative charge and are acidic, while others carry a positive charge and are basic. Some carry no charge. Some examples are given below. For this class, you don't need to memorize amino acid structures or names, but you should appreciate the diversity of amino acids and understand that it is the side chain that makes each different.

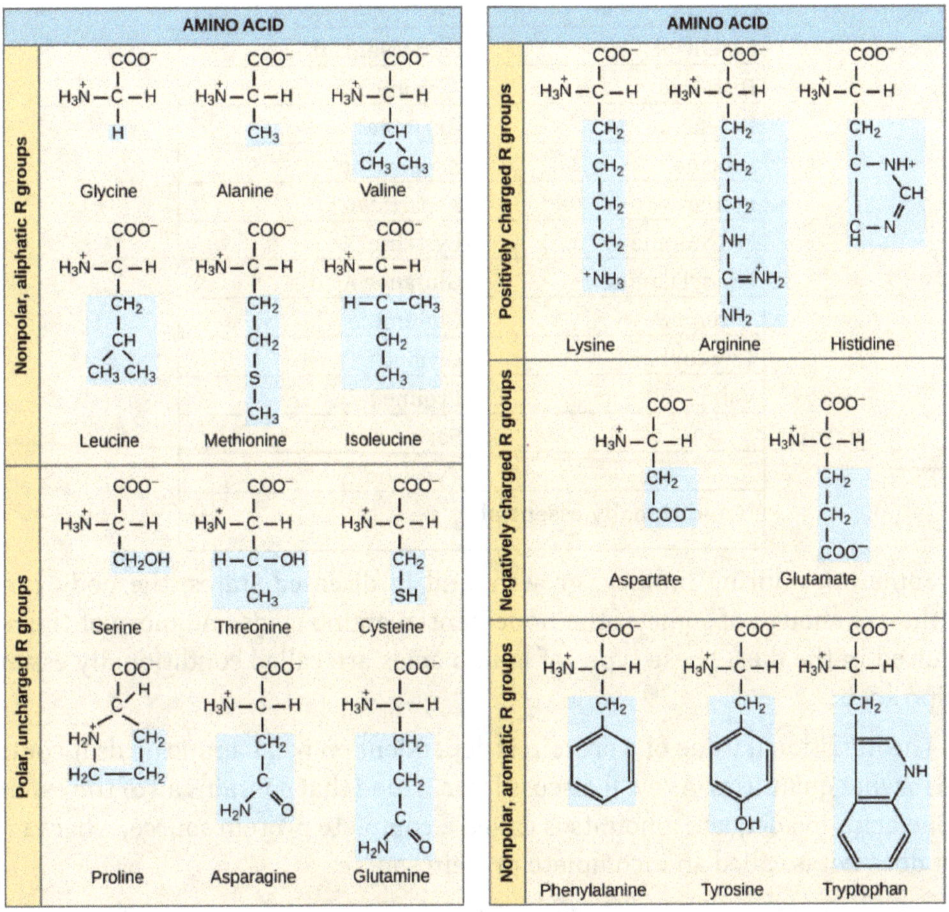

Figure 6.2. Amino acids have different structures and chemical properties, determined by their side chains.

Essential and nonessential Amino Acids

We also classify amino acids based on their nutritional aspects (Table 6.1 "Essential and Nonessential Amino Acids"):

- Nonessential amino acids are not required in the diet, because the body can synthesize them. They're still vital to protein synthesis, and they're still present in food, but because the body can make them, we don't have to worry about nutritional requirements. There are 11 nonessential amino acids.
- Essential amino acids can't be synthesized by the body in sufficient amounts, so they must be obtained in the diet. There are 9 essential amino acids.

Table 6.1. Essential and nonessential amino acids

Essential	Nonessential
Histidine	Alanine
Isoleucine	Arginine*
Leucine	Asparagine
Lysine	Aspartic Acid
Methionine	Cysteine*
Phenylalanine	Glutamic Acid
Threonine	Glutamine
Tryptophan	Glycine*
Valine	Proline*
	Serine
	Tyrosine*
***Conditionally essential**	

Sometimes during infancy, growth, and in diseased states, the body cannot synthesize enough of some of the nonessential amino acids and more of them are required in the diet. These types of amino acids are called conditionally essential amino acids.

The nutritional value of a protein is dependent on what amino acids it contains and in what quantities. As we'll discuss later, a food that contains all of the essential amino acids in adequate amounts is called a complete protein source, whereas one that does not is called an incomplete protein source.

The Many Different Types of Proteins

There are over 100,000 different proteins in the human body. Proteins are similar to carbohydrates and lipids in that they are polymers (simple repeating units); however, proteins are much more structurally complex. In contrast to carbohydrates, which have identical repeating units, proteins are made up of amino acids that are different from one another. Different proteins are produced because there are 20 types of naturally occurring amino acids that are combined in unique sequences.

Additionally, proteins come in many different sizes. The hormone insulin, which regulates blood glucose, is composed of only 51 amino acids. On the other hand, collagen, a protein that acts like glue between cells, consists of more than 1,000 amino acids. Titin is the largest known protein. It accounts for the elasticity of muscles and consists of more than 25,000 amino acids!

The huge diversity of proteins is also due to the unending number of amino acid sequences that can be formed. To understand how so many different proteins can be made from only 20 amino acids, think about music. All of the music that exists in the world has been derived from a basic set of seven notes C, D, E, F, G, A, B (with the addition of sharps and flats), and there is a vast array of music all composed of specific sequences from these basic musical notes. Similarly, the 20 amino acids can be linked together in an extraordinary number of sequences. For example, if an amino acid sequence for a protein is 104 amino acids long, the possible combinations of amino acid sequences is equal to 20^{104}, which is 2 followed by 135 zeros!

Building Proteins with Amino Acids

The decoding of genetic information to synthesize a protein is the central foundation of modern biology. The building of a protein consists of a complex series of chemical reactions that can be summarized into three basic steps: transcription, translation, and protein folding.

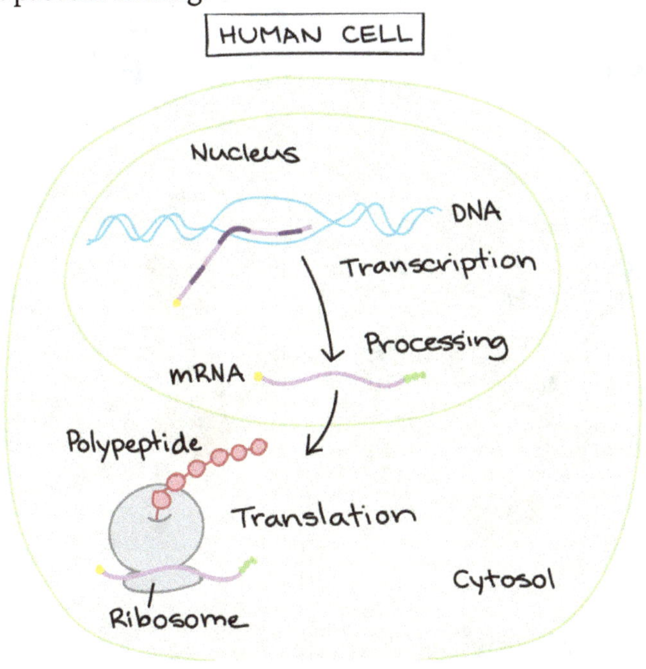

Figure 6.3. Overview of protein synthesis. Protein folding happens after translation.

- Transcription – Deoxyribonucleic acid, or DNA, is the long, double-stranded molecules containing your genome—instructions for making all of the proteins in your body. In the nucleus of the cell, the DNA must be transcribed or copied into the single-stranded messenger ribonucleic acid (mRNA), which carries the genetic instructions into the cell's cytosol for protein synthesis.
- Translation – At the ribosomes in the cell's cytosol, amino acids are linked together in the specific order dictated by the mRNA. Each amino acid is connected to the next amino acid by a special chemical bond called a peptide bond (Figure 6.4). The peptide bond forms between the carboxylic acid group of one amino acid and the amino group of another, releasing a molecule of water. As amino acids are linked sequentially by peptide bonds, following the specific order dictated by the mRNA, the protein chain, also known as a polypeptide chain, is built (Figure 6.5).

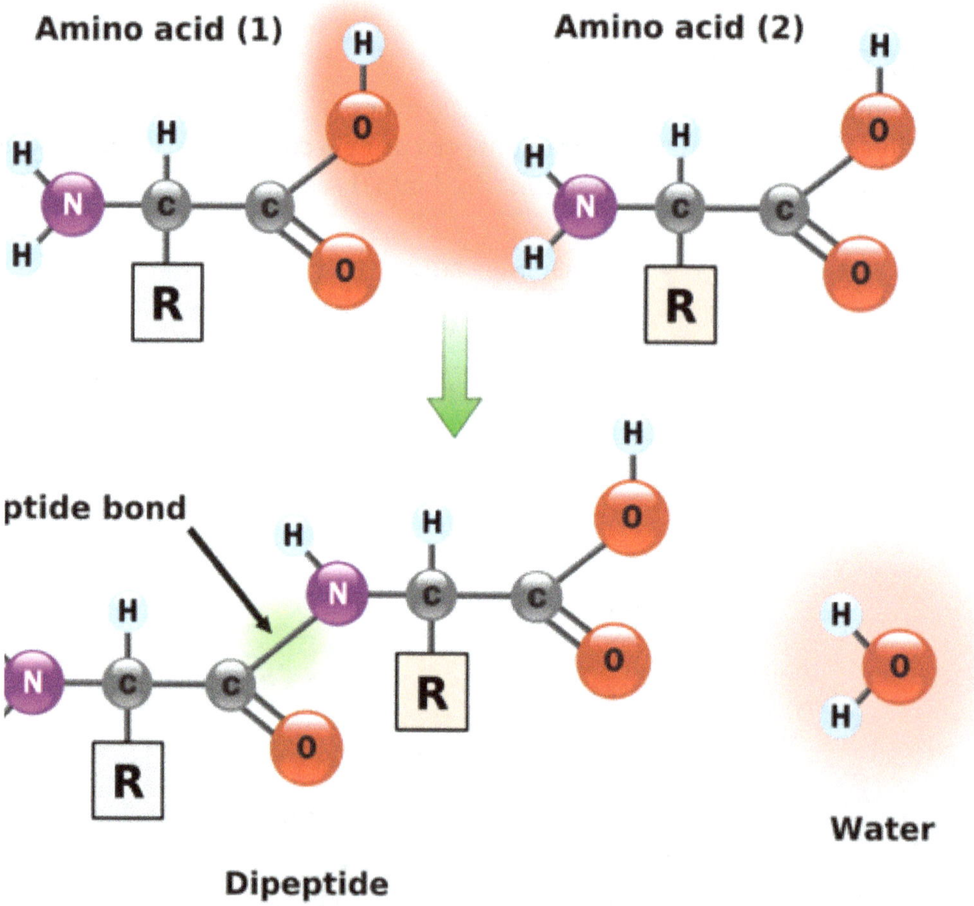

Figure 6.4. Peptide bond formation

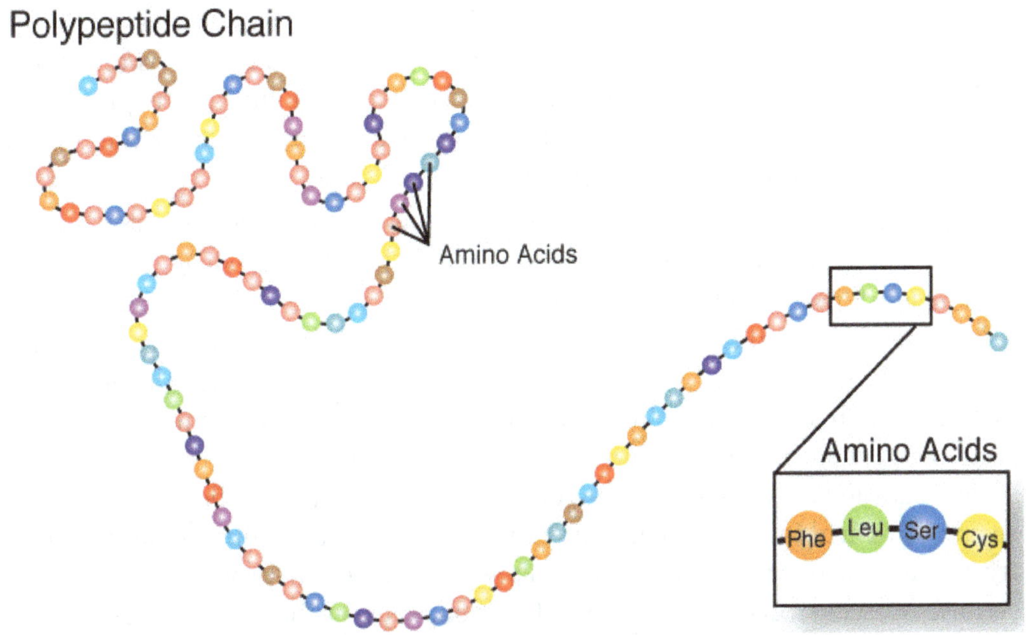

Figure 6.5. A polypeptide chain

3. Protein folding – The polypeptide chain folds into specific three-dimensional shapes, as described in the next section.

Protein organization

Protein's structure enables it to perform a variety of functions. There are four different structural levels of proteins (Figure 6.6.):

- **Primary structure** – This is the *one-dimensional polypeptide chain of amino acids*, held together by peptide bonds.
- **Secondary structure** – The polypeptide chain folds into simple coils (also called helices) and sheets, determined by the chemical interactions between amino acids.
- **Tertiary structure** – This is the unique three-dimensional shape of a protein, formed as the different side chains of amino acids chemically interact, either repelling or attracting each other. Thus, the sequence of amino acids in a protein directs the protein to fold into a specific, organized shape.
- **Quaternary structure** – In some proteins, multiple folded polypeptides called subunits combine to make one larger functional protein. This is called quaternary protein structure. The protein hemoglobin is an example of a protein that has quaternary structure. It is composed of four polypeptides that bond together to form a functional oxygen carrier.

A protein's structure also influences its nutritional quality. Large fibrous protein structures are more difficult to digest than smaller proteins and some, such as keratin, are indigestible. Because digestion of some fibrous proteins is incomplete, not all of the amino acids are absorbed and available for the body to utilize, thereby decreasing their nutritional value.

The specific three-dimensional structure of proteins can be disrupted by changes in their physical environment, causing them to unfold. This is called denaturation, and it results in loss of both structure and function of proteins. Changes in pH (acidic or basic conditions) and exposure to heavy metals, alcohol, and heat can all cause protein denaturation. The proteins in cooked foods are at least partially denatured from the heat of cooking, and denaturation in the stomach is an important part of protein digestion, as we'll discuss later in this unit. We can see everyday examples of denaturation in cooking techniques, like how egg whites become solid and opaque with cooking, and cream becomes fluffy when it's whipped. Both of these are examples of denaturation leading to physical changes in protein structure, and because protein structure determines function, denaturation also causes proteins to lose their function.

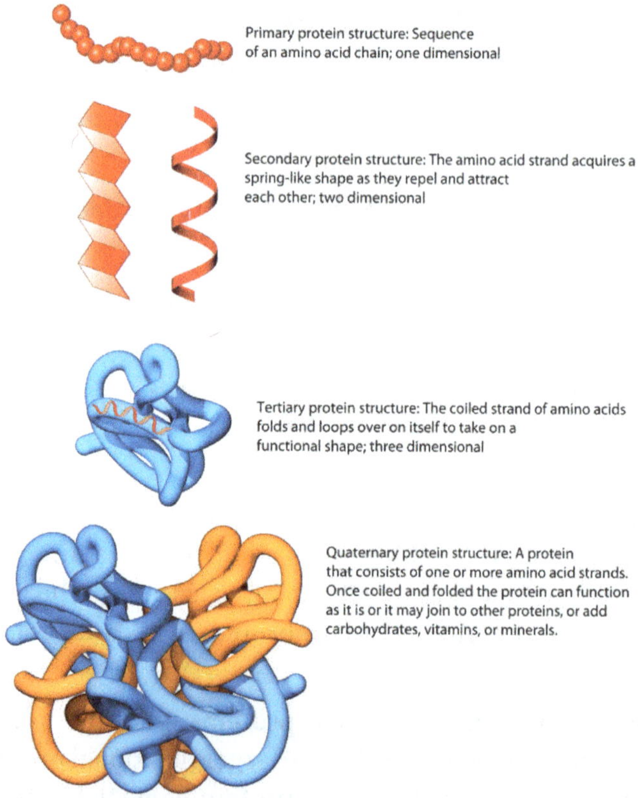

Figure 6.6. A protein has four different structural levels.

Shape Determines Function

An important concept with proteins is that shape determines function. A change in the amino acid sequence will cause a change in protein shape. Each protein in the human body differs in its amino acid sequence and consequently, its shape. The synthesized protein is structured to perform a particular function in a cell. A protein made with an incorrectly placed amino acid may not function properly, and this can sometimes cause disease. An example of this is sickle cell anemia, a genetic disorder. Below is a picture of hemoglobin, a protein with a globular three-dimensional structure. When packed in red blood cells to deliver oxygen, this structure gives red blood cells a donut shape.

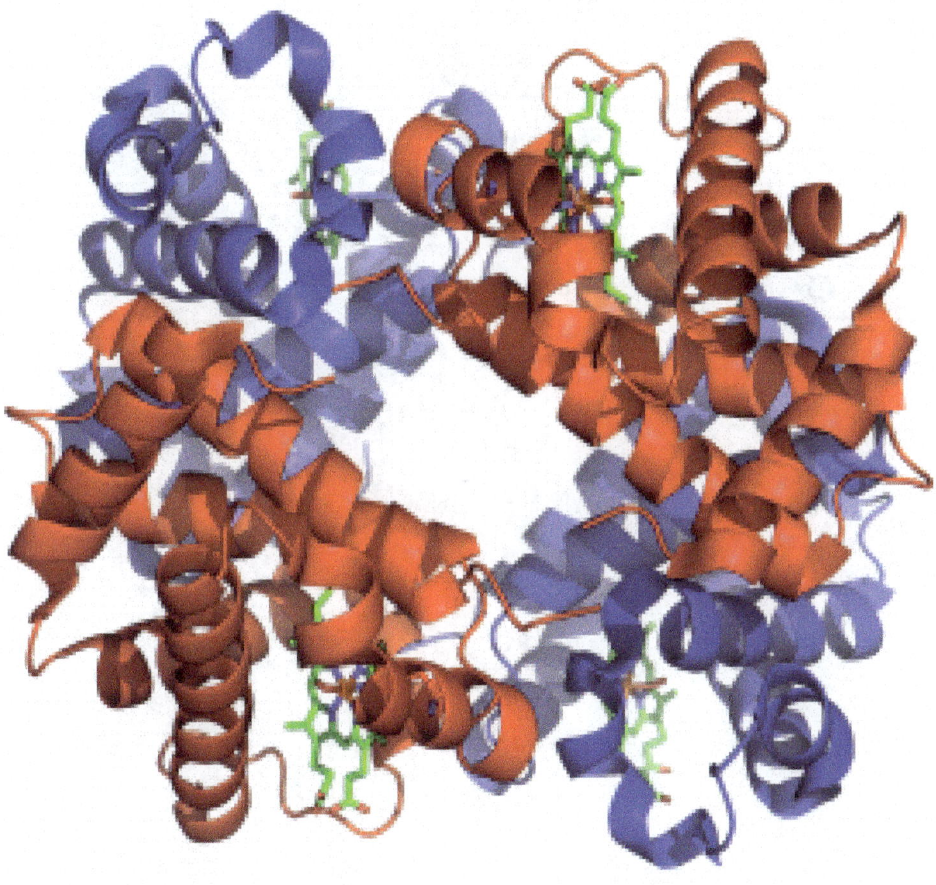

Figure 6.7. Structure of hemoglobin

In people with sickle cell anemia, DNA gives cells the incorrect message when bonding amino acids together to make hemoglobin. The result is crescent-shaped red blood cells that are sticky and do not transport oxygen like normal red blood cells, as illustrated in the figure below.

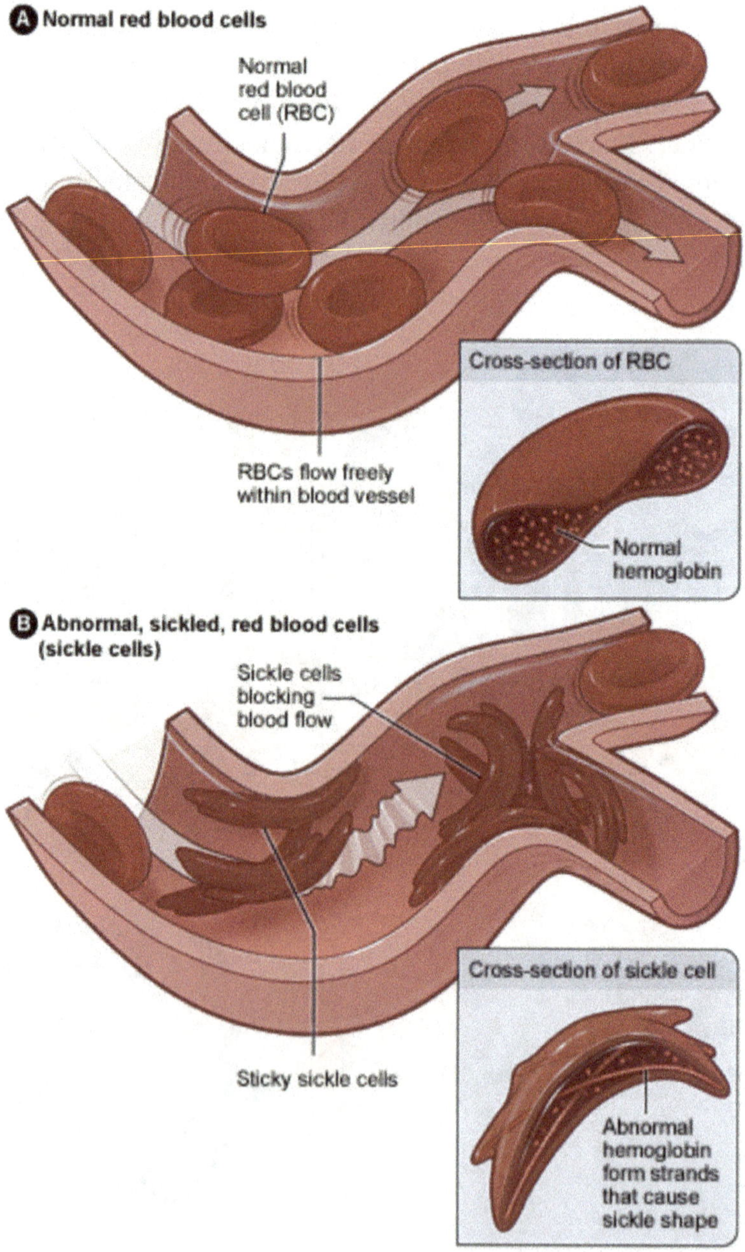

Figure 6.8. Difference in blood cells and blood flow between normal red blood cells and sickle shaped blood cells.

Attributions

- Lindshield, B. L. Kansas State University Human Nutrition (FNDH 400) Flexbook. goo.gl/vOAnR, CC BY-NC-SA 4.0
- "Defining Protein," section 6.1 from the book An Introduction to Nutrition (v. 1.0), CC BY-NC-SA 3.0

FUNCTIONS OF PROTEINS

Proteins are the "workhorses" of the body and participate in many bodily functions. As we've already discussed, proteins come in all sizes and shapes, and each is specifically structured for its particular function. This page describes some of the important functions of proteins. As you read through them, keep in mind that synthesis of all of these different proteins requires adequate amounts of amino acids. As you can imagine, consuming a diet that is deficient in protein and essential amino acids can impair many of the body's functions. (More on that later in the unit.)

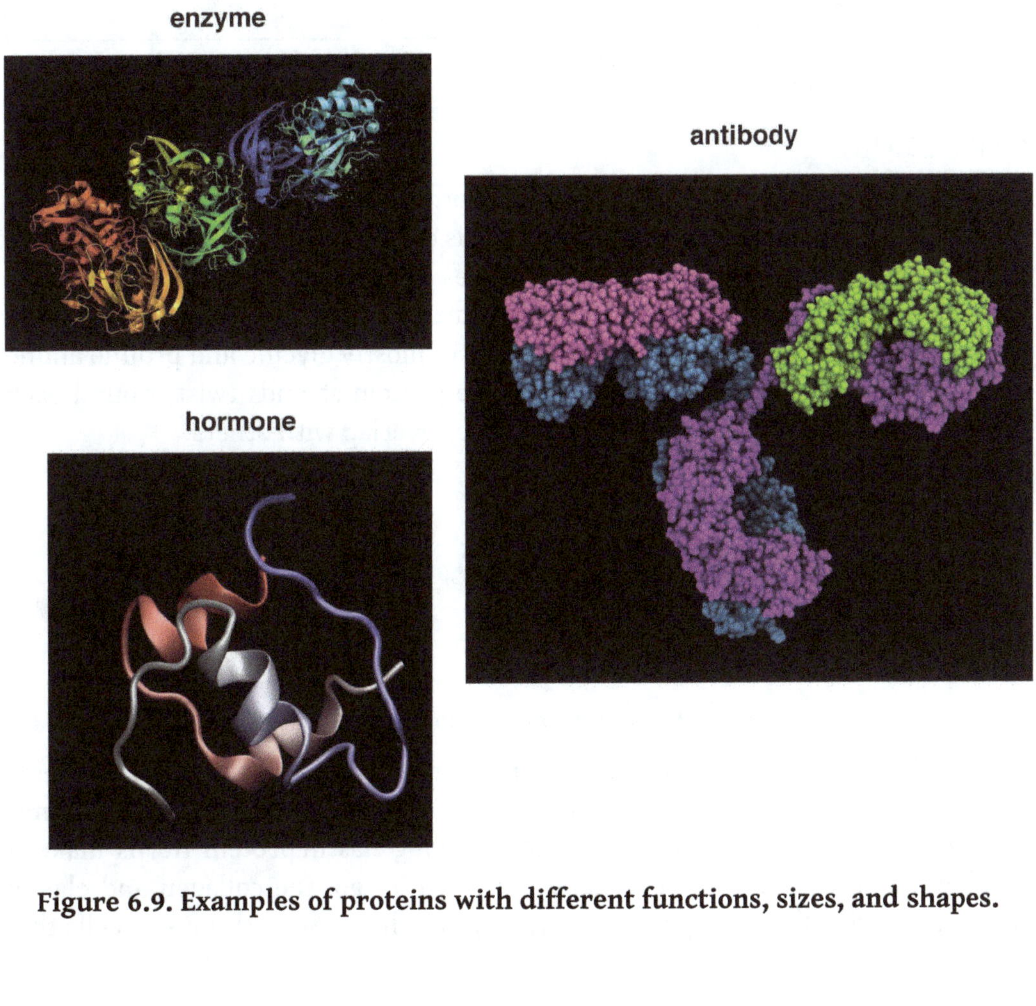

Figure 6.9. Examples of proteins with different functions, sizes, and shapes.

Major types and functions of proteins are summarized in the table below, and the subsequent sections of this page give more detail on each of them.

Table 6.2. Protein types and functions

Protein Types and Functions		
Type	**Examples**	**Functions**
Structure	Actin, myosin, collagen, elastin, keratin	Give tissues (bone, tendons, ligaments, cartilage, skin, muscles) strength and structure
Enzymes	Amylase, lipase, pepsin, lactase	Digest macronutrients into smaller monomers that can be absorbed; performs steps in metabolic pathways to allow for nutrient utilization
Hormones	Insulin, glucagon, thyroxine	Chemical messengers that travel in blood and coordinate processes around the body
Fluid and acid-base balance	Albumin, hemoglobin	Maintains appropriate balance of fluids and pH in different body compartments
Transport	Hemoglobin, albumin, protein channels, carrier proteins	Carry substances around the body in the blood or lymph; help molecules cross cell membranes
Defense	Collagen, lysozyme, antibodies	Protect the body from foreign pathogens

STRUCTURE

More than one hundred different structural proteins have been discovered in the human body, but the most abundant by far is collagen, which makes up about 6 percent of total body weight. Collagen makes up 30 percent of bone tissue and comprises large amounts of tendons, ligaments, cartilage, skin, and muscle. Collagen is a strong, fibrous protein made up of mostly glycine and proline amino acids. Within its quaternary structure, three protein strands twist around each other like a rope and then these collagen ropes overlap with others.

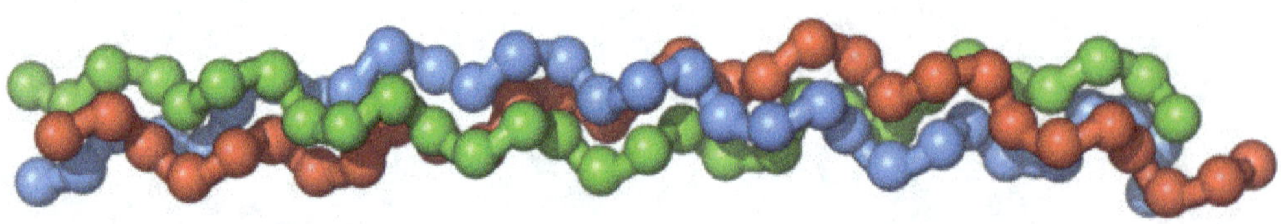

Figure 6.10. Triple-helix structure of collagen

This highly ordered structure is even stronger than steel fibers of the same size. Collagen makes bones strong but flexible. Collagen fibers in the skin's dermis provide it with structure, and the accompanying elastin protein fibrils make it flexible. Pinch the skin on your hand and then let go; the collagen and elastin proteins in skin allow it to go back to its original shape. Smooth-muscle cells that secrete collagen and elastin proteins surround blood vessels, providing the vessels with structure and the ability to stretch back after blood is pumped through them. Another strong, fibrous protein is keratin, an important component of skin, hair, and nails.

ENZYMES

Enzymes are proteins that conduct specific chemical reactions. An enzyme's job is to provide a site for a chemical reaction and to lower the amount of energy and time it takes for that chemical reaction to happen (this is known as "catalysis"). On average, more than 100 chemical reactions occur in cells every single second, and most of them require enzymes. The liver alone contains over 1,000 enzyme systems. Enzymes are specific and will use only particular substrates that fit into their active site, similar to the way a lock can be opened only with a specific key. Fortunately, an enzyme can fulfill its role as a catalyst over and over again, although eventually it is destroyed and rebuilt. All bodily functions, including the breakdown of nutrients in the stomach and small intestine, the transformation of nutrients into molecules a cell can use, and building all macromolecules, including protein itself, involve enzymes.

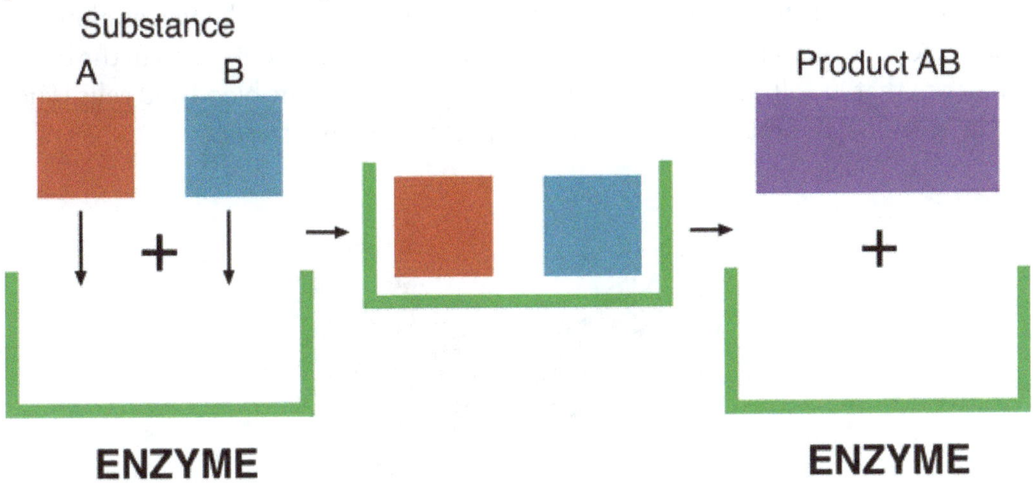

Figure 6.11. Enzymes are proteins. An enzyme's job is to provide a site for substances to chemically react and form a product, and decrease the amount of energy and time it takes for this to happen.

HORMONES

Proteins are responsible for hormone synthesis. Hormonesare the chemical messengers produced by the endocrine glands. When an endocrine gland is stimulated, it releases a hormone. The hormone is then transported in the blood to its target cell, where it communicates a message to initiate a specific reaction or cellular process. For instance, after you eat a meal, your blood glucose levels rise. In response to the increased blood glucose, the pancreas releases the hormone insulin. Insulin tells the cells of the body that glucose is available and to take it up from the blood and store it or use it for making energy or building macromolecules. A major function of hormones is to turn enzymes on and off, so some proteins can even regulate the actions of other proteins. While not all hormones are made from proteins, many of them are.

Fluid and Acid-Base Balance

Adequate protein intake enables the basic biological processes of the body to maintain homeostasis (constant or stable conditions) in a changing environment. One aspect of this is fluid balance, keeping water distributed properly in the different compartments of the body. If too much water suddenly moves from the blood into a tissue, the results are swelling and, potentially, cell death. Water always flows from an area of high concentration to an area of low concentration. As a result, water moves toward areas that have higher concentrations of other solutes, such as proteins and glucose. To keep the water evenly distributed between blood and cells, proteins continuously circulate at high concentrations in the blood. The most abundant protein in blood is the butterfly-shaped protein known as albumin. The presence of albumin in the blood makes the protein concentration in the blood similar to that in cells. Therefore, fluid exchange between the blood and cells is not in the extreme, but rather is minimized to preserve homeostasis.

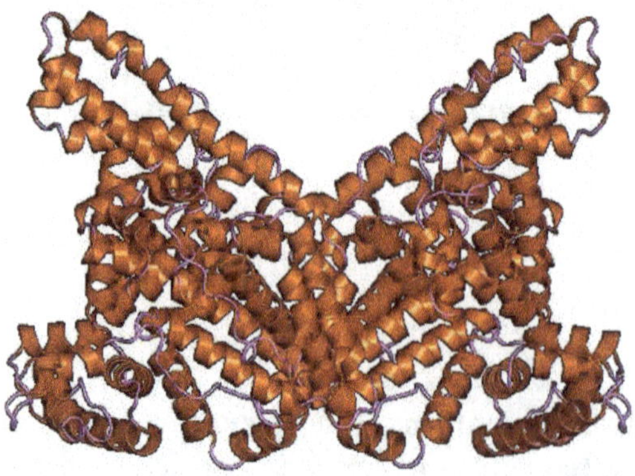

Figure 6.12. The butterfly-shaped protein, albumin, has many functions in the body including maintaining fluid and acid-base balance and transporting molecules.

Protein is also essential in maintaining proper pH balance (the measure of how acidic or basic a substance is) in the blood. Blood pH is maintained between 7.35 and 7.45, which is slightly basic. Even a slight change in blood pH can affect body functions. The body has several systems that hold the blood pH within the normal range to prevent this from happening. One of these is the circulating albumin. Albumin is slightly acidic, and because it is negatively charged it balances the many positively charged molecules circulating in the blood, such as hydrogen protons (H^+), calcium, potassium, and magnesium. Albumin acts as a buffer against abrupt changes in the concentrations of these molecules, thereby balancing blood pH and maintaining homeostasis. The protein hemoglobin also participates in acid-base balance by binding hydrogen protons.

Transport

Proteins also play vital roles in transporting substances around the body. For example, albumin chemically binds to hormones, fatty acids, some vitamins, essential minerals, and drugs, and transports them throughout the circulatory system. Each red blood cell contains millions of hemoglobin molecules that bind oxygen in the lungs and transport it to all the tissues in the body. A cell's plasma membrane is usually not permeable to large polar molecules, so to get the required nutrients and molecules into the cell, many transport proteins exist in the cell membrane. Some of these proteins are channels that allow particular molecules to move in and out of cells. Others act as one-way taxis and require energy to function.

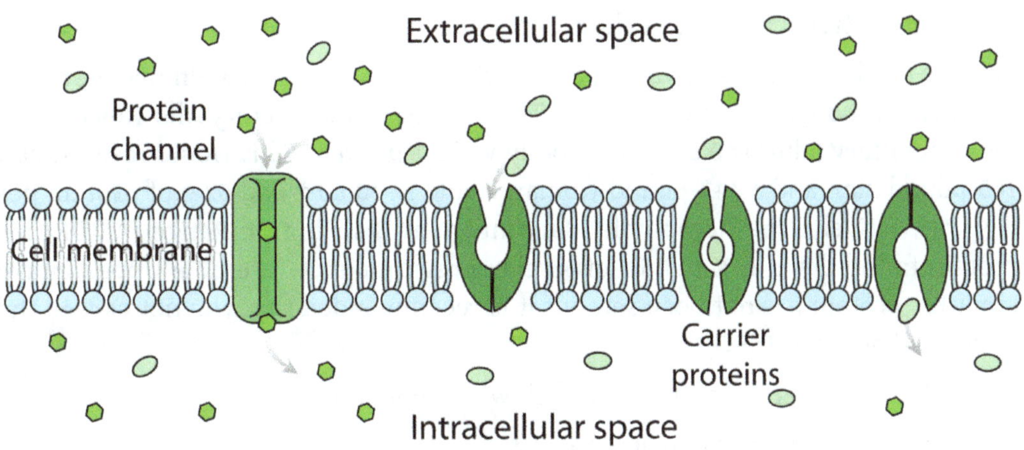

Figure 6.13. Molecules move in and out of cells through transport proteins, which are either channels or carriers.

Immunity

Proteins also play important roles in the body's immune system. The strong collagen fibers in skin provide it with structure and support, but it also serves as a barricade against harmful substances. The immune system's attack and destroy functions are dependent on enzymes and antibodies, which are also proteins. For example, an enzyme called lysozyme is secreted in the saliva and attacks the walls of bacteria, causing them to rupture. Certain proteins circulating in the blood can be directed to build a molecular knife that stabs the cellular membranes of foreign invaders. The antibodies secreted by white blood cells survey the entire circulatory system, looking for harmful bacteria and viruses to surround and destroy. Antibodies also trigger other factors in the immune system to seek and destroy unwanted intruders.

Energy Production

Some of the amino acids in proteins can be disassembled and used to make energy. Only about 10 percent of dietary proteins are catabolized each day to make cellular energy. The liver is able to break down amino acids to the carbon skeleton, which can then be fed into the citric acid or Krebs cycle. This is similar to the way that glucose is used to make ATP. If a person's diet does not contain enough carbohydrates and fats, their body will use more amino acids to make energy, which can compromise the synthesis of new proteins and destroy muscle proteins if calorie intake is also low.

DIETARY RECOMMENDATIONS AND PROTEIN IN FOODS

In this section, we'll discuss how to determine how much protein you need and your many choices in designing an optimal diet with high-quality protein sources.

How much dietary protein does a person need?

Because our bodies are so efficient at recycling amino acids, protein needs are not as high as carbohydrate and fat needs. The Recommended Dietary Allowance (RDA) for a sedentary adult is 0.8 g per kg body weight per day. This would mean that a 165-pound man and a 143-pound woman would need 60 g and 52 g of protein per day, respectively. The Acceptable Macronutrient Distribution Range (AMDR) for protein for adults is 10% to 35% of total energy intake. A Tolerable Upper Intake Limit for protein has not been set, but it is recommended that you not exceed the upper end of the AMDR.

Protein needs are higher for the following populations:

- growing children and adolescents
- women who are pregnant (they're using protein to help grow a fetus)
- lactating women (breast milk has protein in it for the baby's nutrition, so mothers need more protein to synthesize that milk)
- athletes

The Academy of Nutrition and Dietetics, Dietitians of Canada, and the American College of Sports Medicine recommend 1.2 to 2.0 grams of protein per kilogram of body weight per day for athletes, depending on the type of training.[1] Higher intakes may be needed for short periods during intensified training or with reduced energy intake.

Nitrogen Balance To Determine Protein Needs

The appropriate amount of protein in a person's diet is that which maintains a balance between what is taken in and what is used. The RDAs for protein were determined by assessing nitrogen balance. Nitrogen is one of the four basic elements contained in all amino acids. When amino acids are broken down, nitrogen is released. Most nitrogen is excreted as urea in urine, but some urea is also contained in feces. Nitrogen is also lost in sweat and as hair and nails grow. The RDA, therefore, is the amount of protein a person should consume in their diet to balance the amount of protein used by the body, measured as the amount of nitrogen lost from the body. The Institute Of Medicine used data from multiple studies that determined nitrogen balance in people of different age groups to calculate the RDA for protein.

- Nitrogen Balance- A person is said to be in nitrogen balance when the nitrogen consumed equals the amount of nitrogen excreted. Most healthy adults are in nitrogen balance. If more protein is consumed than needed, this extra protein is used for energy, and the nitrogen waste that results is excreted. The lowest amount of protein a person can consume and still remain in nitrogen balance represents that person's minimum protein requirement.

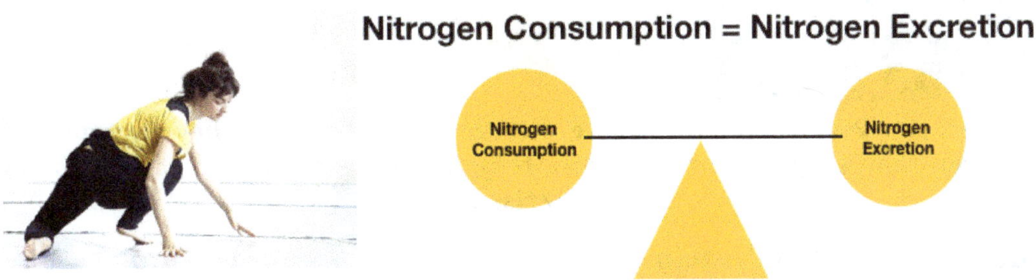

Figure 6.14. People are in nitrogen balance when they excrete as much nitrogen as they consume.

- Negative Nitrogen Balance- A person is in negative nitrogen balance when the amount of excreted nitrogen is greater than that consumed, meaning that the body is breaking down more protein to meet its demands. This state of imbalance can occur in people who have certain diseases, such as cancer or muscular dystrophy. Someone who is eating a low-protein diet may also be in negative nitrogen balance as they are taking in less protein than they actually need.

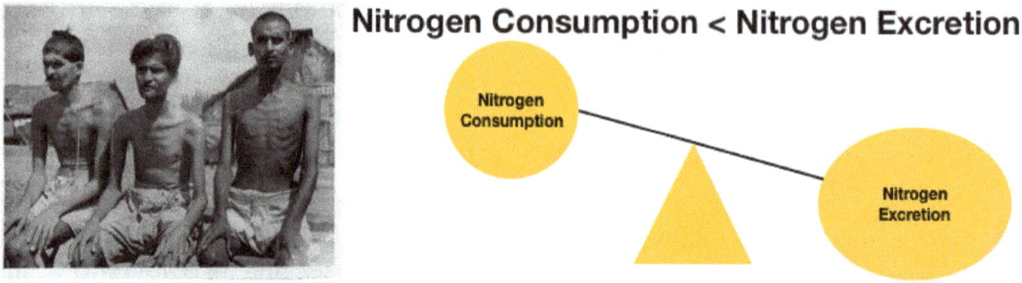

Figure 6.15. People are in negative nitrogen balance when they excrete more nitrogen than they consume, usually because they are not eating enough protein to meet their needs.

- Positive Nitrogen Balance– A person is in positive nitrogen balance when a person excretes less nitrogen than what is taken in by the diet, such as during pregnancy or growth in childhood. At these times the body requires more protein to build new tissues, so more of what gets consumed gets used up and less nitrogen is excreted. A person healing from a severe wound may also be in positive nitrogen balance because protein is being used up to repair tissues.

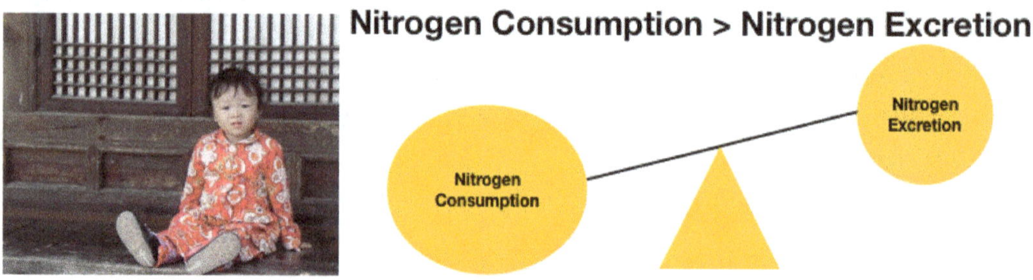

Figure 6.16. People are in positive nitrogen balance when they excrete less nitrogen than they consume, because they are using protein to actively build new tissue.

Dietary Sources Of Protein

Although meat is the typical food that comes to mind when thinking of protein, many other foods are rich in protein as well, including dairy products, eggs, beans, whole grains, and nuts. Table 6.3 lists the grams of protein in a standard serving for a variety of animal and plant foods.

Table 6.3. Protein in common foods2

Animal Sources	Grams of Protein per Standard Serving
Egg White	3 g per 1 large white
Whole Egg	6 g per 1 large egg
Cheddar Cheese	7 g per 1 oz. (30 g)
Milk, 1%	8 g per 1 cup (8 fl oz)
Yogurt	11 g per 8 oz
Greek Yogurt	22 g per 8 oz
Cottage Cheese	15 g per ½ cup
Hamburger	30 g per 4 oz
Chicken	35 g per 4 oz
Tuna	40 g per 6 oz can
Almonds, dried	6 g per 1 oz
Almond milk	1 g per cup (8 fl oz)
Soy milk	8g per cup (8 fl oz)
Peanut butter	4 g per 1 tbsp
Hummus	8 g per ½ cup
Refried beans	6 g per ½ cup
Lentil soup	11 g per 10.5 oz
Tofu, extra firm	11 g per 3.5 oz
Enriched wheat bread	1 g per slice (45 g)
Whole Grain Bread	5g per slice (45 g)
Grape Nuts	7 g per ½ cup

Notice in the table above that whole foods contain more protein than refined foods. When foods are refined—for example, going from a whole almond to almond milk or whole grain to refined grain—protein is lost in that processing. Very refined foods like oil and sugar contain no protein.

The USDA provides some tips for choosing your dietary protein sources. The overall suggestion is to eat a variety of protein-rich foods to benefit health. Examples include:

- Lean meats, such as round steaks, top sirloin, extra lean ground beef, pork loin, and skinless chicken.
- 8 ounces of cooked seafood every week (typically as two 4-ounce servings).
- Choosing to eat beans, peas, or soy products as a main dish. For example, chili with kidney and pinto beans, hummus on pita bread, and black bean enchiladas.

- Enjoy nuts in a variety of ways. Put them on a salad, in a stir-fry, or use them as a topping for steamed vegetables in place of meat or cheese.

Protein Quality

While protein is contained in a wide variety of foods, it differs in quality. High-quality complete proteins contain all nine essential amino acids. Lower-quality incomplete proteins do not contain all nine essential amino acids in proportions needed to support growth and health.

Foods that are complete protein sources include animal foods such as milk, cheese, eggs, fish, poultry, and meat. A few plant foods also are complete proteins, such as soy (soybeans, soy milk, tofu, tempeh) and quinoa.

Most plant-based foods are deficient in at least one essential amino acid and therefore are incomplete protein sources. For example, grains are usually deficient in the amino acid lysine, and legumes are low in methionine and tryptophan. Because grains and legumes are not deficient in the same amino acids, they can complement each other in a diet. When consumed in tandem, they contain all nine essential amino acids at adequate levels, so they are called complementary proteins. Some examples of complementary protein foods are given in Table 6.4. Mutual supplementationis another term used when combining two or more incomplete protein sources to make a complete protein. Complementary protein sources do not have to be consumed at the same time—as long as they are consumed within the same day, you will meet your protein needs. Most people eat complementary proteins without thinking about it, because they go well together. Think of a peanut butter sandwich and beans and rice; these are examples of complementary proteins. So long as you eat a variety of foods, you don't need to worry much about incomplete protein foods. They may be called "lower quality" in terms of protein, but they're still great choices, as long as they're not the only foods you eat!

Table 6.4. Complementary protein sources

Foods	Lacking Amino Acids	Complementary Food	Complementary Menu
Legumes	Methionine, tryptophan	Grains, nuts and seeds	Hummus and whole-wheat pita
Grains	Lysine, isoleucine, threonine	Legumes	Cornbread and kidney bean chili
Nuts and seeds	Lysine, isoleucine	Legumes	Stir-fried tofu with cashews

The second component of protein quality is digestibility, as not all protein sources are equally digested. In general, animal-based proteins are more fully digested than plant-based proteins, because some proteins are contained in the plant's fibrous cell walls and these pass through the digestive tract unabsorbed by the body. Animal proteins tend to be 95 percent or more digestible; soy is estimated at 91 percent; and many grains are around 85 to 88 percent digestible.[3]

REFERENCES:

- Thomas, D. T., Erdman, K. A., & Burke, L. M. (2016). Position of Dietitians of Canada, the Academy of Nutrition and Dietetics and the American College of Sports Medicine: Nutrition and Athletic Performance. Journal of the Academy of Nutrition and Dietetics, 116(3), 501-528.
- USDA National Nutrient Database for Standard Reference, December 2018.
- Tome, D. (2012). Criteria and markers for protein quality assessment – a review. British Journal of Nutrition 108, S222–S229.

DIGESTION AND ABSORPTION OF PROTEINS

When you eat food, the body's digestive system breaks down dietary protein into individual amino acids, which are absorbed and used by cells to build other proteins and a few other macromolecules, such as DNA. Let's follow the path that proteins take down the gastrointestinal tract and into the circulatory system.

Eggs are a good dietary source of protein and will be used as our example as we discuss the processes of digestion and absorption of protein. One egg, whether raw, hard-boiled, scrambled, or fried, supplies about six grams of protein.

In the image below, follow the numbers to see what happens to the protein in our egg at each site of digestion.

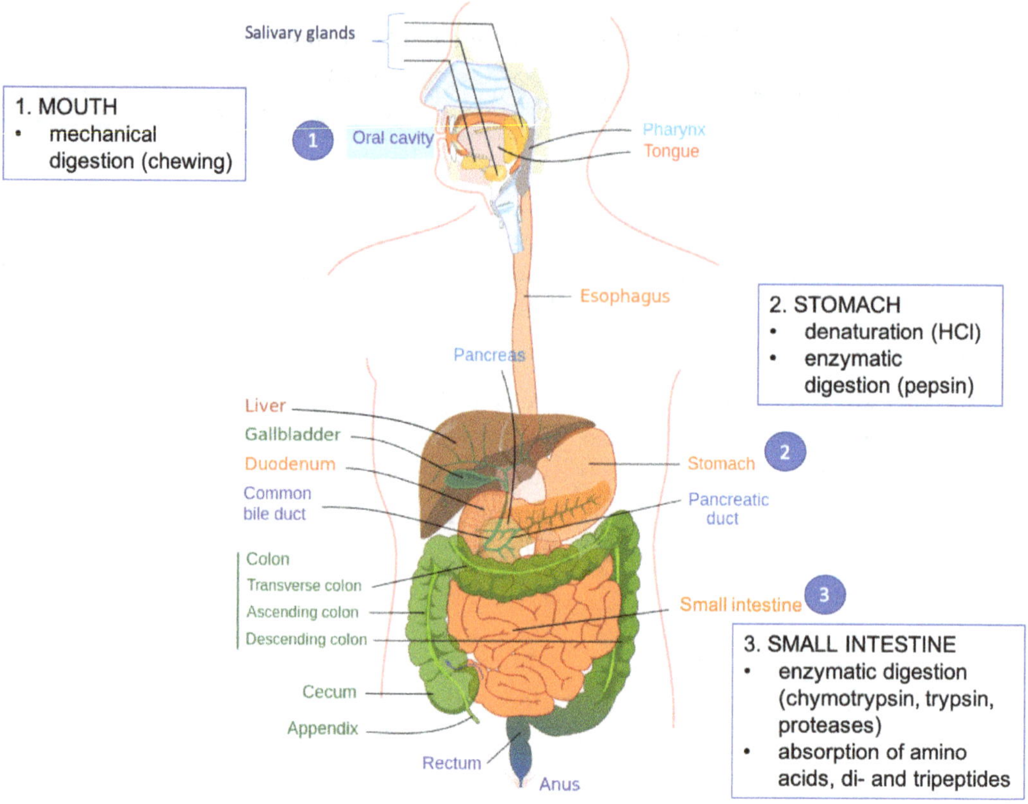

Fig. 6.17. Protein digestion in the human GI tract.

1 – Protein Digestion In The Mouth

Unless you are eating it raw, the first step in digesting an egg (or any other solid food) is chewing. The teeth begin the mechanical breakdown of large egg pieces into smaller pieces that can be swallowed. The salivary glands secrete saliva to aid swallowing and the passage of the partially mashed egg through the esophagus.

2 – Protein Digestion In The Stomach

The mashed egg pieces enter the stomach from the esophagus. As illustrated in the image below, both mechanical and chemical digestion take place in the stomach. The stomach releases gastric juices containing hydrochloric acid and the enzyme, pepsin, which initiate the chemical digestion of protein. Muscular contractions, called peristalsis, also aid in digestion. The powerful stomach contractions churn the partially digested protein into a more uniform mixture, which is called chyme.

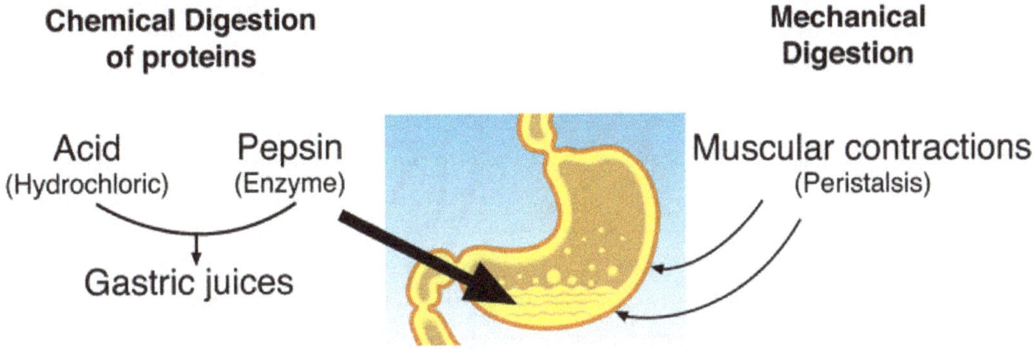

Fig. 6.18. Protein digestion in the stomach

Because of the hydrochloric acid in the stomach, it has a very low pH of 1.5-3.5. The acidity of the stomach causes food proteins to denature, unfolding their three-dimensional structure to reveal just the polypeptide chain. This is the first step of chemical digestion of proteins. Recall that the three-dimensional structure of a protein is essential to its function, so denaturation in the stomach also destroys protein function. (This is why a protein such as insulin can't be taken as an oral medication. Its function is destroyed in the digestive tract, first by denaturation and then further by enzymatic digestion. Instead, it has to be injected so that it is absorbed intact into the bloodstream.)

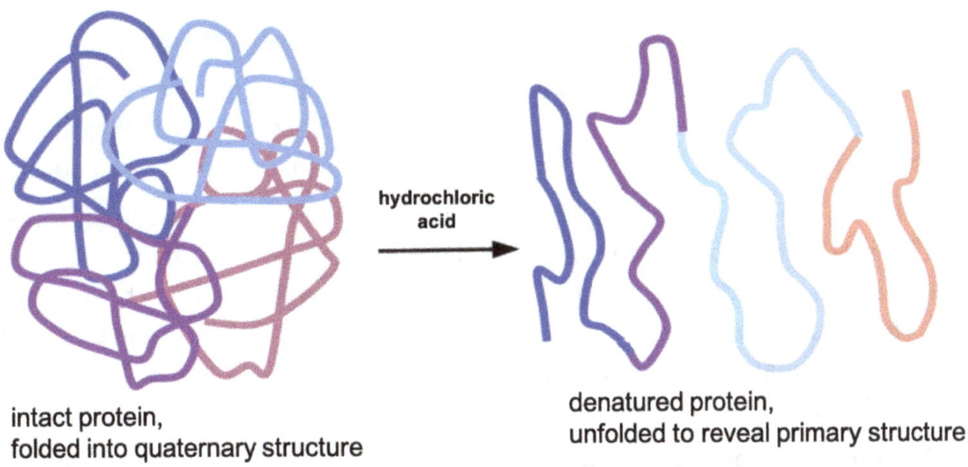

Fig. 6.19. In the stomach, proteins are denatured because of the acidity of hydrochloric acid.

Once proteins are denatured in the stomach, the peptide bonds linking amino acids together are more accessible for enzymatic digestion. That process is started by pepsin, an enzyme that is secreted by the cells that line the stomach and is activated by hydrochloric acid. Pepsin begins breaking peptide bonds, creating shorter polypeptides.

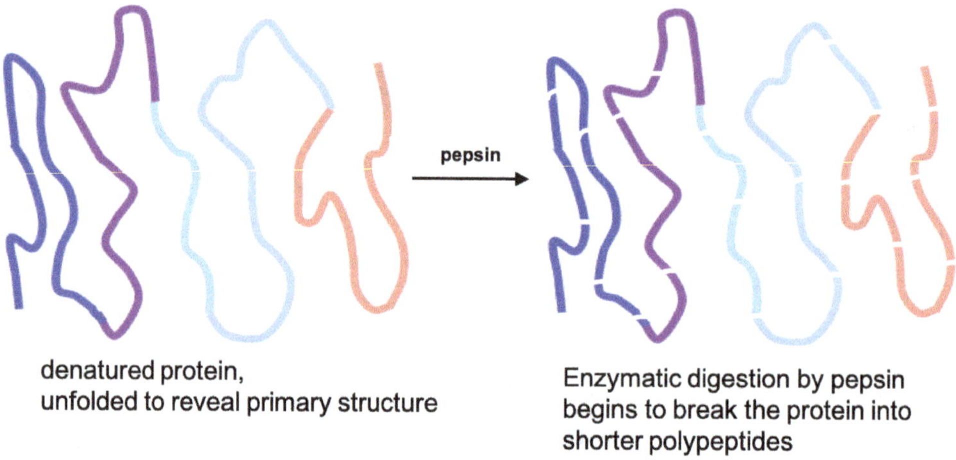

Fig. 6.20. Enzymatic digestion of proteins begins in the stomach with the action of the enzyme pepsin.

Proteins are large globular molecules, and their chemical breakdown requires time and mixing. Protein digestion in the stomach takes a longer time than carbohydrate digestion, but a shorter time than fat digestion. Eating a high-protein meal increases the amount of time required to sufficiently break down the meal in the stomach. Food remains in the stomach longer, making you feel full longer.

3 – Protein Digestion And Absorption In The Small Intestine

The chyme leaves the stomach and enters the small intestine, where the majority of protein digestion occurs. The pancreas secretes digestive juices into the small intestine, and these contain more enzymes to further break down polypeptides.

The two major pancreatic enzymes that digest proteins in the small intestine are chymotrypsin and ***trypsin***. Trypsin activates other protein-digesting enzymes called proteases, and together, these enzymes break proteins down to tripeptides, dipeptides, and individual amino acids. The cells that line the small intestine release additional enzymes that also contribute to the enzymatic digestion of polypeptides.

Tripeptides, dipeptides, and single amino acids enter the enterocytes of the small intestine using active transport systems, which require ATP. Once inside, the tripeptides and dipeptides are all broken down to single amino acids, which are absorbed into the bloodstream. There are several different types of transport systems to accommodate different types of amino acids. Amino acids with structural similarities end up competing to use these transporters. That's not a problem if your protein is coming from food, because it naturally contains a mix of amino acids. However, if you take high doses of amino acid supplements, those could theoretically interfere with absorption of other amino acids.

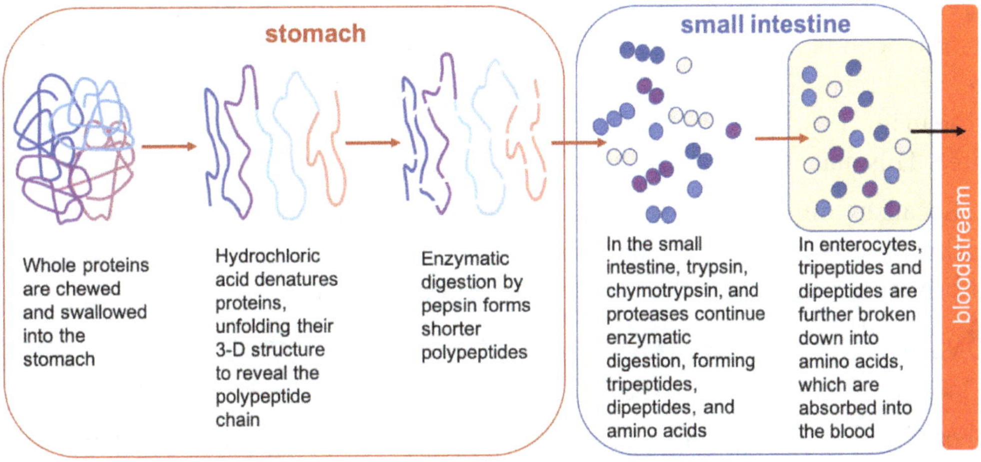

Fig. 6.21. Summary of protein digestion. Note that the lines representing polypeptide chains in the stomach consist of strings of amino acids connected by peptide bonds, even though the individual amino acids aren't shown in this simplified representation.

Proteins that aren't fully digested in the small intestine pass into the large intestine and are eventually excreted in the feces. Recall from the last page that plant-based proteins are a bit less digestible than animal proteins, because some proteins are bound in plant cell walls.

What happens to absorbed amino acids?

Once the amino acids are in the blood, they are transported to the liver. As with other macronutrients, the liver is the checkpoint for amino acid distribution and any further breakdown of amino acids, which is very minimal. Dietary amino acids then become part of the body's amino acid pool.

Assuming the body has enough glucose and other sources of energy, those amino acids will be used in one of the following ways:

- Protein synthesis in cells around the body
- Making nonessential amino acids needed for protein synthesis
- Making other nitrogen-containing compounds
- Rearranged and stored as fat (there is no storage form of protein)

If there is not enough glucose or energy available, amino acids can also be used in one of these ways:

- Rearranged into glucose for fuel for the brain and red blood cells
- Metabolized as fuel, for an immediate source of ATP

In order to use amino acids to make ATP, glucose, or fat, the nitrogen first has to be removed in a process called deamination, which occurs in the liver and kidneys. The nitrogen is initially released as ammonia, and because ammonia is toxic, the liver transforms it into urea. Urea is then transported to the kidneys and excreted in the urine. Urea is a molecule that contains two nitrogens and is highly soluble in water. This makes it ideal for transporting excess nitrogen out of the body.

Because amino acids are building blocks that the body reserves in order to synthesize other proteins, more than 90 percent of the protein ingested does not get broken down further than the amino acid monomers.

CONSEQUENCES FOR HEALTH OF EITHER TOO LITTLE OR TOO MUCH DIETARY PROTEIN

A healthy diet incorporates all nutrients in moderation, meaning that there's neither too little nor too much. As with all nutrients, having too little or too much protein can have health consequences.

The AMDR for protein for adults is between 10 and 35 percent of kilocalories.[1] That's a fairly wide range, and it encompasses typical protein intakes of many traditional human cultures.

Fig. 6.22. Diverse human cultures have survived on different levels of dietary protein. The photo on the left shows Inuit families sharing frozen, aged walrus meat. Their traditional diet is very dependent on meat and high in both protein and fat. On the right is a traditional vegetarian meal in India, high in carbohydrates but still providing adequate levels of protein.

Protein intake below the RDA is inadequate to support the body's needs for synthesis of structural and functional proteins. On the other hand, there are some concerns that high protein intake is associated with chronic disease. However, as we'll discuss, it's not just the quantity of protein that matters, but also the nutritional package that it comes in. According to a 2018 study published in the American Journal of Clinical Nutrition, most Americans get enough protein, averaging about 88 grams per day and 14 to 16 percent of caloric intake. The study also found that diets with protein above 35 percent of caloric intake, the upper end of the AMDR, were extremely rare.

Health Consequences of Protein Deficiency

Although severe protein deficiency is rare in the developed world, it is a leading cause of death in children in many poor, underdeveloped countries. There are two main syndromes associated with protein deficiencies: Kwashiorkor and Marasmus.

Kwashiorkor affects millions of children worldwide. When it was first described in 1935, more than 90 percent of children with Kwashiorkor died. Although the associated mortality is slightly lower now, most children still die after the initiation of treatment.

The name Kwashiorkor comes from a language in Ghana and means, "rejected one." The syndrome was named because it occurs most commonly in children recently weaned from breastfeeding, usually because the mother had a new baby, and the older child is switched to a diet of watery porridge made from low-protein grains. The child may be consuming enough calories, but not enough protein.

Kwashiorkor is characterized by swelling (edema) of the feet and abdomen, poor skin health, poor growth, low muscle mass, and liver malfunction. Recall that one of the roles of protein in the body is fluid balance. Diets extremely low in protein do not provide enough amino acids for the synthesis of the protein albumin. One of the functions of albumin is to hold water in the blood vessels, so having lower concentrations of blood albumin results in water moving out of the blood vessels and into tissues, causing swelling. The primary symptoms of Kwashiorkor include not only swelling, but also diarrhea, fatigue, peeling skin, and irritability. Severe protein deficiency in addition to other micronutrient deficiencies, such as folate, iodine, iron, and vitamin C all contribute to the many health manifestations of this syndrome.

Children and adults with marasmus are protein deficient, but at the same time, they're also not taking in enough calories. Body weights of children with Marasmus may be up to 80 percent less than that of a healthy child of the same age. Marasmus is a Greek word meaning "starvation." The syndrome affects more than fifty million children under age five worldwide. It is characterized by an extreme emaciated appearance, poor skin health, poor growth, and increased risk of infection. The symptoms are acute fatigue, hunger, and diarrhea.

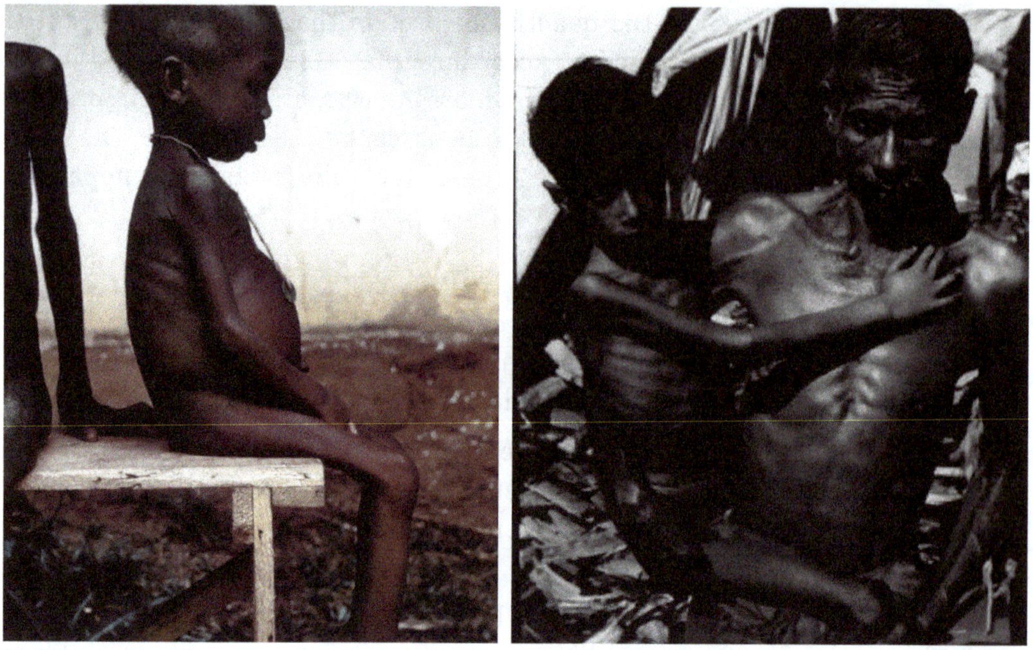

Figure 6.23. The photo on the left shows a child suffering from kwashiorkor (note the swollen belly) in the late 1960s in a Nigerian relief camp during the Nigerian-Biafran War. The photo on the right shows an Indian child suffering from marasmus.

Kwashiorkor and marasmus often coexist as a combined syndrome termed marasmic kwashiorkor. Children with the combined syndrome have variable amounts of edema and the characterizations and symptoms of marasmus. Although organ system function is compromised by undernutrition, the ultimate cause of death is usually infection. Undernutrition is intricately linked with suppression of the immune system at multiple levels, so undernourished children commonly die from severe diarrhea and/or pneumonia resulting from bacterial or viral infection. According to the United Nations Children's Fund (UNICEF), nearly half of all deaths of children under age five are related to malnutrition. That translates to about 3 million child deaths each year.

While severe protein deficiency is rare in the U.S., there are several groups at risk of low protein intake. A 2018 study found that 23 percent of U.S. adolescent girls (aged 14 to 18 years old) and 11 percent of adolescent boys were consuming below the RDA for protein, which may compromise their growth and development.[2] This is thought to be related to the growing independence in food choices and the high prevalence of dieting in this group. Low protein intake is also a concern for the elderly in the U.S. The same 2018 study found that among those 71 years and older, 19 percent of women and 13 percent of men consume less protein than the RDA.[2] This is a particular concern in this age group, as loss of muscle is accelerated with aging, and that can lead to greater frailty, loss of balance, and greater risk of falls. Some researchers argue that older adults actually need more protein than recommended by the RDA in order to maintain muscle mass and function.[4]

HEALTH CONSEQUENCES OF TOO MUCH PROTEIN IN THE DIET

When the Food and Nutrition Board of the Institute of Medicine wrote the DRI for macronutrients, published in 2005, they concluded that there wasn't enough evidence to establish an Upper Limit for protein. The high end of the AMDR, 35 percent of kilocalories for protein, was set in order to allow the total diet to be well-balanced with carbohydrate and fat.[1] Higher levels of protein intake haven't been well-studied, but over the years, there have been many concerns with high protein diets. However, current evidence indicates it's large amounts of animal protein (particularly from red meat or processed meats) that can be problematic, not high amounts of protein per se. For example, a diet containing lots of steak, bacon, and sausage would be high in protein, but it also might be high in saturated fat, cholesterol, salt, and nitrates. Eating more red meat and processed red meat is linked to an increased risk of heart disease, stroke, and cancer (especially colorectal, stomach, pancreatic, prostate, and breast cancers). This link doesn't seem to be caused by the protein but rather the nutritional package that it comes in. In addition, the link to cancer may be related to the carcinogens that can form when meat is cooked at high temperatures, particularly when it's charred by grilling. On the other hand, studies show that when protein comes from lean meat and plant sources, risk of chronic diseases may be reduced. For example, a 2015 study found that frequent consumption of red meat in adolescence was associated with a higher risk of breast cancer later in life, whereas consuming poultry, fish, legumes, and nuts instead lowered risk.[6] Other studies have shown that higher protein diets can reduce the risk of heart disease, provided the protein comes from healthier sources.

Fig. 6.24. Compare several different "protein packages." The steak and bacon provide protein but also large amounts of saturated fat and sodium. Salmon provides as much protein as the steak but with less saturated fat and more polyunsaturated fats. Lentils are a good source of protein, are low in fat, and are a great source of fiber.

Several other concerns about high protein diets haven't turned out to be problematic after all:

- **Osteoporosis** – High protein diets were once thought to increase the risk of osteoporosis, because researchers noticed that urinary calcium excretion increases when people consume high amounts of protein. However, a 2017 systematic review and meta-analysis from the National Osteoporosis Foundation found that this is not a concern, and some studies even show that higher protein intake is associated with greater bone mineral density.[7]
- **Kidney function** – Another concern was that high protein diets would strain the kidneys because of the increased need to filter and excrete nitrogen. A 2018 meta-analysis concluded that this is not a concern. However, people who already have chronic kidney disease should avoid high protein diets and maintain protein intake around the RDA of 0.8 g/kg, a lower intake than many Americans are consuming.8 A 2009 study tested a very low protein diet of 0.6 g/kg protein per day in people with chronic kidney disease and found that it did not prevent the development of kidney failure but did increase the risk of death, underscoring the importance of adequate protein even in people with kidney disease.9

From all of this research, there's little evidence that a high protein diet is inherently harmful, so long as the protein doesn't come packaged with a lot of saturated fat and red meat consumption is limited. Still, there's little research directly testing the health effects of very high protein diets, including those achieved using protein supplements of purified protein, so it's probably wise to keep protein balanced with the other macronutrients, focusing on whole foods from all the food groups.

REFERENCES:

Food and Nutrition Board, & Institute of Medicine. (2005). Dietary Reference Intakes for Energy, Carbohydrate, Fiber, Fat, Fatty Acids, Cholesterol, Protein, and Amino Acids (p. 1359). Washington, D.C.: The National Academies Press.

Berryman, C. E., Lieberman, H. R., Fulgoni, V. L., & Pasiakos, S. M. (2018). Protein intake trends and conformity with the Dietary Reference Intakes in the United States: Analysis of the National Health and Nutrition Examination Survey, 2001-2014. The American Journal of Clinical Nutrition, 108(2), 405–413. https://doi.org/10.1093/ajcn/nqy088

UNICEF. (2019). The State of the World's Children 2019. Children, Food and Nutrition: Growing well in a changing world. New York.

Traylor, D. A., Gorissen, S. H. M., & Phillips, S. M. (2018). Perspective: Protein Requirements and Optimal Intakes in Aging: Are We Ready to Recommend

More Than the Recommended Daily Allowance? Advances in Nutrition (Bethesda, Md.), 9(3), 171–182. https://doi.org/10.1093/advances/nmy003

Harvard T.H. Chan School of Public Health. (n.d.). Protein. Retrieved from https://www.hsph.harvard.edu/nutritionsource/what-should-you-eat/protein/#protein-research

Farvid, M. S., Cho, E., Chen, W. Y., Eliassen, A. H., & Willett, W. C. (2015). Adolescent meat intake and breast cancer risk. International Journal of Cancer, 136(8), 1909–1920. https://doi.org/10.1002/ijc.29218

Shams-White, M. M., Chung, M., Du, M., Fu, Z., Insogna, K. L., Karlsen, M. C., ... Weaver, C. M. (2017). Dietary protein and bone health: A systematic review and meta-analysis from the National Osteoporosis Foundation. The American Journal of Clinical Nutrition, 105(6), 1528–1543. https://doi.org/10.3945/ajcn.116.145110

Devries, M. C., Sithamparapillai, A., Brimble, K. S., Banfield, L., Morton, R. W., & Phillips, S. M. (2018). Changes in Kidney Function Do Not Differ between Healthy Adults Consuming Higher- Compared with Lower- or Normal-Protein Diets: A Systematic Review and Meta-Analysis. The Journal of Nutrition, 148(11), 1760–1775. https://doi.org/10.1093/jn/nxy197

Menon, V., Kopple, J. D., Wang, X., Beck, G. J., Collins, A. J., Kusek, J. W., ... Sarnak, M. J. (2009). Effect of a very low-protein diet on outcomes: Long-term follow-up of the Modification of Diet in Renal Disease (MDRD) Study. American Journal of Kidney Diseases: The Official Journal of the National Kidney Foundation, 53(2), 208–217. https://doi.org/10.1053/j.ajkd.2008.08.009

CHOICES IN PROTEIN FOOD AND SUSTAINABILITY

Before it gets to our plates, every food has a story. Maybe it started as a seed, planted in soil and nurtured to maturity with water, sunlight, and fertilizer. Or maybe it came from an animal, one raised for its meat or to produce milk or eggs. We choose foods based on their taste, price, convenience, and nutritional value, but it's also worth considering their backstories. This is particularly true for protein foods, because animal protein production generally consumes more resources and is less sustainable than plant protein sources. Agricultural animals need care, feeding, housing, disposal of their waste, and sometimes medication use throughout their lives. It's worth considering where our protein comes from and how our choices affect the planet, especially since most of us consume more protein than we need.

Animal Agriculture and Resource Use

The World Resources Institute, a global research non-profit organization with a mission "to move human society to live in ways that protect Earth's environment" has compiled data on the environmental impact of protein choices. In the graphic below, you'll see that in terms of greenhouse gas emissions, protein sources from plants have a much lower impact than protein sources from animals. Most plant proteins, with the exception of nuts, are also less expensive. Beef, lamb, and goat meat come at a higher cost to the environment and your wallet.

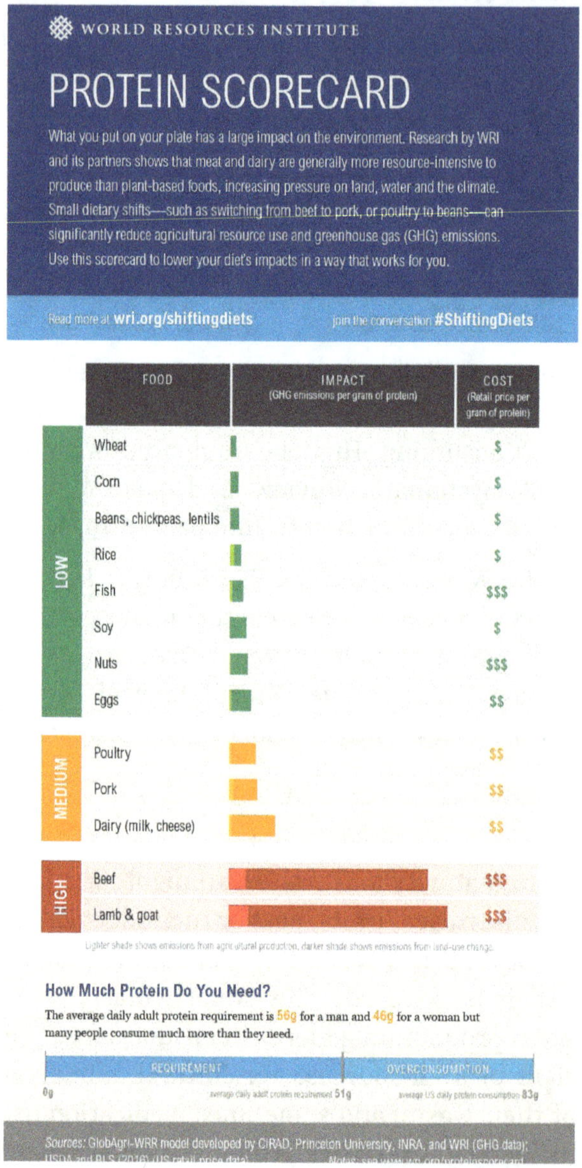

Fig. 6.25. Protein Scorecard from the World Resources Institute. Source: https://www.wri.org/resources/data-visualizations/protein-scorecard

Beef is among the most resource-intensive sources of protein. A 2014 study published in the Proceedings of the National Academy of Sciences calculated that beef production uses 28 times more land and 11 times more irrigation water than the average of dairy, poultry, pork, and egg production.[1]

It's important to point out that sustainable animal agriculture does fill some important roles that plants can't. For example, much of the world's pasture land is on steep terrain that wouldn't work well for growing food crops. And animal waste—in the form of manure—is an important fertilizer, including in organic food systems. So, animal agriculture and eating meat aren't inherently bad for the environment, but it would probably be good for the planet if we ate less meat, as shown in the graphic below.

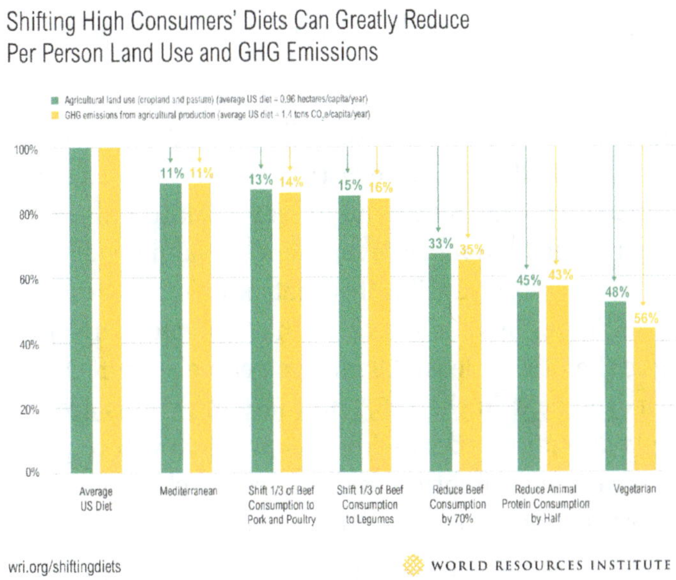

Fig. 6.26. Shifting High Consumers' Diets Can Greatly Reduce Per Person Land Use and GHG Emissions from the World Resources Institute. Source: https://www.wri.org/resources/charts-graphs/animal-based-foods-are-more-resource-intensive-plant-based-foods

In terms of environmental impact, making small shifts can have a significant impact. Consider the following approaches[2]:

- If you eat meat every day, try adding a "meatless Monday" into your week and experiment with some vegetarian recipes. Once you've adapted to that, try adding another day.
- Replace some of your beef meals with dishes featuring chicken, pork, eggs, fish, or legumes.
- Eat smaller portions of meat and add more plant foods to your plate. For example, if you enjoy spaghetti and meat sauce, try using less meat in your

sauce and adding in vegetables like mushrooms, bell peppers, and carrots. Your meal will be more nutrient-dense and maybe even more flavorful.

When you consider that moderate shifts like these would not only be good for the planet but also good for our health, then they don't seem like much of a sacrifice. A 2016 study published in the Proceedings of the National Academy of Sciences concluded that just following standard dietary guidelines (which recommend a variety of protein sources, including plant proteins, and eating more whole grains, fruits and vegetables) could reduce mortality by 6 to 10 percent *and* cut greenhouse gas emissions by 29 to 70 percent. This page from the World Resources Institute provides more information: Sustainable Diets: What You Need to Know in 12 Charts, by Janet Ranganathan and Richard Waite, April 20, 2016.

Animal Agriculture And Antibiotic Resistance

One of the biggest current threats to public health is antibiotic resistance. Antibiotics are life-saving drugs, but over time, bacteria can develop resistance to them. This means that the antibiotics no longer work to kill the bacteria causing infections, leaving people with more severe illnesses and fewer treatment options, often needing to try different antibiotics that have more side effects. There are now some bacterial infections for which we have no working antibiotics to treat them. According to the CDC, at least 2.8 million people are infected with antibiotic-resistant bacteria each year in the U.S, and these infections are thought to kill at least 35,000 people annually.[5] Addressing this problem will require us to be more careful about how we use antibiotics, invest in research to develop new ones, and to develop other ways of preventing bacterial disease, such as new vaccines.

Antibiotics are important to both human and animal medicine. When we're sick with a bacterial illness, we may need antibiotics to treat it, and the same is true of animals, whether they're raised for agriculture or part of our families as our pets. The problem is that the more we use antibiotics, the more chances bacteria have to evolve resistance to them, and the less effective those antibiotics become.

The overuse of antibiotics in both human medicine and animal agriculture has contributed to the growth of antibiotic-resistant bacteria. For example, taking antibiotics for an illness caused by a virus, such as the common cold or the flu, won't make you better and just gives harmful bacteria chances to evolve resistance. Historically, antibiotics were also routinely given to food production animals to make them fatter, and that allowed for the growth of antibiotic resistance. As of 2017, the FDA ruled that antibiotics can no longer be used for growth promotion in animal agriculture, a significant step in reducing the overuse of these drugs that are important to both humans and animals. Data released by the FDA in December 2018 show that antibiotic sales for farm animals have dropped significantly after this rules change.

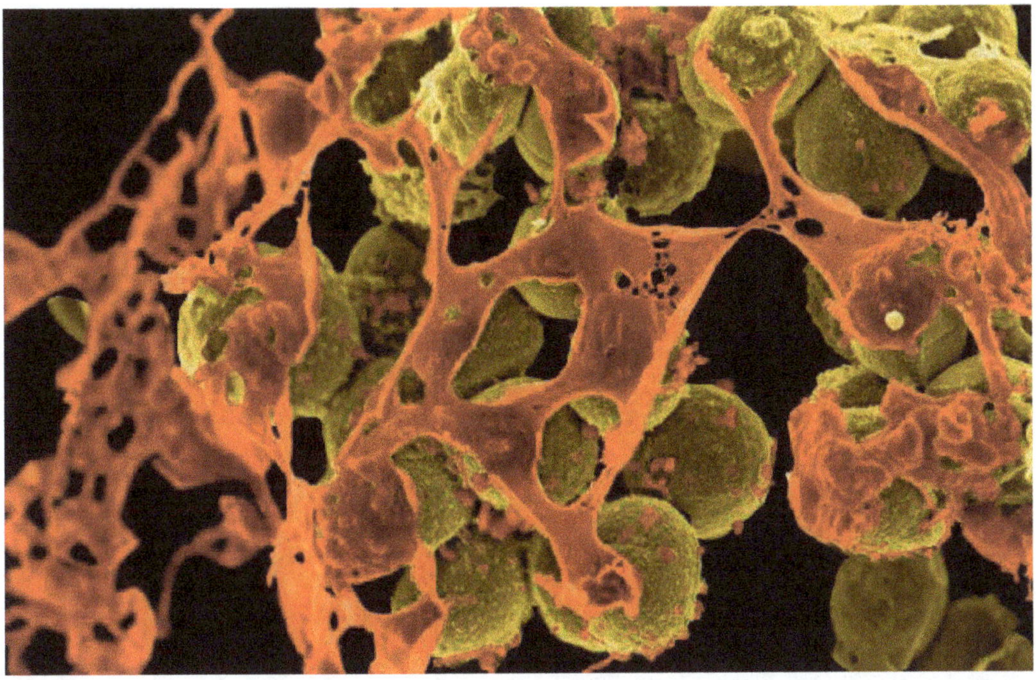

Fig. 6.27. Scanning electron micrograph of methicillin-resistant Staphylococcus aureus (MRSA, brown) surrounded by cellular debris. MRSA resists treatment with many antibiotics. Credit: NIAID

Antibiotics can still be used to treat sick farm animals or stop the spread of disease, which is important for animal health and welfare. However, antibiotics can also be used to *prevent* disease in animals that might become sick. Many experts argue that allowing antibiotics to be used for disease prevention leaves a loophole for large amounts of antibiotics to continue to be used, especially in farming systems where animals are crowded and diseases can spread quickly. The World Health Organization has called for this practice to stop, reserving antibiotics only for use in animals that are already sick, not healthy animals. The goal is to reduce antibiotic use in order to reduce the development of antibiotic-resistant bacteria, so that we can still use these valuable medicines to treat sick animals and humans when needed.

When antibiotics are used in food animals, bacteria can evolve resistance to those antibiotics. Those antibiotics-resistant bacteria can then be present in your meat, and they can spread in the environment from animal feces, including into the water used to irrigate fruits and vegetables. Humans exposed to these bacteria by handling or eating contaminated food can then become sick with infections that are resistant to antibiotic treatment, as shown in this infographic from the CDC:

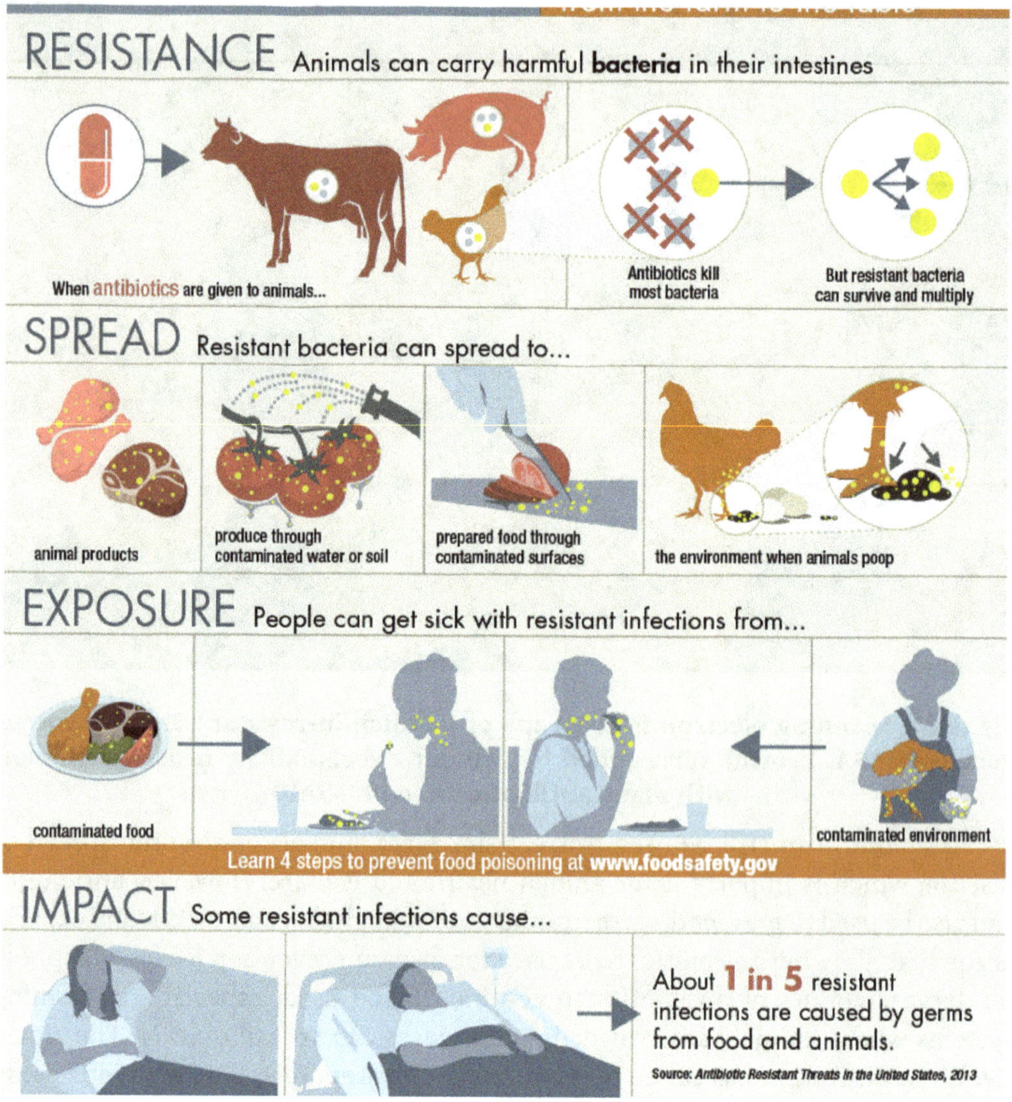

Fig. 6.28. Antibiotic Resistance from Farm to Table infographic from the CDC. Source: https://www.cdc.gov/foodsafety/challenges/from-farm-to-table.html

What can you do to prevent yourself and your family from getting sick with antibiotic-resistant infections? The CDC offers these tips, most of which are useful for reducing your risk of foodborne illness in general:

- Take antibiotics only when needed.
- Follow simple Food Safety Tips:
- COOK. Use a food thermometer to ensure that foods are cooked to a safe internal temperature: 145°F for whole beef, pork, lamb, and veal (allowing the meat to rest for 3 minutes before carving or consuming), 160°F for

ground meats, and 165°F for all poultry, including ground chicken and turkey.

- CLEAN. Wash your hands after touching raw meat, poultry, and seafood. Also wash your work surfaces, cutting boards, utensils, and grill before and after cooking.
- CHILL. Keep your refrigerator below °40F and refrigerate foods within 2 hours of cooking (1 hour during the summer heat).
- SEPARATE. Germs from raw meat, poultry, seafood, and eggs can spread to produce and ready-to-eat foods unless you keep them separate. Use different cutting boards to prepare raw meats and any food that will be eaten without cooking.
- Wash your hands after contact with poop, animals, or animal environments.
- Report suspected outbreaks of illness from food to your local health department.
- Review CDC's Traveler's Health recommendations when preparing to travel to a foreign country.

When you shop for meat, milk, and eggs, you'll see lots of different types of labels making claims about how the animals were raised. Does any of this matter when it comes to antibiotic resistance? First, it's important to note that no animal products—however they're labeled—should ever contain antibiotics, as required by federal law. If an animal is treated with an antibiotic, it can't be sold for slaughter or have its milk sold until the antibiotic is cleared from its system. However, if animals were routinely treated with antibiotics earlier in their lives, those practices could have contributed to the growing problem of antibiotic resistance.

Choosing products that are certified organic ensures that antibiotics weren't used in their production, because organic farms are not allowed to use antibiotics even for sick animals.[7] (If an animal becomes sick and requires antibiotic treatment to get better, its milk, meat, or eggs can no longer be sold as organic, but it can be sold to a conventional farm.)

You'll also see labels stating "raised without antibiotics," and buying these products helps to support farmers and companies that have committed to reducing antibiotic use in their production systems, even if they aren't certified organic. However, antibiotic resistant bacteria (bacteria that have evolved resistance to antibiotics so could cause hard-to-treat infections) may still be present in products that are labeled certified organic or "raised without antibiotics" (the bacteria could have spread to these animals from somewhere else), so follow the food safety rules no matter where your meat comes from.

Figure 6.29. These eggs are certified organic, so you can be confident that antibiotics weren't used in their production. The sausage is not organic, but it is made from chickens raised without antibiotics, so its production is unlikely to have contributed to the problem of antibiotic resistance.

More and more companies are also recognizing how the overuse of antibiotics can contribute to antibiotic resistance, and they're changing their practices. In December 2018, a large consortium of companies and industry groups, including Walmart, McDonald's, and Tyson Foods, committed to a framework for more responsible use of antibiotics.[9]

Issues Of Fish Sustainability

Fish are a good source of protein and healthful polyunsaturated fats, as well as micronutrients like vitamin D, so they're often mentioned as a good choice. From the charts at the top of this page, you can also see that fish are a relatively sustainable source of protein in terms of using little land and freshwater and producing low levels of greenhouse gases.

However, the oceans have been overfished, and global supplies of wild-caught fish are dwindling. Aquaculture, or fish farming, has also created new environmental challenges. Both of these issues are being solved with good management, like careful limits on wild-caught fishing and new management practices for fish farming. You can encourage these practices by purchasing sustainably-sourced seafood.[10] The Monterey Bay Aquarium Seafood Watch program can help with this. You can download their app to help with buying decisions in the grocery store and find more information on their website: Monterey Bay Aquarium Seafood Watch.

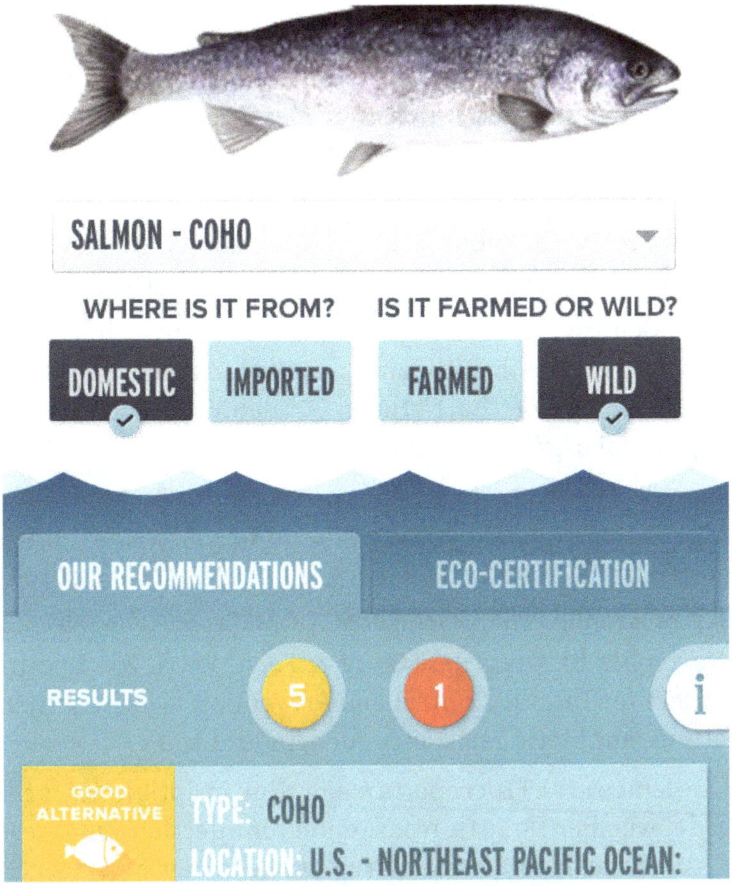

Fig. 6.30. A screenshot from the Seafood Watch app, showing how it can help you make sense of seafood buying options.

You may have also heard that fish can contain dangerous levels of mercury. This is true of some large species of fish that are higher up the food chain, because they accumulate mercury from their smaller prey and then we get a big dose when we eat them. Fish that have dangerous levels of mercury include king mackerel, marlin, orange roughy, shark, swordfish, tilefish, and bigeye tuna. Pregnant women and growing children in particular should take care to avoid these types of fish, because mercury can interfere with brain development. However, the same groups also stand to benefit from healthful omega-3 fatty acids, like DHA and EPA, which are *helpful* for brain development. Thus, it's good for pregnant women and children to eat fish, so long as they avoid the ones with high levels of mercury. Most common types of fish have lower levels of mercury and can be eaten at least once per week, if not two or three times per week[11]. Learn more at this FDA page: Eating Fish: What Pregnant Women and Parents Should Know.

REFERENCES

Eshel, G., Shepon, A., Makov, T., & Milo, R. (2014). Land, irrigation water, greenhouse gas, and reactive nitrogen burdens of meat, eggs, and dairy production in the United States. Proceedings of the National Academy of Sciences, 111(33), 11996–12001. https://doi.org/10.1073/pnas.1402183111

Harvard T.H. Chan School of Public Health. Sustainability. Retrieved from https://www.hsph.harvard.edu/nutritionsource/sustainability/#plate-and-planet

Springmann, M., Godfray, H. C. J., Rayner, M., & Scarborough, P. (2016). Analysis and valuation of the health and climate change cobenefits of dietary change. Proceedings of the National Academy of Sciences, 113(15), 4146–4151. https://doi.org/10.1073/pnas.1523119113

Aleksandrowicz, L., Green, R., Joy, E. J. M., Smith, P., & Haines, A. (2016). The Impacts of Dietary Change on Greenhouse Gas Emissions, Land Use, Water Use, and Health: A Systematic Review. PLOS ONE, 11(11), e0165797. https://doi.org/10.1371/journal.pone.0165797

CDC. (2018, November 8). Drug Resistance & Food. Retrieved December 4, 2018, from Centers for Disease Control and Prevention website: https://www.cdc.gov/features/antibiotic-resistance-food/index.html

Dall, C. (2018, December 19). FDA reports major drop in antibiotics for food animals. Retrieved November 18, 2019, from Center for Infectious Disease Research and Policy website: http://www.cidrap.umn.edu/news-perspective/2018/12/fda-reports-major-drop-antibiotics-food-animals

7United States Department of Agriculture. (2103, July). Organic Livestock Requirements. Retrieved from https://www.ams.usda.gov/sites/default/files/media/Organic%20Livestock%20Requirements.pdf

Smith, T. C. (n.d.). What does 'meat raised without antibiotics' mean—And why is it important? Washington Post. Retrieved from https://www.washingtonpost.com/news/speaking-of-science/wp/2015/10/28/what-does-raised-without-antibiotics-mean-and-why-is-it-important/

Monterey Bay Aquarium Seafood Watch Program. Ocean Issues. Retrieved November 18, 2019, from https://www.seafoodwatch.org/ocean-issues.

U.S. Food and Drug Administration. (2019, October 15). Advice about Eating Fish. Retrieved November 18, 2019, from http://www.fda.gov/food/consumers/advice-about-eating-fish

UNIT-7

HEALTH AND BODY WEIGHT

BODY WEIGHT AND HEALTH: AN INTRODUCTION

On December 26, 2018, 33-year-old Colin O'Brady of Portland, Oregon, became the first person to cross the landmass of Antarctica solo, unassisted, and without any resupply shipments. Others had crossed the continent with the help of a wind sail to propel them over the ice or with resupply drops along the way—both ways of saving precious energy—but O'Brady completed the 926-mile trek only on skis, pulling a sled packed with food, fuel, and supplies the entire way. Speed was important, because he was racing another man, Louis Rudd, a 49-year-old British Army captain. And of course, he didn't want to run out of food hundreds of miles from the finish. Rudd finished the Antarctica crossing just two days after O'Brady.

Figure 7.1. Endurance athlete Colin O'Brady, photographed in March 2016.

To prepare for the expedition, O'Brady and his team had to make careful calculations to estimate his nutrient and caloric needs. He'd be skiing all day, every day in below zero temperatures and against constant wind for about two months. O'Brady estimated that he'd burn about 10,000 calories per day on his journey, and he knew that if he didn't pack enough food, he wouldn't have the strength to complete this epic test of endurance in extreme conditions. Previous explorers died in the Antarctic because they didn't pack enough food.

Figure 7.2. O'Brady and Rudd raced across a landscape similar to that shown in this photo from Antarctica—a polar desert and the coldest, windiest, driest continent on earth.

But O'Brady also knew that the more food he packed, the heavier his sled would be—ironically making him burn more calories, plus slowing him down and prolonging his trip. So he focused on choosing foods that were calorie- and nutrient-dense but lightweight: oatmeal with added oil and protein powder; freeze-dried dinners reconstituted with melted snow; and 4,500-calorie slabs of a custom-made "Colin bar" made from coconut oil, nuts, seeds, and dried fruit.

At the start of his journey, O'Brady's sled weighed 375 pounds and contained enough food to provide him with 8,000 calories per day. That was a bit short of the 10,000 calories he estimated he'd burn every day, so to build up some additional energy stores, he gained 15 pounds prior to his trip. In the end, after 54 days of skiing through ice and snow, he lost 25 pounds during his Antarctic crossing. He was successful, and while his fitness level and determination were surely critical to his feat, the trip would have been impossible without an adequate supply of calories.

In our daily lives, we need far fewer calories than an Antarctic adventurer, and we don't need to schlep a two-month supply of food on our backs wherever we go. Thankfully, we also get to enjoy fresher and more interesting food options. But each of us, every day, whether we're aware of it or not, is attempting to balance the calories we consume with the calories we burn, just like Colin O'Brady. This is the concept of energy balance—one we'll be exploring throughout this unit. If adults eat roughly the same number of calories as they burn each day, their body weight will generally stay very constant. If they burn more calories than they eat, like

O'Brady on his expedition, they'll lose weight. If they consume more calories than they burn, they'll gain weight.

Energy balance may seem like a simple concept, but in practice, how many calories a person eats and expends each day is influenced by so many different factors that it can be frustratingly difficult to apply. The human body has many sophisticated mechanisms that have evolved to help us protect and maintain our body weight, and in the history of our species, these mechanisms have been vital to our survival in times of food shortages. But today, we live in a world where highly palatable food is easily available and convenient, and we're constantly bombarded with marketing messages enticing us to eat more of it. At the same time, the diet industry and media send messages saying that being thin is the same as being healthy and that restrictive dieting is the virtuous path to wellness, which ironically may cause weight gain and worsen people's mental and physical health. Meanwhile, disparities in access to healthy food, safe places for physical activity, and healthcare have grown, all of which are associated with body weight and health. Many other factors, including the use of some medications, exposure to certain chemicals in the food supply and environment, inadequate sleep, excess stress, microbiome changes, and experiencing traumatic events, have been linked to body weight.

In this unit, we'll explore these complex concepts and concerns and seek some answers, but we acknowledge that we raise more questions than we resolve. The truth is that researchers don't fully understand why body weight has increased over the last several decades, much less how to change this trend. What is clear is that body weight is extraordinarily complex, and people have long been harmed by oversimplified messages to "eat less and move more" to reach a "normal" or "healthy" body weight. This approach implies that obesity is a failure of personal responsibility, but the last several decades have shown that blaming individuals for the size and shape of their bodies has not changed the trajectory of increasing body weight, and it has likely worsened people's health by harming their relationships with food and perpetuating weight stigma.

We have done our best to approach this topic with care, recognizing that it can be very personal. We know that the language used to describe people's bodies can be hurtful, and that encouraging people to engage in restrictive dieting practices for the purposes of making their bodies smaller can cause harm. Researchers' understanding of body weight regulation is incomplete and still evolving, and we have tried to include different perspectives in this unit. We hope you'll come away with an appreciation of the complexities, nuances, and uncertainties of this topic.

ENERGY BALANCE: ENERGY IN VS. ENERGY OUT

The concept of energy balance seems simple: Balance the calories you consume with the calories you expend. But many factors play a role in energy intake and energy expenditure. Some of these factors are under our control and others are not. In this section, we define energy balance, look at the different components of energy expenditure, and discuss the variables that influence energy expenditure. We'll also consider some of the factors that can affect energy intake and consider why energy balance is more complex than it seems.

Energy Balance

Our body weight is influenced by our energy intake (calories we consume) and our energy output (energy we expend during rest and physical activity). This relationship is defined by the energy balance equation:

Energy Balance = energy intake – energy expenditure

When an individual is in energy balance, energy intake equals energy expenditure, and weight should remain stable.

Energy intake is made up of the calories we consume from food and beverages. These calories come from the macronutrients (carbohydrates, proteins, and fats) and alcohol. Remember that when the body has a surplus of energy, this energy can be stored as fat. When the body has an inadequate supply of food calories to match energy expenditure, it will turn to stored energy (i.e., adipose tissue, glycogen, and some muscle protein) to meet energy demands, resulting in weight loss.

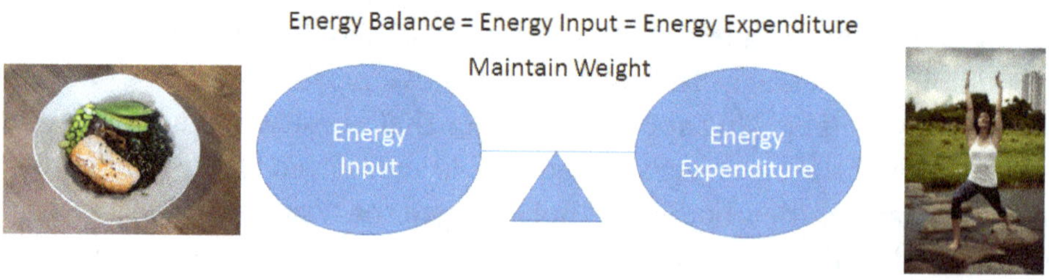

Positive energy balance **occurs when energy intake is greater than energy expenditure, usually resulting in weight gain.**

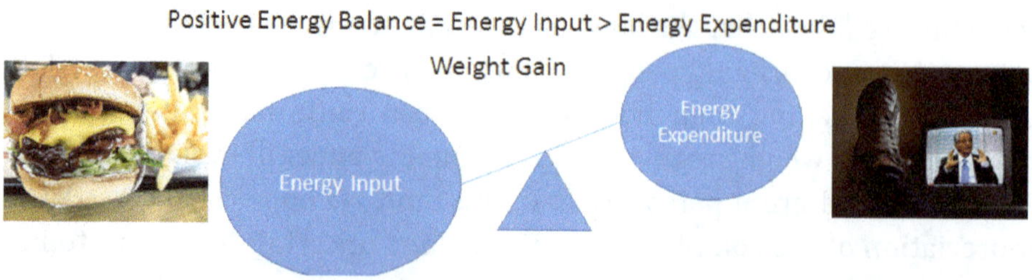

Negative energy balance **is when energy intake is less than energy expenditure, usually resulting in weight loss.**

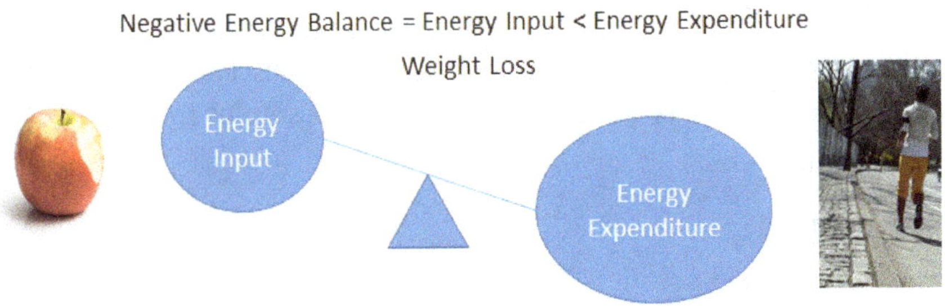

You may see simple rules about energy balance, such as the idea that a pound of fat contains 3,500 calories, so cutting 3,500 calories from your diet will result in a pound of weight loss. This is sometimes extrapolated to say that cutting your caloric intake by 500 calories per day will lead to one pound of weight loss per week, or 52 pounds of weight loss in one year. However, this is a myth, because the body responds to caloric restriction by making adjustments to energy expenditure and resisting changes in body weight, especially over long periods of time.[1]

Energy balance is complex, dynamic, and variable between individuals—something we'll explore a bit more later on this page—but it is still a vital concept in understanding body weight. Next, let's look at the energy expenditure side of the energy balance equation, to see the components that make up energy expenditure and the factors that influence them.

Components Of Energy Expenditure

The sum of caloric expenditure is referred to as total energy expenditure (TEE). There are three main components of TEE:

- Basal metabolic rate (BMR)
- Thermic effect of food (TEF)
- Physical activity

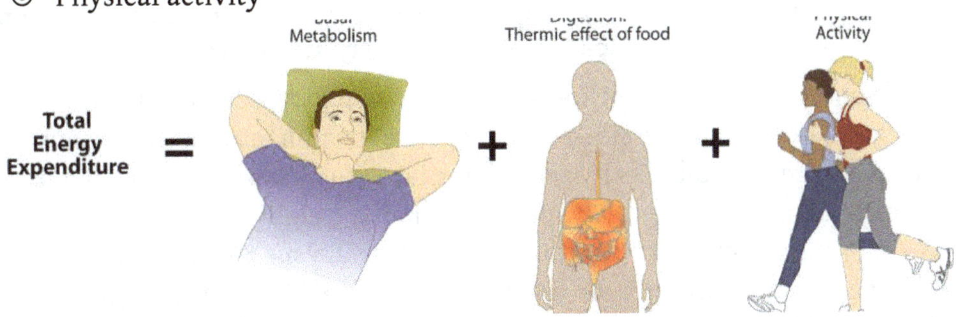

Figure 7.3. Components of total energy expenditure include basal metabolism, the thermic effect of food, and physical activity.

1. Basal Metabolic Rate (BMR)

BMR is the energy expended by the body when at rest. These are the behind-the-scenes activities that are required to sustain life. Examples include:

- respiration
- circulation
- nervous system activity
- protein synthesis
- temperature regulation

Basal metabolic rate does not include the energy required for digestion or physical activity.

BMR is usually the largest component of energy expenditure, making up about 60 to 75 percent of total energy output. For example, a sedentary person might need about 1800 calories in a day, with about 1200 of them being for BMR.

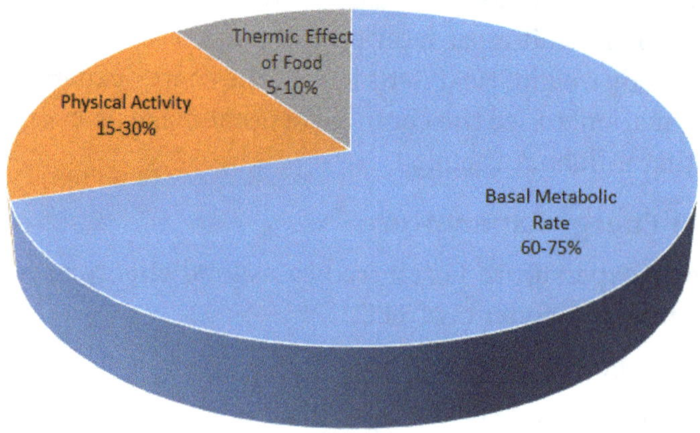

Figure 7.4. Components of energy expenditure and their percent contribution to the total in sedentary to moderately active people.

BMR can vary widely among individuals. An individual's lean body mass—made up of organs, bone, and muscle—is the biggest determinant of BMR, because lean body tissue is more metabolically active than fat tissue. This means that a muscular person expends more energy than a person of similar weight with more fat. Likewise, increasing your muscle mass can cause an increase in your BMR. However, skeletal muscle at rest only accounts for about 18 percent of the total energy expended by lean mass. Most is used to meet the energy needs of vital organs. The liver and brain, for example, together account for nearly half of the energy expenditure by lean mass.

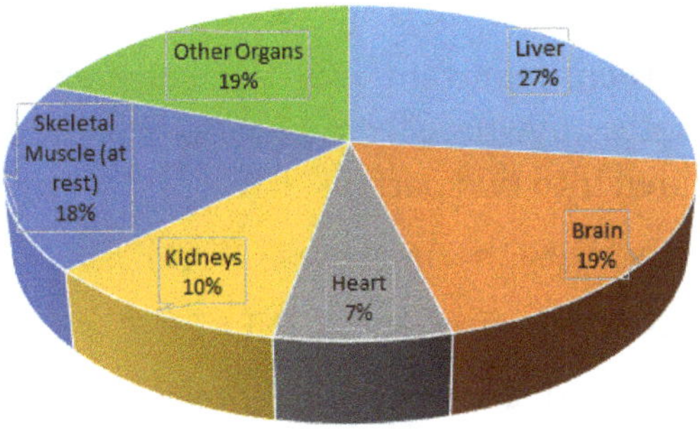

Figure 7.5. Energy expenditure of organs.

BMR depends not only on body composition but also on body size, sex, age, nutritional status, genetics, body temperature, and hormones (Table 7.1). People with a larger frame size have a higher BMR simply because they have more mass. On average, women have a lower BMR than men, because they typically have a smaller frame size and less muscle mass. As we get older, muscle mass declines, and therefore BMR declines as well.

Nutritional status also affects basal metabolism. If someone is fasting or starving, or even just cutting their caloric intake for a diet, their BMR will decrease. This is because the body attempts to maintain homeostasis and adapts by slowing down its basic functions (BMR) to help preserve energy and balance the decrease in energy intake. This is a protective mechanism during times of food shortages, but it also makes intentional weight loss more difficult.

Table 7.1. Factors that impact BMR.

Factors That Increase BMR	Factors That Decrease BMR
Higher lean body mass	Lower lean body mass
Larger frame size	Smaller frame size
Younger age	Older age
Male sex	Female sex
Stress, fever, illness	Starvation or fasting
Elevated levels of thyroid hormone	Lower levels of thyroid hormone
Pregnancy or lactation	
Stimulants such as caffeine and tobacco	

2. Thermic Effect of Food (TEF)

This is the energy needed to digest, absorb, and metabolize the nutrients in foods. It accounts for 5 to 10 percent of total energy expenditure and does not vary greatly amongst individuals.

3. Physical activity

Physical activity is another important way the body expends energy. Physical activity usually contributes anywhere from 15 to 30 percent of energy expenditure and can be further divided into two parts:

- Exercise-related activity thermogenesis (EAT)
- Non-exercise activity thermogenesis (NEAT)

EAT is planned, structured, and repetitive physical activity with the objective of improving health (participating in a sport like soccer or strength training at the gym, for example).

NEAT is the energy expenditure for unstructured and unplanned activities. This includes daily-living activities like cleaning the house, yard work, shopping, and occupational activities. NEAT also includes the energy required to maintain posture and spontaneous movements such as fidgeting and pacing.[2]

NEAT can vary by up to 2,000 calories a day for two people of similar size, according to Dr. James Levine, the Mayo Clinic researcher who first coined the term. NEAT may be an important component of obesity prevention and is currently an area of research.

Factors Affecting Energy Intake

Given the importance of energy's role in sustaining life, it's not surprising that energy balance is tightly regulated by complex physiological processes. A region in the brain called the hypothalamus is the main control center for hunger and satiety. There is a constant dialogue between our brains and gastrointestinal tracts through hormonal and neural signals, which determine if we feel hungry or full.

The hypothalamus can sense nutrient levels in the blood, and when nutrient levels are low, the hunger center is stimulated. Conversely, when nutrient levels are high, the satiety center is stimulated.

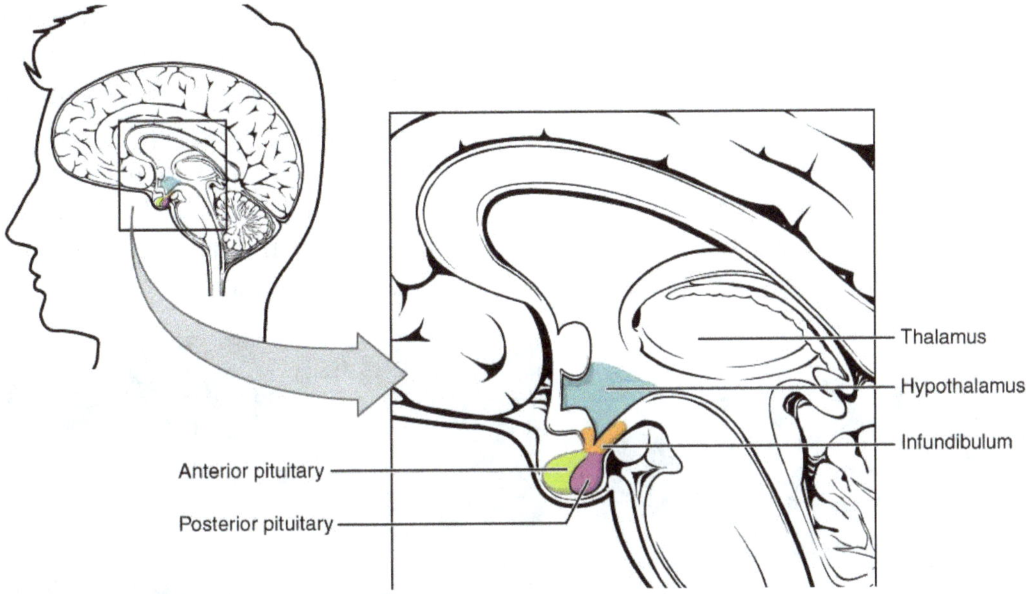

Figure 7.6. The hypothalamus, shown in blue, is about the size of an almond and serves as the hunger center of the brain, receiving signals from the gastrointestinal tract, adipose tissue, and blood and signaling hunger and satiety.

Hunger is the physiological need to eat. When the stomach is empty, it contracts and starts to grumble and growl. The stomach's mechanical movements relay neural signals to the hypothalamus. (Of course, the stomach also contracts when it's full and hard at work digesting food, but we can't hear these movements as well because the stomach's contents muffle the noise.) The stomach is also the main organ that produces and secretes the "hunger hormone," ***ghrelin***, the only gut hormone found to increase hunger. Ghrelin levels are high before a meal and fall quickly once nutrients are absorbed.

Appetite is the psychological desire to eat. Satiety is the sensation of feeling full. After you eat a meal, the stomach stretches and sends a neural signal to the brain stimulating the sensation of satiety and relaying the message to stop eating. There are many hormones that are associated with satiety, and various organs secrete these hormones, including the gastrointestinal tract, pancreas, and adipose tissue. Cholecystokinin (CCK) is an example of one of these satiety hormones and is secreted in response to nutrients in the gut, especially fat and protein. In addition to inhibiting food intake, CCK stimulates pancreatic secretions, gall bladder contractions, and intestinal motility—all of which aide in the digestion of nutrients.

Adipose tissue also plays a role in regulating food intake. Adipose is the primary organ that produces the hormone ***leptin***, and as fat stores increase, more leptin is produced. Higher levels of leptin communicate to the satiety center in the hypothalamus that the body is in positive energy balance. Leptin acts on the brain to suppress hunger and increase energy expenditure. The discovery of leptin's functions sparked excitement in the research world and the weight loss industry, as it was hypothesized that leptin might be used as a weight loss drug to decrease food intake. In several clinical trials, it was found that people who are overweight or obese are actually resistant to the hormone, meaning their brain does not respond as well to it. Therefore, when you administer leptin to an overweight or obese person, there is generally no sustained effect on food intake.[4]

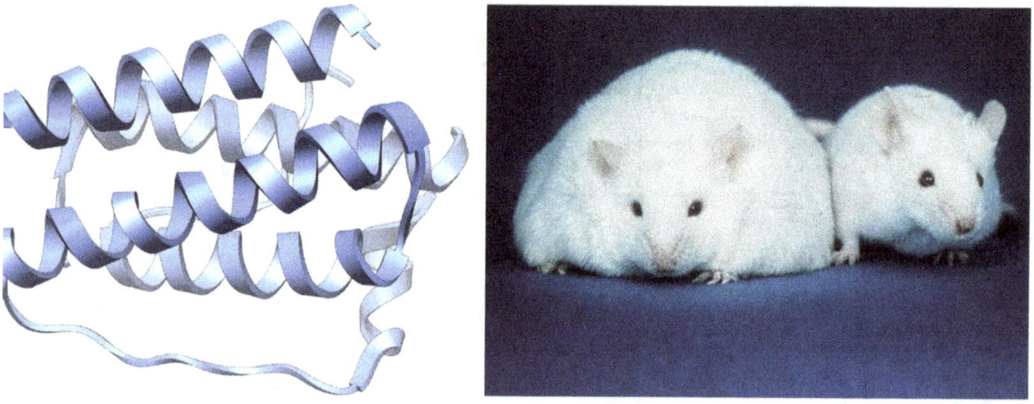

Figure 7.7. The structure of the hormone leptin (left), which is primarily produced by adipose tissue. The obese mouse in the photo has a gene mutation that makes it unable to produce leptin, resulting in constant hunger, lethargy, and severe obesity. For comparison, a mouse with normal leptin production is also shown. Such gene mutations are rare, but they serve as a dramatic illustration of the importance of the hormone in signaling energy balance.

The Complexity Of Energy Balance

Energy balance seems like it should be a simple math problem, and in fact, it is based on a fundamental truth in physics—the first law of thermodynamics. This law states that energy can't be created or destroyed; it can only change form. That is, calories that are consumed must go somewhere, and if they aren't metabolized (which converts caloric energy to heat and work energy), they'll have to be stored, usually in the form of adipose tissue. What makes energy balance challenging is the reality that both energy intake and energy expenditure are dynamic variables that are constantly changing, in response to each other and overall energy balance.

Let's first look at the energy intake side. As we've already discussed, how much food we eat each day is not just a matter of willpower or self-control. It's the result of powerful physiological and psychological forces that tell us if we need to eat, or if we've had enough. Our brains are hard-wired to seek food if we're in negative energy balance, an instinct required for survival. This means that if you start to exercise more—increasing your energy expenditure—you will also feel hungrier, because your body needs more fuel to support the increase in physical activity. If you eat fewer calories, perhaps in an effort to lose weight, your stomach will produce more ghrelin, and your adipose tissue will produce less leptin. These and other shifting hormone levels work together to increase hunger and make you focus on obtaining more calories. People who try to gain weight run into the opposite problem. Their leptin levels increase, suppressing hunger. It's also uncomfortable to eat beyond satiety, and food doesn't taste as good once you're full.

Even measuring how much energy is consumed is not as simple as you might think. We can measure the caloric content of food from a chemical standpoint, but we can only *estimate* how much energy a person will absorb from that food. This will depend on how well the food is digested and how well the macronutrients are absorbed—factors which vary depending on the food itself, the digestion efficiency of the person eating it, and even the microbes living in their gut. Two people may eat the exact same meal, yet not absorb the same number of calories.

Energy expenditure is also dynamic and changes under different conditions, including increased or decreased caloric intake. Decreased caloric intake and negative energy balance cause a drop in BMR to conserve energy. Muscles also become more efficient, requiring less energy to work, and without realizing it, people in negative energy balance often decrease their NEAT activity level. These adaptations help to conserve body weight and make it more difficult to stay in negative energy balance. People may still be able to lose weight despite their bodies working to prevent it, but maintaining a new, lower weight requires constant vigilance, and weight regain is common.

Research has also shown that people respond differently to positive energy balance. When a group of people are overfed, the amount of weight gained amongst study participants varies widely. In a study of identical twins who were given an extra 1,000 calories per day for 100 days, weight gain varied between 10 and 30 pounds among participants. Weight gain between twins was more similar (though not exactly the same), which may be attributed to genetic factors.[7] People gain and lose weight differently; we don't necessarily follow formulas.

When people say that the answer to weight gain is to eat less and move more, they may be partially correct. But this is also an oversimplified answer, because of all the complexities underlying energy intake and energy expenditure.

REFERENCES

"Balancing Energy Input with Energy Output", section 11.2 from the book An Introduction to Nutrition (v. 1.0), CC BY-NC-SA 3.0

Webb, D. (2014, November). Farewell to the 3,500-Calorie Rule. Today's Dietitian, 26(11), 36. https://www.todaysdietitian.com/newarchives/111114p36.shtml.

Chun, N., Park, M., Kim, J., Park, H., Hwang, H., Lee, C., Han, J., So, J., Park, J., & Lim, K. (2018). Non-exercise activity thermogenesis (NEAT): a component of total daily expenditure. J Exerc Nutrition Biochem, 22(2), 23–30. doi: 10.20463/jenb.2018.0013.

Austin J., & Marks, D. (2008). Hormonal Regulators of Appetite. Int J Pediatr Endocrinol. 2009. doi: 10.1155/2009/141753.

Dardeno, T. A. et al. (2010). Leptin in Human Physiology and Therapeutics. Front Neuroendocrinol, 31 (3), 377–93. http://www.ncbi.nlm.nih.gov/pmc/articles/PMC2916735/?tool=pubmed.

Hall, K. D., & Guo, J. (2017). Obesity Energetics: Body Weight Regulation and the Effects of Diet Composition. Gastroenterology, 152(7), 1718-1727.e3. https://doi.org/10.1053/j.gastro.2017.01.052.

Hall, K. D., Heymsfield, S. B., Kemnitz, J. W., Klein, S., Schoeller, D. A., & Speakman, J. R. (2012). Energy balance and its components: Implications for body weight regulation123. The American Journal of Clinical Nutrition, 95(4), 989–994. https://doi.org/10.3945/ajcn.112.036350.

Bouchard, C., Tremblan, A.,...Fournier, G. (1990). The response to long-term overfeeding in identical twins. N. Engl. J. Med. 322, 1477-1482.

Fig 7.4. "Components of energy expenditure and the percentage they contribute" by Tamberly Powell is licensed under CC BY-NC-SA 2.0.

Fig 7.5. "Energy Expenditure of Organs" by Tamberly Powell is licensed under CC BY-NC-SA 2.0.

Table 7.1. Factors that Impact BMR by Tamberly Powell is licensed under CC BY-NC-SA 2.0.

"Raking Alternative" by Jack Zalium is licensed under CC BY-ND 2.0.

Fig 7.6."The Hypothalamus-Pituitary Complex" by OpenStax College is licensed under CC BY-SA 3.0.

Fig 7.7. "Leptin" by Vossman is licensed under CC BY-SA 3.0; "Fatmouse" by Human Genome wall for SC99 is in the Public Domain.

METHODS OF MEASUREMENT OF BODY SIZE, BODY COMPOSITION, AND METABOLIC HEALTH

Healthy Body Size is Personal

Ask a room full of people what the term "healthy body weight" means, and you'll probably get lots of different answers. This isn't surprising given that people naturally come in different shapes and sizes. In addition, body weight or fatness does not determine a person's health, and health encompasses not just physical health but also mental health and social well-being.

Here's how the authors of the text "Sport Nutrition for Health and Performance" describe a healthy body weight[1]:

- A weight that is appropriate for your age and physical development
- A weight you can achieve and sustain without severely curtailing your food intake and constantly dieting
- A weight that is compatible with normal blood pressure, lipid levels, and glucose tolerance (in other words, you are metabolically fit)
- A weight that is based on your genetic background and family history of body shape and weight (after all the apple doesn't fall too far from the tree)
- A weight that promotes good eating habits and allows you to participate in regular physical activity
- A weight that is acceptable to you

Another way of describing healthy body weight is "best weight," suggested in the 2020 Canadian Adult Obesity Clinical Practice Guidelines as a way for physicians working with patients who want to lose weight to help center their goals. "Best weight" is described as "the weight that a person can achieve and maintain while living their healthiest and happiest life."2 It is thus not based on a number on the scale or a metric of body composition, but on individual bodies, lifestyle, and family history. It requires a truly individual assessment, and it isn't dictated by a formula, fad, or societal expectations.

Yet, the truth is that simple metrics are often used to assess body size and composition. We find them in online calculators, on scales at the gym, and in our doctors' offices. None of these metrics give anywhere near a complete picture of an individual's health or whether they are at their best weight. But it's worth understanding some of these measures so that you know what they mean when you encounter them, how they may be linked to health, and what their limitations are.

Body Mass Index

Body Mass Index (BMI) is a simple calculation used in clinical and research settings to assess body weight relative to height. BMI is easy and inexpensive to measure, and because it takes height into account, it is more predictive of how much body fat a person has than weight alone. However, BMI is not a direct measure of body fat, and it shouldn't be used on its own to make assumptions about the health of an individual.[3]

BMI is calculated using the following equations:

$$BMI = [\text{weight (kg)}/\text{height (m}^2)]$$

OR

$$BMI = [\text{weight (lb)}/\text{height (in}^2)] \times 703$$

The Centers for Disease Control and Prevention has a BMI calculator on its website.

For adults, BMI ranges are divided into four main categories:

- Underweight – BMI < 18.5
- Normal weight – BMI from 18.5-24.9
- Overweight – BMI from 25-29.9
- Obesity – BMI from 30 or higher

Obesity is frequently subdivided into categories:

- Class 1 obesity: BMI of 30 to < 35
- Class 2 obesity: BMI of 35 to < 40
- Class 3 obesity: BMI of 40 or higher

BMI values are interpreted differently for children, because body fatness changes so much during growth and varies with sex.[3]

BMI Interpretation in Populations

Because of its simplicity, BMI is a useful metric for researchers studying large groups of people. For example, BMI is used to track changes in body size within a population over time. It is also used in research studies looking at correlations between BMI and health conditions.

In general, studies have consistently shown that having a very low BMI or high BMI is associated with greater health risks, because there are health risks to carrying too little or too much body fat. When researchers have looked at BMI and health in large groups of people, they have often found that the lowest risk of dying younger is in the range of BMI of 20 to 25.4-6 BMI values in the overweight and especially the obesity range have generally been found to be associated with higher rates of type 2 diabetes, cardiovascular disease, stroke, and some types of cancer, as well as other complications of obesity, such as osteoarthritis.6-10

However, which BMI range is associated with the best health outcomes is somewhat controversial, and different studies have come to different conclusions. For example, a meta-analysis published by CDC researchers in 2013 found that a BMI of 25 to 30 (classified as overweight) was associated with the lowest risk of mortality, and there was no difference between normal BMI or class 1 obesity in mortality risk.11 We see a similar pattern in studies of specific health conditions. A 2022 study published in the Journal of the American Heart Association reported that among patients with coronary artery disease followed for 3 to 5 years, the risk of death was lowest in those with BMI of 25 to 35 (overweight to class 1 obesity).12 In another recent study of people with type 2 diabetes, the risk of dying of cardiovascular disease or of having a heart attack or stroke was also lowest in people with BMI classified as overweight or class 1 obesity.13

The finding that overweight or mild obesity based on BMI can be associated with better health outcomes is often called the "obesity paradox," but some scholars argue that it is simply a reflection of our evolving understanding of the complex relationships between body size and health.14 Regardless, it is clear that health risks associated with different body sizes do not always align into neat BMI categories, and simple labels such as "normal" and "overweight" may not be very meaningful.

Limitations of BMI for Individuals

We've discussed how BMI is an imperfect metric of health in population studies, but it has even more limitations when it comes to indicating an individual's body composition or health. For one thing, since BMI is based only on weight and height, it doesn't distinguish between muscle mass and adipose tissue. It's not unusual for muscular athletes to be classified as overweight based on BMI, yet have relatively little body fat. On the flip side, BMI can underestimate body fatness in someone with very low muscle mass, such as a person who is elderly and frail. In addition, BMI can't tell us where body fat is located in the body, and as we'll learn, this is a major factor determining its impact on health. BMI is also not useful for estimating body fatness in a pregnant person, because pregnancy weight gain includes placental and fetal tissues.3

BMI's accuracy in predicting body fatness is also affected by biological sex and race/ethnicity—not surprising given the natural diversity in human bodies. At the same BMI, women tend to carry more body fat than men. Also at the same BMI, a Black person tends to have less body fat, and a person of Asian descent tends to have more body fat, compared to a white person. This means that a high BMI may overestimate health risk in a Black person and underestimate health risk in a person of Asian descent.3

This natural variation means that using BMI as a universal metric to screen for body fatness and health risk can result in misleading conclusions. For example, Black Americans have, on average, higher BMI values and higher rates of obesity (defined by BMI) than white Americans, and this is especially true for women. However, the relationship between high BMI and mortality is weaker in Black Americans; in Black women, having a higher BMI has been found to be associated with a lower death rate.16 Some scholars argue that using a crude metric like BMI in doctors' offices disproportionately harms Black women, who may be more likely to be told that weight is the cause of their health problems without acknowledging the gender- and race-related differences in BMI or the systemic issues — such as racism, poverty, pollution, and stress — that may be more important factors affecting their health.17

All of this means that BMI is not particularly useful for comparing one individual to another or even one population to another. And for an individual, one BMI measurement at a single point in time is not all that meaningful; other measurements and clinical assessments are needed to evaluate disease risk, as we'll discuss in a moment.

BMI is perhaps most useful for tracking changes in body composition over time, whether of a population or an individual. For example, the data on average BMI in the U.S. show a clear increase over the last several decades, and the most likely explanation for this trend is not that people in the U.S. are all gaining several pounds of muscle each year, but that we're carrying more fat. On the other hand, someone may have a BMI classified as overweight, but if they've been at that BMI their entire adult life and are active and metabolically healthy, they may be at their "best weight," and labeling them as "overweight" may be misleading and stigmatizing.

Measuring Body Composition

A person's body mass is made up of water, lean body mass (including organs, bone, and muscle), fat, and other components like minerals. The weight on the scale does not distinguish between these different components, but body composition measurements can.

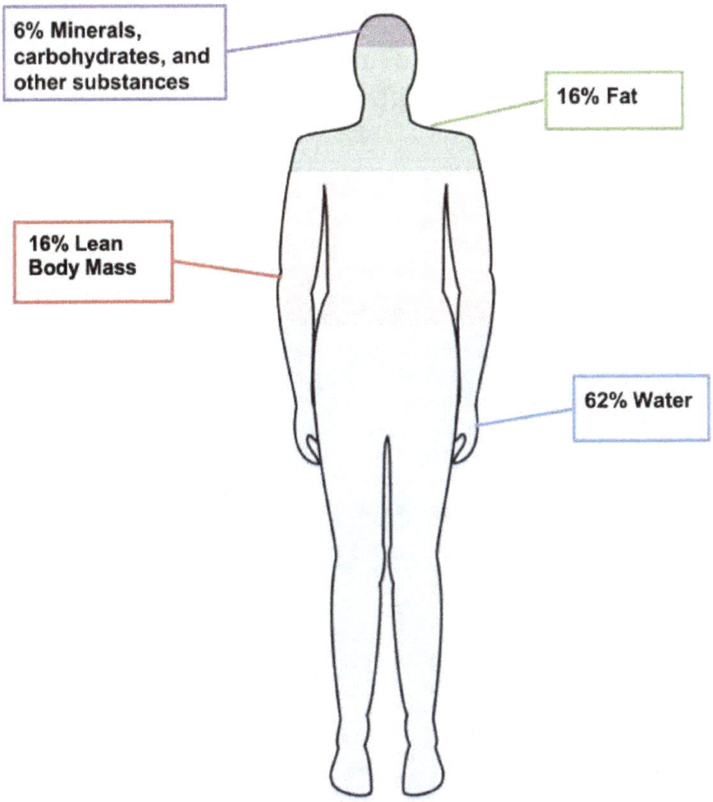

Figure 7.8. The four main components of body weight are water, fat, lean body mass, and other components like minerals.

Body composition measurements are sometimes used by individuals and researchers to determine how much of a person's weight is made up of body fat and lean body mass. For example, an individual might use body composition measurements to track their progress in building muscle with a new strength training program. Since increased body fat can be a risk factor for diseases like cardiovascular disease and diabetes, researchers are often interested in this type of data. There are several different methods used to measure body composition, each with advantages and limitations.[18]

- Skinfold Test. This is a simple, non-invasive, and low-cost way to assess fat mass. Calipers are used to measure the thickness of skin on three to seven different parts of the body, and these numbers are then entered into a conversion equation. The accuracy of the skinfold test depends on the skill of the person taking the measurements, the accuracy of the calipers, and the appropriateness of the conversion equations. Best practice is for the same person to take repeated measurements if using them to monitor changes over time. Repeated measurements by different technicians, using different calipers, and different conversion equations will yield very different results.

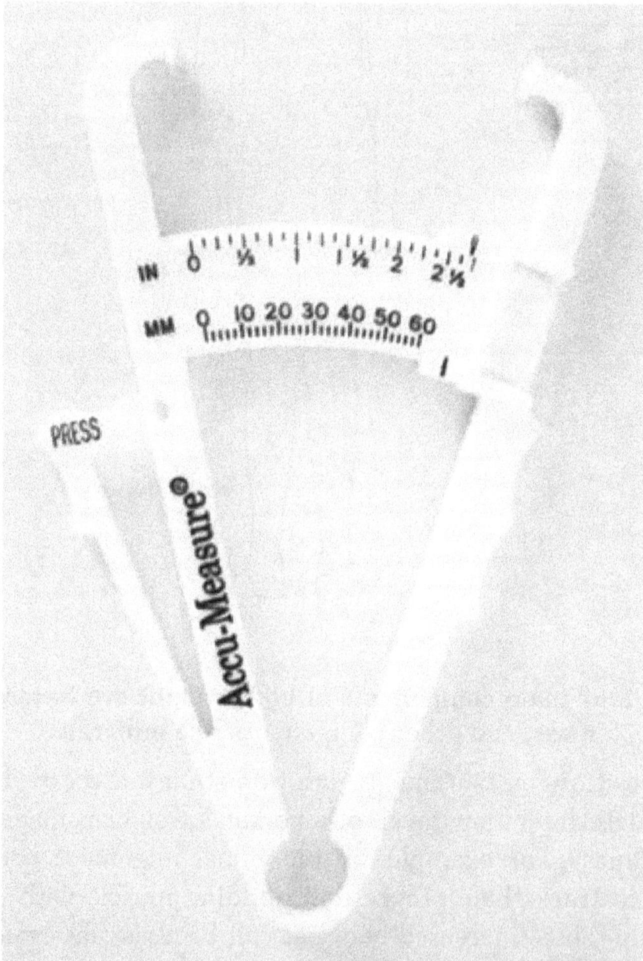

Figure 7.9. Calipers used to assess body fat during skinfold testing.

- Bioelectric Impedance Analysis (BIA). This is a simple, non-invasive, quick tool that does not require extensive training. BIA estimates body composition by sending a small amount of electricity through the body. Since water is a good conductor of electricity, and lean body mass contains more water than fat, the rate at which the current travels can be used to estimate percent body fat. Body fluid levels must be normal with BIA, which is a limitation, since hydration can be impacted by exercise, alcohol, and menstrual cycles.

Figure 7.10. BIA hand device.

- Air Displacement Plethysmography (ADP). This is a non-invasive, quick tool that does not take extensive training. It's more accurate but also more expensive than BIA. While a person sits inside an enclosed chamber (usually called a Bod Pod), changes in air pressure are used to determine the amount of air that is displaced in the test chamber, which can determine body volume. These measurements are then translated into percent body fat. Hydration status can affect the accuracy of this test. This test also needs to be conducted in a facility with a Bod Pod, so it is not as accessible as the skinfold test or the BIA.

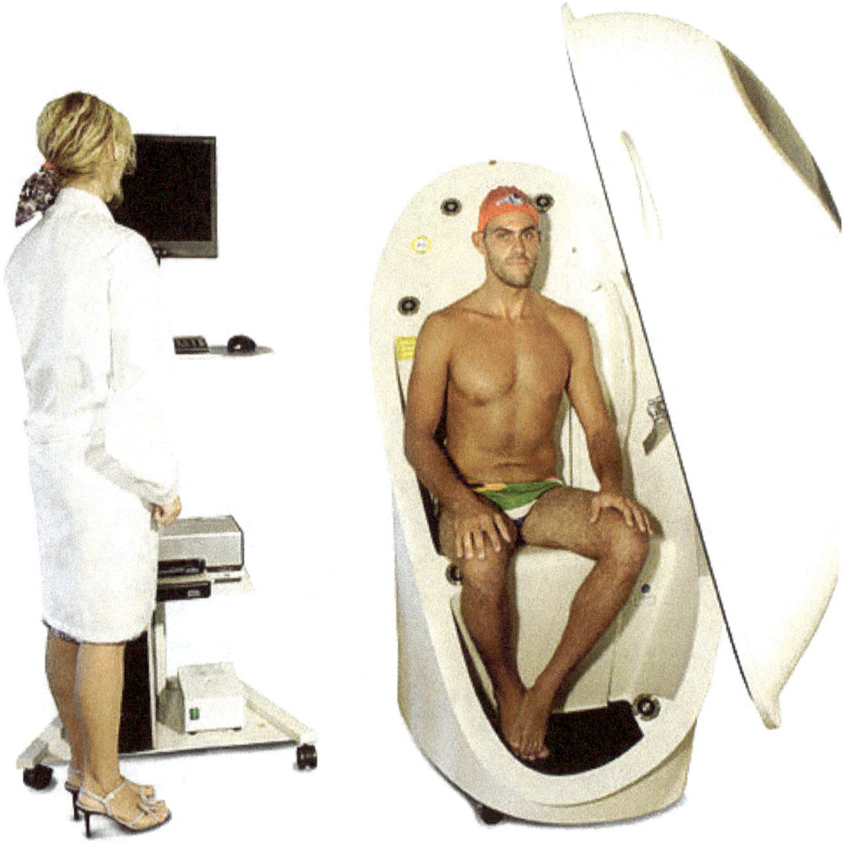

Figure 7.11. Body composition measurement with whole-body air displacement plethysmography (ADP) technology or BodPod

- Dual-energy X-ray Absorptiometry (DXA). This method directs two low-dose X-ray beams through the body and determines the amount of energy absorbed from the beams. The amount of energy absorbed is dependent on the body's content of bone, lean tissue mass, and fat mass. Using standard mathematical formulas, fat content and bone density can be accurately estimated. Although this is one of the most accurate methods of measuring fat mass, it is expensive and mostly used in research. It also requires low doses of radiation to the subject being tested, and is not appropriate for pregnant women.

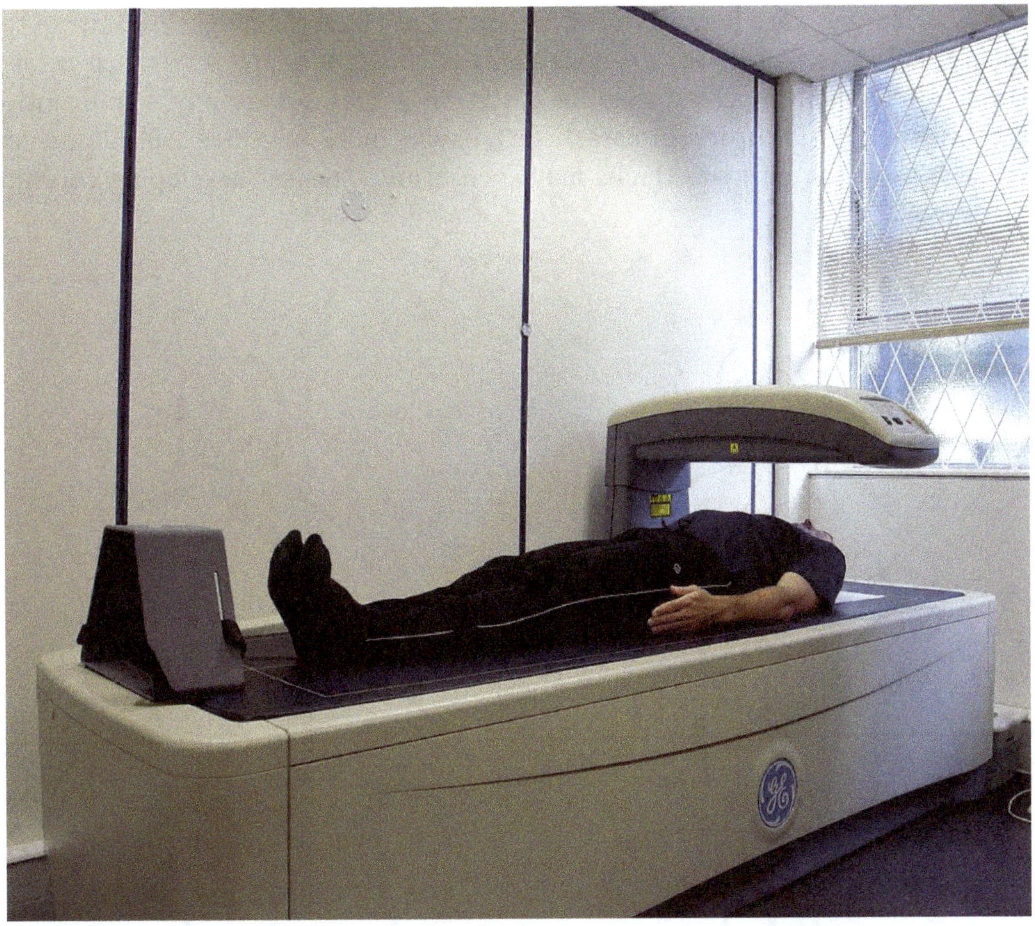

Figure 7.12. Dual-Energy X-ray Absorptiometry (DXA)

Keep in mind that body composition can be hard to measure accurately when using inexpensive and accessible techniques like skinfold testing and BIA. Your best bet is to pick one method and use it over time to compare numbers and see how they change. Just don't get too hung up on the numbers, as their accuracy will be questionable depending on the method chosen. And just as there is no one body weight or BMI that is ideal for everybody, there is no one ideal body composition. Striving to hit a particular goal on the scale or on body composition measurements can lead to an unhealthy fixation on the numbers.

Measuring Fat Distribution

How fat is distributed in the body is important, because people who store fat more centrally (apple-shaped) have a higher risk of chronic diseases like cardiovascular disease and type 2 diabetes compared to people who store fat in the hips, thighs, and buttocks (pear-shaped). This is because visceral fat that surrounds vital organs (common in central obesity or apple-shaped fat patterning) is more metabolically active, releasing more hormones and inflammatory factors thought to contribute to disease risk compared to subcutaneous fat. Subcutaneous fat stored just below the skin (common in pear-shaped fat patterning) does not seem to significantly increase the risk for chronic disease. In fact, having more lower body subcutaneous fat such as in the hips and thighs may even protect against developing chronic diseases.[19]

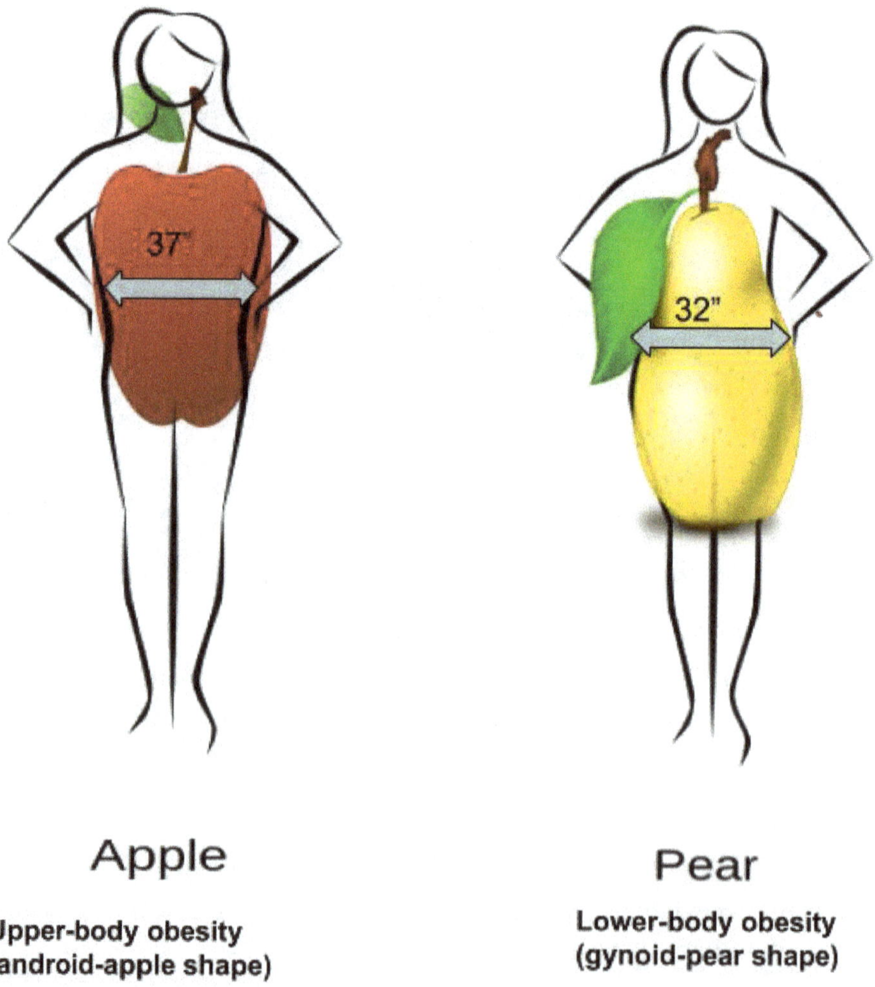

Figure 7.13. Fat can be located in the abdominal region (apple shape) or hips, thighs, and buttocks (pear shape).

Body fat distribution can be measured by waist circumference and waist-to-hip ratio, both of which only require a measuring tape.

- Waist circumference is measured just above the hip bone, level with your belly button. Men with a waist circumference greater than 40 inches and women with a waist circumference greater than 35 inches are predicted to face greater health risks.
- Waist-to-hip ratio is calculated by measuring waist circumference and hip circumference (at its widest part) and dividing the two values. Abdominal obesity is defined by the World Health Organization as a waist-to-hip ratio above 0.90 for men and above 0.85 for women.

Indicators Of Metabolic Health

Metabolic health refers to the body's ability to maintain normal homeostasis and effectively regulate measures like blood pressure, blood lipids, and blood glucose. It is often assumed that when an individual is classified as overweight and obese based on measurements of BMI, body composition, and fat distribution, metabolic health is negatively impacted. However, some individuals who meet the criteria for obesity do not experience an increased risk of metabolic health.[20] These individuals are known as metabolically healthy obese (MHO). Metabolically healthy can be described as an absence of insulin resistance, type 2 diabetes, dyslipidemia, and hypertension.

Along the same lines, there are also individuals who are classified as normal weight (BMI < 25) but show an increased metabolic risk. These individuals are known as metabolically obese normal weight (MONW).

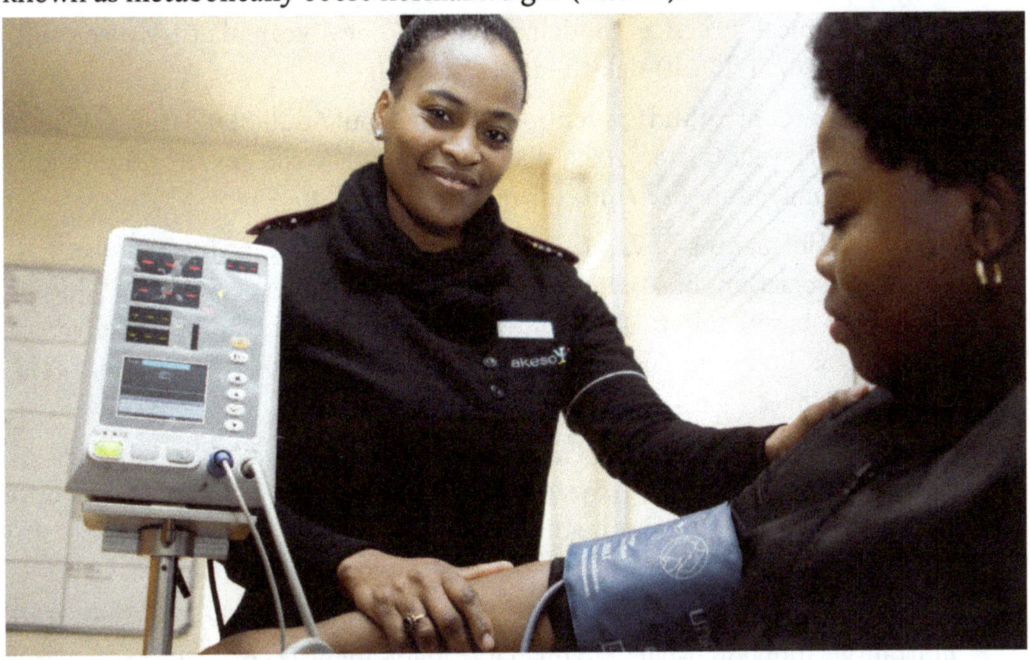

In a 2016 study, researchers analyzed BMI and measures of metabolic health (blood pressure, triglyceride and cholesterol levels, blood glucose, insulin resistance, and a marker of inflammation) for more than 40,000 Americans. Among the group classified as normal weight based on BMI, 69% were metabolically healthy based on measures of metabolic health, and 31% were metabolically unhealthy. In the higher BMI categories, the proportion of metabolically healthy people declined, but a significant number of individuals with larger body sizes had normal values for measures of metabolic health; 47% of people classified as overweight, 29% of those classified as having obesity, and 16% of those classified as having class 2 or 3 obesity were metabolically healthy.[21] In other words, a person's body size is not an accurate indicator of their metabolic health, and it's important to measure and track more direct indicators of health and disease risk such as blood pressure, blood glucose, and blood lipid levels.

These variations may challenge the assumptions we hold about body fatness. It can not be assumed that thinness equals health or that fatness indicates a lack of health.

REFERENCES

Manore, M.M., Meyer, N.L., & Thompson, J.L. (2009). Sport Nutrition for Health and Performance, 2nd ed. Champaign, IL: Human Kinetics.

Obesity Canada and the Canadian Association of Bariatric Physicians and Surgeons (2020). Clinical Recommendations: Quick Guide; Canadian Adult Obesity Clinical Practice Guidelines. http://obesitycanada.ca/wp-content/uploads/2020/11/CPG-Quick-Guide-English.pdf

Centers for Disease Control and Prevention. (2021). About Adult BMI | Healthy Weight | CDC. RetrievedMay 6, 2022, from https://www.cdc.gov/healthyweight/assessing/bmi/adult_bmi/index.html

Bhaskaran, K., dos-Santos-Silva, I., Leon, D. A., Douglas, I. J., & Smeeth, L. (2018). Association of BMI with overall and cause-specific mortality: A population-based cohort study of 3·6 million adults in the UK. The Lancet. Diabetes & Endocrinology, 6(12), 944–953. https://doi.org/10.1016/S2213-8587(18)30288-2

Aune, D., Sen, A., Prasad, M., Norat, T., Janszky, I., Tonstad, S., ... Vatten, L. J. (2016). BMI and all cause mortality: Systematic review and non-linear dose-response meta-analysis of 230 cohort studies with 3.74 million deaths among 30.3 million participants. The BMJ, 353. https://doi.org/10.1136/bmj.i2156

The Global BMI Mortality Collaboration. (2016). Body-mass index and all-cause mortality: Individual-participant-data meta-analysis of 239 prospective studies in four continents. Lancet, 388(10046), 776–786.

7Jayedi, A., Soltani, S., Motlagh, S. Z., Emadi, A., Shahinfar, H., Moosavi, H., & Shab-Bidar, S. (2022). Anthropometric and adiposity indicators and risk of type 2 diabetes: systematic review and dose-response meta-analysis of cohort studies. BMJ (Clinical research ed.), 376, e067516. https://doi.org/10.1136/bmj-2021-067516

Kim, M. S., Kim, W. J., Khera, A. V., Kim, J. Y., Yon, D. K., Lee, S. W., Shin, J. I., & Won, H. H. (2021). Association between adiposity and cardiovascular outcomes: an umbrella review and meta-analysis of observational and Mendelian randomization studies. European heart journal, 42(34), 3388–3403. https://doi.org/10.1093/eurheartj/ehab454

Liu X, Zhang D, Liu Y, Sun X, Hou Y, Wang B, Ren Y, Zhao Y, Han C, Cheng C, Liu F, Shi Y, Chen X, Liu L, Chen G, Hong S, Zhang M, Hu D. A J-shaped relation of BMI and stroke: Systematic review and dose-response meta-analysis of 4.43 million participants. Nutr Metab Cardiovasc Dis. 2018 Nov;28(11):1092-1099. doi: 10.1016/j.numecd.2018.07.004.

Zheng, H., & Chen, C. (2015). Body mass index and risk of knee osteoarthritis: systematic review and meta-analysis of prospective studies. BMJ open, 5(12), e007568. https://doi.org/10.1136/bmjopen-2014-007568

Flegal K.M., Kit B.K., Orpana H., & Graubard B.I. (2013) Association of All-Cause Mortality With Overweight and Obesity Using Standard Body Mass Index Categories: A Systematic Review and Meta-analysis. JAMA, 309(1):71–82. doi:10.1001/jama.2012.113905

White, H. D., & Wallentin, L. (2022). Body Mass Index and Association With Cardiovascular Outcomes in Patients With Stable Coronary Heart Disease – A STABILITY Substudy. Journal of the American Heart Association, 11(3), e023667. https://doi.org/10.1161/JAHA.121.023667

Pagidipati, N. J., Zheng, Y., Green, J. B., McGuire, D. K., Mentz, R. J., Shah, S., Aschner, P., Delibasi, T., Rodbard, H. W., Westerhout, C. M., Holman, R. R., Peterson, E. D., & TECOS Study Group (2020). Association of obesity with cardiovascular outcomes in patients with type 2 diabetes and cardiovascular disease: Insights from TECOS. American Heart Journal, 219, 47–57. https://doi.org/10.1016/j.ahj.2019.09.016

Harper, K. (2021, November 18). The weight game: How body-size bias can hold back health science. Scientific American. https://www.scientificamerican.com/article/the-weight-game-how-body-size-bias-can-hold-back-health-science/

Centers for Disease Control and Prevention (2021). Adult Obesity Facts. Retrieved May 6, 2022 from https://www.cdc.gov/obesity/data/adult.html.

Jackson, C. L., Wang, N. Y., Yeh, H. C., Szklo, M., Dray-Spira, R., & Brancati, F. L. (2014). Body-mass index and mortality risk in U.S. blacks compared to whites. Obesity (Silver Spring, Md.), 22(3), 842–851. https://doi.org/10.1002/oby.20471

Strings, S. & L. Bacon (2020). The racist roots of fighting obesity. Scientific American. https://www.scientificamerican.com/article/the-racist-roots-of-fighting-obesity2/

Lemos, T., & Gallagher, D.(2017). Current body composition measurement techniques. Current Opinion in Endocrinology & Diabetes and Obesity, 24(5), 310–314. doi: 10.1097/MED.0000000000000360

Vasan, S. K., Osmond, C., Canoy, D., Christodoulides, C., Neville, M. J., Di Gravio, C., Fall, C., & Karpe, F. (2018). Comparison of regional fat measurements by dual-energy X-ray absorptiometry and conventional anthropometry and their association with markers of diabetes and cardiovascular disease risk. International journal of obesity (2005), 42(4), 850–857. https://doi.org/10.1038/ijo.2017.289

Matthew, H., Farr, O.M., & Mantzoros, S.C. (2016). Metabolic Health and Weight: Understanding metabolically unhealthy normal weight or metabolically healthy obese patients. Metabolism, 65(1), 73-80. doi: 10.1016/j.metabol.2015.10.019

Tomiyama, A. J., Hunger, J. M., Nguyen-Cuu, J., & Wells, C. (2016). Misclassification of cardiometabolic health when using body mass index categories in NHANES 2005-2012. International journal of obesity (2005), 40(5), 883–886. https://doi.org/10.1038/ijo.2016.17

BODY FAT RISKS OF TOO MUCH AND TOO LITTLE

As the previous section illustrated, most measures of body composition are imperfect estimates, and they are often not accurate indicators of health. However, in general, carrying too little and too much body fat are both correlated with greater health issues. We see this pattern in studies of mortality risk and BMI, which usually report a J-shaped curve indicating a greater risk of dying sooner at low and high BMI. The lowest average mortality risk is usually found somewhere between BMI values of about 20 and 30, corresponding to normal and overweight BMI categories. Which BMI category is correlated to the lowest mortality risk is controversial, and different studies have reached different conclusions depending on the population studied and the statistical methods used. Most studies find the lowest mortality rate in the normal range with little to no increase in mortality in the overweight range for BMI, and some find that the overweight group has the lowest mortality rate.

As an example, the results of a 2018 study of 3.6 million adults in the UK are shown below.[1] You can see that the relationship between BMI and mortality rate varies by biological sex and age group, but the general pattern holds.

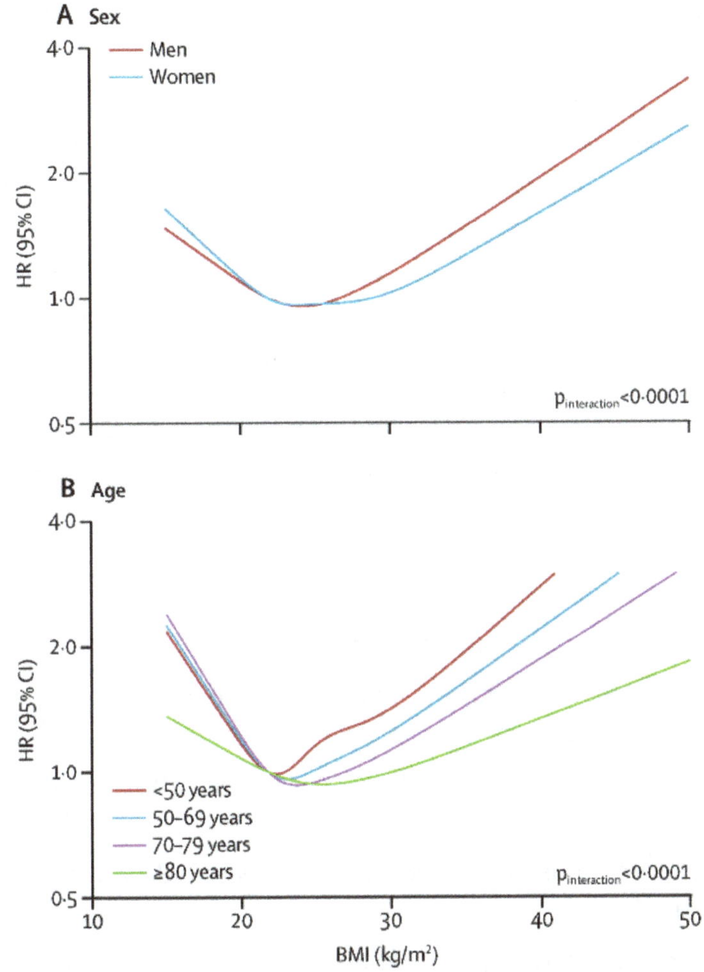

Fig. 7.14. The relationship between BMI and mortality risk (higher HR on the y-axis indicates higher risk of dying during the study) forms a J-shaped curve, which can vary depending on factors like sex and age. This example is from Bhaskaran et al. (2018) and illustrates the typical pattern seen in similar studies, although the exact placement of the curve along the BMI scale varies between studies.

What are the specific risks associated with carrying too little or too much body fat? Let's take a closer look at each of these situations.

Health Risks of Being Underweight

The 2017–2018 National Health and Nutrition Examination Survey (NHANES) estimated that 1.6 percent of adults and 4.1 percent of children and adolescents in the United States are underweight.[7,8] Underweight individuals represent a small portion of Americans, yet the health risks associated with being underweight are an important part of the discussion on nutrition and health. Globally, the World Health Organization estimated that in 2014, approximately 462 million adults were underweight.[9]

Keep in mind that BMI is used to classify underweight, and just as it's a crude measure for estimating health risks associated with higher body fat, the same can be true for underweight. Some people are genetically predisposed to be lean, and they may be at their "best weight" even if their BMI classifies them as underweight.

Adipose tissue plays important roles in the body. It is our largest form of energy storage, helping us to survive during periods of low food supply and storing extra energy during periods of abundance. Adipose also provides insulation and padding, and it produces hormones and other signals that regulate metabolism and support immune function. It is critical for reproductive function, including normal secretion of sex and reproductive hormones, pubertal development, pregnancy, and lactation.[10]

Being underweight has been shown to be associated with an increased risk of several health conditions:[11-16]

- Osteoporosis and low bone mineral density
- Increased risk of infection, delayed wound healing, and greater post-surgical complications
- Cardiovascular disease, including stroke, heart attack, and coronary heart disease
- Some cancers, plus poorer response to treatment and survival rates after diagnosis
- Irregular menstrual cycles, decreased fertility, and poorer pregnancy outcomes
- Depression in women
- Decreased semen quality in men
- Stunted growth in children

The most common underlying cause of underweight in the U.S. is inadequate nutrition, which may occur because of lack of access to enough food; difficulty eating or digesting food, which is common in older adults; or disordered eating. Note that eating disorders do not always cause underweight and can be present in people of all body sizes. Eating disorders are covered in greater depth on the next page. Other causes of underweight include wasting diseases like cancer, multiple sclerosis, and tuberculosis. People with wasting diseases are encouraged to seek nutritional counseling, as a healthy diet greatly affects survival and improves responses to disease treatments.

Health Risks of Excess Body Weight

In 2017-2018, it was estimated that 42 percent of adults and 19 percent of children in the United States had obesity. Globally, the prevalence of obesity was estimated to be 13 percent of adults and 7 percent of children and adolescents in 2016. According to the World Health Organization (WHO), overweight and obesity are linked to more deaths worldwide than underweight.

Carrying too much body fat can affect health through several mechanisms. Fat cells are not lifeless storage tanks—they're dynamic, metabolically-active tissue that secrete a number of different hormones and hormone-like messengers, causing low-grade inflammation that's believed to contribute to chronic disease development such as type 2 diabetes, cardiovascular disease, and some types of cancer. Carrying more body weight can also contribute to joint problems and difficulty with mobility.

According to the Centers for Disease Control and Prevention, obesity is associated with an increased risk of the following health conditions[21]:

- High blood pressure and high cholesterol, which are risk factors for heart disease
- Type 2 diabetes
- Breathing problems such as asthma and sleep apnea
- Joint problems such as osteoarthritis and musculoskeletal discomfort
- Gallstones and gallbladder disease
- Stroke
- Many types of cancer
- Depression and anxiety

Children with obesity face many of these same health risks, and they are more likely to have obesity as adults. Obesity in children is also associated with lower self-esteem and facing greater social challenges such as bullying and stigma.

As we discussed on the previous page, a significant number of people living in larger bodies—who might be classified as overweight or obese based on BMI—are metabolically healthy.[22] Health issues may be more common in people carrying more fat, but many other factors are at play, including genetics, the location of body fat (subcutaneous or visceral), diet quality, physical activity, socioeconomic factors, and environmental factors. Most of the studies on obesity and health conditions are observational, and as we covered in Unit 2, observational research can only show correlations, not causation. And with so many overlapping factors involved, it can be difficult to tease apart correlation and causation. For example, a person with a family history and genetic predisposition for diabetes or heart disease might also have a genetic predisposition for higher body weight, so it is difficult to say whether or how much higher body weight contributed to the development of disease. Likewise, as we'll discuss later in the unit, factors such as access to healthcare and education, food security, poverty, neighborhood safety, housing stability, stress, and racism can affect both body weight and health. These factors are called ***social determinants of health*** and are defined by the CDC as "conditions in the places where people live, learn, work, and play that affect a wide range of health risks and outcomes." If a person living in a larger body has health problems, we shouldn't assume that those problems are caused by their weight, as there could be numerous contributing factors. In addition, many of the health problems associated with obesity can be mitigated by lifestyle changes such as increased physical activity or improved diet quality, so just as excess weight may not be the sole cause of associated health issues, weight loss is also not the only answer.

To further add to the complexity of this discussion, studies of the relationship between weight and health rarely account for the effects of weight stigma. As we'll discuss in more detail below, research has shown that experiencing weight stigma may contribute to both weight gain and poorer health outcomes. In this sense, some of the health problems associated with higher body weight may not be because of the weight itself but because of how others treat people with larger bodies.

All of that said, recent studies have used innovative study designs that incorporate genetic data to better understand the complex relationship between body weight and health conditions. These studies generally find that, although the relationship is neither simple nor the effects universal, excess weight does contribute to increased risk of many health conditions.

Effects of Weight Bias and Stigma

According to the 2020 Joint International Consensus Statement for Ending Stigma of Obesity, "***weight stigma*** refers to social devaluation and denigration of individuals because of their excess body weight, and can lead to negative attitudes, stereotypes, prejudice, and discrimination." The statement, which was authored by 36 experts from around the world, is a call to action for healthcare professionals, policymakers, and the public to end weight stigma.

It begins:

- "Individuals with obesity face not only increased risk of serious medical complications but also a pervasive, resilient form of social stigma.

Often perceived (without evidence) as lazy, gluttonous, lacking will power and self-discipline, individuals with overweight or obesity are vulnerable to stigma and discrimination in the workplace, education, healthcare settings, and society in general. Weight stigma can cause considerable harm to affected individuals, including physical and psychological consequences. The damaging impact of weight stigma, however, extends beyond harm to individuals. The prevailing view that obesity is a choice and that it can be entirely reversed by voluntary decisions to eat less and exercise more can exert negative influences on public health policies, access to treatments, and research."

- "About 40% of the general population reports that it has experienced some type of weight stigma—whether it be weight-based teasing, unfair treatment, or discrimination," said Rebecca Puhl, PhD, deputy director of the Rudd Center for Food Policy & Obesity at the University of Connecticut, in Today's Dietitian.[31] Weight stigma can be experienced in nearly every aspect of life, including the workplace, healthcare, social settings, and even within a person's own family.

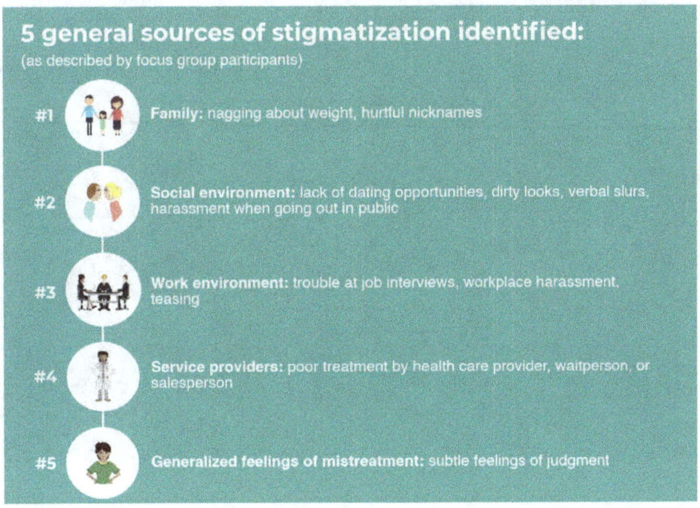

Fig. 7.15. Common sources of weight stigmatization identified by individuals who are overweight or obese include families, social environments, work environments, and service providers, as well as generalized feelings of judgment from others. Source: Cossrow, N. H., Jeffery, R. W., & McGuire, M. T. (2001). Understanding weight stigmatization: A focus group study. Journal of nutrition education, 33(4), 208-214.

Experiencing weight stigma has been associated with many negative physical and mental health outcomes[30]:

- Lower self-esteem and greater levels of depression, anxiety, social isolation, stress, and substance use
- Unhealthy eating and weight control behaviors, such as binge eating
- Exercise avoidance and lower levels of physical activity
- Greater weight gain over time
- Higher levels of inflammatory markers and stress hormones
- Greater long-term cardiometabolic risk and increased mortality
- Lower quality medical care, poorer treatment outcomes, and avoidance of medical care

It is important to note that these effects have been shown to be associated with experiencing weight stigma, above and beyond the consequences of excess body weight. And as noted above, this research points to weight stigma as one of the factors that contributes to poorer health among people with larger bodies.

Combatting weight stigma and discrimination will require change on many levels. Governments can include weight as a category covered in anti-discrimination laws. Anti-bullying efforts at the school level should include policies on harassment and bullying related to weight. Healthcare should include reimbursement for obesity treatment, and weight bias training should be required for healthcare providers. Providers should ensure that their facilities are comfortable and welcoming to people of all sizes, and that they offer evidence-based care and treatment options to all patients, regardless of their size. And education and awareness about weight bias on an individual level will help change negative attitudes toward overweight and obesity.[30,32]

REFERENCES:

Bhaskaran, K., dos-Santos-Silva, I., Leon, D. A., Douglas, I. J., & Smeeth, L. (2018). Association of BMI with overall and cause-specific mortality: A population-based cohort study of 3·6 million adults in the UK. Lancet. Diabetes & Endocrinology, 6(12), 944–953. https://doi.org/10.1016/S2213-8587(18)30288-2

Aune, D., Sen, A., Prasad, M., Norat, T., Janszky, I., Tonstad, S., ... Vatten, L. J. (2016). BMI and all cause mortality: Systematic review and non-linear dose-response meta-analysis of 230 cohort studies with 3.74 million deaths among 30.3 million participants. The BMJ, 353. https://doi.org/10.1136/bmj.i2156

The Global BMI Mortality Collaboration. (2016). Body-mass index and all-cause mortality: Individual-participant-data meta-analysis of 239 prospective studies in four continents. Lancet, 388(10046), 776–786. https://doi.org/10.1016/S0140-6736(16)30175-1

Flegal K.M., Kit B.K., Orpana H., & Graubard B.I. (2013) Association of All-Cause Mortality With Overweight and Obesity Using Standard Body Mass Index Categories: A Systematic Review and Meta-analysis. JAMA, 309(1):71–82. doi:10.1001/jama.2012.113905

Romero-Corral, A., Montori, V. M., Somers, V. K., Korinek, J., Thomas, R. J., Allison, T. G., Mookadam, F., & Lopez-Jimenez, F. (2006). Association of bodyweight with total mortality and with cardiovascular events in coronary artery disease: a systematic review of cohort studies. Lancet (London, England), 368(9536), 666–678. https://doi.org/10.1016/S0140-6736(06)69251-9

Orpana, H. M., Berthelot, J. M., Kaplan, M. S., Feeny, D. H., McFarland, B., & Ross, N. A. (2010). BMI and mortality: results from a national longitudinal study of Canadian adults. Obesity (Silver Spring, Md.), 18(1), 214–218. https://doi.org/10.1038/oby.2009.191

Centers for Disease Control and Prevention. (2020, December 16). Prevalence of underweight among adults aged 20 and over: United States, 1960–1962 through 2017–2018. Centers for Disease Control and Prevention. Retrieved May 13, 2022, from https://www.cdc.gov/nchs/data/hestat/underweight-adult-17-18/underweight-adult.htm

Centers for Disease Control and Prevention. (2020, December 16). Prevalence of underweight among children and adolescents aged 2–19 years: United States, 1963–1965 through 2017–2018. Centers for Disease Control and Prevention. Retrieved May 13, 2022, from https://www.cdc.gov/nchs/data/hestat/underweight-child-17-18/underweight-child.htm

World Health Organization. (n.d.). Fact sheets – malnutrition. World Health Organization. Retrieved May 13, 2022, from https://www.who.int/news-room/fact-sheets/detail/malnutrition

Richard AJ, White U, Elks CM, et al. Adipose Tissue: Physiology to Metabolic Dysfunction. [Updated 2020 Apr 4]. In: Feingold KR, Anawalt B, Boyce A, et al., editors. Endotext [Internet]. South Dartmouth (MA): MDText.com, Inc. Available from: https://www.ncbi.nlm.nih.gov/books/NBK555602/

Golubnitschaja, O., Liskova, A., Koklesova, L., et al. (2021). Caution, "normal" BMI: health risks associated with potentially masked individual underweight-EPMA Position Paper 2021. The EPMA journal, 12(3), 243–264. https://doi.org/10.1007/s13167-021-00251-4

Dobner, J., & Kaser, S. (2018). Body mass index and the risk of infection – from underweight to obesity. Clinical microbiology and infection : the official publication of the European Society of Clinical Microbiology and Infectious Diseases, 24(1), 24–28. https://doi.org/10.1016/j.cmi.2017.02.013

Park, D., Lee, J.H., & Han, S. (2017). Underweight: another risk factor for cardiovascular disease?: A cross-sectional 2013 Behavioral Risk Factor Surveillance System (BRFSS) study of 491,773 individuals in the USA. Medicine, 96(48), e8769. https://doi.org/10.1097/MD.0000000000008769

Underweight. Office on Women's Health. (2021). Retrieved May 13, 2022, from https://www.womenshealth.gov/healthy-weight/underweight

Han, Z., Mulla, S., Beyene, J., Liao, G., McDonald, S. D., & Knowledge Synthesis Group (2011). Maternal underweight and the risk of preterm birth and low birth weight: a systematic review and meta-analyses. International journal of epidemiology, 40(1), 65–101. https://doi.org/10.1093/ije/dyq195

Guo, D., Xu, M., Zhou, Q., Wu, C., Ju, R., & Dai, J. (2019). Is low body mass index a risk factor for semen quality? A PRISMA-compliant meta-analysis. Medicine, 98(32), e16677. https://doi.org/10.1097/MD.0000000000016677

Centers for Disease Control and Prevention. (2021, September 30). Adult obesity facts. Centers for Disease Control and Prevention. Retrieved May 13, 2022, from https://www.cdc.gov/obesity/data/adult.html

Centers for Disease Control and Prevention. (2021, April 5). Childhood obesity facts. Centers for Disease Control and Prevention. Retrieved May 13, 2022, from https://www.cdc.gov/obesity/data/childhood.html

World Health Organization. (n.d.). Obesity and overweight. World Health Organization. Retrieved May 13, 2022, from https://www.who.int/news-room/fact-sheets/detail/obesity-and-overweight

Harvey, I., Boudreau, A., & Stephens, J. M. (2020). Adipose tissue in health and disease. Open biology, 10(12), 200291. https://doi.org/10.1098/rsob.200291

Centers for Disease Control and Prevention. (n.d.). Consequences of obesity. Centers for Disease Control and Prevention. Retrieved May 13, 2022, from https://www.cdc.gov/obesity/basics/consequences.html

Zembic, A., Eckel, N., Stefan, N., Baudry, J., & Schulze, M. B. (2021). An Empirically Derived Definition of Metabolically Healthy Obesity Based on Risk of Cardiovascular and Total Mortality. JAMA Network Open, 4(5), e218505. https://doi.org/10.1001/jamanetworkopen.2021.8505

Nuttall F. Q. (2015). Body Mass Index: Obesity, BMI, and Health: A Critical Review. Nutrition Today, 50(3), 117–128. https://doi.org/10.1097/NT.0000000000000092

24Koolhaas, C. M., Dhana, K., Schoufour, J. D., Ikram, M. A., Kavousi, M., & Franco, O. H. (2017). Impact of physical activity on the association of overweight and obesity with cardiovascular disease: The Rotterdam Study. European Journal of Preventive Cardiology, 24(9), 934–941. https://doi.org/10.1177/2047487317693952

Puhl, R. M., Himmelstein, M. S., & Pearl, R. L. (2020). Weight stigma as a psychosocial contributor to obesity. The American Psychologist, 75(2), 274–289. https://doi.org/10.1037/amp0000538

Tomiyama, A. J., Carr, D., Granberg, E. M., Major, B., Robinson, E., Sutin, A. R., & Brewis, A. (2018). How and why weight stigma drives the obesity 'epidemic' and harms health. BMC Medicine, 16(1), 123. https://doi.org/10.1186/s12916-018-1116-5

Kim, M. S., Kim, W. J., Khera, A. V., Kim, J. Y., Yon, D. K., Lee, S. W., Shin, J. I., & Won, H. H. (2021). Association between adiposity and cardiovascular outcomes: an umbrella review and meta-analysis of observational and Mendelian randomization studies. European Heart Journal, 42(34), 3388–3403. https://doi.org/10.1093/eurheartj/ehab454

Corbin, L. J., & Timpson, N. J. (2016). Body mass index: Has epidemiology started to break down causal contributions to health and disease?. Obesity (Silver Spring, Md.), 24(8), 1630–1638. https://doi.org/10.1002/oby.21554

Franks, P. W., & Atabaki-Pasdar, N. (2017). Causal inference in obesity research. Journal of internal medicine, 281(3), 222–232. https://doi.org/10.1111/joim.12577

30Rubino, F., Puhl, R. M., Cummings, D. E., Eckel, R. H., Ryan, D. H., Mechanick, J. I., Nadglowski, J., Ramos Salas, X., Schauer, P. R., Twenefour, D., Apovian,

C. M., Aronne, L. J., Batterham, R. L., Berthoud, H. R., Boza, C., Busetto, L., Dicker, D., De Groot, M., Eisenberg, D., Flint, S. W., ... Dixon, J. B. (2020). Joint international consensus statement for ending stigma of obesity. Nature Medicine, 26(4), 485–497. https://doi.org/10.1038/s41591-020-0803-x

Dennett, C. The Health Impact of Weight Stigma. Today's Dietitian Vol. 20, No. 1, P. 24. Retrieved November 7, 2019, from https://www.todaysdietitian.com/newarchives/0118p24.shtml.

Understanding obesity stigma. (2019). Obesity Action Coalition. Retrieved November 7, 2019, from https://www.obesityaction.org/get-educated/public-resources/brochures-guides/understanding-obesity-stigma-brochure/.

DISORDERS OF EATING

In any conversation about weight and health, it is important to understand the risk and prevalence of eating disorders. The National Institute of Mental Health (NIMH) defines eating disorders as "serious and sometimes fatal illnesses that cause severe disturbances to a person's eating behaviors."[1] People with eating disorders often experience a preoccupation with food choices and body weight, and they frequently have a distorted body image, believing that self-worth is tied to body size and shape.[2] Eating disorders are serious and complex health conditions, and their prevention, proper diagnosis, and treatment must involve a multi-faceted approach, including physical, emotional, and social issues related to each individual's needs.[3]

The short video below summarizes the main types of eating disorders, and you'll find sections with more details on each type further down this page.

Researchers have estimated that about 9 percent of people living in the United States, or 28.8 million Americans, will have an eating disorder in their lifetime.[4] The prevalence of eating disorders also seems to be increasing.[5] Eating disorders can be accompanied by many health complications, and among all mental illnesses, eating disorders have one of the highest fatality rates. Anorexia nervosa in particular can be deadly; it has the second highest mortality rate of mental illnesses, outranked only by substance use disorder.[6] Eating disorders cause approximately 10,200 deaths each year, or one death every 52 minutes in the United States.[4] In a large study of U.S. college students published in 2020, those with the greatest level of eating disorder symptoms were 11 times more likely to have attempted suicide compared with those who did not have an eating disorder.[7]

Historically, there has been a stereotype of eating disorders typically affecting young, white, straight women and resulting in a visibly underweight body. But this is a false stereotype. Eating disorders can affect anyone, regardless of their background, body size, gender, sexuality, race, ethnicity, or age. This is an important point because misconceptions about who is affected by eating disorders have resulted in these illnesses being underrecognized, undiagnosed, and untreated in people who don't fit the stereotype.[8]

Risk factors for eating disorders are multifactorial. The National Eating Disorders Foundation lists the following risk factors for eating disorders[9]:

- Biological Factors
- Having a close relative with an eating disorder or other mental health condition
- History of dieting or negative energy balance
- Having type 1 diabetes
- Psychological Factors
- Perfectionism
- Body image dissatisfaction
- Personal history of an anxiety disorder
- Behavioral inflexibility (always following the rules or feeling there is one "right way" to do things)
- Social Factors
- Experiencing weight stigma
- Experiencing teasing or bullying
- Believing there is an "ideal body"
- Having a limited social network
- Acculturation, such as when people of different ethnicities are introduced to Western media that emphasizes diet culture and an ideal body shape
- Having experienced historical or intergenerational trauma, such as Holocaust survivors and Indigenous groups that experienced colonization

It is especially important to note how dieting and weight stigma are associated with the risk of developing an eating disorder. For example, in one U.S. survey, adults who reported experiencing weight stigma were more likely to have disordered eating, regardless of BMI.[10] In another study, -14year-old girls who were told by a family member or another person in their life that they were "too fat" were, at age 19, more likely to be dissatisfied with their body and to be using unhealthy weight control behaviors, such as vomiting, taking diet pills, and using laxatives. These effects were seen regardless of BMI, meaning that these unhealthy outcomes also occurred in girls with a normal weight BMI who were labeled by others as "too fat."[11] In one cohort of teens followed for three years, those who dieted were five times more likely to develop eating disorders, and those who were on severely restrictive diets were 18 times more likely to develop eating disorders.[12] It's also worth noting that dieting in young people is actually linked to greater weight gain over time, not less. For these reasons, the American Academy of Pediatrics discourages dieting in young people and recommends that families concerned about a child's body weight focus on adopting healthy and sustainable eating and physical activity patterns rather than trying to change weight.[13] When we talk about weight management interventions or public health policies, it's important to be mindful of the risks of overly restrictive diets or practices that stigmatize people based on weight.

With proper diagnosis and treatment, people can recover completely from eating disorders. Treatment can involve individual, group, and/or family psychotherapy; medical care and monitoring; nutrition counseling; and medications such as antidepressants and mood stabilizers. A registered dietitian is often a vital part of the treatment team.[1]

ANOREXIA NERVOSA

Anorexia nervosa, often referred to as simply "anorexia," is an eating disorder characterized by severe restriction of the amount and often types of foods eaten due to an obsession about body weight, fear of gaining weight, and/or a distorted body image. Anorexia results in extreme nutrient inadequacy, weight loss and/or failure to grow normally, and eventually organ malfunction. A 2019 review of studies estimated that 1.4 percent of women and 0.2 percent of men will have anorexia at some point in their lifetime.[14] Anorexia often begins during adolescence, but it can emerge much later in adulthood. In addition to severe food restriction, some people with anorexia also may exercise compulsively and have binge-eating and purging episodes, during which they eat large amounts of food and then purge it by vomiting or using laxatives or diuretics.[1] Though some people with anorexia are noticeably underweight, the disorder can also affect people in larger bodies, who may be less likely to be diagnosed because of their size.[15]

Anorexia can affect mental and physical health in many ways. It can cause excessive weight loss, osteoporosis, anemia, muscle wasting and weakness, brittle hair and nails, skin abnormalities, growth of fine hair all over the body, severe constipation, low blood pressure, slowed pulse and breathing rate, low body temperature, fatigue, infertility, menstrual irregularities or loss of periods, heart and brain damage, and organ failure.[1] The goals of treatment for anorexia are to restore a healthy body weight and significantly reduce the weight obsession and behaviors associated with the eating disorder.

BULIMIA

Bulimia nervosa is characterized by episodes of eating large amounts of food followed by purging, which is accomplished by vomiting or the use of laxatives and diuretics. People with bulimia often feel a lack of control over these episodes. The prevalence of bulimia is estimated to be 1.9 percent in women and 0.6 percent in men.[14] People with bulimia typically have a BMI in the normal weight or overweight range; underweight is less common.

Signs and symptoms of bulimia include a chronically inflamed and sore throat, swollen salivary glands, gastric reflux, severe erosion of tooth enamel due to exposure to stomach acid, intestinal distress, dehydration, and electrolyte imbalances, which can lead to stroke or heart attack.[1]

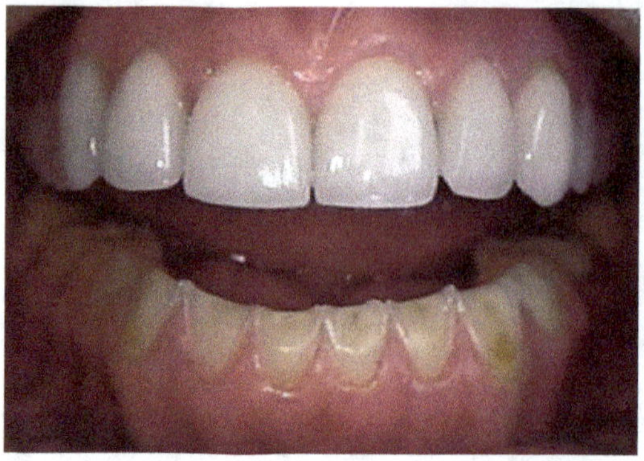

Fig. 7.16. This photo shows the erosion on the lower teeth caused by bulimia. For comparison, the upper teeth were restored with porcelain veneers.

Binge-Eating Disorder (Bed)

Like those with anorexia or bulimia, people with binge-eating disorder (BED) have lost control over their eating. They will periodically overeat to the extreme, often quickly and to the point of discomfort, but their loss of control over eating is not followed by fasting, purging, or compulsive exercise. BED may be diagnosed at any weight, but compared with other eating disorders, is more often seen in people living in larger bodies.[1] BED is the most common eating disorder in the U.S.; it has been estimated to affect 2.6 percent of people in their lifetime. Unlike bulimia and anorexia, which often begin in adolescence, the average age of onset of BED is 23 years of age.[17] There may be a link between dieting and BED, as many people with BED also have a history of dieting, and restricting food intake may increase the urge to binge. The health consequences of BED are those linked to weight stigma and having a higher body weight, such as hypertension, cardiovascular disease, type 2 diabetes, and chronic pain. Additionally, people with BED often experience guilt, shame, and depression, and BED is also associated with phobias, social anxiety, posttraumatic stress disorder, and alcohol use disorder.[18] Treatment often involves psychiatric and nutritional counseling as well as medications to treat depression or other mental health conditions.[1]

Orthorexia

Orthorexia is a newer disordered eating behavior defined as an obsession with healthy eating. The term "orthorexia" was first defined in 1998, but it has yet to be formally classified as an eating disorder, making it difficult to determine how prevalent it is. Research suggests it may be identified as a form of obsessive-compulsive disorder. While focusing on a healthy diet isn't inherently a bad thing, a person with orthorexia takes the emphasis on healthy or "clean" eating to the extreme, so much so that it becomes a fixation, putting their health at risk. Signs of orthorexia include compulsively reading food labels, cutting several food groups out of the diet, spending an unusual amount of time focusing on what foods may be available at upcoming events, and experiencing a high level of stress when healthy foods are not available. The obsession with healthfulness comes with a high social cost as it is often difficult to enjoy eating out or sharing meals with friends and family. There is no formal treatment plan for orthorexia, but many eating disorder experts treat it similarly to anorexia and obsessive-compulsive disorder.[19]

If you think you or someone close to you may have an eating disorder and you want to learn more or find help, we've provided links to more resources below. Note that there are several other types of eating disorders not covered here, so if you're struggling with your relationship with food but your experience doesn't match one of the disorders described above, it may still be worth learning more and speaking with a professional to seek help.

- The National Eating Disorder Association offers many resources, including a short online screening tool that "can help determine if it's time to seek professional help."
- ANAD is a nonprofit providing free support services, including a helpline, recovery mentorship, and support groups, to anyone struggling with an eating disorder, disordered eating, and body image.
- Project HEAL is a nonprofit that offers free, 1-hour phone assessments for eating disorders, as well as free and sliding scale treatment, with a focus on increasing access for people identifying as BIPOC, LGBTQ+, and from other marginalized communities.

REFERENCES

Eating disorders. (2021). The National Institute of Mental Health. Retrieved June 10, 2022, from https://www.nimh.nih.gov/health/statistics/eating-disorders.shtml#part_155061.

Body image. (2019). National Eating Disorder Association. Retrieved June 10, 2022, from https://www.nationaleatingdisorders.org/toolkit/parent-toolkit/body-image.

Treatment. (2017). National Eating Disorders Association. Retrieved June 10, 2022, from https://www.nationaleatingdisorders.org/treatment

Report: Economic Costs of Eating Disorders. (2020, June 11). T.H. Chan Harvard School of Public Health. Retrieved June 10, 2022, from https://www.hsph.harvard.edu/striped/report-economic-costs-of-eating-disorders/

Zipfel, S., Schmidt, U., & Giel, K. E. (2022). The hidden burden of eating disorders during the COVID-19 pandemic. The Lancet Psychiatry, 9(1), 9–11. https://doi.org/10.1016/S2215-0366(21)00435-1

Chesney, E., Goodwin, G. M., & Fazel, S. (2014). Risks of all-cause and suicide mortality in mental disorders: A meta-review. World Psychiatry: Official Journal of the World Psychiatric Association (WPA), 13(2), 153–160. https://doi.org/10.1002/wps.20128

Lipson, S. K., & Sonneville, K. R. (2020). Understanding suicide risk and eating disorders in college student populations: Results from a National Study. The International Journal of Eating Disorders, 53(2), 229–238. https://doi.org/10.1002/eat.23188

Identity & Eating Disorders. (2017). National Eating Disorders Association. Retrieved June 10, 2022, from https://www.nationaleatingdisorders.org/identity-eating-disorders

Risk Factors. (2017, February 21). National Eating Disorders Association. https://www.nationaleatingdisorders.org/risk-factors

Lee, K. M., Hunger, J. M., & Tomiyama, A. J. (2021). Weight stigma and health behaviors: Evidence from the Eating in America Study. International Journal of Obesity (2005), 45(7), 1499–1509. https://doi.org/10.1038/s41366-021-00814-5

Hunger, J. M., & Tomiyama, A. J. (2018). Weight Labeling and Disordered Eating Among Adolescent Girls: Longitudinal Evidence From the National Heart, Lung, and Blood Institute Growth and Health Study. The Journal of Adolescent Health: Official Publication of the Society for Adolescent Medicine, 63(3), 360–362. https://doi.org/10.1016/j.jadohealth.2017.12.016

Patton, G. C., Selzer, R., Coffey, C., Carlin, J. B., & Wolfe, R. (1999). Onset of adolescent eating disorders: Population based cohort study over 3 years. BMJ (Clinical Research Ed.), 318(7186), 765–768. https://doi.org/10.1136/bmj.318.7186.765

Golden, N. H., Schneider, M., Wood, C., Committee on Nutrition, Committee on Adolescence, & Section on Obesity. (2016). Preventing Obesity and Eating Disorders in Adolescents. Pediatrics, 138(3), e20161649. https://doi.org/10.1542/peds.2016-1649

Galmiche, M., Déchelotte, P., Lambert, G., & Tavolacci, M. P. (2019). Prevalence of eating disorders over the 2000–2018 period: A systematic literature review. The American Journal of Clinical Nutrition, 109(5), 1402–1413. https://doi.org/10.1093/ajcn/nqy342

Anorexia Nervosa. (2017, February 25). National Eating Disorders Association. https://www.nationaleatingdisorders.org/learn/by-eating-disorder/anorexia

Bulimia Nervosa. (2017, February 26). National Eating Disorders Association. https://www.nationaleatingdisorders.org/learn/by-eating-disorder/bulimia

Kessler, R. C., Berglund, P. A., Chiu, W. T., Deitz, A. C., Hudson, J. I., Shahly, V., Aguilar-Gaxiola, S., Alonso, J., Angermeyer, M. C., Benjet, C., Bruffaerts, R., de Girolamo, G., de Graaf, R., Maria Haro, J., Kovess-Masfety, V., O'Neill, S., Posada-Villa, J., Sasu, C., Scott, K., ... Xavier, M. (2013). The prevalence and correlates of binge eating disorder in the World Health Organization World Mental Health Surveys. Biological Psychiatry, 73(9), 904–914. https://doi.org/10.1016/j.biopsych.2012.11.020

Binge-eating disorder—Symptoms and causes. Mayo Clinic. Retrieved July 14, 2022, from https://www.mayoclinic.org/diseases-conditions/binge-eating-disorder/symptoms-causes/syc-20353627

Yager, J. (2022). Eating disorders: Overview of epidemiology, clinical features, and diagnosis. UpToDate. https://www.uptodate.com/contents/eating-disorders-overview-of-epidemiology-clinical-features-and-diagnosis

Orthorexia. (2018). National Eating Disorder Association. Retrieved JUne 14, 2022, from https://www.nationaleatingdisorders.org/learn/by-eating-disorder/other/orthorexia.

CAUSES OF EXCESS WEIGHT THAT ARE COMPLEX

Since the 1980s, the prevalence of obesity (defined as BMI of 30 or greater) in the United States has increased dramatically. Data collected by the Centers for Disease Control and Prevention show rising rates of obesity across the nation, state-by-state.[1]

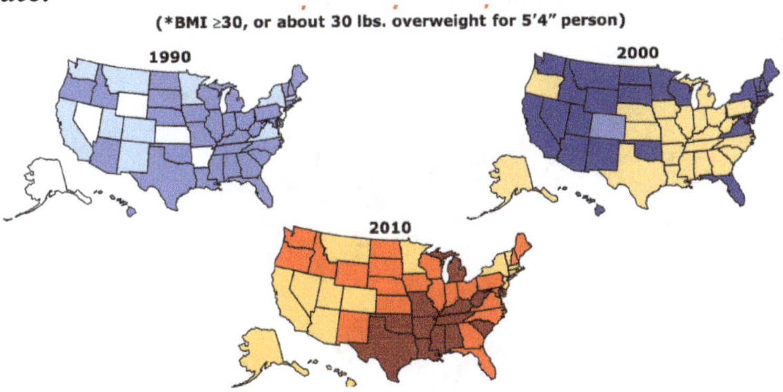

Fig 7.17. Each year since 1990, the CDC has published maps of the United States in which states are color-coded based on the percentage of their population estimated to be obese. The maps show a clear increase in the prevalence of obesity between 1990 and 2010.

The methods used by the CDC to collect the data changed in 2011, so we can't make direct comparisons between the periods before and after that change, but the trend has continued. Every year, more and more people in the U.S. report having a

BMI that classifies them as having obesity, a trend that public health officials have labeled the obesity epidemic.

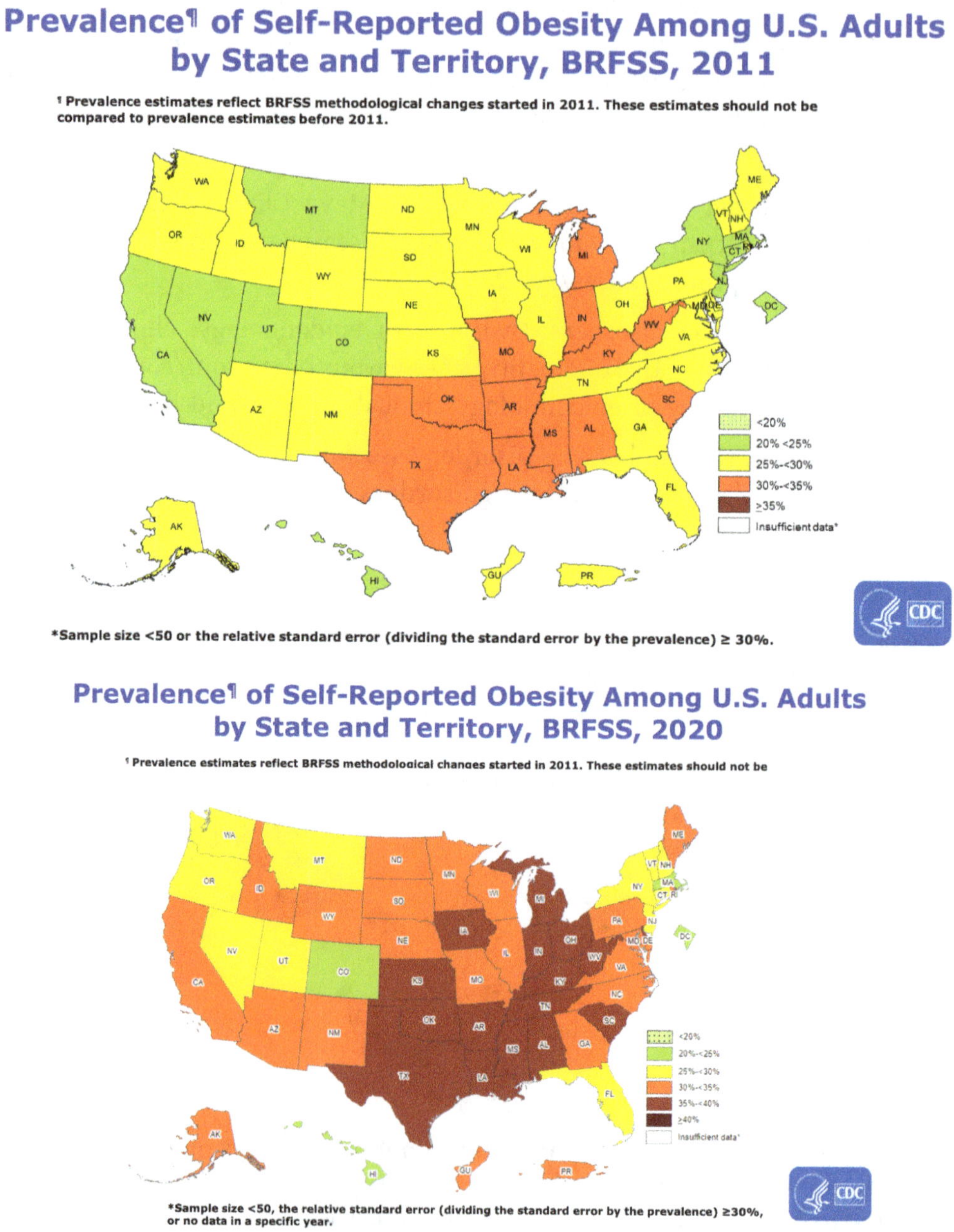

Fig 7.18. The prevalence of obesity among U.S. adults has continued to rise between 2011 and 2020.

These trends are unmistakable, and they're not just occurring in adults. Childhood obesity has seen similar increases over the last few decades. This is a concern because the metabolic and health effects of carrying too much weight can be compounded over a person's lifetime, and people with obesity in childhood are more likely to also have obesity as adults.[2]

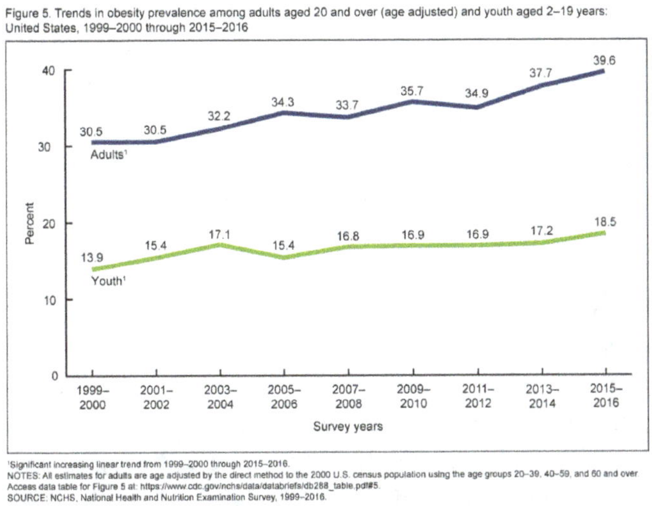

Fig 7.19. Between 1999 and 2016, the prevalence of obesity in both children and adults has risen steadily.

While obesity is a problem across the United States, it affects some groups of people more than others. Based on 2017-2018 data, the prevalence of obesity is 50 percent among Black adults, 45 percent among Hispanic adults, 42 percent among white adults, and 17 percent among Asian adults in the U.S. Overall, people who are college-educated and have a higher income are less likely to be obese.3 These health disparities point to the importance of looking at social determinants of health when examining causes and solutions. Not everyone has the same opportunity for good health, or an equal ability to make changes to their circumstances, because of factors like poverty and longstanding inequities in how resources are invested in communities.4

In addition to the health issues that can be associated with obesity, excess weight is also associated with greater costs of medical care. Cost estimates vary widely, in part because it's very difficult to tease out health issues caused by obesity versus those correlated with it. A 2016 review of studies found that, on average, a person with obesity had $1,901 greater medical costs than a person classified as normal weight, adding up to $149 billion in the United States.5 A 2021 study reported similar cost estimates.6 The obesity epidemic is also not unique to the United States. Obesity is rising around the globe, and in 2015, it was estimated to affect 2 billion people worldwide, making it one of the largest factors affecting poor health in most countries.7 Previously, overweight and obesity mainly affected high-income countries, but some of the most dramatic increases in childhood overweight over the last decade have been in low income countries, such as those in Africa and South Asia.

Despite the gravity of the problem, no country has yet been able to implement policies that have reversed the trend and brought about a decrease in obesity. This represents "one of the biggest population health failures of our time," wrote an international group of researchers in the journal The Lancet in 2019.9 Turning the tide will require an understanding of the complex causes of the rise in excess weight. As we'll see, these are interwoven into our environments, economies, societies, and policies, and we can thus assume that effective solutions will need to go far beyond the simplistic advice that people should "eat less and move more."

Possible Causes of Excess Weight

What has caused the rise in obesity around the world? This question is still an active area of debate among researchers. The predominant explanation comes back to the concept of energy balance: A chronic shift towards positive energy balance—consuming more energy or calories than expended—has led to a gradual but persistent increase in body weight. This shift is thought to be the result of changes in the environments in which we live, which include fewer opportunities for physical activity and greater availability and marketing of inexpensive, convenient, energy-dense foods.[10]

Putting the focus on energy balance often leads people to assume that obesity is an individual problem—that those who weigh more simply need to change their behavior to eat less and exercise more, and if this doesn't work, it must be because of a personal failing, such as a lack of self-control or motivation. However, science tells us that the reality is far more complex. Many factors can contribute to energy balance, and neurological and hormonal control of food intake and energy expenditure mean that it's difficult to maintain a negative energy balance and weight loss. Some researchers also argue that food intake and physical activity can't explain the continued increase in body weight, and we need to look to alternative explanations.11 In the sections below, we discuss some of the hypothesized causes for the rise in obesity, acknowledging there are likely multiple causes to this complex issue.

Environment and Physical Activity

Many of us live in what researchers and public health experts call "obesogenic environments." That is, the ways in which our neighborhoods are built and our lives are structured influence our physical activity and food intake to encourage weight gain.12 Human physiology and metabolism evolved in a world where obtaining enough food for survival required significant energy investment in hunting or gathering—very different from today's world where more people earn their living in sedentary occupations. From household chores, to workplace productivity, to daily transportation, getting things done requires fewer calories than it did in past generations.

Fig 7.20. Some elements of our environment that may make it easier to gain weight include sedentary jobs, easy access to inexpensive calories, and cities built more for car travel than for physical activity.

Our jobs have become more and more sedentary, with fewer opportunities for non-exercise thermogenesis (NEAT) throughout the day. There's less time in the school day for recess and physical activity, and lack of neighborhood safety can limit kids' ability to get out and play after the school day is over. Our towns and cities are built more for cars than for walking or biking. We can't turn back the clock on human progress, and finding ways to stay active in modern environments is a significant challenge.

That said, recent research has called into question the role of physical activity in driving the rise in obesity. Studies of modern hunter-gatherer societies on the savannas of Africa have found that these very active people burn a similar number of calories each day as adults living more sedentary lifestyles in modern cities.13 In addition, even as rates of obesity have continued to increase, research suggests that physical activity among U.S. adults has also been increasing.11 Also, some studies find that people with higher BMI are more physically active than those in the normal weight category. "These findings contrast with conventional views that primarily attribute the cause of obesity to sedentary lifestyles and suggest that compensatory metabolic adaptations maintain total energy expenditure relatively constant among human populations and across various levels of physical activities," write the authors of the Joint International Consensus Statement for Ending Stigma of Obesity.14 Being physically active benefits our health in a multitude of ways, including preventing chronic diseases often associated with higher body weight, but it's not clear how much moving more affects long term energy balance or weight.

Food Intake And Diet Quality

Our environments can also impact our food choices. We're surrounded by vending machines, fast food restaurants, coffeeshops, and convenience stores that offer quick and inexpensive access to calories. These foods are also heavily advertised, and especially when people are stretched thin by working long hours or multiple jobs, they can be a welcome convenience. However, they tend to be more calorie-dense (and less nutrient-dense) and more heavily processed compared with home-cooked food. In addition, portion sizes at restaurants, especially fast food chains, have increased over the decades, and people are eating at restaurants more and cooking at home less. A randomized controlled trial found that participants had a greater caloric intake when eating processed foods compared with unprocessed foods, and eating more fruits and vegetables and less processed foods has been shown to be associated with less weight gain over time.10,15

Social Determinants of Health

As mentioned earlier in this unit, social determinants of health are the conditions in the places where people live, learn, work, and play that can affect health. They include factors related to socioeconomic instability, education, healthcare access and quality, neighborhood and built environment, and social and community context.4 In the United States, people with more disadvantaged social determinants of health have a greater incidence of obesity, emphasizing the importance of addressing systemic inequities to improve health of all Americans.16

Social determinants such as poverty and socioeconomic instability decrease opportunities for health in many ways. People living in poorer neighborhoods often lack grocery stores where they can purchase fresh fruits and vegetables and basic ingredients for cooking at home. Such areas are called food deserts—where healthy foods simply aren't available or easily accessible. In these neighborhoods, people rely more on packaged and convenience foods that are more highly processed and may lead to greater energy intake.

People living with lower and less stable incomes also have to contend with food insecurity. Food security means "access by all people at all times to enough food for an active, healthy life."16 Food insecurity means an inability to consistently obtain adequate food. It may seem counter-intuitive, but in the United States, food insecurity is linked to obesity. That is, people who have difficulty obtaining enough food are more likely to become obese and to suffer from diabetes and hypertension. This is likely related to the fact that inexpensive foods tend to be high in calories but low in nutrients, and when these foods form the foundation of a person's diet, they can cause both obesity and nutrient deficiencies. It's estimated that 12 percent of U.S. households are food insecure, and food insecurity is higher among Black (22 percent) and Latino (18 percent) households.

Poor neighborhoods are also sometimes not safe or pleasant places to walk, play, or exercise. They may have busy traffic and polluted air, and they may lack sidewalks, green spaces, and playgrounds. A person living in this type of neighborhood will find it much more challenging to incorporate physical activity into their day compared with someone living in a neighborhood where it's safe to walk to school or work, play at a park, ride a bike, or go for a run.

Social determinants of health likely drive and compound the racial and ethnic differences in obesity prevalence. Studies show that U.S. regions with more racial inequality in poverty, unemployment, and homeownership—all factors driven by historic and present-day inequities—have greater rates of obesity. In addition, experiencing racism and bias results in chronic stress, which is associated with greater weight gain.18-19

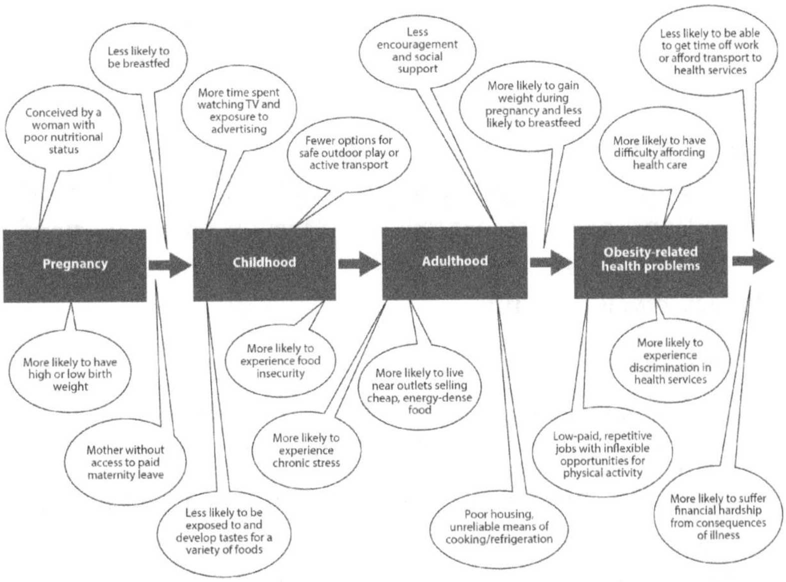

Figure 7.21. Social and economic inequities result in accumulated conditions that can increase risk of obesity and other health conditions.

Genetics

What about genetics? While it's true that our genes can influence our susceptibility to weight gain, researchers say genes can't be the sole cause of the increasing prevalence of obesity. Genes take many generations to evolve, and the rise in obesity has occurred over just the last 40 to 50 years—only a few generations. When our grandparents were children, they were much less likely to become obese than our own children. That's not because their genes were different, but more likely because they grew up in different conditions. However, it is true that a person's genes can influence their susceptibility to weight gain in this environment, and obesity is more prevalent in some families. Research shows that about 50 to 75 percent of the variation in BMI is explained by genetic makeup.[10]

Dieting And Weight Stigma

We live in a society saturated by diet culture: Beliefs continuously promoted by diet companies, media, and influencers that center the idea that being thin is the same as being healthy, and that anyone who applies themselves by eating the right foods in the right amounts can achieve that ideal. But ironically, diet culture may contribute to weight gain. Studies show that dieting during adolescence is associated with greater risk of becoming overweight and of developing an eating disorder.[20] Having a greater "drive for thinness" in adolescence is associated with similar attitudes and weight gain in adulthood.[21] There's also some research that finds that experiencing cycles of weight loss and weight regain (i.e., weight cycling or "yo-yo dieting") may cause people to gain even more weight over time, although the evidence on this point is somewhat mixed and controversial.[22,23]

In addition, living in a larger body and experiencing weight stigma, which has been shown to affect people at home, in social settings, in the workplace, and in healthcare settings, may also contribute to weight gain. Experiencing weight stigma is associated with greater stress, less access to healthcare, less physical activity, and greater weight gain.[24]

Other Possible Causes

Many other possible causes of weight gain have been investigated and likely contribute on some level to the increasing prevalence of obesity:

- Many medications that may improve our health are also known to cause weight gain. These include some medications used to treat common conditions such as diabetes, hypertension, autoimmune disorders, and depression.14
- Greater exposure to endocrine disrupting chemicals, such as those found in food packaging and pollutants, is associated with weight gain.14
- Not getting enough sleep or being under greater stress is associated with weight gain.
- Changes in the gut microbiome (the bacteria living in our gastrointestinal tract), caused by changing dietary patterns and environmental exposures to microbes, may predispose people to weight gain.11
- Having had adverse childhood experiences (ACEs), defined as intensive stressors and traumatic events occuring in the first 18 years of life, is associated with a greater risk of obesity in adolescence and adulthood. ACEs include "recurring childhood maltreatment (physical, emotional and sexual abuse, and neglect), household dysfunction (witnessing violence between parents or caregivers, parental separation, incarceration or substance

abuse of a household member, mental illness, and attempted or complete suicide by a family member), and community violence that deprive the sense of stability, safety and development of a child." Experiencing ACEs can cause changes in the brain, hormone levels, inflammation, sleep quality, and immune function, all of which can affect a person's risk of developing obesity.25

SOLUTIONS FOR HEALTH IMPROVEMENT

According to a 2019 study, 42 percent of adults in the United States had recently attempted to lose weight, primarily through reduced food consumption and exercise.[1] This is not a surprise, given the growing prevalence of obesity and the constant messages of diet culture, which promote thinness as being equivalent to wellness and the latest diet fads as the keys to unlocking good health.

But, as we have learned, the causes of obesity are complex and interwoven into our society, and solutions for improving health will need to be equally far-reaching, going beyond individual efforts at weight loss. On this page, we look at potential solutions to health concerns related to body weight, both at the population and individual level.

Population Level Solutions

Because obesity can affect people over the course of their lifespan and is difficult to reverse, the focus of many of these efforts is on prevention, starting as early as the first years of life. Here, we'll review some strategies happening in schools, communities, and at state and federal levels.

Focus on Health Equity

Social determinants of health have a strong impact on obesity risk, so prevention efforts understandably target areas such as access to healthy foods and safe and affordable opportunities for physical activity. These are important, but fully addressing them will likely require societies to dig deeper into the root causes of existing disparities, to look at how factors such as poverty, racism, and discrimination affect people's opportunities for health. This means looking at how communities and governments can work toward ***health equity***, defined as "the fair and just opportunity for all to be as healthy as possible," requiring removal of obstacles to health such as poverty, discrimination, and their consequences.

Consider low income neighborhoods that lack grocery stores, leaving residents without access to fresh foods and cooking ingredients. Policies and programs that bring groceries into these neighborhoods are important, but they may not do much to improve the nutrition or the food security of residents if they can't afford to shop there or have no time to cook because they need to work multiple jobs to support their families. Likewise, programs that give children bicycles or teach them about the importance of physical activity may not have an impact if there are no safe places to ride a bike or play in their neighborhood. Looking at this issue through the lens of health equity means tackling even bigger issues of poverty and access to education, resources, and safety. In this sense, policies and programs at community, state, and national levels that seek to improve people's lives by reducing income inequality and improving access to resources can help to target some of the core causes of obesity in populations.

An example of community-based interventions that aim to reduce health disparities can be found in the Racial and Ethnic Approaches to Community Health across the United States (REACH US) project, launched by the CDC in 2007. REACH US invested in many projects across the U.S., including 14 projects in predominantly Black communities aiming to reduce cardiovascular disease or diabetes. Each community project was unique and depended on local needs and preferences, but all included community coalitions, involving local leaders and residents, and targeted access to healthy food and/or improving physical activity opportunities. Between 2009 and 2012, the prevalence of obesity in REACH US communities dropped by 5 percent, while it increased by 2 percent in nearby comparison communities without a REACH program.[4]

End Weight Stigma

Weight stigma is a significant barrier to health, particularly for people living in larger bodies. It is harmful to both mental and physical health and counter-productive to the goal of reducing obesity and improving health. The following is the executive summary of the Joint International Consensus Statement for Ending Stigma of Obesity[5]:

- Weight stigma is reinforced by misconceived ideas about body-weight regulation and lack of awareness of current scientific evidence. Weight stigma is unacceptable in modern societies, as it undermines human rights, social rights, and the health of afflicted individuals.
- Research indicates that weight stigma can cause significant harm to affected individuals. Individuals who experience it suffer from both physical and psychological consequences, and are less likely to seek and receive adequate healthcare.
- Despite scientific evidence to the contrary, the prevailing view in society is that obesity is a choice that can be reversed by voluntary decisions to eat less and exercise more. These assumptions mislead public health policies, confuse messages in popular media, undermine access to evidence-based treatments, and compromise advances in research.
- For the reasons above, weight stigma represents a major obstacle in efforts to effectively prevent and treat obesity and type 2 diabetes. Tackling stigma is not only a matter of human rights and social justice, but also a way to advance prevention and treatment of these diseases.
- Academic institutions, professional organizations, media, public health authorities, and government should encourage education about weight

stigma and facilitate a new public narrative of obesity, coherent with modern scientific knowledge.

Support Healthy Dietary Patterns

Interventions that support healthy dietary patterns, especially among people more vulnerable because of food insecurity or poverty, may reduce obesity. Here are some examples:

- Implement and support better nutrition standards for childcare, schools, hospitals, and worksites. The National School Lunch Program in particular provides meals for approximately 30 million children from low-income families each school day, and recent changes in school lunch standards have improved dietary quality for students consuming school lunches.3,6
- Limit or end marketing of processed foods and sugar sweetened beverages, especially ads that disproportionately target children of color.7
- Provide incentives for supermarkets and farmers markets to establish businesses in underserved areas.7

Figure 7.22. *Farmers markets can expand healthy food options for neighborhoods and build connections between consumers and local farmers.*

- Increase access to food assistance programs and align them with nutrition recommendations. For example, in 2009, the U.S. Department of Agriculture revised the food packages for the Women, Infants, and Children (WIC) program to better align with the Dietary Guidelines for Americans. The new packages emphasized more fruits, vegetables, whole grains, and low-fat dairy and decreased the availability of juice. After this change, between 2010 and 2016, there was a decrease in the obesity rate of children in the WIC program.8 Similar progress may be made by increasing access to the

Supplemental Nutrition Assistance Program (SNAP) in order to reduce food insecurity, and providing incentives to use SNAP benefits on foods like fruits and vegetables. Many farmers' markets now accept SNAP benefits for the purchase of fresh fruit and vegetables.7

Fig 7.23. As of 2018, restaurant chains and some other food vendors are required to list calorie counts on their menus. Would these make you pause before ordering?

- Tax sugary drinks, such as soda and sports drinks, which contribute significant empty calories to the U.S. diet and are associated with childhood obesity. Local taxes on soda and other sugary drinks are often controversial, and soda companies lobby to prevent them from passing. However, research from U.S. cities with soda taxes show that they do work to decrease soda consumption, and one study concluded that a national soda tax may be the most cost-effective obesity prevention measure, estimating that it could prevent more than 500,000 cases of childhood obesity over 10 years.7 In the U.S., soda has only been taxed at the local level, and the tax has been paid by consumers. The United Kingdom has taken a different approach: They started taxing soft drink manufacturers for the sugar content of the products they sell. Between 2015 and 2018, the average sugar content of soda sold in the U.K. dropped by 29 percent.9

- A controversial policy intervention has been the requirement that restaurant menus include calorie information. Beginning in 2018, as part of the Affordable Care Act, chain restaurants in the U.S. with more than 20 locations were required to add calorie information to their menus, and some had already done so voluntarily. Overall, the research on menu labeling has shown it may slightly reduce the calorie count of restaurant purchases, but it doesn't seem to reduce calories consumed in a sitting, and there's no evidence that menu labeling affects total calories consumed in the long term, dietary quality, or risk of obesity.

Support Greater Physical Activity

Physical activity can help maintain energy balance and promote health in many ways, including preventing chronic diseases often associated with higher body weight. The CDC recommends several evidence-based strategies to increase physical activity in communities[12,13]:

- Activity-friendly routes to everyday destinations – This strategy focuses on creating easier, more appealing, and more accessible opportunities for everyday physical activity within the community. It includes adding sidewalks, bike lanes and paths, hiking trails, and public transit to connect homes, schools, green spaces, parks, worksites, stores, restaurants, and libraries. This infrastructure allows for more physical activity throughout the day.
- Access to places for physical activity – This strategy provides more access (free or reduced-cost) to spaces for physical activity, such as parks, recreation centers, and gyms. It includes workplaces that provide gym facilities and/or time off during the day for physical activity.
- School and youth programs – This includes providing physical activity programs in schools, enhancing physical activity during recess periods, and before- and after-school activities, such as sports and activity clubs.
- Community-wide campaigns – These programs encourage community-wide participation in physical activity through the use of media coverage and promotion, community events, classes, contests, and educational opportunities.
- Social supports – This strategy focuses on building social groups centered on physical activity, such as through walking, cycling, or hiking clubs.
- Individual supports – Worksites, schools, organizations, and churches can offer individuals support in the form of education and counseling to build skills and set activity goals.

- Prompts to encourage physical activity – These include elements such as signs that encourage taking the stairs over using an elevator or choosing walking and biking routes through neighborhoods.

Fig 7.24. There are many ways to increase physical activity, including walking to work, playing with friends, and going for a bike ride.

Individual Level Solutions

While we acknowledge that population level solutions are likely most important for improving the health of communities, we know that individuals may also be concerned about their weight or health. For people who believe that their weight is affecting their health or quality of life, what steps can they take? There is no one-size-fits-all approach. In this section, we'll discuss lifestyle changes that can improve health (whether or not they result in weight loss), the evidence for weight loss diets, weight loss medications, bariatric surgery, and non-diet approaches.

Biology of weight loss and maintenance

As we discussed earlier in this unit, energy balance determines whether we are gaining, losing, or maintaining body weight. To lose weight, we need to shift into a negative energy balance, so that we are consuming fewer calories than we are burning each day. Most people find that they are able to do this by restricting their calorie or macronutrient intake (i.e., dieting) for a short period of time, perhaps 6 months or a year, and they may lose 5 to 10 percent of their body weight. However, it is very common for people to experience gradual weight regain within a year or two. This fact is what fuels the diet industry; diet companies market their plans based on short-term success stories and frame the longer term reality of weight regain as personal failing or lack of willpower. Even scientific studies of weight loss rarely follow people for more than 12 months, which can lead to reporting of overly optimistic results.

Why is it so difficult to maintain a reduced body weight? Keep in mind that weight, like so many other parts of our physiology, is tightly regulated. Our bodies can make adjustments to both sides of the energy balance equation to help us maintain a steady weight. During periods of negative energy balance, the body signals weight loss as a threat to survival, and these regulatory mechanisms kick in to defend the original weight. Basal metabolic rate drops to conserve energy and protect against starvation. Skeletal muscles also become more efficient, so that they need less energy to fuel their movement. Additionally, after weight loss, there is less physical mass of the body that has to be moved from place to place throughout the day, resulting in fewer calories burned through physical movement and activity, and less metabolically active tissue using calories for fuel throughout the day. These metabolic changes conserve energy, reducing the "energy out" side of the energy balance equation and favoring weight regain.[16,17] In individuals maintaining a %10 or greater weight loss, these changes combine to account for an estimated decrease of 400-300 calories in energy expenditure per day beyond what is expected due to the change in body composition alone.[18]

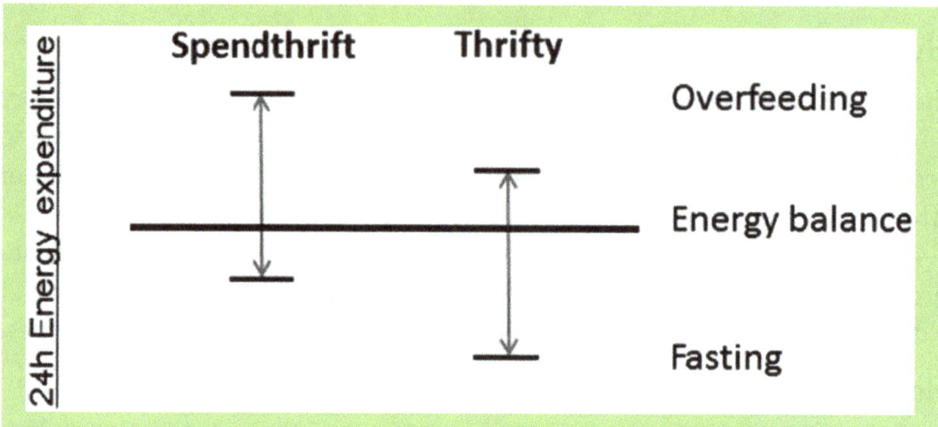

Fig. 7.25. Illustration of the concept of spendthrift and thrifty metabolism, characterized by response to overfeeding and fasting.

On the "energy in" side of the equation, neurological and hormonal changes also increase appetite, lower satiety, and increase the brain's focus on obtaining food. Researchers have estimated that weight loss of about 2 pounds increases appetite by about 95 calories per day.[16,17] These biological factors and their influence on weight are discussed further in the video below.

Biological differences in individual metabolism may also impact weight loss and maintenance. Researchers have found that some individuals have a "thrifty" metabolism, meaning that they have a lower metabolic rate and will expend significantly fewer calories when in a fasting or calorie-restricted state, resulting in less weight loss. In contrast, individuals with a "spendthrift" metabolism tend to have a higher metabolic rate in a fasting state, burning more calories and thus yielding bigger weight loss results.[19] This research explains why some people may be able to lose weight and maintain weight loss, while others will struggle to do so, even with substantial effort.

Setting Goals and Expectations

It's clear that sustained weight loss is difficult and complex. But weight loss doesn't need to be the primary goal to improve health. There are many ways to improve your health that aren't tied to changing your weight or BMI. If you're at a higher weight and concerned about your weight or health, a good starting point may be thinking of steps you can take to improve your health that are in your immediate control. For example, are you getting enough sleep? What strategies can you take to reduce your stress level? How can you improve your eating habits to better nourish your body? Can you incorporate more enjoyable forms of physical activity into your daily routine? Setting goals around behavior and health rather than a number on the scale improves your likelihood of feeling successful and making sustainable changes. And since many of these daily habits are also thought to affect weight, making positive changes may (but may not) lead to some weight loss. In addition, talk with your doctor about any medications you are using that could cause weight gain.

If weight loss is one of your goals, it's important to have realistic expectations. The goal of weight loss efforts should be to "prevent, treat, or reverse the complications of obesity and improve the quality of life."[20] Specific goals and approaches will look different for everyone depending on their individual risk factors and lifestyle. Research shows that losing just 5 percent of body weight is associated with significant health benefits such as lower blood pressure and improved blood cholesterol levels, and changing diet or exercise habits typically can reduce body weight by about 5 to 7 percent.[20] However, many people set weight loss goals well beyond this mark, leading to disappointment and frustration if they fall short. It may be helpful to set a goal of reaching your "best weight," defined in the 2020 Canadian Adult Obesity Clinical Practice Guidelines as "the weight that a person can achieve and maintain while living their healthiest and happiest life."[21]

Even more difficult than losing weight is maintaining a reduced weight. Weight regain is common, so at the outset, it's important to think about how weight loss strategies can be sustained for the long term. Weight control registries from around the world have tracked thousands of people who have successfully lost weight and maintained a reduced weight over time. While these registries include self-selected samples and thus are not representative of all people attempting weight loss, their findings provide some examples of the types of habits associated with maintaining individual weight loss. For example, more than 80 percent of global registry members report consuming breakfast daily, increasing their consumption of vegetables and high fiber foods, eating regular meals, watching their consumption of foods higher in fat and sugar, and keeping healthy foods on hand at home. Physical activity is also an important part of their ongoing efforts;

most registry members reported exercising regularly, many for about an hour per day.

Evidence-Based Dietary Approaches

There are many dietary approaches to weight loss, and research shows that there is no single dietary strategy that is superior to others. Dietary approaches that have been shown to be effective, at least in the short-term, include the Mediterranean diet, the Dietary Approaches to Stop Hypertension (DASH) eating plan, vegetarian diet, and a variety of lower-calorie macronutrient distribution diets, such as low-fat and low- carbohydrate approaches.

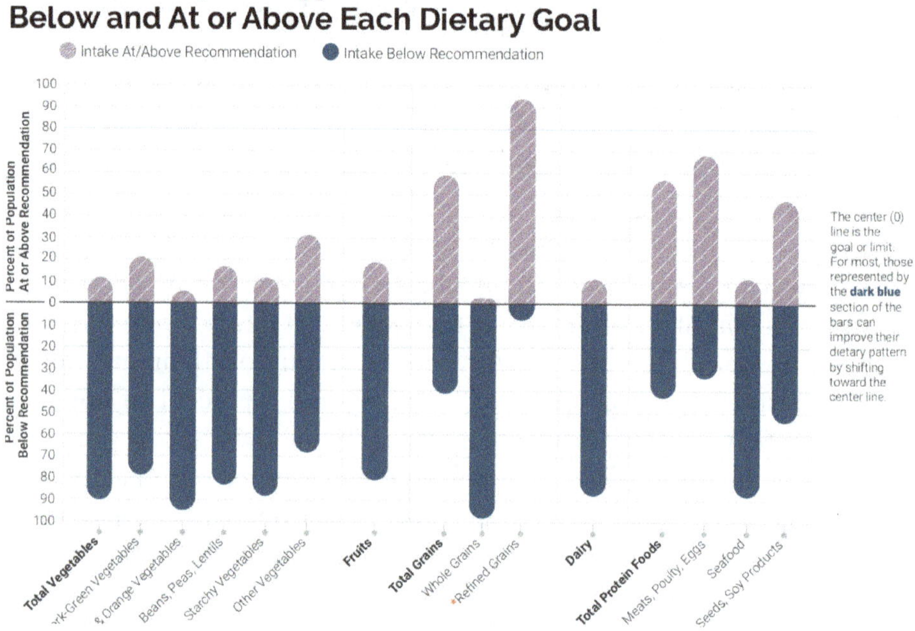

Figure 7.26. This graph indicates the percentage of the U.S. population ages 1 year and older with intakes below the recommendation or above the limit for different food groups and dietary components.

While not specifically about weight loss, the 2020 Dietary Guidelines for Americans also offers specific, evidence-based recommendations for dietary habits that provide a good roadmap for healthy eating. These guidelines are aimed at keeping calorie intake in balance with physical activity, which is key for weight management. These recommendations include following a healthy eating pattern that accounts for all foods and beverages within an appropriate calorie level. Unit 1: Tools for Achieving a Healthy Diet includes a more detailed look at these guidelines.

Most Americans do not meet the recommendations of the Dietary Guidelines, indicating room for improvement in dietary quality. Figure 7.26 shows how Americans are falling short of meeting the recommendations for vegetables, fruit, whole grains, dairy, and seafood and consume well over the recommended amount for refined grains. Meanwhile, many Americans also exceed the recommended limits for added sugars, saturated fats, sodium, and alcohol. Shifting towards more nutrient-dense choices, as recommended in the Dietary Guidelines, can help balance caloric intake and better meet nutrient needs for optimal health. In addition, evidence-based guidelines for obesity emphasize the importance of an individualized eating approach that is culturally acceptable, affordable, and prioritizes individual preferences to promote long-term adherence.[24]

Evidence-Based Physical Activity Recommendations

The other part of the energy balance equation is physical activity. The 2018 Physical Activity Guidelines for Americans, issued by the Department of Health and Human Services (HHS) provide evidence-based guidelines for appropriate physical activity levels. These guidelines provide recommendations to Americans aged three and older about how to improve health and reduce chronic disease risk through physical activity. Increased physical activity has been found to lower the risk of heart disease, stroke, high blood pressure, Type 2 diabetes, some forms of cancer, falls and fractures, depression, and early death. Increased physical activity not only reduces disease risk but also improves overall health by increasing cardiovascular and muscular fitness, increasing bone density and strength, improving cognitive function, and assisting in weight loss and weight maintenance. In addition, physical activity has been shown to improve mood disorders, body image, and overall health-related quality of life. These benefits of physical activity can help counter some of the potential effects of carrying excess weight, and some research suggests that increasing physical activity and fitness levels is more effective than weight loss when it comes to reducing the risk of heart disease and premature death.[26] To be most effective, individuals should find activities they enjoy so they can sustain their activity level for lifelong health.

Key guidelines for physical activity are summarized in Unit 10: Essential Elements and Benefits of Physical Fitness, but a few worth mentioning specific to weight include the following:

- For substantial health benefits, adults should do at least 150 minutes (2 hours and 30 minutes) to 300 minutes (5 hours) per week of moderate-intensity aerobic activity, or 75 minutes (1 hour and 15 minutes) to 150 minutes (2 hours and 30 minutes) per week of vigorous-intensity aerobic physical activity, or an equivalent combination of moderate- and vigorous-intensity aerobic activity.
- Engaging in physical activity beyond the equivalent of 300 minutes (5 hours) of moderate-intensity physical activity per week can result in additional health benefits and may help with weight loss and weight loss maintenance.

- Adults should also do muscle-strengthening activities of at least moderate intensity that involve all major muscle groups on 2 or more days per week, as these activities provide additional health benefits and promote weight maintenance. Exercises such as push-ups, sit-ups, squats, and lifting weights are all examples of muscle-strengthening activities.

Despite the indisputable benefits of regular physical activity, just 23 percent of Americans met the national physical activity guidelines in 2018, although this number has gradually increased over the last several decades.[28] Given the number of Americans that are falling short on both nutrition and physical activity recommendations, both of these areas are of primary interest in improving health in the United States.

Evidence-Based Behavioral Recommendations

Behavioral weight loss interventions have been described as approaches "used to help individuals develop a set of skills to achieve a healthier weight. It is more than helping people to decide what to change; it is helping them identify how to change."[29] Cornerstones for these interventions typically include self-monitoring through daily recording of food intake and exercise, nutrition education and dietary changes, physical activity goals, and behavior modification.[30] Research shows that these types of interventions can result in weight loss and a lower risk for type 2 diabetes, and similar maintenance strategies lead to less weight regained later.[31]

Behavioral interventions have been shown to help individuals achieve and maintain weight loss of at least 5 percent from baseline weight. The Food and Drug Administration (FDA) considers a 5 percent weight loss to be clinically significant, as this level of weight loss has been shown to improve cardiometabolic risk factors such as blood lipid levels and insulin sensitivity.[32,33] The behavioral intervention team often includes primary care clinicians, dietitians, psychologists, behavioral therapists, exercise physiologists, and lifestyle coaches.[32] These programs may include a variety of delivery methods, often through group classes of 20-10 participants both in-person and online, and may use print-based or technology-based materials and resources. The interventions usually span one to two years with more frequent contact in the initial months (weekly to bi-monthly), followed by less frequent contact (monthly) in the latter months or maintenance phase.[32] A variety of behavioral topics are covered over the course of the program and range from nutrition education and goal-setting, to problem-solving and assertiveness. Relapse prevention is included as participants move into the maintenance phase.[31]

Behavioral Weight- Loss Treatment Topics

- Self-monitoring
- Healthy food choices
- Physical activity
- Stimulus control
- Eating patterns
- Lifestyle activity
- Thoughts on weight control
- Changing the quality of your diet: fat and fiber
- Problem solving
- Eating in social situations
- High-risk situations
- Restaurant eating
- Assertiveness training
- Recipe modification
- Stress management
- Motivation enhancement
- Relapse prevention

Fig. 7.27. Common topics included in behavioral interventions for weight loss, adapted from Smith, C. E., & Wing, R. R. (2000). New directions in behavioral weight-loss programs. Diabetes Spectrum, 13(3), 142-148.

One example of a comprehensive behavior intervention is the Diabetes Prevention Program (DPP). This was a high quality randomized controlled trial that enrolled participants at high risk for developing type 2 diabetes and included an intervention group that made dietary and physical activity changes with the support of frequent meetings with researchers to implement behavioral changes. In the DPP study, participants lost an average of 15 pounds and reduced their risk of developing type 2 diabetes by 58 percent in the first year. They gradually regained an average of 10 pounds by 10 years later. However, 10 years after the start of the study, intervention participants still had a 34 percent reduced risk in developing type 2 diabetes, indicating significant health benefits of the program.

The DPP yielded such significant improvements in risk of developing diabetes that it was expanded to a national program available in communities throughout the U.S. To find a DPP intervention program near you, visit the National Diabetes Prevention Program website.

Pharmacotherapy and Bariatric Surgery

When lifestyle changes in diet, exercise, and behavior modification do not produce or sustain meaningful levels of weight loss, the use of medications may be considered to improve weight loss and health outcomes. While continuing with a healthy lifestyle, medications are typically considered for individuals with a BMI over 30, or BMI over 27 with at least one coexisting condition, such as heart disease, type 2 diabetes, or hypertension.

A newer class of weight loss medications that act similarly to the body's protein glucagon-like peptide-1 (GLP-1) and were originally developed as treatments for type 2 diabetes have been shown to be very effective for weight loss and generally have fewer side effects than older medications. As with many drugs used to treat chronic diseases, people generally need to continue using these medications for the long term to avoid weight regain. Over-the-counter weight loss supplements are not monitored by the FDA and are not recommended due to safety concerns.[34,35]

Surgical interventions may be appropriate for people with a BMI over 40 or BMI over 35 with obesity-related coexisting conditions, so long as they're motivated to lose weight and behavioral interventions (with or without medication) have not been effective. Potential candidates for surgery should be referred to an experienced bariatric surgeon for consultation and evaluation.

Non-Diet Approaches

In addition to weight management approaches that focus on dietary changes, physical activity programs, and behavioral interventions, there is a growing movement for non-diet approaches for a healthier mentality toward weight, food, and body image. These approaches focus on establishing healthier relationships with food and more body acceptance and positivity regardless of shape and size. They seek to normalize relationships with food, make eating an enjoyable experience focused on well-being rather than dieting, do away with shame or guilt often associated with failed weight loss efforts, and promote respect and inclusivity for all people regardless of weight or size. Mindful eating or intuitive eating are common components of these approaches. These non-diet approaches have been shown to improve quality of life, general well-being, body image, cardiovascular outcomes, physical activity, and eating behaviors regardless of changes in weight status.

One of these non-diet approaches, the Satter Eating Competence Model, is based on four components: eating attitudes, food acceptance, regulation of food intake and body weight, and management of the eating context. According to Ellyn Satter, a registered dietitian, family therapist, and the founder of the model, competent eaters are "confident, comfortable, and flexible with eating and are matter-of-fact and reliable about getting enough to eat of enjoyable and nourishing food." This approach enhances "the importance of eating by making it positive, joyful, and intrinsically rewarding."[36] This model emphasizes that by developing a healthier relationship with food, individuals will yield the following benefits:[37]

- Have better diets
- Feel more positive about food and eating
- Have better overall health
- Have the same or lower BMI
- Sleep better
- Be more active
- Have better physical self-acceptance
- Be more trusting of themselves and others

Health at Every Size (HAES) is another non-diet movement started by the Association for Size Diversity and Health (ASDAH) organization as an alternative to weight-centered health models. HAES aims to decrease our culture's obsession with body size and weight, decrease weight discrimination and stigma, and instead promote size acceptance and inclusivity.[38] Key principles of the HAES approach include:

- Acceptance and respect for the inherent diversity of body shapes and sizes
- Health enhancement through policies and services that promote well-being in all aspects of health, including physical, economic, social, emotional, and spiritual needs
- Respectful care and elimination of weight bias and discrimination through proper education and training
- Eating behaviors driven by hunger, satiety, nutritional needs, and pleasure instead of external regulation by diets and eating plans
- Physical activity through life-enhancing movement for all sizes and abilities

To learn more about non-dieting approaches for a healthy lifestyle, check out the following video.

It's clear that obesity is a complex disease that must be managed through a focus on overall health and well-being, not just by changing a number on the scale. Issues around weight bias and stigma must be addressed. Our current health system must improve to support individuals regardless of body shape and size. And current cultural narratives around obesity must change to remove assumptions of weakness or lack of willpower as causes of obesity and eliminate attitudes that cast blame and shame on people living with obesity.[24]

REFERENCES:

Han L, You D, Zeng F, et al. Trends in Self-perceived Weight Status, Weight Loss Attempts, and Weight Loss Strategies Among Adults in the United States, 1999-2006. (2019). JAMA Network Open. doi: https://doi.org/10.1001/jamanetworkopen.2019.15219

National Academies of Sciences, Engineering, and Medicine. (2019). A Health Equity Approach to Obesity Efforts: Proceedings of a Workshop in Brief. The National Academies Press. https://doi.org/10.17226/25496

Kumanyika, S. K. (2022). Advancing Health Equity Efforts to Reduce Obesity: Changing the Course. Annual Review of Nutrition, 42(1), null. https://doi.org/10.1146/annurev-nutr-092021-050805

Pan, L. (2019). State-Specific Prevalence of Obesity Among Children Aged 2–4 Years Enrolled in the Special Supplemental Nutrition Program for Women, Infants, and Children—United States, 2010–2016. MMWR. Morbidity and Mortality Weekly Report, 68. https://doi.org/10.15585/mmwr.mm6846a3

Rubino, F., Puhl, R. M., Cummings, D. E., Eckel, R. H., Ryan, D. H., Mechanick, J. I., Nadglowski, J., Ramos Salas, X., Schauer, P. R., Twenefour, D., Apovian, C. M.,

Aronne, L. J., Batterham, R. L., Berthoud, H. R., Boza, C., Busetto, L., Dicker, D., De Groot, M., Eisenberg, D., Flint, S. W., ... Dixon, J. B. (2020). Joint international consensus statement for ending stigma of obesity. Nature Medicine, 26(4), 485–497. https://doi.org/10.1038/s41591-020-0803-x

Kinderknecht, K., Harris, C., & Jones-Smith, J. (2020). Association of the Healthy, Hunger-Free Kids Act With Dietary Quality Among Children in the US National School Lunch Program. JAMA, 324(4), 359–368. https://doi.org/10.1001/jama.2020.9517

Trust for America's Health. (2021). The State of Obesity: Better Policies for a Healthier America. Retrieved from https://www.tfah.org/wp-content/uploads/2021/09/2021ObesityReport_Fnl.pdf

Pan, L. (2019). State-Specific Prevalence of Obesity Among Children Aged 2–4 Years Enrolled in the Special Supplemental Nutrition Program for Women, Infants, and Children—United States, 2010–2016. MMWR. Morbidity and Mortality Weekly Report, 68. https://doi.org/10.15585/mmwr.mm6846a3

Public Health England. (2019). Sugar reduction: Report on progress between 2015 and 2018. Retrieved from https://assets.publishing.service.gov.uk/government/uploads/system/uploads/attachment_data/file/832182/Sugar_reduction__Yr2_progress_report.pdf

Crockett, R. A., King, S. E., Marteau, T. M., Prevost, A. T., Bignardi, G., Roberts, N. W., Stubbs, B., Hollands, G. J., & Jebb, S. A. (2018). Nutritional labeling for healthier food or non alcoholic drink purchasing and consumption. Cochrane Database of Systematic Reviews, 2. https://doi.org/10.1002/14651858.CD009315.pub2

McGeown, L. (2019). The calorie counter-intuitive effect of restaurant menu calorie labelling. Canadian Journal of Public Health, 110(6), 816–820. https://doi.org/10.17269/s41997-019-00183-7

CDC. (2022, May 9). What Works: Strategies to Increase Physical Activity. Centers for Disease Control and Prevention. https://www.cdc.gov/physicalactivity/activepeoplehealthynation/strategies-to-increase-physical-activity/index.html

Schmid, T. L., Fulton, J. E., McMahon, J. M., Devlin, H. M., Rose, K. M., & Petersen, R. (2021). Delivering Physical Activity Strategies That Work: Active People, Healthy NationSM. Journal of Physical Activity and Health, 18(4), 352–356. https://doi.org/10.1123/jpah.2020-0656

Brown, J., Clarke, C., Stoklossa, C. J., & Sievenpiper, J. (2020). Medical Nutrition Therapy in Obesity Management. In Canadian Adult Obesity Clinical Practice Guidelines: Medical Nutrition Therapy in Obesity Management (p. 25). Retrieved July 15, 2022, from https://obesitycanada.ca/guidelines/nutrition/

Mann, T., Tomiyama, A. J., Westling, E., Lew, A.-M., Samuels, B., & Chatman, J. (2007). Medicare's search for effective obesity treatments: Diets are not the answer. The American Psychologist, 62(3), 220–233. https://doi.org/10.1037/0003-066X.62.3.220

Aronne, L. J., Hall, K. D., M. Jakicic, J., Leibel, R. L., Lowe, M. R., Rosenbaum, M., & Klein, S. (2021). Describing the Weight-Reduced State: Physiology, Behavior, and Interventions. Obesity, 29(S1), S9–S24. https://doi.org/10.1002/oby.23086

Hall, K. D., & Kahan, S. (2018). Maintenance of Lost Weight and Long-Term Management of Obesity. The Medical Clinics of North America, 102(1), 183–197. https://doi.org/10.1016/j.mcna.2017.08.012

Rosenbaum, M., Kissileff, H. R., Mayer, L. E. S., Hirsch, J., & Leibel, R. L. (2010). Energy Intake in Weight-Reduced Humans. Brain Research, 1350, 95–102. https://doi.org/10.1016/j.brainres.2010.05.062

Piaggi, P., Vinales, K. L., Basolo, A., Santini, F., & Krakoff, J. (2018). Energy expenditure in the etiology of human obesity: Spendthrift and thrifty metabolic phenotypes and energy-sensing mechanisms. Journal of Endocrinological Investigation, 41(1), 83–89. https://doi.org/10.1007/s40618-017-0732-9

Perreault, L., & Apovian, C. (2019). Obesity in adults: Overview of management. UpToDate, Kunins L. Accessed on July 7, 2022.

Obesity Canada and the Canadian Association of Bariatric Physicians and Surgeons (2020). Clinical Recommendations: Quick Guide; Canadian Adult Obesity Clinical Practice Guidelines. http://obesitycanada.ca/wp-content/uploads/2020/11/CPG-Quick-Guide-English.pdf

Paixão, C., Dias, C. M., Jorge, R., Carraça, E. V., Yannakoulia, M., de Zwaan, M., ... & Santos, I. (2020). Successful weight loss maintenance: a systematic review of weight control registries. Obesity Reviews, 21(5), e13003.

Research Findings. The National Weight Control Registry. Retrieved July 15, 2022, from http://www.nwcr.ws/Research/default.htm

Wharton, S., Lau, D. C., Vallis, M., Sharma, A. M., Biertho, L., Campbell-Scherer, D., ... & Wicklum, S. (2020). Obesity in adults: a clinical practice guideline. CMAJ, 192(31), E875-E891.

2018 Physical Activity Guidelines for Americans. US Department of Health and Human Services. Retrieved November 11, 2019, from https://health.gov/paguidelines/second-edition/.

Boulé NG, Prud'homme D. (2020). Canadian Adult Obesity Clinical Practice Guidelines: Physical Activity in Obesity Management. Available from: https://obesitycanada.ca/guidelines/physicalactivity. Accessed on July 7, 2022.

Gaesser, G. A., & Angadi, S. S. (2021). Obesity treatment: Weight loss versus increasing fitness and physical activity for reducing health risks. IScience, 24(10). https://doi.org/10.1016/j.isci.2021.102995

FastStats. (2021, June 11). https://www.cdc.gov/nchs/fastats/exercise.htm

Yanovski, S. (2017, December). The challenge of treating obesity and overweight: Proceedings of a workshop. National Academies of Sciences, Engineering, and Medicine; Health and Medicine Division; Food and Nutrition Board. Roundtable on Obesity Solutions, Washington (DC). Retrieved November 13, 2019, from https://www.ncbi.nlm.nih.gov/books/NBK475856/.

Smith, C. E., & Wing, R. R. (2000). New directions in behavioral weight-loss programs. Diabetes Spectrum, 13(3), 142-148. Retrieved November 13, 2019, from http://journal.diabetes.org/diabetesspectrum/00v13n3/pg142.htm.

Curry, S. J., Krist, A. H., Owens, D. K., Barry, M. J., Caughey, A. B., Davidson, K. W., ... & Kubik, M. (2018). Behavioral weight loss interventions to prevent obesity-related morbidity and mortality in adults: US preventive services task force recommendation statement. Jama, 320(11), 1163-1171.

Donnelly, J. E., Blair, S. N., Jakicic, J. M., Manore, M. M., Rankin, J. W., & Smith, B. K. (2009). Appropriate physical activity intervention strategies for weight loss and prevention of weight regain for adults. Medicine & Science in Sports & Exercise, 41(2), 459-471.

Knowler, W. C., Fowler, S. E., Hamman, R. F., Christophi, C. A., Hoffman, H. J., Brenneman, A. T., ... & Nathan, D. M. (2009). 10-year follow-up of diabetes incidence and weight loss in the Diabetes Prevention Program Outcomes Study. Lancet (London, England), 374(9702), 1677-1686.

jensen, M. D., Ryan, D. H., Apovian, C. M., Ard, J. D., Comuzzie, A. G., Donato, K. A., ... & Loria, C. M. (2014). 2013 AHA/ACC/TOS guideline for the management of overweight and obesity in adults: a report of the American College of Cardiology/American Heart Association Task Force on Practice Guidelines and The Obesity Society. Journal of the American College of Cardiology, 63(25 Part B), 2985-3023.

Perreault, L. (2022). Obesity in adults: Drug therapy—UpToDate. UpToDate. https://www.uptodate.com/contents/obesity-in-adults-drug-therapy

Satter, E. (2007). Eating competence: Nutrition education with the Satter eating competence model. Journal of Nutrition Education and Behavior, 39(5), S189-S194. Retrieved November 12, 2019, from https://www.jneb.org/article/S1499-4046(07)00467-8/abstract.

The Satter Eating Competence Model. Retrieved November 10, 2019, from https://www.ellynsatterinstitute.org/satter-eating-competence-model/.

The Health at Every Size Approach. Association for Size Diversity and Health. Retrieved November 10, 2019, from https://www.sizediversityandhealth.org/content.asp?id=152.

UNIT-8

VITAMINS AND MINERALS-I

VITAMINS AND MINERALS: AN INTRODUCTION

Vitamins and minerals are micronutrients, and by definition, they make up a relatively small part of our diet. However, when it comes to vitamins and minerals, a little bit goes a long way. They have many essential jobs in our bodies.

For example, if you've taken a drink of water today, you can thank the minerals that serve as electrolytes, helping to balance fluids in the body. If you've taken a breath of air, you can thank the vitamins and minerals that act as antioxidants, protecting vital molecules from free radical damage. If you've taken a step, you can thank the vitamin D, calcium, and other minerals that make your bones strong. If you've moved a muscle, you can thank the many vitamins and minerals that serve as cofactors in metabolic reactions, which unlock the energy contained in nutrients so that your body can use it.

There are some 13 vitamins and 16 minerals important to human nutrition, and each serves multiple functions in the body. Entire books have been written about each one, and we could easily spend a whole term learning about all of these amazing nutrients. But as this is an introductory course, we'll use the next two units to introduce you to some of the most interesting vitamins and minerals, with a focus on those that are commonly limiting in the human diet.

We'll begin this unit with a general introduction to vitamins and minerals, and we'll consider the role of dietary supplements in meeting our vitamin and mineral requirements. Then, we'll spend the remainder of this unit and the next exploring major functions of vitamins and minerals, where we find them in food, and what happens if we consume too little or too much of each.

Unit Learning Objectives

After completing this unit, you should be able to:

- Classify the vitamins as fat-soluble or water-soluble, including differences in absorption, storage, and toxicity.
- Identify the major minerals and trace minerals, including factors that impact absorption and bioavailability.
- Identify common food sources of vitamins and minerals and how processing affects nutrient retention in foods.
- Describe how vitamins can be made in the body through provitamins and intestinal bacteria.
- Define dietary supplements and describe how supplements are regulated and the concerns with their safety and efficacy.
- Identify guidelines and recommendations for choosing nutrition supplements and for their appropriate use.
- Describe the role of electrolytes in fluid balance, as well as the more specific functions, food sources, and effects of deficiency and toxicity for sodium, potassium, and chloride.
- Describe the general function of antioxidants, as well as the more specific functions, food sources, and effects of deficiency and toxicity for vitamin E, vitamin C, and selenium.
- Describe how vitamin A and beta-carotene contribute to normal vision, and know common food sources and effects of deficiency and toxicity of vitamin A.

VITAMINS AND MINERALS CLASSIFICATION

Vitamins and minerals are only needed in small quantities in the body, but their role is essential to overall health and proper functioning of all body systems. And while many vitamins and minerals work together to perform various functions in the body, they are classified based on their independent characteristics. These characteristics impact not only how we obtain them in our diets, but also how we absorb them and store them, as well as how we experience deficiencies or toxicities when too little or too much is consumed. After we review the classifications for vitamins and minerals, we will examine key vitamins and minerals based on their similar functions to further highlight the importance of how these micronutrients work together.

Vitamins

The name "vitamin" comes from Casimir Funk, who in 1912 thought "vital amines" (similar to amino acids) were responsible for preventing what we know now as vitamin deficiencies. He coined the term "vitamines" to describe these organic substances that were recognized as essential for life, yet unlike other organic nutrients (carbohydrates, protein, and fat), do not provide energy to the body. Eventually, when scientists discovered that these compounds were not amines, the 'e' was dropped to form the term "vitamins."[1]

Vitamins Classification

Vitamins are essential, non-caloric, organic micronutrients. There is energy contained in the chemical bonds of vitamin molecules, but our bodies don't make the enzymes to break these bonds and release their energy; instead, vitamins serve other essential functions in the body. Vitamins are traditionally categorized into two groups: water-soluble or fat-soluble. Whether vitamins are water-soluble or fat-soluble can affect their functions and sites of action. For example, water-soluble vitamins often act in the cytosol of cells (the fluid inside of cells) or in extracellular fluids such as blood, while fat-soluble vitamins play roles such as protecting cell membranes from free radical damage or acting within the cell's nucleus to influence gene expression.

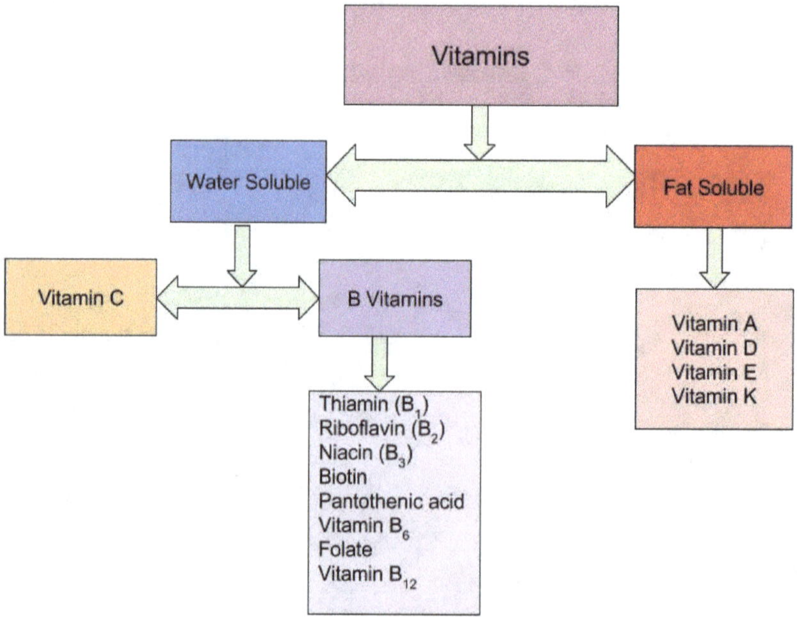

Figure 8.1. Classification of vitamins as water-soluble or fat-soluble.

One major difference between water-soluble and fat-soluble vitamins is the way they are absorbed in the body. Water-soluble vitamins are absorbed directly from the small intestine into the bloodstream. Fat-soluble vitamins are first incorporated into chylomicrons, along with fatty acids, and transported through the lymphatic system to the bloodstream and then on to the liver. The bioavailability (i.e., the amount that gets absorbed) of these vitamins is dependent on the food composition of the diet. Because fat-soluble vitamins are absorbed along with dietary fat, if a meal is very low in fat, the absorption of the fat-soluble vitamins in that meal may be impaired.

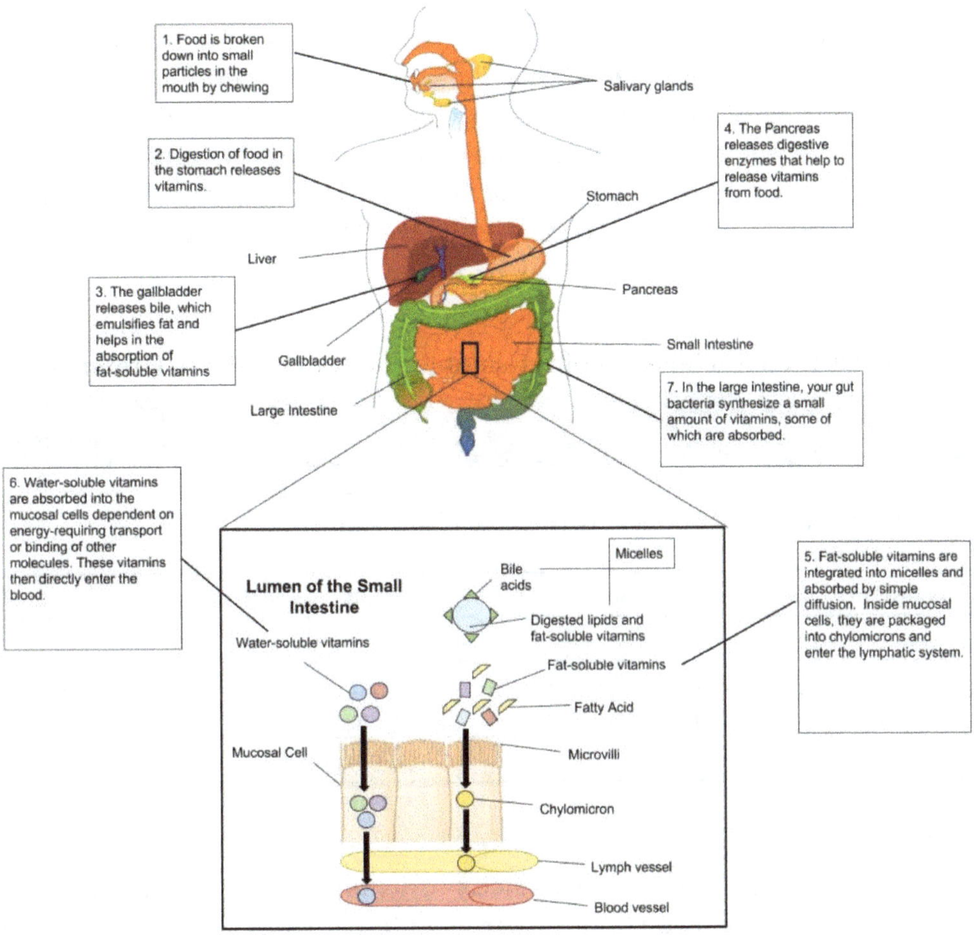

Figure 8.2. "Absorption of Fat-Soluble and Water-Soluble Vitamins."

Fat-soluble and water-soluble vitamins also differ in how they are stored in the body. The fat-soluble vitamins—vitamins A, D, E, and K—can be stored in the liver and the fatty tissues of the body. The ability to store these vitamins allows the body to draw on these stores when dietary intake is low, so deficiencies of fat-soluble vitamins may take months to develop as the body stores become depleted. On the flip side, the body's storage capacity for fat-soluble vitamins increases the risk for toxicity. While toxic levels are typically only achieved through vitamin supplements, if large quantities of fat-soluble vitamins are consumed, either through foods or supplements, vitamin levels can build up in the liver and fatty tissues, leading to symptoms of toxicity.

There is limited storage capacity in the body for water-soluble vitamins, thus making it important to consume these vitamins on a daily basis. Deficiency of water-soluble vitamins is more common than fat-soluble vitamin deficiency because of this lack of storage. That also means toxicity of water-soluble vitamins is rare. Because of their solubility in water, intake of these vitamins in amounts above what is needed by the body can, to some extent, be excreted in the urine, leading to a lower risk of toxicity. Similar to fat-soluble vitamins, a toxic intake of water-soluble vitamins is not common through food sources, but is most frequently seen due to supplement use.

Table 8.1. Characteristics of fat-soluble and water-soluble vitamins.

Characteristics of Fat-Soluble Vitamins	Characteristics of Water-Soluble Vitamins
Protect cell membranes from free radical damage; act within the cell's nucleus to influence gene expression	Act in the cytosol of cells or in extracellular fluids such as blood
Absorbed into lymph with fats from foods	Absorbed directly into blood
Large storage capacity in fatty tissues	Little to no storage capacity
Do not need to be consumed daily to prevent deficiency (may take months to develop)	Need to be consumed regularly to prevent deficiency
Toxicity is more likely	Toxicity is rare

Minerals

Similarly to vitamins, minerals are micronutrients that are essential to human health and can be obtained in our diet from different types of food. Minerals are abundant in our everyday lives. From the soil in your front yard to the jewelry you wear on your body, we interact with minerals constantly. Minerals are inorganic elements in their simplest form, originating from the Earth. They can't be broken down or used as an energy source, but like vitamins, serve essential functions based on their individual characteristics. Living organisms can't make minerals, so the minerals our bodies need must come from the diet. Plants obtain minerals from the soil they grow in. Humans obtain minerals from eating plants, as well as indirectly from eating animal products (because the animal consumed minerals in the plants it ate). We also get minerals from the water we drink. The mineral content of soil and water varies from place to place, so the mineral composition of foods and water differs based on geographic location.

Minerals Classification

Minerals are classified as either major minerals or trace minerals, depending on the amount needed in the body. Major minerals are those that are required in the diet in amounts larger than 100 milligrams each day. These include sodium, potassium, chloride, calcium, phosphorus, magnesium, and sulfur. These major minerals can be found in many foods. While deficiencies are possible with minerals, consuming a varied diet significantly improves an individual's ability to meet their nutrient needs. We'll discuss the concern of both deficiencies and toxicities of specific minerals later in this unit.

Trace minerals are classified as minerals required in the diet in smaller amounts, specifically 100 milligrams or less per day. These include iron, copper, zinc, selenium, iodine, chromium, fluoride, manganese, and molybdenum. Although trace minerals are needed in smaller amounts, a deficiency of a trace mineral can be just as detrimental to your health as a major mineral deficiency.

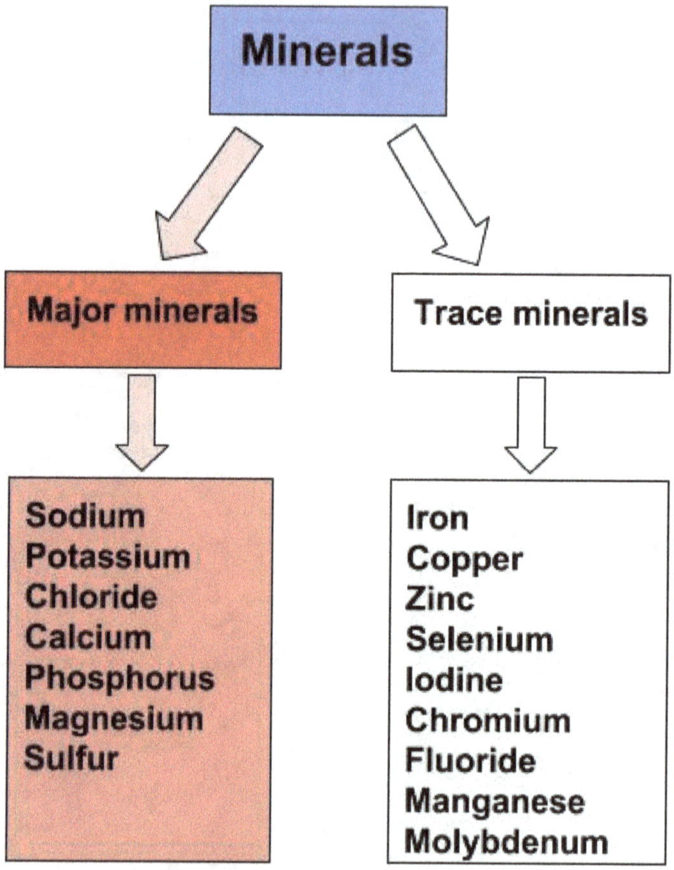

Figure 8.3. The classification of minerals as either major minerals or trace minerals.

Minerals are water-soluble and do not require enzymatic digestion. They are absorbed directly into the bloodstream, although some minerals need the assistance of transport proteins for absorption and transport in blood. Minerals are not as efficiently absorbed as most vitamins, and many factors influence their bioavailability:

- Minerals are generally better absorbed from animal-based foods. Plant-based foods often contain compounds that can bind to minerals and inhibit their absorption (e.g., oxalates, phytates).
- In most cases, if dietary intake of a particular mineral is increased, absorption will decrease.
- Some minerals influence the absorption of others. For instance, excess zinc in the diet can impair iron and copper absorption. Conversely, certain vitamins enhance mineral absorption. For example, vitamin C boosts iron

absorption, and vitamin D boosts calcium and magnesium absorption.

- As is the case with vitamins, mineral absorption can be impaired by certain gastrointestinal disorders and other diseases, such as Crohn's disease and kidney disease, as well as the aging process. Thus, people with malabsorption conditions and the elderly are at higher risk for mineral deficiencies.

REFERENCES

- 1Carpenter, K. J. (2003). A short history of nutritional science: part 3 (1912–1944). The Journal of nutrition, 133(10), 3023-3032.
- 2Linus Pauling Institute-Micronutrient Information Center. (2020). Minerals. https://lpi.oregonstate.edu/mic/minerals

VITAMINS AND MINERALS SOURCES

Vitamins And Minerals In Food

Eating a variety of foods from all food groups is the best way to ensure you are getting all the micronutrients needed for a healthy diet. Each food group lends itself to specific vitamins and minerals (see Figure 8.4). Keep in mind that whole foods (e.g., fresh fruits and vegetables, whole grains, lean meats, and low-fat dairy foods such as milk and cheese) contain more vitamins and minerals than their processed counterparts. A whole baked potato contains more vitamin C, folate, and potassium than a potato that is cut, soaked, and fried to make a french fry.

Figure 8.4. Common vitamins and minerals found in each food group.

Effects of Processing on Nutrient Content in Foods

The nutrient content of foods is typically highest when foods are allowed to ripen on the plant, allowing the plant to fully develop the nutrients and phytochemicals it needs to sustain life. Harvesting plants in the peak state of ripeness helps to ensure maximum vitamin and mineral content, and consuming freshly-picked or harvested produce usually maximizes how much of those nutrients make it into our bodies to be put to use by cells. But not all foods can be consumed immediately after harvest. How foods are handled, processed, stored, and prepared can impact how much of that peak nutrient level remains in the food.

Processing of food is an important step in our food supply. Harvesting and transporting foods to communities increases access to a variety of foods. Preservation techniques like canning and freezing extend the shelf life of foods and increase their availability outside of their peak harvest season. In fact, because fresh foods can deteriorate rapidly, food processing techniques may result in better nutrient retention over time when compared to fresh items. However, some vitamins are more stable than others, and the amount retained depends on the specific vitamin and the processing technique. Water-soluble vitamins are the most susceptible to the effects of processing, though other nutrients can be impacted as well.

Table 8.2. The effects of processing on nutrient retention.

Method of Processing	Effect on Nutrient Retention	How to Minimize Nutrient Loss
Air exposure and time	Enzymes present in foods and exposure to air can destroy nutrients, because as soon as the food is harvested, the food begins to slowly decompose.	Purchase fresh items in quantities that can be used as soon as possible. Cut up foods only when ready to use. Buy local produce to cut back on transport time and air exposure.
Temperature	Cooking helps kill bacteria, makes foods more appealing, and in some situations improves bioavailability of nutrients. But high temperatures for prolonged amounts of time can destroy some vitamins.	Use fast cooking methods like microwaving, steaming, or stir-frying.
Water	Minerals and water-soluble vitamins can leach into the water.	Don't soak produce in water. Limit the amount of water used to cook foods (e.g., steam vegetables rather than boil them). Use cooking water in food preparation.

Method of Processing	Effect on Nutrient Retention	How to Minimize Nutrient Loss
Canning	High temperatures may be used, which can destroy water-soluble vitamins, but commercial techniques usually use rapid heating, which helps to reduce nutrient loss.	Choose a variety of canned goods that don't have added sugars or sodium to maximize nutrient density.
Freezing	Freezing does not reduce nutrient content, but if foods are blanched prior to freezing it may slightly reduce levels of water-soluble vitamins.	Choose a variety of frozen goods that don't have added sugars, syrups, or sauces to maximize nutrient density.
Refining of grains	Many B vitamins, minerals, and phytochemicals, as well as fiber, are lost when whole grains are refined.	Choose whole grains whenever possible.

There is a notable exception to the effects of processing described here. In contrast to most vitamins, the bioavailability of beta-carotene, a precursor to vitamin A, and similar phytochemicals called carotenoids is actually increased by the cooking process, because cooking, chopping, and homogenizing releases carotenoids from the plant matrix. Thus, cooked carrots can be a better source of vitamin A than raw carrots. However, overcooking transforms some of the carotenoids into inactive products, and in general it is best to chop and lightly steam vegetables containing carotenoids to maximize their availability from foods.

Vitamin Synthesis In The Body

In addition to getting vitamins from the foods we eat, there are some vitamins that can be synthesized in the body. There are two ways the body can make vitamins: certain vitamins can be made from a provitamin, or a *precursor* substance that can be converted into the active form of a vitamin; other vitamins can be synthesized by bacteria living in the intestinal tract.

Vitamins Made from Precursors

Vitamins made in the body from precursors include vitamin A, vitamin D, and niacin, one of the B vitamins.

- The active form of vitamin A, called retinol, is found in animal foods, but plants contain beta-carotene, a provitamin that can be converted to vitamin A in the body. This red-orange pigment found in fruits and vegetables is converted to vitamin A primarily in the small intestine.[2] We will discuss this conversion in more depth later in this unit.

- Vitamin D can be made when ultraviolet light from sunlight strikes cholesterol in the skin. Cholesterol, which our body can make, is a precursor for vitamin D. This process of making vitamin D from cholesterol is limited by geographic location (both latitude and altitude) and seasonal changes, both of which influence the quality, quantity, and intensity of ultraviolet rays that reach the skin.[3] We will discuss this conversion in more depth later in this unit.
- Niacin can be made in the liver from the amino acid tryptophan, when tryptophan is available in quantities greater than needed for protein synthesis. The efficiency in which tryptophan is converted to niacin varies greatly in individuals.[4]

Vitamins Made by Intestinal Bacteria

Some vitamins can be synthesized not by our bodies, but by the helpful bacteria living within us. Bacteria in the gut can make some of the vitamin K and B vitamins that our body needs, though we still need to get these vitamins from food as well.

- Bacteria that colonize the large intestine can synthesize one form of vitamin K, although the total amount made in the large intestine is not clear.[5]
- Gut bacteria are also able to make all B vitamins, though the amount synthesized of each vitamin is dependent on the composition of each individual's microbiome.[6,7] Dietary choices (e.g. intake of high fiber foods or probiotics) and medication use can alter a person's microbiome, possibly promoting or inhibiting the production of vitamins in the large intestine.[8]

REFERENCES

Devi, R. (2005). Food processing and impact on nutrition. Research Scholar, Department of Economics, Kurukshetra University, Kurukshetra, Haryana, India. Retrieved from http://saspjournals.com/wp-content/uploads/2015/08/SJAVS-24A304-311.pdf

National Institutes of Health Office of Dietary Supplements. (February 14, 2020). Vitamin A. Retrieved April 11, 2020, from https://ods.od.nih.gov/factsheets/VitaminA-HealthProfessional/

Vitamin D reference:Linus Pauling Institute-Micronutrient Information Center. (2020). https://lpi.oregonstate.edu/mic/vitamins/vitamin-D

National Institutes of Health Office of Dietary Supplements. (March 6, 2020). Niacin. Retrieved April 11, 2020, from https://ods.od.nih.gov/factsheets/Niacin-HealthProfessional/

Linus Pauling Institute-Micronutrient Information Center. (2020). Vitamin K. https://lpi.oregonstate.edu/mic/vitamins/vitamin-K

Yoshii, K., Hosomi, K., Sawane, K., & Kunisawa, J. (2019). Metabolism of dietary and microbial vitamin B family in the regulation of host immunity. Frontiers in nutrition, 6, 48. https://www.frontiersin.org/articles/10.3389/fnut.2019.00048/full#h4

Gu, Q., & Li, P. (2016). Biosynthesis of vitamins by probiotic bacteria. Probiotics and prebiotics in human nutrition and health. https://www.intechopen.com/books/probiotics-and-prebiotics-in-human-nutrition-and-health/biosynthesis-of-vitamins-by-probiotic-bacteria

Valdes, A. M., Walter, J., Segal, E., & Spector, T. D. (2018). Role of the gut microbiota in nutrition and health. Bmj, 361, k2179. https://www.bmj.com/content/361/bmj.k2179

DIETARY SUPPLEMENTS

By now you know that a balanced, nutritious diet is important for good health. But you may also wonder if your diet is adequate, or if you could benefit from taking a vitamin or mineral supplement. You may also be curious about other supplements that claim to help you lose weight, build muscle, ease joint pain, or help build good bacteria in your gut.

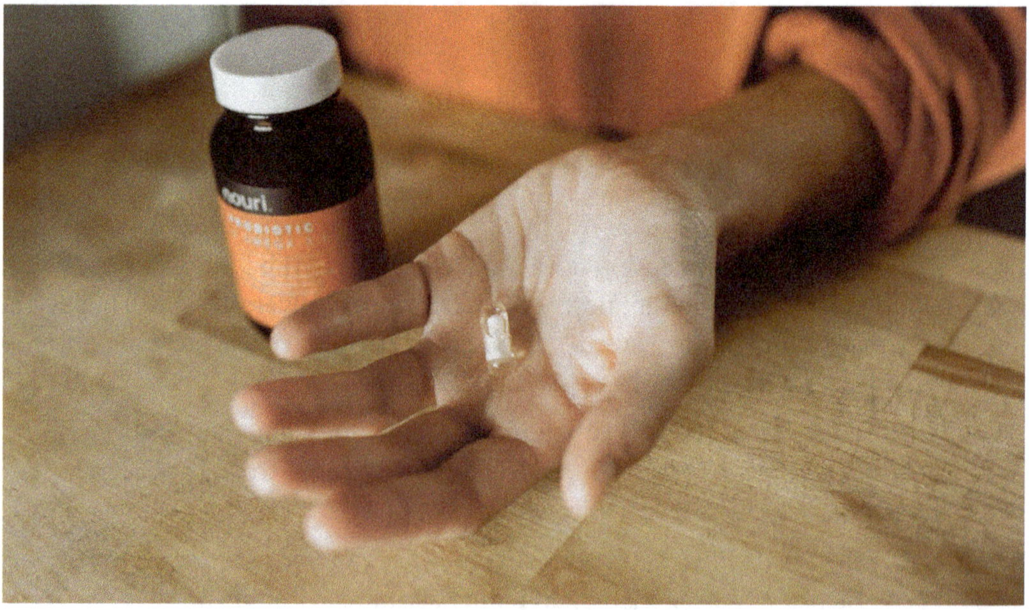

According to the results of a large national survey published in JAMA in ,2016 %52 of U.S. adults reported using supplements in [2].2012-2011 With over half of Americans reporting dietary supplement use, it's important to have a conversation about the safety and efficacy of these products. Are dietary supplements safe? Are they effective? Do Americans need a multivitamin/mineral supplement? If so, what are the guidelines and recommendations for choosing a supplement? We will explore these important questions in this section.

Regulation Of Dietary Supplements

A dietary supplement is defined as a product that:[1]

- Is intended to supplement the diet
- Contains one or more dietary ingredients (e.g., vitamins, minerals, herbs or other botanicals, amino acids, and enzymes)
- Can be taken by mouth as a pill, capsule, tablet, or liquid
- Is labeled as being a dietary supplement

The Food and Drug Administration (FDA) regulates dietary supplements, but under a different set of regulations than foods, pharmaceuticals, and over-the-counter drugs. In 1994, the Dietary Supplement Health and Education Act (DSHEA) created new regulations for the labeling and safety of dietary supplements. Under these rules, the "FDA is not authorized to review dietary supplement products for safety and effectiveness before they are marketed."[3] Therefore, the responsibility falls on the manufacturers and distributors of dietary supplements to ensure their products are safe before they go to consumers. The FDA has to prove that the product is unsafe in order to remove it from the market. This is a big contrast from pharmaceuticals, which must obtain approval from the FDA, showing substantial evidence that their drugs are safe and effective before reaching the marketplace.

The FDA issued Good Manufacturing Practices (GMPs) for dietary supplements in 2007. GMPs are a set of requirements and expectations by which dietary supplements must be manufactured, prepared, and stored to ensure quality.[4] Manufacturers are expected to guarantee the identity, purity, strength, and composition of their dietary supplements.

Once a dietary supplement is on the market, the FDA tracks side effects reported by consumers, supplement companies, and others. If the FDA finds a product to be unsafe, it can take legal action against the manufacturer or distributor and may issue a warning or require that the product be removed from the marketplace. However, the FDA says it can't test all products marketed as dietary supplements that may have potentially harmful hidden ingredients.

Safety Of Dietary Supplements

With current regulations, the safety of dietary supplements is the manufacturers' responsibility. Unfortunately, manufacturers don't always have the public's best interest in mind, especially when there is a profit to be gained.

According to the FDA website, the "FDA has found that manufacturing problems have been associated with dietary supplements. Products have been recalled because of microbiological, pesticide, and heavy metal contamination and because they do not contain the dietary ingredients they are represented to contain or they contain more or less than the amount of the dietary ingredient claimed on the label."[2]

What this means for consumers is that they should use caution when considering whether to use a dietary supplement. Keep in mind the following information from the Center for Complementary and Integrative Health at the National Institutes of Health:[1]

- What's on the label may not be what is in the product. For example, the FDA has found prescription drugs, including anticoagulants (e.g., warfarin), anticonvulsants (e.g., phenytoin), and others, in products being sold as dietary supplements. You can see a list of some of those products on the FDA's Tainted Supplements webpage.
- Supplement labels may make illegal claims to make their product more appealing. A 2012 Government study of 127 dietary supplements marketed for weight loss or to support the immune system found that 20 percent made illegal claims.
- Dietary supplements can interact with other medications, and cause harm. For example, the herbal supplement St. John's wort makes many medications less effective.

- The term natural does not always mean safe. Ephedra, an evergreen plant native to central Asia is associated with heart problems and risk of death. In 2004, the FDA banned the sale of ephedrine in dietary supplements for these reasons. Supplements can contain natural herbs and other plant-based ingredients that have not been adequately studied. We don't know if supplement ingredients are dangerous until people end up really sick or even die from them. Dietary supplements result in an estimated 23,000 emergency room visits every year in the United States, according to a 2015 study. Many of the patients are young adults having heart problems from weight-loss or energy products and older adults having swallowing problems from taking large vitamin pills.
- The term "standardized" (or "verified" or "certified") on a supplement does not always guarantee product quality or safety. These are terms used by manufacturers to sell their product and have not been legally defined.

You can report safety concerns about a dietary supplement through the U.S. Health and Human Services Safety Reporting Portal. For more information on contaminants in dietary supplements, visit the FDA's Dietary Supplement Products & Ingredients webpage.

Efficacy of Dietary Supplements

The amount of scientific evidence on dietary supplements varies widely—there is a lot of information on some and very little on others. The Center for Complementary and Integrative Health at the National Institutes of Health offers these key points about efficacy of dietary supplements:

- Dietary supplements can't be marketed with claims that they can diagnose, treat, cure, mitigate, or prevent any disease; such claims would require the product to be approved by the FDA as a pharmaceutical. Instead, dietary supplements are marketed with health claims or structure/function claims, similar to claims on food labels. Recall from Unit 1 that structure/function claims (e.g., "builds strong bones," or "boosts immunity") are intentionally vague and require no evidence to support them. Supplements are often labeled with claims that have little to no scientific basis.
- Studies have found that some dietary supplements may have benefits, such as melatonin for jet lag. Others may have little or no benefit, such as ginkgo for dementia.[1] Many dietary supplements haven't been studied at all in humans.
- Studies of many supplements haven't supported claims made about them. For example, in several studies, echinacea didn't help cure colds and Ginkgo biloba wasn't useful for dementia—but you can still find Ginkgo

biloba supplements with claims that they improve memory and echinacea supplements with claims of providing "immune support." Many times the research on a dietary supplement is conflicting, such as whether the supplements glucosamine and chondroitin improve symptoms of osteoarthritis. Research design and interpretation can also be biased when funded by the supplement industry.

- Most research shows that taking multivitamin/mineral (MVM) supplements doesn't result in living longer, slowing cognitive decline, or lowering the chance of getting cancer, heart disease, or diabetes. However, taking a multivitamin is unlikely to pose health risks, providing you follow the guidelines below for choosing supplements.

Guidelines for Choosing Supplements

Throughout this text, we have discussed the importance of whole foods. As you might suspect, supplements can not replace real, whole food. Marion Nestle, professor emerita at New York University and author of many books about nutrition, wrote eloquently about the benefits of getting nutrients from food instead of supplements in a 2006 blog post:

"Unless you have been diagnosed with a vitamin or mineral deficiency and need to replenish that nutrient in a great big hurry, it is always better to get nutrients from foods—the way nature intended. I can think of three benefits of whole foods as compared to supplements:

- You get the full variety of nutrients—vitamins, minerals, antioxidants, etc–in that food, not just the one nutrient in the supplement;
- The amounts of the various nutrients are balanced so they don't interfere with each other's digestion, absorption, or metabolism; and
- There is no possibility of harm from taking nutrients from foods (OK. Polar bear liver is an exception; its level of vitamin A is toxic).

In contrast, high doses of single nutrients not only fail to improve health but also can make things worse, as has been shown in some clinical trials of the effects of beta-carotene, vitamin E, and folic acid, for example, on heart disease or cancer. And foods taste a whole lot better, of course."5

However, there are certain populations that might be at risk for developing nutrient deficiencies, and they may benefit from a MVM supplement or supplements of specific nutrients. These groups include the elderly, strict vegetarians or vegans, people restricting their caloric intake, pregnant women, or individuals with food insecurity. If you choose to take supplements, keep moderation in mind, and use the following guidelines to help you choose a supplement.

- Don't substitute for whole foods. According to the 2020 Dietary Guidelines for Americans, "Because foods provide an array of nutrients and other components that have benefits for health, nutritional needs should be met primarily through foods."[7] Whole foods are complex and not only contain essential vitamins and minerals, but also dietary fiber and phytochemicals that may have positive health benefits. As their name suggests, supplements should never act as replacements for whole food, but rather as supplements to fill in some nutritional gaps.
- Check the dose carefully. Since dietary supplements are not regulated before they hit the market, it is not uncommon to find nutrient levels that exceed the upper intake level (UL). A good rule-of-thumb is to choose a supplement that keeps the dose close to %100 of the Daily Value (unless advised by a doctor to help correct a deficiency) and definitely no more than the UL.
- If getting supplements from multiple sources, make sure you add together the doses. Supplements not only come in pill form, but also powder and liquid form. Vitamin water, protein powder, and other products like Emergen-C are often fortified with large amounts of vitamins and minerals.
- Be skeptical of product claims. Remember that supplement manufacturers promote their products with structure/function claims, and they don't have to provide any evidence that the product actually does what it claims to do. If something sounds too good to be true, it probably is.
- Look for third-party testing when purchasing a supplement. ConsumerLab.com, NSF International, U.S. Pharmacopeia (USP), and UL are all companies that do third-party testing on dietary supplements.[8] If you see these companies' stamps (see Figure 8.5) on a supplement bottle, it means that the product is periodically tested to check that it:
- Contains the ingredients listed on the label, in the declared potency and amounts.
- Does not contain harmful levels of specified contaminants (e.g., heavy metals and pesticides).
- Will break down and release into the body within a specified amount of time. (If a supplement does not break down properly to allow its ingredients to be available for absorption in the body, the consumer will not get the full benefit of its contents.)

Figure 8.5. The U.S. Pharmacopeia (USP) verification mark. USP is a nonprofit organization that does third-party testing on dietary supplements. This mark helps assure consumers that the product has been tested for quality, purity, potency, performance, and consistency.

- Choose a MVM supplement that is tailored to your age, sex, and other characteristics (e.g., pregnancy). This is important because different populations have different nutrient needs. MVMs for seniors typically provide more calcium and vitamin D for bone health than MVMs for younger adults. MVMs for women contain more iron than MVMs for men. Prenatal supplements generally provide no vitamin A as retinol, and most children's MVMs provide age-appropriate amounts of nutrients.10
- Check with your healthcare provider to ensure the supplement you are considering is safe for you. Supplements can interact with both prescription medications and over-the-counter medications potentially causing life-threatening complications.

REFERENCES

National Institutes of Health Center for Complementary and Integrative Health. Using Dietary Supplements Wisely. Retrieved April 1, 2020, from https://www.nccih.nih.gov/health/using-dietary-supplements-wisely

Kantor E.D., Rehm C.D., Du M., White E., Giovannucci E.L. (2016). Trends in dietary supplement use among US adults from 1999–2012. JAMA, 316, 1464–1474. doi: 10.1001/jama.2016.14

Information for Consumers Using Dietary Supplements. U.S. Food and Drug Administration. Retrieved April 1, 2020, from https://www.fda.gov/food/dietary-supplements/inf ormation-consumers-using-dietary-supplements

What You Need to Know About Dietary Supplements. U.S. Food and Drug Administration. Retrieved April 1, 2020 from https://www.fda.gov/food/buy-store-serve-safe-food/what-you-need-know-about-dietary-supplements?utm_campaign=buffer&utm_content=buffer6d184&utm_medium=social&utm_source=facebook.com

Nestle, M. (2007). Food vs. Supplements. Food Politics. Retrieved from https://www.foodpolitics.com/2007/06/foods-vs-supplements/

Vitamins Minerals and Supplements: Do You Need to Take Them? Academy of Nutrition and Dietetics. Retrieved April 1, 2020, from https://www.eatright.org/food/vitamins-and-supplements/dietary-supplements/vitamins-minerals-and-supplements-do-you-need-to-take-them

U.S. Department of Agriculture and U.S. Department of Health and Human Services. (2020). Dietary Guidelines for Americans, 2020-2025, 9th Edition. Retrieved from https://www.dietaryguidelines.gov/

Loria, K. (2019, October 30). How To Choose Supplements Wisely. Consumer Reports. Retrieved from https://www.consumerreports.org/supplements/how-to-choose-supplements-wisely/

Dietary Supplements Verification Program. USP. Retrieved April 3, 2020, from https://www.usp.org/verification-services/dietary-supplements-verification-program

Multivitamin/Mineral Supplements. National Institutes of Health Office of Dietary Supplements. Retrieved April 3, 2020, from https://ods.od.nih.gov/factsheets/MVMS-HealthProfessional/

VITAMINS AND MINERALS NECESSARY FOR MAINTAINING FLUID AND ELECTROLYTE BALANCE

Water is the foundation of all life. The surface of the earth is 70% water, and human beings are mostly water, ranging from about 75% of body mass in infants, 50–60% in adults, and as low as 45% in old age. (The percent of body water changes with development, because the proportions of muscle, fat, bone, and other tissues change from infancy to adulthood.) Of all the nutrients, water is the most critical, as its absence proves lethal within a few days. The importance of water in the human body can be loosely categorized into four basic functions: transportation vehicle, medium for chemical reactions, lubricant/shock absorber, and temperature regulator. Maintaining the right level of water in your body is crucial to survival, as either too little or too much will result in less-than-optimal functioning. Several minerals are key to regulating water balance in different compartments of the body; the most important of these are sodium, potassium, and chloride.

Water Distribution And Composition

In the human body, water is distributed into two compartments: inside cells, called intracellular fluid (ICF), and outside cells, called extracellular fluid (ECF). Extracellular fluid includes both the fluid component of the blood (called plasma) and the interstitial fluid (IF) that surrounds all cells not in the blood (Figure 8.6).

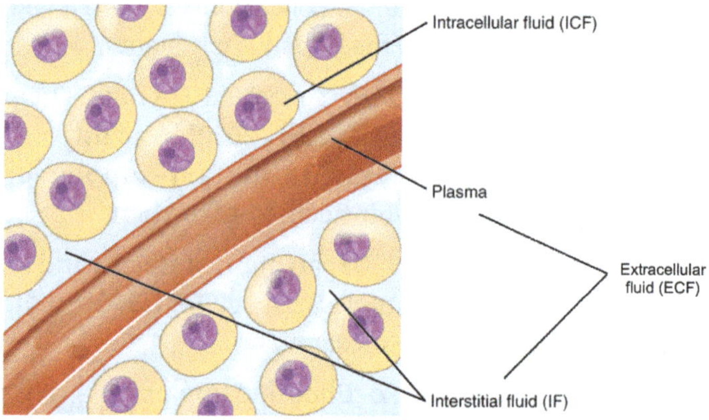

Figure 8.6. Fluid compartments in the human body. The intracellular fluid (ICF) is the fluid within cells. The extracellular fluid (ECF) includes both the blood plasma and the interstitial fluid (IF) between the cells.

Although water makes up the largest percentage of body volume, it is not actually pure water, but rather a mixture of dissolved substances (solutes) that are critical to life. These solutes include electrolytes, substances that dissociate into charged ions when dissolved in water. For example, sodium chloride (the chemical name for table salt) dissociates into sodium (Na^+) and chloride (Cl^-) in water. In extracellular fluid, sodium is the major positively-charged electrolyte (or cation), and chloride (Cl^-) is the major negatively-charged electrolyte (or anion). Potassium (K^+) is the major cation inside cells. Together, these electrolytes are involved in many body functions, including water balance, acid-base balance, and assisting in the transmission of electrical impulses along cell membranes in nerves and muscles.

Fluid And Electrolyte Balance

One of the essential homeostatic functions of the body is to maintain fluid and electrolyte balance within cells and their surrounding environment. Cell membranes are selectively permeable: Water can move freely through the cell membrane, while other substances, such as electrolytes, require special transport proteins, channels, and often energy. The movement of water between the intracellular and extracellular fluid happens by osmosis, which is simply the movement of water through a selectively permeable membrane from an area where solutes are less concentrated to an area where solutes are more concentrated (Figure 8.7).

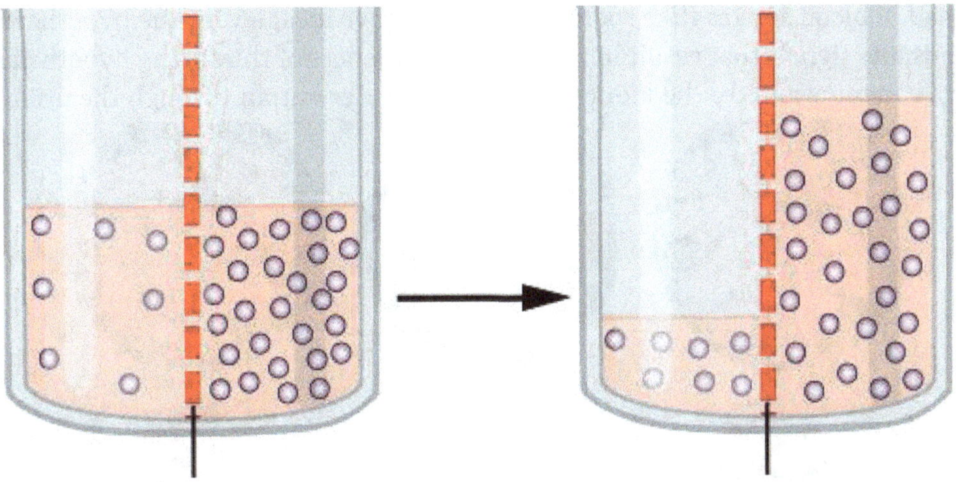

Figure 8.7. Osmosis is the diffusion of water through a semipermeable membrane towards higher solute concentration. If a membrane is permeable to water but not a solute, water will equalize its own concentration by diffusing to the side of lower water concentration (and thus the side of higher solute concentration). In the beaker on the left, the solution on the right side of the membrane is more concentrated with solutes; therefore, water diffuses to the right side of the beaker to equalize its concentration.

To maintain water and electrolyte balance, cells control the movement of electrolytes across their membranes, and water follows the electrolytes by osmosis. The health of the cell depends on proper fluid and electrolyte balance. If the body's fluid and electrolyte levels change too rapidly, cells can struggle to correct the imbalance quickly enough. For example, consider a person exercising strenuously, losing water and electrolytes in the form of sweat, and drinking excessive amounts of water. The excess water dilutes the sodium in the blood, leading to hyponatremia, or low blood sodium concentrations. Sodium levels within the cells are now more concentrated, leading water to enter the cells by osmosis. As a result, the cells swell with water and can burst if the imbalance is severe and prolonged.

In contrast, the opposite situation can occur in a person exercising strenuously for a long duration with inadequate fluid intake. This can lead to dehydration and hypernatremia, or elevated blood sodium levels. The high concentration of sodium in the extracellular fluid causes water to leave cells by osmosis, making them shrink (Figure 8.8). This scenario can also occur anytime a person is dehydrated because of significant fluid loss, such as from diarrhea and/or vomiting caused by illness. When a person becomes dehydrated, and solutes like sodium become too concentrated in the blood, the thirst response is triggered. Sensory receptors in the thirst center in the hypothalamus monitor the concentration of solutes of the blood. If blood solutes (like sodium) increase above ideal levels, the hypothalamus transmits signals that result in a conscious awareness of thirst. The hypothalamus also communicates to the kidneys to decrease water output through the urine.

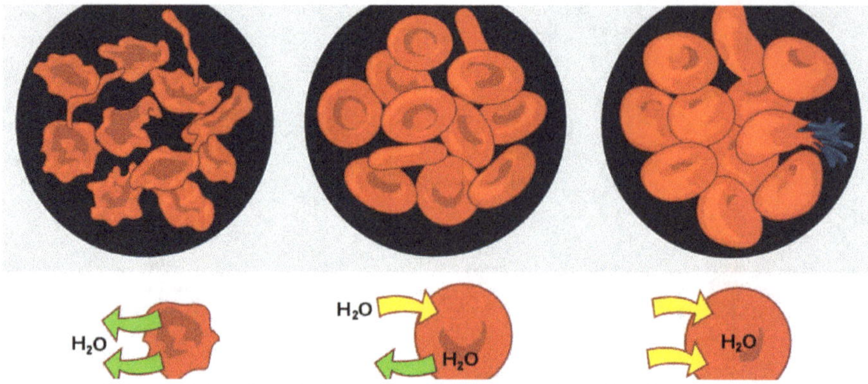

Figure 8.8. Effect of fluid imbalance on cells. With dehydration, the concentration of electrolytes becomes greater outside of cells, leading to water leaving cells and making them shrink. In fluid balance, electrolyte concentrations are in balance inside and outside of cells, so water is in balance too. During overhydration, electrolyte concentrations are low outside the cell relative to inside the cell (like in the situation of hyponatremia), so water moves into the cells, making them swell.

The cell is able to control the movement of the two major cations, sodium and potassium, with a sodium-potassium pump (Na^+/K^+ pump). This pump transports sodium out of cells while moving potassium into cells.

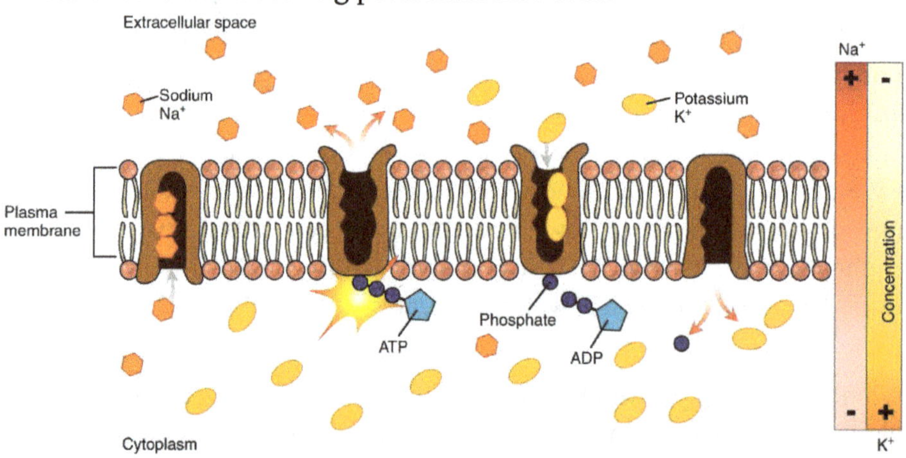

Figure 8.9. The sodium-potassium pump is found in many cell (plasma) membranes. Powered by ATP, the pump moves sodium and potassium ions in opposite directions, each against its concentration gradient. In a single cycle of the pump, three sodium ions are extruded from and two potassium ions are imported into the cell.

The Na^+/K^+ pump is an important ion pump found in the membranes of many types of cells and is particularly abundant in nerve cells. When a nerve cell is stimulated (e.g., the touch of a hand), there is an influx of sodium ions into the nerve cell. Similar to how a current moves along a wire, a sodium current moves along a nerve cell. Stimulating a muscle contraction also involves the movement of sodium ions. For a muscle to contract, a nerve impulse travels to a muscle. The movement of the sodium current in the nerve signals the muscle cell membrane to open and sodium rushes in, creating another current that travels along the muscle and eventually leading to muscle contraction. In both nerve and muscle cells, the sodium that went in during a stimulus now has to be moved out by the sodium-potassium pump in order for the nerve and muscle cell to be stimulated again.

Sodium

Although sodium often gets vilianized because of its link to hypertension, it is an essential nutrient that is vital for survival. As previously discussed, it is not only important for fluid balance, but also nerve impulse transmission and muscle contraction.

Food Sources of Sodium

Sodium can be found naturally in a variety of whole foods, but most sodium in the typical American diet comes from processed and prepared foods. Manufacturers add salt to foods to improve texture and flavor, and also to act as a preservative. Even foods that you wouldn't consider to be salty, like breakfasts cereals, can have greater than 10% of the DV for sodium. Most Americans exceed the adequate intake recommendation of 1500 mg per day, averaging 3,393 mg per day.[1] The sources of sodium in the American diet are shown below.

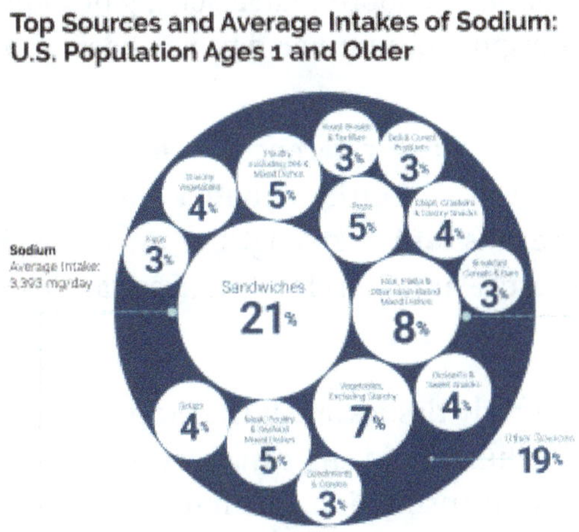

Figure 8.10. Top sources and average intake of sodium in the U.S. population, ages 1 year and older.6

This slideshow from WebMD, "Sources of Salt and How to Cut Back," offers some tools for reducing dietary sodium.

Sodium Deficiency and Toxicity

Deficiencies of sodium are extremely rare since sodium is so prevalent in the American diet. It is too much sodium that is the main concern. High dietary intake of sodium is one risk factor for hypertension, or high blood pressure. In many people with hypertension, cutting salt intake can help reduce their blood pressure. However, studies have shown that this isn't always the case. According to Harvard Medical School, "About 60% of people with high blood pressure are thought to be salt-sensitive — [a trait that means your blood pressure increases with a high-sodium diet]. So are about a quarter of people with normal blood pressure, although they may develop high blood pressure later, since salt sensitivity increases with age and weight gain."[2] Genetics, race, sex, weight, and physical activity level are determinants of salt sensitivity. African Americans, women, and overweight individuals are more salt-sensitive than others.

The Dietary Approaches to Stop Hypertension (DASH) is an eating pattern that has been tested in randomized controlled trials and shown to reduce blood pressure and LDL cholesterol levels, resulting in decreased cardiovascular disease risk. The DASH plan recommends focusing on eating vegetables, fruits, and whole grains, as well as including fat-free or low-fat dairy products, fish, poultry, beans, nuts, and vegetable oils; together, these foods provide a diet rich in key nutrients, including potassium, calcium, magnesium, fiber, and protein. DASH also recommends limiting foods high in saturated fat (e.g., fatty meats, full-fat dairy products, and tropical oils such as coconut or palm oils), sugar-sweetened beverages, and sweets. DASH also suggests consuming no more than 2,300 mg of sodium per day and notes that reduction to 1,500 mg of sodium per day has been shown to further lower blood pressure.[1]

Although the updated dietary reference intake (DRI) for sodium does not include an upper intake level (UL), the updated adequate intake (AI) considers chronic disease risk.[3] There is a high strength of evidence that reducing sodium intake reduces blood pressure and therefore reduces cardiovascular disease risk.[4]

Potassium

Potassium is present in all body tissues and is the most abundant positively charged electrolyte in the intracellular fluid. As discussed previously, it is required for proper fluid balance, nerve transmission, and muscle contraction.[5]

Food Sources of Potassium

Potassium is found in a wide variety of fresh plant and animal foods. Fresh fruits and vegetables are excellent sources of potassium, as well as dairy products (e.g., milk and yogurt), beans (e.g., lentils and soybeans), and meat (e.g., salmon and beef).

The 2020-2025 Dietary Guidelines for Americans identifies potassium as a "dietary component of public health concern," because dietary surveys consistently show that people in the United States consume less potassium than is recommended.[6] This is a nutritional gap that must be corrected through food since most dietary supplements do not contain significant amounts of potassium.

Figure 8.11. Dietary sources of potassium. Source: Dietary Guidelines for Americans, 2015-2020

Potassium Deficiency and Toxicity

Low potassium intake may have negative health implications on blood pressure, kidney stone formation, bone mineral density, and type 2 diabetes risk. Although there is a large body of evidence that has found a low potassium intake increases the risk of hypertension, especially when combined with high sodium intake, and higher potassium intake may help decrease blood pressure, especially in salt-sensitive individuals, the body of evidence to support a cause-and-effect relationship is limited and inconclusive. However, it is important to remember that a lack of evidence does not mean there is a lack of effect of potassium intake on chronic disease outcomes. This is an area that needs more research to determine the effect dietary potassium has on chronic disease risk.

There is no UL set for potassium since healthy people with normal kidney function can excrete excess potassium in the urine, and therefore high *dietary* intakes of potassium do not pose a health risk.[7] However, the absence of a UL does not mean that there is no risk from excessive *supplemental* potassium intake, and caution is warranted against taking high levels of supplemental potassium.

Chloride

Chloride helps with fluid balance, acid-base balance, and nerve cell transmission. It is also a component of hydrochloric acid, which aids digestion in the stomach.[9]

Table salt is 60% chloride, so most chloride in the diet comes from salt. Each teaspoon of salt contains 3.4 grams of chloride. The chloride AI for adults is 2.3 grams. Therefore, the chloride requirement can be met with less than a teaspoon of salt each day. Other dietary sources of chloride include tomatoes, lettuce, olives, celery, rye, whole-grain foods, and seafood.

Chloride deficiency is rare since most foods containing sodium also provide chloride, and sodium intake in the American diet is high.[9]

REFERENCES:

National Heart, Lung, and Blood Institute. DASH Eating Plan. Retrieved February 9, 2021, from https://www.nhlbi.nih.gov/health-topics/dash-eating-plan

Harvard Health Publishing. (August 2019). Salt Sensitivity: Sorting out the science. Retrieved from https://www.health.harvard.edu/heart-health/salt-sensitivity-sorting-out-the-science

The National Academies of Sciences, Engineering and Medicine. Dietary Reference Intakes for Sodium and Potassium. (2019). Chapter 14 – Sodium: Dietary Reference Intakes for Toxicity. Retrieved from https://www.nap.edu/read/25353/chapter/14

The National Academies of Sciences, Engineering and Medicine. Dietary Reference Intakes for Sodium and Potassium. (2019). Chapter 13 – Sodium: Dietary Reference Intakes for Adequacy. Retrieved from https://www.nap.edu/read/25353/chapter/13

National Institutes of Health of Dietary Supplements. Potassium – Health Professional Fact Sheet. Retrieved April 6, 2020 from https://ods.od.nih.gov/factsheets/Potassium-HealthProfessional/

U.S. Department of Agriculture and U.S. Department of Health and Human Services. (2020). Dietary Guidelines for Americans, 2020-2025, 9th Edition. Retrieved from https://www.dietaryguidelines.gov/

The National Academies of Sciences, Engineering and Medicine. Dietary Reference Intakes for Sodium and Potassium. (2019). Chapter 10 – Potassium: Dietary Reference Intakes Based on Chronic Disease. Retrieved from https://www.nap.edu/read/25353/chapter/10

The National Academies of Sciences, Engineering and Medicine. Dietary Reference Intakes for Sodium and Potassium. (2019). Chapter 9 – Potassium: Dietary

Reference Intakes for Toxicity. Retrieved from https://www.nap.edu/read/25353/chapter/9

The National Academies of Sciences, Engineering and Medicine. Dietary Reference Intakes for Water, Potassium, Sodium Chloride and Sulfate. (2005). Chapter 6 – Sodium and Chloride. Retrieved from https://www.nap.edu/read/10925/chapter/8#272

AS ANTIOXIDANTS, VITAMINS AND MINERALS

What are antioxidants, and why do we need them?

Recall from Unit 1 that atoms are composed of a nucleus, which contains neutrons and protons, and electrons, which orbit the nucleus. Atoms are most stable when they have an even number of electrons, so that they can orbit the nucleus in pairs.

An atom or group of atoms with an unpaired electron is called a free radical. Free radicals are inherently unstable and highly reactive. They steal electrons from other molecules in order to stabilize themselves, but in doing so, they create additional free radicals. This electron-grabbing is called oxidation and can set up a chain reaction, creating new free radicals and damaging important molecules along the way, similar to how one falling domino can bring down countless more.

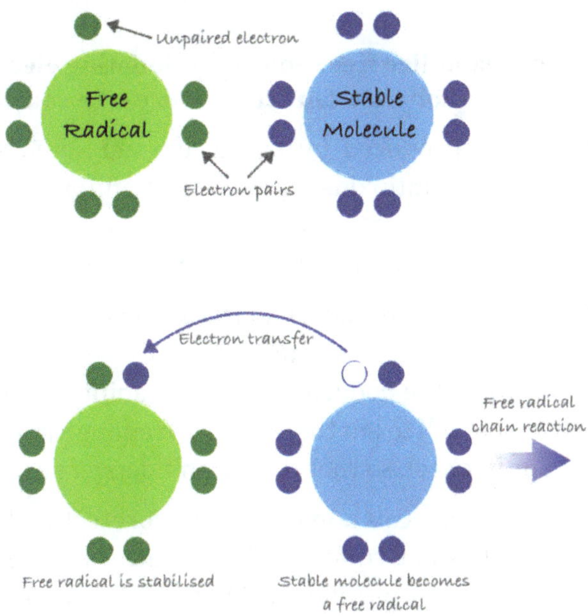

Figure 8.12. A free radical is a molecule with an unpaired electron. It can steal electrons from a stable molecule, creating a new free radical and initiating a chain reaction.

Antioxidants are molecules that can donate an electron to stabilize and neutralize free radicals. Like a domino that refuses to fall, an antioxidant can stop the free radical chain reaction in its tracks. In donating an electron, the antioxidant itself becomes a free radical. However, antioxidants are special in that they are not very reactive themselves and have processes for quick stabilization.

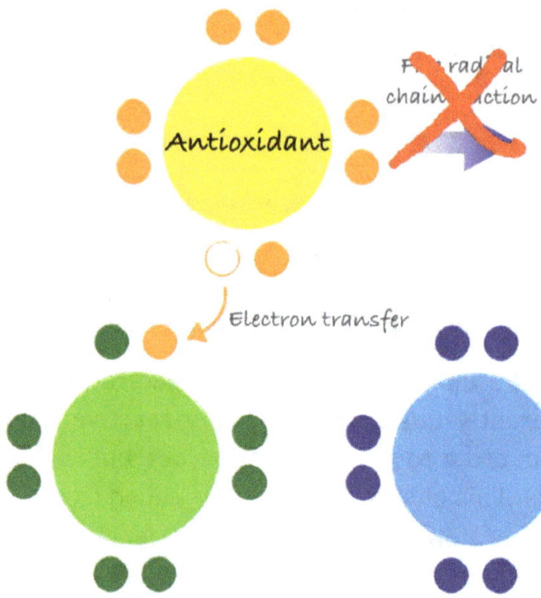

Figure 8.13. Antioxidants stabilize free radicals by donating electrons, preventing the chain reaction that can create more free radicals.

Some antioxidants are produced by the body, and some are consumed in the diet. Examples of dietary antioxidants include vitamins C and E, the mineral selenium, and phytochemicals, such as beta-carotene. We'll focus on vitamins C and E and selenium on this page, and we'll discuss beta-carotene on the next page.

Free radicals are a natural byproduct of metabolic reactions and of exercise, and it's normal to have low levels of free radicals in the body. In fact, free radicals play a role in normal functioning of the body, including the ability to fight off pathogens and to send signals from one cell to another. And with enough antioxidants present, free radicals can be kept in check so that they aren't dangerous.[1]

However, too many free radicals and not enough protection from antioxidants creates a situation called oxidative stress. Free radical-induced damage, when left unrepaired, destroys lipids, proteins, RNA, and DNA, and can contribute to disease. Oxidative stress has been implicated as a contributing factor to cancer, cardiovascular disease, arthritis, diabetes, kidney disease, Alzheimer's disease, Parkinson's disease, and eye diseases such as cataracts.[2]

Substances and energy sources from the environment can add to or accelerate the production of free radicals within the body. Exposure to UV radiation (e.g., from sunlight), air pollution, tobacco smoke, heavy metals, ionizing radiation, asbestos, and other toxic chemicals increase the amount of free radicals in the body. They do so by being free radicals themselves or by adding energy that provokes electrons to move between atoms. Excessive exposure to environmental sources of free radicals can contribute to disease by overwhelming the free radical detoxifying systems and those processes involved in repairing oxidative damage.

Oxidative stress is associated with the development of chronic diseases, and eating a diet rich in antioxidant-containing foods like fruits, vegetables, and whole grains seems to protect against many of these same diseases. It was thus natural for researchers to hypothesize that taking antioxidants in supplement form might also offer protection from these diseases. However, several decades of research investigating this hypothesis have revealed disappointing results. Not only have these studies shown that antioxidant supplements generally aren't beneficial, some have shown that they can cause health risks. For example, high doses of beta-carotene increased the risk of lung cancer in smokers, and high doses of vitamin E increased the risk of prostate cancer in men. Antioxidant supplements may also interact with other medications, further emphasizing the importance of talking with your doctor before taking supplements.[2]

Researchers aren't sure why some antioxidant supplements have turned out to be dangerous. It may come back to the fact that free radicals play important roles in the body, and adding high doses of antioxidant supplements overwhelms the normal balance of free radicals and does more harm than good. It may also be that the benefits of eating antioxidant-rich foods come not just from the antioxidants but from the entire package of nutrients, like fiber and phytochemicals in the whole foods, a combination that simply can't be replicated in a pill. Regardless, you can obtain adequate levels of dietary antioxidants simply by eating a healthy diet.[1,2]

Let's take a closer look at several of the most important dietary antioxidants: vitamin E, vitamin C, and selenium.

Vitamin E

When we talk about vitamin E, we're actually referring to 8 chemically similar substances, of which alpha-tocopherol appears to be the most potent antioxidant. Because vitamin E is fat-soluble, its antioxidant capacity is especially important to lipids, including those in cell membranes and lipoproteins. For example, free radicals can oxidize LDL cholesterol (stealing an electron from it), and it is this damaged LDL that lodges in blood vessels and forms the fatty plaques characteristic of atherosclerosis, increasing the risk of heart attack, stroke, and other complications of cardiovascular disease.

After alpha-tocopherol interacts with a free radical it is no longer capable of acting as an antioxidant unless it is enzymatically regenerated. Vitamin C helps to regenerate some of the alpha-tocopherol, but the remainder is eliminated from the body. Therefore, to maintain vitamin E levels, you ingest it as part of your diet.

In addition to its antioxidant functions, vitamin E, mainly as alpha-tocopherol, plays a role in the immune system, regulation of gene expression, and cell signaling. It also enhances the dilation of blood vessels and inhibits blood clot formation.

Food Sources of Vitamin E

Excellent dietary sources of vitamin E include nuts, seeds, and vegetable oils, with additional amounts provided by green leafy vegetables and fortified cereals.[3]

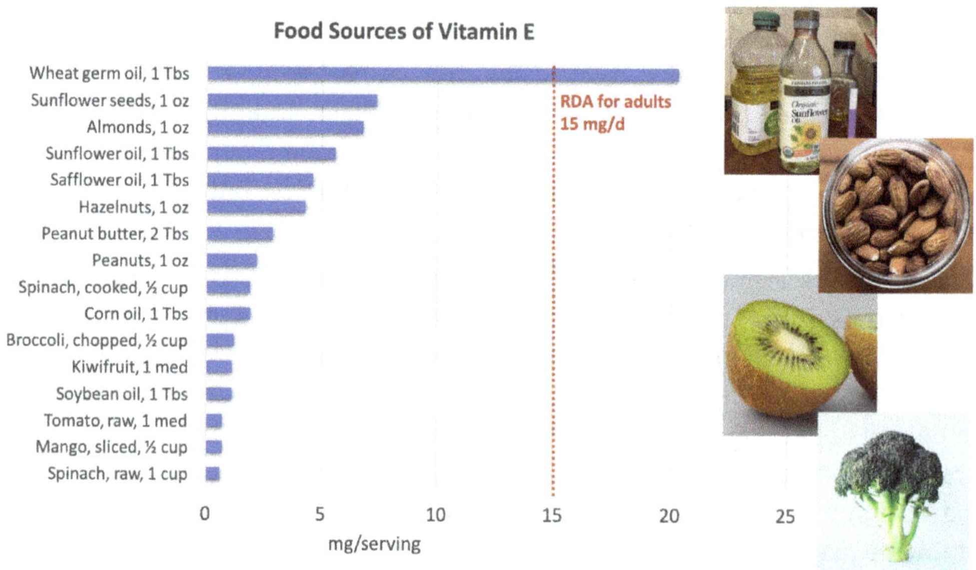

Figure 8.14. Dietary sources of vitamin E. Source: NIH Office of Dietary Supplements

Surveys of Americans' diets often find that they provide less than the RDA for vitamin E. However, these studies may underestimate the amount of vitamin E in the diet because they don't fully account for vegetable oils in the diet, as these are rich sources of vitamin E. Vitamin E can be destroyed at high temperatures, especially when reheated repeatedly, so oils used in deep frying are not good sources of the vitamin.

Vitamin E Deficiency and Toxicity

Outright vitamin E deficiency with obvious deficiency symptoms is very rare in healthy people. It most often occurs in people with an underlying disorder that impairs the digestion and absorption of fat. Symptoms of vitamin E deficiency include nerve and muscle damage, vision problems, and a weakened immune system.[3]

Studies have not found any risks of consuming vitamin E in foods. The UL for vitamin E is set at 1,000 mg for adults, far above the RDA of 15 mg and far higher than could naturally be obtained from food. These amounts are available in supplement form, however. As mentioned, high-dose vitamin E supplements were shown to increase the risk of prostate cancer in men. Other studies have found that high-dose vitamin E supplements are associated with an increased risk of hemorrhage, stroke, and death.

Vitamin C

Vitamin C, also called ascorbic acid, is a water-soluble vitamin essential in the diet for humans. Interestingly, most other mammals can readily synthesize vitamin C and don't require it in their diets. Vitamin C's ability to easily donate electrons makes it a highly effective antioxidant. Since it is water-soluble, it acts both inside and outside cells to protect molecules in aqueous environments. Vitamin C also plays a vital role in regenerating vitamin E after it has acted as an antioxidant, allowing it to be recycled and used again.

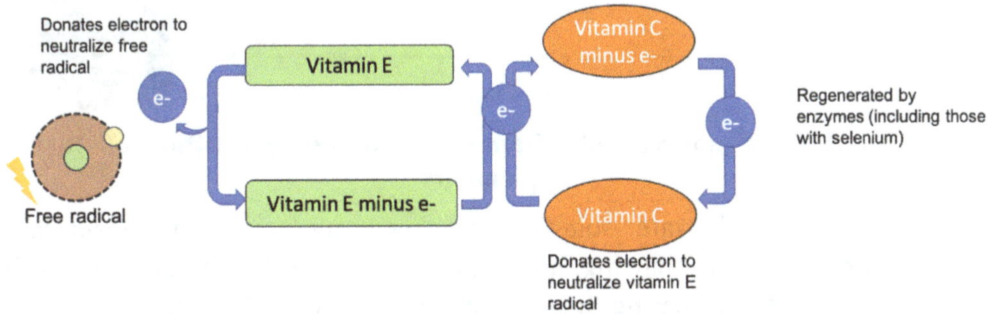

Figure 8.15. After vitamin E donates an electron to neutralize a free radical, it can be regenerated by an electron from vitamin C. Vitamin C is then regenerated by antioxidant enzymes.

In addition to its role as an antioxidant, vitamin C is a required part of several enzymes involved in the synthesis of collagen, a protein important to the strength and structure of muscles, bones, tendons, ligaments, connective tissue, and skin. Vitamin C is also required to synthesize neurotransmitters important for signaling in the brain, some hormones, and amino acids. It also plays a role in immune function and improves the absorption of dietary iron.

The body's vitamin C status is tightly controlled to maintain steady tissue and plasma concentrations. This means that if you consume high doses of vitamin C, you'll absorb less from the intestine and excrete more in urine to prevent excessive concentrations in the body. Vitamin C is not stored in any significant amount in the body, but once it has reduced a free radical, it is very effectively regenerated and therefore can exist in the body as a functioning antioxidant for many weeks.

Food Sources of Vitamin C

Fruits and vegetables are great sources of vitamin C. Some of the best sources include bell pepper, citrus, broccoli, strawberries, Brussels sprouts, and cantaloupe.[4]

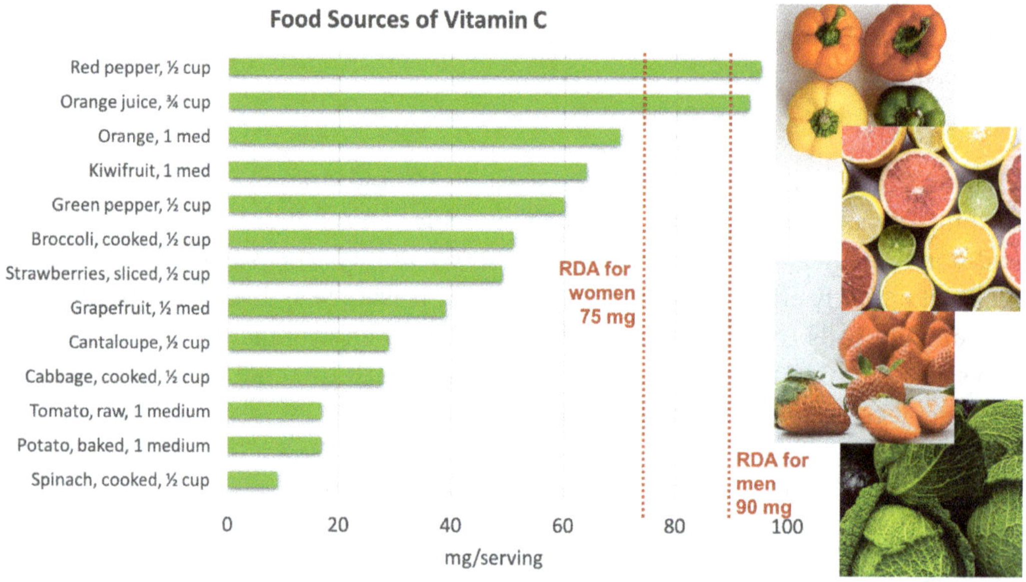

Figure 8.16. Dietary sources of vitamin C. Source: NIH Office of Dietary Supplements

Because vitamin C is water-soluble, it leaches away from foods considerably during cooking, freezing, thawing, and canning. Up to 50% of vitamin C can be boiled away. Therefore, eating fruits and vegetables raw or lightly steamed maximizes the vitamin C value of these foods.

Vitamin C Deficiency and Toxicity

The classic condition caused by vitamin C deficiency is scurvy. The signs and symptoms of scurvy include skin disorders, bleeding gums, joint pain, and weakness—all of which may be related to vitamin C's role in collagen synthesis. Additional symptoms of scurvy include abnormally-thickened skin, fatigue, depression, iron deficiency anemia, and increased susceptibility to infections.

In the past, scurvy was common among sailors on long ocean voyages, whose diets were completely lacking in fruits and vegetables for many months. In the mid-1700s, British Navy surgeon Sir James Lind's experiments revealed that citrus fruits and juices could prevent scurvy in sailors. British sailors were often referred to as "limeys," as they carried sacks of limes onto ships to prevent scurvy. It was not until 1932 that scientists showed that vitamin C was the essential nutrient involved in this cure.

Figure 8.17. This drawing of a sailor with scurvy, from the 1929 edition of "Kranken-Physiognomik" by von K. H. Baumgärtner, shows common effects of the disease, including sunken eyes and skin lesions.

Scurvy is prevented by even a low intake of fruits and vegetables, and it takes at least a month of consuming very little or no vitamin C for scurvy symptoms to develop. Thus, scurvy is rare in developed countries today. Diet surveys show that most Americans meet the RDA for vitamin C. When vitamin C deficiency occurs today, it is in people who consume very limited food variety, such as those with mental illness, people who abuse alcohol or drugs, people on very restrictive diets, and impoverished people with limited fruit and vegetable access. People who smoke also require more vitamin C to counter the free radicals generated by smoking.

The risk of vitamin C toxicity from foods is essentially nonexistent, because the body can adjust intestinal absorption and urinary excretion to maintain a healthy vitamin C level. However, high doses of vitamin C from supplements have been reported to cause numerous problems, including gastrointestinal upset and diarrhea. To prevent these discomforts, the UL for vitamin C is set at 2,000 milligrams per day for adults, more than twenty times the RDA.

At very high doses in combination with iron, vitamin C has sometimes been found to increase oxidative stress, reaffirming that getting your antioxidants from foods is better than getting them from supplements. There is also some evidence that taking vitamin C supplements at high doses increases the likelihood of developing kidney stones; however, this effect is most often observed in people that already have multiple risk factors for kidney stones.

Can Vitamin C Supplements Prevent the Common Cold?

Many people believe that taking a vitamin C supplement can prevent the common cold or decrease its symptoms. This idea was popularized by Linus Pauling in the 1970s, and it's continuously promoted today in the form of over-the-counter supplements such as Emergen-C and Airborne. These typically contain doses in the range of 1000 mg of vitamin C, far higher than normal levels of vitamin C in the diet and enough to reach the UL of 2000 mg if a person takes two doses per day.

Do these high-dose vitamin C supplements do anything to prevent or treat the misery of the common cold? A systematic review and meta-analysis published in 2013 by the Cochrane Collaboration summarized the results of 29 studies conducted on this question. The review concluded that for most people, these supplements don't prevent the common cold but can reduce the duration of symptoms by 8% in adults and 14% in children—amounting to a day or two of relief—but only if they're taken consistently every day and before cold symptoms begin. If taken after the onset of symptoms, a vitamin C supplement does not seem to reduce the duration or severity of symptoms. Some research shows that vitamin C supplements may be more effective in cold prevention in athletes and those in extreme physical conditions, such as marathon runners, endurance skiers, and soldiers.[5]

Selenium

Selenium is an essential trace mineral. It is part of the structure of at least 25 proteins in the body, with functions in reproduction, thyroid hormone metabolism, DNA synthesis, and antioxidant and immune protection.[6] As part of antioxidant enzymes, selenium helps to regenerate other antioxidants, including vitamin C. These enzymes also protect lipids from free radicals, and, in doing so, spare vitamin E. This illustrates how antioxidants work together to protect the body against free radical-induced damage.

Food Sources of Selenium

Organ meats, muscle meats, and seafood have the highest selenium content. Grains and some nuts contain selenium when grown in selenium-containing soils. The selenium content of the soils used to grow animal feed can also affect the selenium content of animal products.

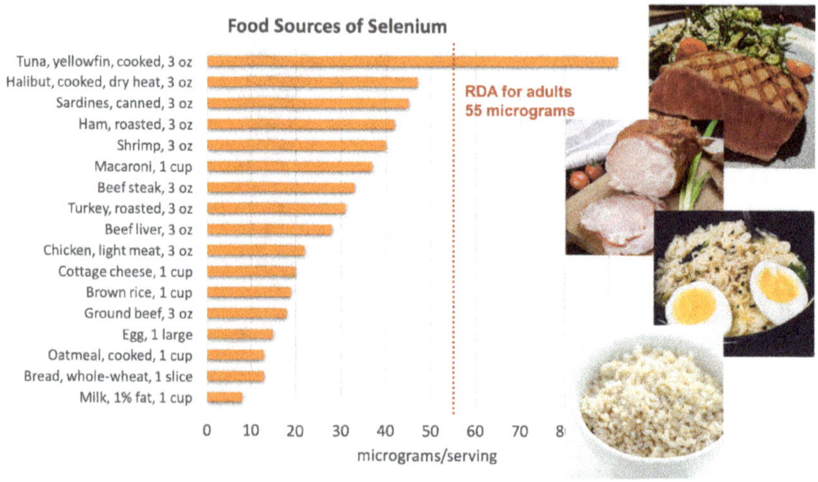

Figure 8.18. Dietary sources of selenium. Source: NIH Office of Dietary Supplements

Selenium Deficiency and Toxicity

Selenium deficiency is very rare in the United States and other developed countries. Worldwide, people with a primarily vegetarian diet in areas with low soil selenium levels, including parts of China and Europe, may be at risk for selenium deficiency.[6]

Chronic exposure to foods grown in soils containing high levels of selenium (above the UL of 400 micrograms per day) can cause brittle hair and nails, gastrointestinal discomfort, skin rashes, halitosis, fatigue, and irritability. Brazil nuts contain very high levels of selenium, so if eaten regularly could cause selenium toxicity. Selenium at doses several thousand times the RDA can cause acute toxicity, and when ingested in gram quantities, can be fatal.

REFERENCES:

Adcock, J. (2018, January 10). What are antioxidants? And are they truly good for us? The Conversation. https://theconversation.com/what-are-antioxidants-and-are-they-truly-good-for-us-86062

National Center for Complementary and Integrative Health. (n.d.). Antioxidants: In Depth. Retrieved April 2, 2020, from https://www.nccih.nih.gov/health/antioxidants-in-depth

National Institutes of Health Office of Dietary Supplements. (n.d.). Vitamin E – Health Professional Fact Sheet. Retrieved April 2, 2020, from https://ods.od.nih.gov/factsheets/VitaminE-HealthProfessional/

National Institutes of Health Office of Dietary Supplements. (n.d.). Vitamin C – Health Professional Fact Sheet. Retrieved April 2, 2020, from https://ods.od.nih.gov/factsheets/VitaminC-HealthProfessional/

Hemila, H., & Chalker, E. (2013). Vitamin C for preventing and treating the common cold. Cochrane Database of Systematic Reviews, CD000980. https://www.ncbi.nlm.nih.gov/pubmed/23440782

National Institutes of Health Office of Dietary Supplements. (n.d.). Selenium-Health Professional Fact Sheet. Retrieved April 2, 2020, from https://ods.od.nih.gov/factsheets/Selenium-HealthProfessional/

ESSENTIAL VITAMINS FOR VISION

Vision is one of our five basic senses, and unless you've experienced vision loss, it's hard to imagine life without sight. Vitamin A plays a key role in vision and eye health. On this page, we'll cover the functions, food sources, and signs of deficiency and toxicity for vitamin A and carotenoids such as beta-carotene, which can be converted to vitamin A.

VITAMIN A

Vitamin A is a generic term for a group of similar compounds called retinoids. Retinol is the form of vitamin A found in animal-derived foods, and it is converted in the body to the biologically active forms of vitamin A: retinal and retinoic acid. The dietary retinol found in animal-derived foods is called preformed or active vitamin A.

This differentiates it from beta-carotene and other carotenoid compounds, which are brightly-colored yellow, orange, and red pigments synthesized by plants. These are called provitamin A carotenoids, because they can be converted to vitamin A in the body. Among the carotenoids, beta-carotene is most efficiently converted to vitamin A. If you look at the beta-carotene molecule, you can see that it looks like two retinol molecules connected together. Enzymes in the intestine and liver can cleave the beta-carotene molecule in half, creating two new molecules of vitamin A.

As with other fat-soluble vitamins, vitamin A is packaged into chylomicrons in the enterocytes of the small intestine and then transported to the liver. The liver stores and exports vitamin A as needed; it is released into the blood bound to a retinol-binding protein, which transports it to cells. Beta-carotene can be converted into vitamin A in the intestine, or it can be absorbed intact, packaged in chylomicrons, and then transported around the body in lipoproteins. Beta-carotene and other carotenoids that aren't converted to vitamin A can also act as powerful antioxidants and have other helpful functions in the body, which may in part explain the health benefits of a diet rich in fruits and vegetables.

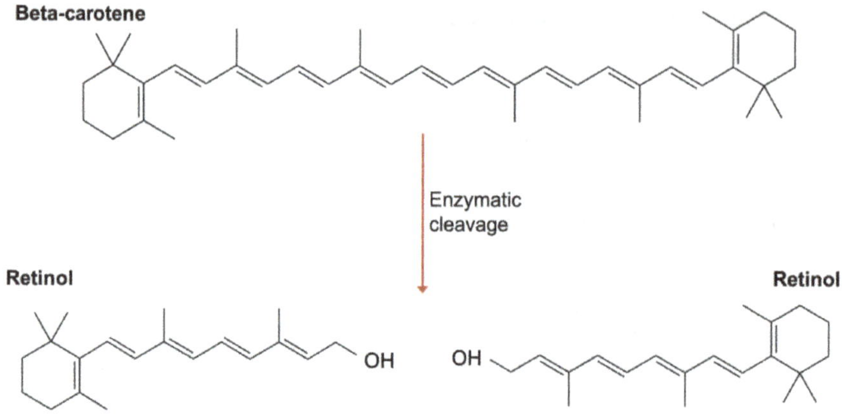

Figure 8.19. One molecule of beta-carotene can be enzymatically cleaved to 2 molecules of vitamin A.

The retinoids are aptly named, as their most notable function is in the retina of the eye. Retinol that is circulating in the blood is taken up by cells in the retina, where it is converted to retinal and is used as part of the pigment rhodopsin. Rhodopsin is especially important to our ability to see in low-light conditions. When light hits rhodopsin in the eye, a nerve signal is sent to the brain, allowing us to detect that light. A person that is deficient in vitamin A has less rhodopsin pigment in the eye and is therefore less able to detect low-level light. This makes it more difficult to see at night, a condition referred to as night blindness, and this is one of the first signs that a person is deficient in vitamin A.

Figure 8.20. A depiction of the vision of a person suffering from night blindness, in which there is inadequate rhodopsin for detection of light.

Vitamin A is also required for normal cellular differentiation, the process by which cells change from stem cells to more specialized cells with specific structure and function. Cellular differentiation is important in every tissue of the body, but if there is a shortage of vitamin A, the eye is one of the first areas to be impacted.

Specialized cells in the lining of the eyes produce mucus and tears, which keep eyes moist and lubricated. When the mucus-secreting cells die, they need to be replaced with new cells. If the body is deficient in vitamin A, those new cells don't differentiate normally, resulting in dry eyes, a condition called xerophthalmia. Instead of producing mucus, these dysfunctional cells produce a protein called keratin. Keratin is a hard, structural protein that is found in nails, hair, and the outer layer of skin, and you can imagine the problems it causes when it accumulates in the eye. Instead of a moist, well-lubricated eye, keratin makes the eye hard and dry, resulting in clouded vision.

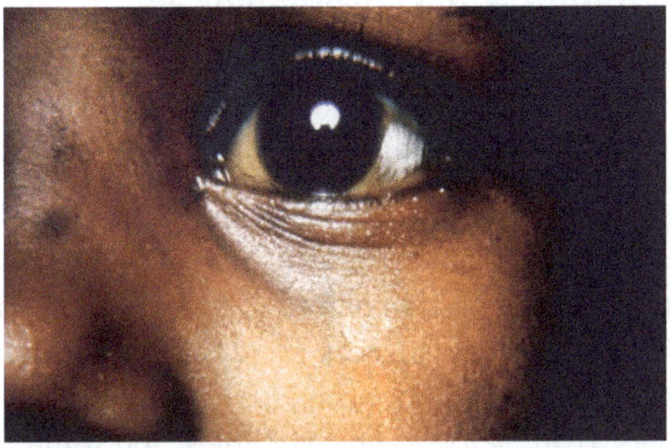

Figure 8.21. A child with xerophthalmia caused by vitamin A deficiency, demonstrating cloudiness on the surface of the eye.

A deficiency in vitamin A can thus impair vision in two ways:

- Development of night blindness due to a lack of the pigment rhodopsin
- Development of xerophthalmia, or dry eyes, caused by abnormal cellular differentiation

Night blindness is usually the first sign of a vitamin A deficiency, followed by xerophthalmia and clouded vision. If the deficiency persists, the damage from keratin in the lining of the eye can cause permanent blindness.

Vitamin A's role in cellular differentiation also makes it critical to cells around the body involved in normal growth, development, reproduction, and immune function. All of these processes require cells to develop in specific ways at specific times, and vitamin A helps to orchestrate these processes. For example, embryonic development requires stem cells to differentiate into specific types of cells to form new organs, and timing is critical.

Vitamin A also helps the immune system produce different types of immune cells, and without adequate vitamin A, a person is more susceptible to infections. The common occurrence of severe xerophthalmia in children who died from infectious diseases led scientists to hypothesize that supplementing vitamin A in the diets of children with xerophthalmia might reduce disease-related mortality. In Asia in the late 1980s, researchers administered vitamin A supplements to targeted populations of children, and their death rates from measles and diarrhea declined by up to 50%. Vitamin A supplementation in these deficient populations did not reduce the number of children who contracted these diseases, but it did decrease the severity of the diseases so that they were less likely to be fatal. Since this discovery, providing vitamin A supplementation to children in the developing world has been a major effort of the World Health Organization and UNICEF.[1]

Food Sources of Vitamin A and Carotenoids

Preformed vitamin A is found only in animal-derived foods. The best food sources are liver and fish oils, as vitamin A is fat-soluble and stored in fatty tissues. Smaller amounts can be found in other animal products, such as meat, eggs, and dairy products.

Provitamin A carotenoids such as beta-carotene are mostly found in fruits and vegetables. Carotenoids are brightly-colored pigments, so vibrant color is a good indicator of their presence. Top sources include orange and yellow vegetables such as carrots, sweet potatoes, and pumpkins (beta-carotene is a bright orange pigment), bell peppers, fruit, leafy green vegetables, and some vegetable oils. Some carotenoids can also be found in animal-derived foods. For example, the yellow color of egg yolk and butter comes from carotenoids absorbed from the diets of the hens and cows, respectively.

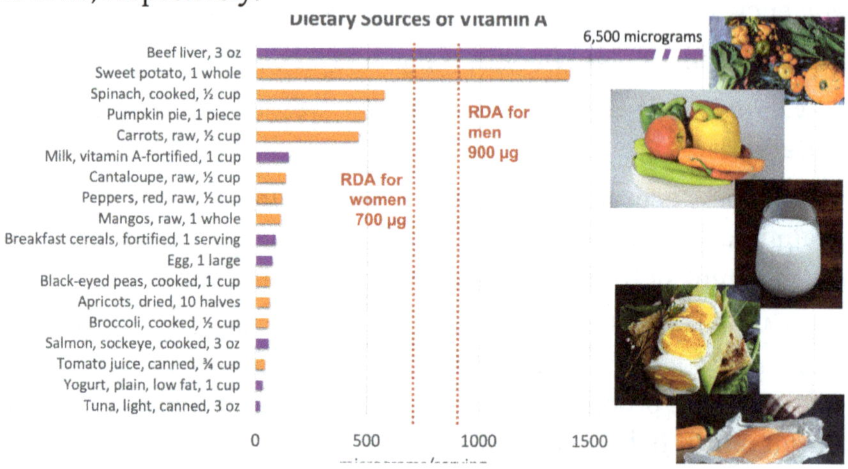

Figure 8.22. Food sources of vitamin A. Food sources that only contain provitamin A carotenoids are represented with orange bars. Those containing preformed or active vitamin A (animal-derived or fortified foods) are represented with purple bars. Source: NIH Office of Dietary Supplements.

Deficiency and Toxicity of Vitamin A and Carotenoids

The main symptoms of vitamin A deficiency are xerophthalmia, night blindness, and increased susceptibility to infections. Another symptom is hyperkeratosis, which occurs when cells in the skin overproduce the protein keratin (similar to what happens in the eye with xerophthalmia) causing the skin to become rough and irritated.

Vitamin A deficiency is rare in developed countries, but globally, it is the leading cause of preventable blindness in children. It is caused by malnutrition related to consumption of inadequate diets predominantly based on staple grains and lacking in animal products, fruits, vegetables, and fat, which increases absorption of vitamin A. According to the World Health Organization, an estimated 250,000 to 500,000 children lose their sight each year due to vitamin A deficiency, and half of them die within a year of developing blindness, likely due to infection. According to UNICEF, 30% of children under 5 have vitamin A deficiency in the world, but rates are as high as 44% and 48% in south Asia and sub-Saharan Africa, respectively.

Vitamin A toxicity causes dry, itchy skin, loss of appetite, dizziness, nausea, swelling of the brain, and joint pain. In the most severe cases, it can cause liver damage, coma, and death. Vitamin A toxicity is almost always caused by taking supplements in doses above the UL of 3,000 micrograms per day for substantial periods. These high doses would not be found in a normal diet, although vitamin A toxicity has been observed in Arctic explorers who ate large amounts of bear and seal liver.

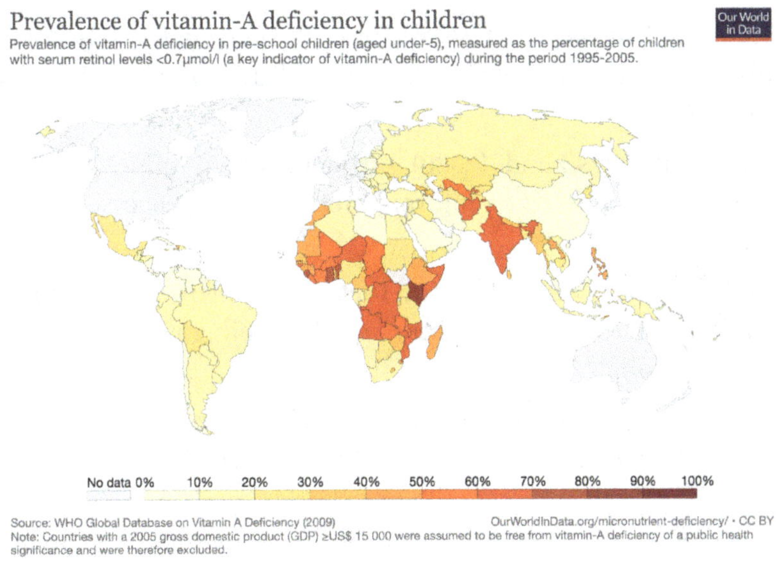

Figure 8.23. The worldwide prevalence of vitamin A deficiency in 1995-2005, based on World Health Organization data.

Consuming excessive amounts of vitamin A during pregnancy can also cause birth defects, so pregnant people should pay close attention to vitamin A contained in supplements. In addition, some synthetic forms of vitamin A (Retin-A and Accutane, for example) are used as acne treatments and should never be used during pregnancy due to the risk of birth defects.

Unlike preformed vitamin A, beta-carotene and other carotenoids do not seem to cause birth defects or other major toxicity effects in high doses. This is because the body doesn't convert beta-carotene to vitamin A if it already has excessive amounts of vitamin A. Because it doesn't cause toxicity, beta-carotene is usually used as the source of vitamin A in prenatal multivitamin supplements.

Beta-carotene that isn't converted to vitamin A is absorbed intact in the intestine. When high levels of beta-carotene are consumed in the diet, it can have the unusual effect of making a person's skin appear to be yellow or orange. The color change doesn't seem to be harmful, and normal skin tone returns once the person stops consuming so much beta-carotene. However, studies have shown that long-term consumption of high-dose beta-carotene supplements have been linked to increased rates of cancer and death, so it's best to get beta-carotene from food rather than supplements.

REFERENCES

Sommer, A. (2008). Vitamin a deficiency and clinical disease: An historical overview. The Journal of Nutrition, 138(10), 1835–1839. https://doi.org/10.1093/jn/138.10.1835

World Health Organization. (n.d.). Micronutrient deficiencies. WHO; World Health Organization. Retrieved April 7, 2020, from https://www.who.int/nutrition/topics/vad/en/

United Nations Children's Fund. (2020). Estimates of Vitamin A Supplementation Coverage in Preschool-age Children: Methods and processes for the UNICEF global database. UNICEF.

National Institutes of Health Office of Dietary Supplements. (n.d.). Vitamin A. Retrieved April 7, 2020, from https://ods.od.nih.gov/factsheets/VitaminA-HealthProfessional/

Higdon, J., Drake, V. J., & Delage, B. (2014, April 28). Carotenoids. Linus Pauling Institute. https://lpi.oregonstate.edu/mic/dietary-factors/phytochemicals/carotenoids.

UNIT-9

VITAMINS AND MINERALS-II

VITAMINS AND MINERALS PART 2: AN INTRODUCTION

In Unit 8, we began our study of vitamins and minerals, covering the basic classification and sources of vitamins and minerals, as well as examining in detail the vitamins and minerals involved in fluid and electrolyte balance, antioxidant function, and vision. In Unit 9, we'll continue our study of vitamins and minerals, focusing on those micronutrients involved in bone health, energy metabolism, and blood health.

BONE HEALTH: AN INTRODUCTION

The human skeleton consists of 206 bones and other connective tissues that together support and protect many organs, produce red and white blood cells, and act as a storage depot for minerals such as calcium, phosphorus, and magnesium. Although bones may look inactive at first glance, they are living tissues that are dynamic and in a constant state of breaking down and rebuilding to withstand mechanical forces. Bones also contain a complex network of canals, blood vessels, and nerves that allow for nutrient transport and communication with other organ systems.

Nutrition influences all body systems, and the skeletal system is no exception. Our lifestyle choices impact the health of our bones. In this section, we will look at how bone forms across the lifecycle and discuss the complex interactions of nutrients, hormones, genetics, and environmental factors that impact bone health.

What is bone?

Bone is a living tissue, made mostly of collagen, a protein that provides a soft framework, and minerals like calcium phosphate that form tiny crystals (called hydroxyapatite) around the collagen fibers. These inorganic minerals harden the collagen framework and provide strength. The combination of collagen and minerals makes bone both flexible and strong, which allows it to withstand stress.

Most bones contain two types of tissue, compact and spongy tissue, but their distribution and concentration vary based on the bone's function. Spongy bone (also known as trabecular bone) is 50 to 90 percent porous and appears as a lattice-like structure under a microscope. It makes up about 20 percent of the adult skeleton and is found at the ends of long bones, in the cores of vertebrae, and in the pelvis, as it supports shifts in weight distribution. Compact bone (also known as cortical bone) is dense so that it can withstand compressive forces. It is only 10 percent porous, and it looks similar to the rings in a tree trunk, with many concentric circles sandwiched together. Compact bone tissue makes up approximately 80 percent of the adult skeleton and surrounds all spongy tissue.

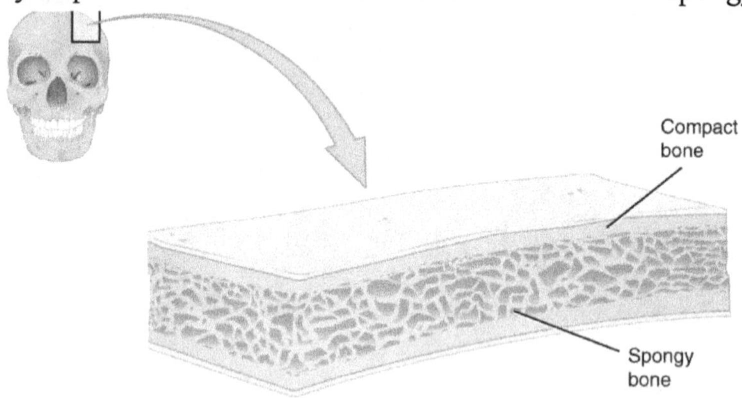

Figure 9.1. This cross-section of a flat bone from the skull shows the spongy bone lined on either side by a layer of compact bone.

Bone Growth, Modeling and Remodeling

Bones change in shape, size, and position throughout the life cycle. During infancy, childhood, and adolescence, bones are continuously growing and changing shape through two processes: growth (or ossification) and modeling. In the process of modeling, bone tissue is dismantled at one site and built up at a *different* site, which influences the *shape* of the bone. During childhood and adolescence, more bone is deposited than dismantled, so bones grow in both size and density, reaching 90 percent of peak bone mass by age 18 in girls and age 20 in boys.[2] Peak bone mass is reached by age 30, at which point bones have reached their maximum strength and density. Factors affecting peak bone mass include sex, race, hormones (e.g., estrogen and testosterone), nutrition (e.g., calcium and vitamin D intake), physical activity, and behavioral factors like smoking. These factors will be discussed in more detail when we discuss osteoporosis.

In adulthood, our bones stop growing and modeling, but they continue to go through a process of bone remodeling, in which bone tissue is degraded and built up at the *same* location. About 10 percent of bone tissue is remodeled each year in adults. Bones adapt their structure to the forces acting upon them, even in adulthood. This is why physical activity increases bone strength, especially when it involves weight-bearing activities. For example, tennis players can have measurably higher bone mass in the arm they use for play compared with the other arm.[3] Ultimately, bones adapt their shape and size to accommodate function.

The dynamic nature of bone means that new tissue is constantly formed, and old, injured, or unnecessary bone is dissolved for repair or for calcium release. The cell type responsible for bone resorption, or breakdown, is the osteoclast. Osteoclasts are continually breaking down old bone tissue. Another type of cell, called osteoblasts, are continually forming new bone. The ongoing balance between osteoblasts and osteoclasts is responsible for the constant but subtle reshaping of bone. The decline in bone mass after age 40 occurs because the rate of bone loss is greater than the rate of bone formation. This means that osteoclast-mediated bone degradation exceeds that of the bone-building activity of osteoblasts. How much bone is lost in adulthood depends on peak bone mass reached in early adulthood and other risk factors, as we'll discuss next.

Osteoporosis

Osteoporosis is a bone disease that occurs when bone density or bone mass decreases. The bone becomes thinner and more porous and is therefore more susceptible to breaking. According to the National Institute of Arthritis and Musculoskeletal and Skin Diseases, more than 53 million people in the U.S. either have osteoporosis already or are at high risk of developing it due to low bone mass.

Bone loss usually occurs without symptoms, so osteoporosis is often called a silent disease. It can go undetected until bones become so weak that they fracture due to a sudden strain, bump, or fall.

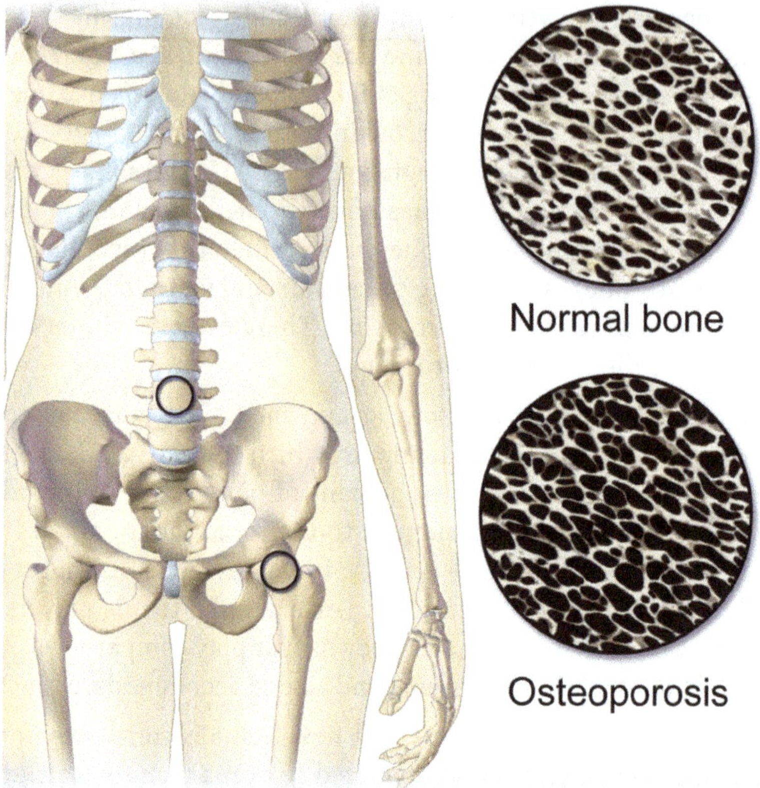

Figure 9.2. Osteoporosis. This illustration shows the difference between the structure of normal bone, which is less porous, and bone with osteoporosis, which is more porous. The two circles located on the spine and hip represent the location of the images.

One way bone health can be assessed is by measuring bone mineral density. A bone mineral density (BMD) test can detect osteoporosis and predict the risk of bone fracture. The most common tool used to measure BMD is called dual energy X-ray absorptiometry. This method can measure bone density over the entire body, but most often the DXA scan focuses on measuring BMD in the hip and the spine. These measurements are then used as indicators of overall bone strength and health. DXA is painless, non-invasive, uses low doses of radiation, and is the most accurate way to measure BMD.

An individual's chances of developing osteoporosis depend on several risk factors, some of which are controllable and some of which are not. It is thought that genetic factors (such as sex and race) may account for up to 75 percent of bone mass, and lifestyle factors (such as diet and exercise habits) account for the remaining 25 percent.

Osteoporosis risk factors that are biological and can't be controlled:

- **Body frame size**- People with small frames are at higher risk for osteoporosis.
- **Race**- Caucasian and Asian populations are at higher risk of osteoporosis compared to African American and Hispanic populations, which are at lower risk.
- **Family history**– Having a family member with osteoporosis may increase risk, as heredity seems to play a part in the development of osteoporosis.
- Age- After age 40, bone mass declines due to bone breakdown exceeding bone formation. Therefore, any person over the age of 40 has an increased likelihood of developing osteoporosis compared with a younger person. Starting out with a higher peak bone mass in early adulthood enables you to lose more bone during the aging process and not develop osteoporosis.
- **Sex**- Females, on average, have a lower peak bone mass compared with males (see Figure 9.3) and a much greater risk of developing osteoporosis, in part because of hormone levels.
- **Hormones**- The female hormone estrogen and the male hormone testosterone both help to increase peak bone mass. Estrogen is the primary female reproductive hormone, and it stimulates bone building and reduces bone breakdown. When women go through menopause (usually around age 50), they experience a natural decline in estrogen levels, which accelerates bone loss and increases the risk of developing osteoporosis (Figure 9.3).

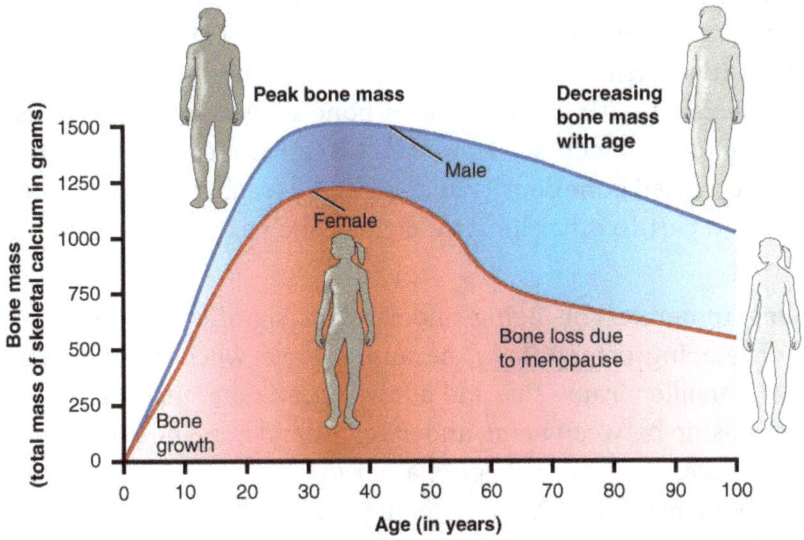

Figure 9.3. Age and bone mass. Bone density peaks at about 30 years of age, and women lose bone mass more rapidly than men, particularly around menopause.

- Risk factors that can be controlled:
- **Physical inactivity**- Physical inactivity lowers peak bone mass, decreases BMD at all ages, and is linked to an increase in fracture risk, especially in the elderly. Regular exercise can help individuals achieve greater peak bone mass, prevents bone loss for women and men age 30 and older, and maintains strength and balance to help prevent falls later in life. The best activities for stimulating new bone are weight-bearing exercise, such as walking, hiking, and dancing, and resistance exercises like weight lifting.
- **Nutrition**- Ensuring adequate nutrition is a key component in maintaining bone health. Having low dietary intakes of calcium and vitamin D are strong risk factors for developing osteoporosis. Protein is also important during childhood and adolescence for proper bone development, and in older age to preserve bone mass.4
- **Smoking**- Smoking cigarettes has long been known to correlate to a decrease in bone mass and an increased risk of osteoporosis and fractures. However, because people who smoke are more likely to be physically inactive and have poor diets, it is difficult to determine whether smoking itself causes osteoporosis. Smoking is also linked to earlier menopause, and therefore the increased risk of developing osteoporosis among female smokers may also be attributed, at least in part, to having reduced estrogen production at an earlier age. However, studies have also shown that tobacco smoke and nicotine can directly impact bone metabolism.5
- **Alcohol intake**- Alcohol intake may also affect bone health, although this seems to depend on the amount consumed. Light to moderate alcohol intake (two drinks or less per day) has been shown in some studies to be associated with an increase in bone density and a decreased risk of developing osteoporosis. However, excessive alcohol intake is associated with decreased bone density and increased fracture risk, although this may be due in part to other lifestyle factors, such as poor diet and less physical activity.
- **Being underweight**- Being underweight significantly increases the risk of developing osteoporosis, because people who are underweight often have a smaller frame size and a lower peak bone mass. The most striking relationship between being underweight and bone health is seen in people with anorexia nervosa. Anorexia nervosa is strongly correlated with low peak bone mass, and more than 50 percent of men and women who have this illness develop osteoporosis, often very early in life.

The changeable risk factors for osteoporosis provide ways for people to improve their bone health, even though some people may be genetically predisposed to the disease. Prevention of osteoporosis begins early in life since this is a critical time of bone growth. Eating a balanced diet that provides adequate amounts of calcium, vitamin D, and protein is important for bone health throughout the life cycle. Participating in exercise such as walking, hiking, and weight lifting, and refraining from risky behaviors like smoking and excessive drinking are all behaviors that will help protect bones.

RESOURCES

National Institute of Arthritis and Musculoskeletal and Skin Diseases. Osteoporosis Overview. Retrieved April 22, 2020 from https://www.bones.nih.gov/health-info/bone/osteoporosis/overview

National Institute of Arthritis and Musculoskeletal and Skin Diseases. Osteoporosis: Peak Bone Mass in Women. Retrieved April 22, 2020 from https://www.bones.nih.gov/health-info/bone/osteoporosis/bone-mass

Kontulainen S., Sievanen H., Kannus P., Pasanen M., Vuori I. (2002). Effect of long-term impact-loading on mass, size, and estimated strength of humerus and radius of female racquet-sports players: a peripheral quantitative computed tomography study between young and old starters and controls. J Bone Miner Res 17: 2281–2289.

International Osteoporosis Foundation. Nutrition. Retrieved April 28, 2020 from https://www.iofbonehealth.org/nutrition

Al-Bashaireh, A. M., Haddad, L. G., Weaver, M., Chengguo, X., Kelly, D. L., & Yoon, S. (2018). The Effect of Tobacco Smoking on Bone Mass: An Overview of Pathophysiologic Mechanisms. Journal of osteoporosis, 2018, 1206235. https://doi.org/10.1155/2018/1206235

Gaddini, G. W., Turner, R. T., Grant, K. A., & Iwaniec, U. T. (2016). Alcohol: A Simple Nutrient with Complex Actions on Bone in the Adult Skeleton. Alcoholism, clinical and experimental research, 40(4), 657–671. https://doi.org/10.1111/acer.13000

CALCIUM IS ESSENTIAL FOR THE BONES AND THE ENTIRE BODY.

As discussed on the previous page, bones are made up of two components: inorganic minerals and a protein matrix. Minerals, which make up 65% of bone tissue, are what gives bones their hardness. Calcium and phosphorus together form hydroxyapatite crystals, the main mineral component of bone. Other minerals, including magnesium, fluoride, sodium, and potassium, play supporting roles. This page focuses on calcium, and we'll cover potassium, magnesium, and fluoride on the following page. (Sodium and potassium are covered in the electrolyte section of this text.)

Calcium Functions and Regulation

Calcium is the most abundant mineral in the body. Most of the body's calcium—more than 99% of it—is stored in bone, where it's important for bone strength and structure. The remaining 1% of the body's calcium is found in the blood and soft tissues, but it is here that calcium performs its most critical functions. For example, calcium is required for the transmission of every nerve impulse, electrical signals sent from one nerve cell to another. It's also required for every cycle of muscle contraction and relaxation. With inadequate calcium, muscles can't relax, and instead become stiff and contract involuntarily, a condition known as tetany. Calcium also plays vital roles in blood pressure regulation, blood clotting, enzyme activation, hormone secretion, and signaling between cells.[1]

The many roles of calcium around the body are critical to daily survival, so maintaining homeostasis, or a steady state, of blood calcium levels is a high priority. The body rigorously controls blood calcium levels in a very tight range. If blood calcium drops, your body initiates several mechanisms to restore homeostasis, including drawing calcium from the bone. While the calcium stored in bone is important for long-term strength and structure of bone, it also serves as a calcium reserve that can be drawn upon to support the vital functions of calcium in the body, should blood calcium drop too low.

Two endocrine glands are key players in the regulation of blood calcium concentrations: the thyroid gland and parathyroid glands. The thyroid gland is a small, butterfly-shaped gland located at the base of the neck. It secretes a hormone called calcitonin. There are four parathyroid glands, each about the size of a pea and located at the back of the thyroid gland. They secrete a hormone called parathyroid hormone (PTH).

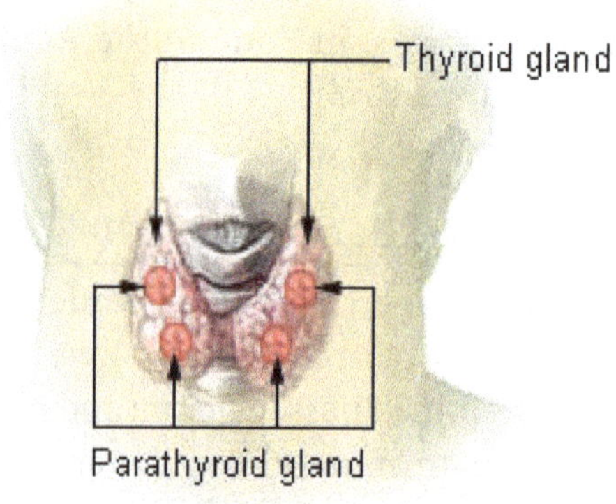

Figure 9.4. The thyroid and parathyroid glands are located at the back of the neck.

Let's take a closer look at how the body regulates blood calcium levels. If the blood calcium concentration drops too low, the parathyroid glands release parathyroid hormone, or PTH. PTH then acts in several ways to increase blood calcium levels:

PTH stimulates the activity of osteoclasts to release calcium from bone.

PTH acts on the kidney to reduce the amount of calcium lost in the urine, returning more to circulation.

PTH stimulates enzymes in the kidney that convert vitamin D to its active form, also called calcitriol. Activated vitamin D acts on the intestine to increase the absorption of calcium. Vitamin D also works together with PTH to stimulate release of calcium from the bone and reduce calcium loss in urine.

Once blood calcium levels are normal, PTH levels drop, turning off all of these mechanisms of increasing calcium.

On the other hand, if blood calcium levels become too high, the thyroid gland releases calcitonin, which inhibits the release of calcium from the bone and increases calcium excretion from the kidneys. These mechanisms help to restore normal blood calcium concentrations, after which calcitonin levels drop again.

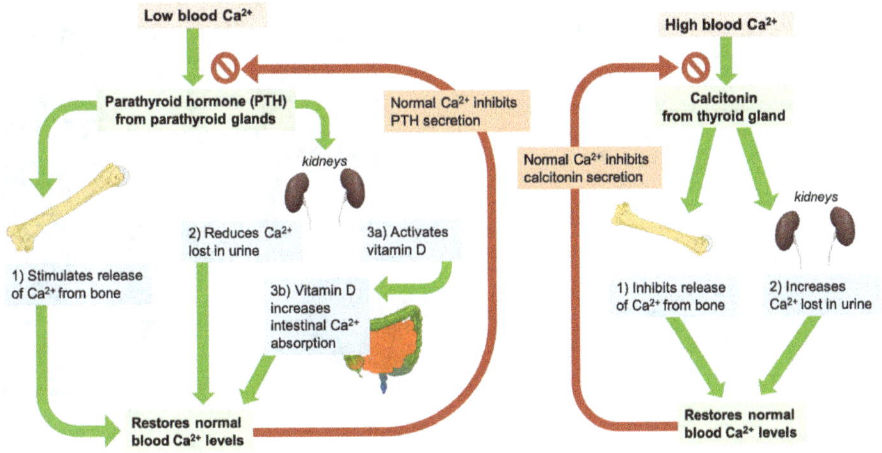

Figure 9.5. Blood calcium (Ca2+) levels are tightly regulated by PTH, vitamin D, and calcitonin.

Through these two opposing pathways—PTH and vitamin D for raising blood calcium and calcitonin for lowering blood calcium—the body can very effectively maintain blood calcium homeostasis. This system is dependent upon stored calcium in bone, which is sacrificed when needed to ensure adequate blood calcium. In the short-term, this isn't a problem, because bone remodeling allows you to replace calcium in the bone. However, in the long-term, inadequate dietary calcium means you continuously draw down the calcium stores in your bones, resulting in declining bone mineral density and increased risk of fracture.

Calcium requirements are highest for children and adolescents, who are growing and building bones, and for older adults, who are losing bone density. The RDA for calcium for children 9 to 13 years old and teens 14 to 18 years old is 1,300 milligrams per day. The RDA for adults is 1,000 mg per day but increases to 1,200 mg per day for women ages 51 and up and for men age 71 and older.

Dietary Sources of Calcium

In the typical American diet, calcium is obtained mostly from dairy products. A slice of cheddar or Swiss cheese contains over 200 milligrams of calcium. One cup of milk contains approximately 300 milligrams of calcium, which is about a third of the RDA for calcium for most adults. Foods fortified with calcium such as cereals, soy milk and other plant-based beverages, and orange juice also provide one third or greater of the calcium RDA. Smaller amounts of calcium are naturally present in plant-based foods such as legumes, leafy greens, and nuts, and with careful planning, adequate calcium can be obtained from non-dairy sources.

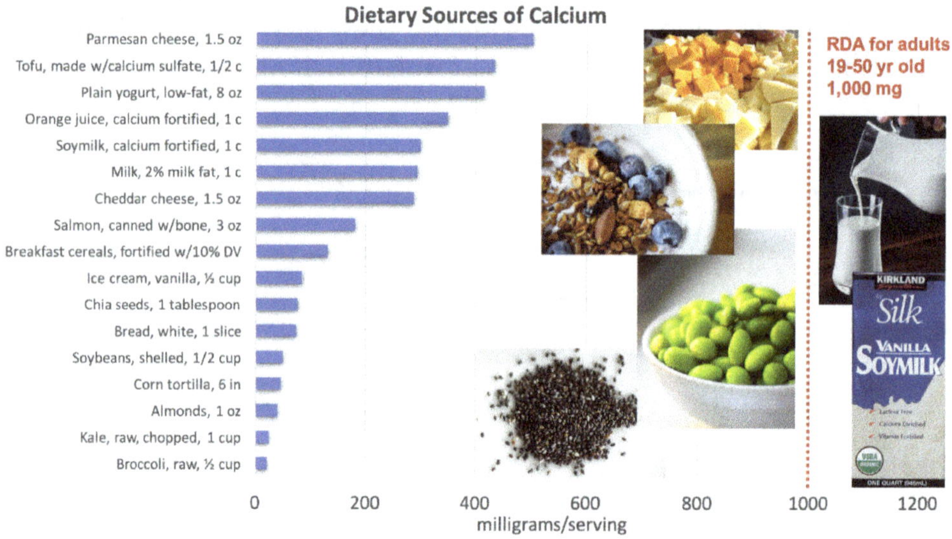

Figure 9.6. Dietary sources of calcium. Examples of good sources pictured include cheese, milk, fortified soymilk, yogurt (with almond granola), edamame, and chia seeds. Source: NIH Office of Dietary Supplements and Dietary Guidelines for Americans, 2015-2020.

Calcium bioavailability, or the amount of dietary calcium that is absorbed from the intestine into the bloodstream, can vary significantly. In general, calcium absorption is highest in infants and young children—who need relatively high amounts of calcium for building bone—and declines with age. With higher calcium intake, especially from supplements, bioavailability decreases in order to prevent excessive calcium absorption. Some chemical components of plant foods, including phytic acid (found in whole grains, beans, seeds, soy, and nuts) and oxalic acid (found in spinach, collard greens, sweet potatoes, rhubarb, and beans), bind to calcium and reduce bioavailability. Despite reduced absorption, these foods can still provide a significant amount of calcium.[1]

Calcium Deficiency and Toxicity

In the short-term, there are no obvious signs of calcium deficiency. This is because the body stores so much calcium in bones, and just 1% of total body calcium is required for daily functioning. If low blood calcium does occur, symptoms include muscle cramping, numbness and tingling in fingers, convulsions, lethargy, poor appetite, and abnormal heart rhythms. Without treatment, low blood calcium can lead to death.[1]

Much more common is a long-term calcium deficiency, resulting from a continuous draw of calcium stores from the bone. This causes osteopenia, or low bone mass, which can lead to osteoporosis if untreated. Osteoporosis significantly increases a person's risk of fractures. Nutrition surveys in the United States show that groups at greatest risk of dietary calcium inadequacy include adolescents and older adults, especially female teens and older women.[1]

Too much calcium can also cause problems, although this is rarely caused by excessive intake of calcium from foods. Abnormally high activity of the parathyroid gland or a parathyroid tumor can cause high blood calcium, leading to calcification or hardening of blood vessels and soft tissue and the formation of kidney stones. High calcium intake from supplements has also been associated with increased risk of kidney stones, and in some studies, increased risk of cardiovascular disease. It can also cause constipation.[1]

REFERENCES

National Institutes of Health Office of Dietary Supplements. (n.d.). *Calcium*. Retrieved April 23, 2020, from https://ods.od.nih.gov/factsheets/Calcium-HealthProfessional/

OTHER MINERALES ESSENTIAL TO BONES' HEALTH

Phosphorus

Phosphorus is the second most abundant mineral in the human body, and 85% of the body's phosphorus is housed in the skeleton. In addition, phosphorus (in the form of phosphate) is a component of the backbones of RNA and DNA, adenosine triphosphate (ATP), and phospholipids. Phosphate also plays important roles in regulating cell signaling, enzyme activity, and acid-base balance, as well as being part of creatine phosphate, which is an energy source for muscles during exercise.

Because phosphorus is present with calcium in mineralized bone, it is somewhat regulated in parallel to calcium. PTH and activated vitamin D stimulate bone resorption, increasing not only blood levels of calcium, but also blood phosphate levels. However, in contrast to the effect of PTH on calcium reabsorption by the kidney, PTH stimulates the renal excretion of phosphate so that it does not accumulate to toxic levels.

Dietary Sources of Phosphorus

In comparison to calcium, most people in the U.S. are not at risk for inadequate intake of phosphorus. Phosphorus is present in many foods, including meat, fish, dairy products, potatoes, nuts, beans, and whole grains. Phosphorus is also added to soft drinks and many processed foods, because it acts as an emulsifying agent, prevents clumping, improves texture and taste, and extends shelf life.

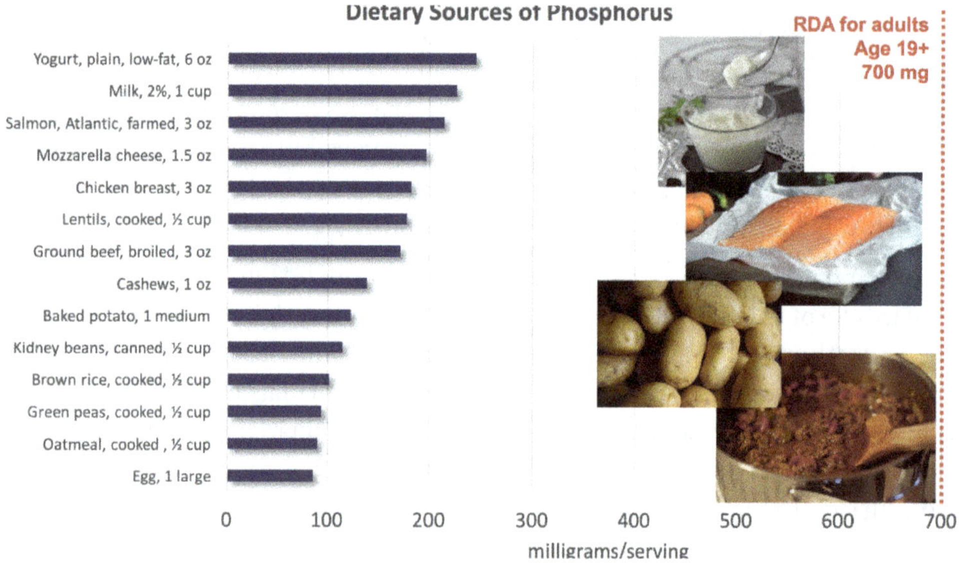

Figure 9.7. Food sources of phosphorus. Examples of good sources pictured include yogurt, salmon, potatoes, and chili (made with ground beef and kidney beans). Source: NIH Office of Dietary Supplements

Phosphorus Deficiency and Toxicity

Both deficiency and toxicity of phosphorus are rare. The average intake of phosphorus in U.S. adults ranges between 1,000 and 1,500 milligrams per day, well above the RDA of 700 milligrams per day. The UL set for phosphorus is 4,000 milligrams per day for adults and 3,000 milligrams per day for people over age 70. Very high doses of phosphorus taken in supplement form can interfere with calcium regulation and cause calcification, or hardening, of soft tissues, especially the kidneys.

Magnesium

Approximately 60% of magnesium in the human body is stored in the skeleton, making up about 1% of mineralized bone tissue. In addition to contributing to bone maintenance, magnesium has several other functions in the body. It is required in every reaction involving ATP, amounting to more than three hundred enzymatic reactions. Magnesium plays a role in the synthesis of DNA and RNA, carbohydrates, and lipids, and it's essential for nerve conduction, muscle contraction, and normal heart rhythm.

Dietary Sources of Magnesium

Magnesium is part of the green pigment, chlorophyll, which is vital for photosynthesis in plants; therefore green leafy vegetables are good dietary sources of magnesium. Magnesium is also found in high concentrations in nuts, whole grains, legumes, potatoes, dairy products, fish, and meats. Most foods that are high in fiber are good sources of magnesium, and it is added to some fortified foods, such as breakfast cereal. Additionally, chocolate, coffee, and hard water contain a good amount of magnesium.

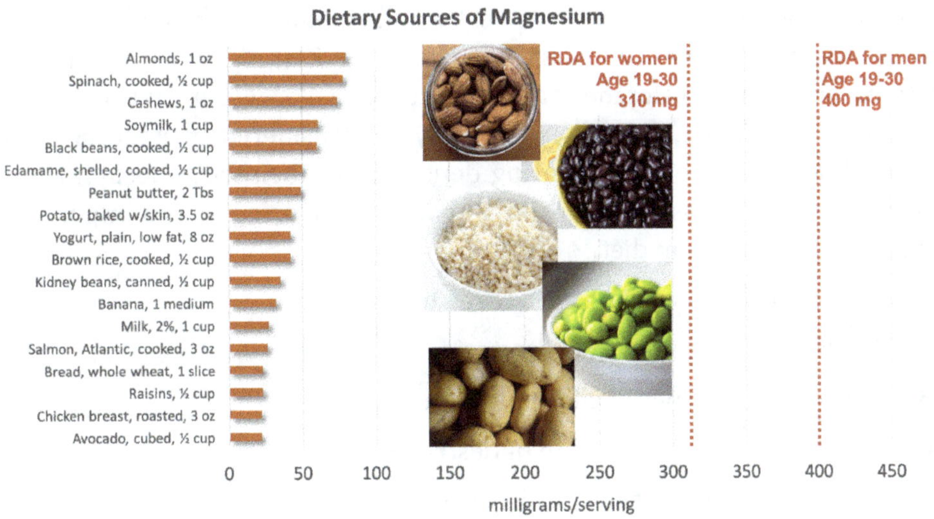

Figure 9.8. Dietary sources of magnesium. Examples of good sources pictured include almonds, black beans, brown rice, edamame, and potatoes. Source: NIH Office of Dietary Supplements

Magnesium Deficiency and Toxicity

Most people in the U.S. do not meet the RDA for magnesium, and studies indicate that consuming adequate magnesium may improve health. For example, people with higher dietary intakes of magnesium tend to have lower rates of cardiovascular disease, type 2 diabetes, and osteoporosis, leading some researchers to hypothesize that low intake of magnesium may increase the risk of these chronic diseases. More studies are needed to determine whether magnesium supplements may help prevent these diseases. However, since magnesium is present in many healthful whole foods, improving magnesium intake through diet may bring multiple benefits.

Obvious magnesium deficiency due to low dietary intake is rare in healthy people, because the kidneys can decrease urinary excretion of this mineral when intake is inadequate. People at greater risk of magnesium deficiency include those with type 2 diabetes, gastrointestinal diseases like Crohn's and celiac, chronic alcoholism, and older adults. A magnesium deficiency can cause decreased appetite, nausea, vomiting, fatigue, and weakness. If extreme, it can cause personality changes, muscle cramps, numbness, tingling, seizures, and an abnormal heart rhythm.[3]

Excessive intake of magnesium from foods is not a risk, as the kidneys can effectively excrete it if it's in excess. However, people should take care not to consume more than the UL for magnesium from supplements and medications.3

Fluoride

Fluoride is a trace mineral needed in very small amounts in the body. It is known mostly as the mineral that combats tooth decay, but it also plays a role in assisting with tooth and bone development and maintenance. Because it isn't necessary for growth or to sustain life, fluoride is generally not considered an essential mineral. However, fluoride's role in preventing dental caries (i.e., tooth decay), the most prevalent chronic disease in children and adults, underscores the importance of this mineral in the human diet.[4]

Fluoride combats tooth decay via three mechanisms:

- Blocking acid formation by bacteria
- Preventing demineralization of teeth
- Enhancing remineralization of destroyed enamel

As a natural mineral, fluoride is present in the soil and water in varying concentrations depending on geographical location. In the 1930s, researchers observed that children living in areas with naturally higher fluoride concentrations in their drinking water had a lower incidence of cavities, leading to the idea that adding fluoride to municipal water supplies could benefit public health. Fluoride was first added to drinking water in 1945 in Grand Rapids, Michigan; now over 60 percent of the U.S. population consumes drinking water that has been supplemented with fluoride to provide amounts that support dental health. The Centers for Disease Control and Prevention (CDC) has reported that fluoridation of water reduces cavities by 25 percent in children and adults and considers water fluoridation one of the ten great public health achievements in the twentieth century. Fluoride's benefits to mineralized tissues of the teeth are well substantiated, but fluoride also plays an important role in the mineralization of bones, increasing their structural stability. Fluoride is currently being researched as a potential treatment for osteoporosis. The data are inconsistent on whether consuming fluoridated water reduces the incidence of osteoporosis and fracture risk. Fluoride does stimulate osteoblast bone building activity, and fluoride therapy in patients with osteoporosis has been shown to increase bone mineral density. In general, it appears that at low doses, fluoride treatment increases bone mineral density in people with osteoporosis and is more effective in increasing bone quality when the intakes of calcium and vitamin D are adequate. The Food and Drug Administration has not approved fluoride for the treatment of osteoporosis, mainly because its benefits are not sufficiently known. It also has several side effects, including frequent stomach upset and joint pain. The doses of fluoride used to treat osteoporosis are much greater than that in fluoridated water.

Dietary Sources of Fluoride

Fluoride is not widely found in the food supply. In communities with municipal water fluoridation, greater than 70 percent of fluoride intake comes from drinking water. In communities without fluoridated water, intake depends on how much fluoride occurs naturally in the water, but in most areas, natural levels fall below amounts recommended for cavity prevention. Other beverages with a high amount of fluoride include teas and grape juice. Solid foods do not generally contain a large amount of fluoride, although this depends on the fluoride level of the soil and water it was grown in and whether it was cooked with fluoridated water. Canned meats and fish that contain bones do contain some fluoride. Other good non-dietary sources are fluoridated toothpaste and dental rinses.

Figure 9.9. Dietary sources of fluoride include water, tea, shellfish, and fluoridated dental products such as toothpaste.

Fluoride Deficiency and Toxicity

Since it is a nonessential mineral, there is no defined fluoride requirement, but lower levels are associated with higher rates of dental cavities in adults and children. This connection is why so many water supplies are fluoridated.

However, as with all minerals, fluoride can also be quite toxic if consumed in excessive amounts. Acute toxicity symptoms from large intakes of fluoride include nausea, vomiting, diarrhea, and convulsions. Chronic toxicity results in an irreversible condition known as fluorosis, characterized by the mottling (i.e., white speckling) and pitting of the teeth (see Figure 9.10). Fluorosis is primarily a risk in children, because mineralization of permanent teeth has typically occurred by age 8.[6]

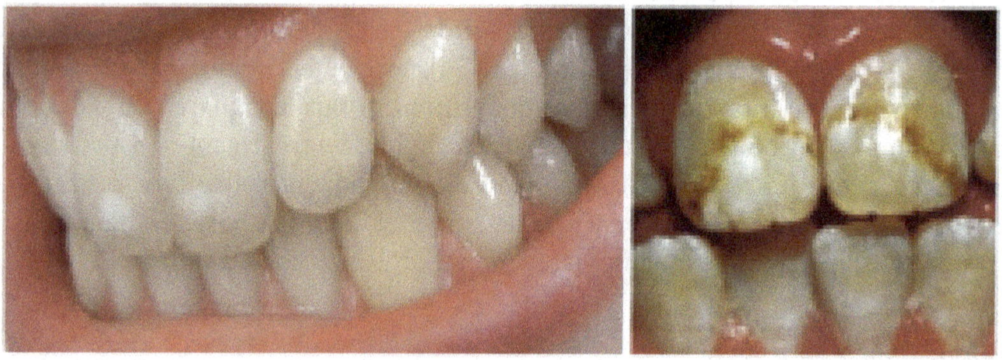

Figure 9.10. A mild case of fluorosis (left) vs. a severe case of fluorosis (right).

Because fluoridated oral care products often taste good, making them appealing to young children, it is important to make sure infants and children do not consume too much fluoride by swallowing toothpaste or other oral care products. Recommendations for managing fluoride intake in children include the following:[7]

- Do not use any fluoride supplement without talking to your health care provider.
- Consider a prescription fluoride supplement for children who live in communities without fluoridated water or rely on a well water supply that is not fluoridated.
- Use only a smear (rice-sized) amount of fluoride toothpaste as soon as the first tooth erupts in infants up through age 2 years.
- Use only a pea-sized amount of fluoride toothpaste in children ages 3 to 6 years.
- Encourage children to spit out toothpaste instead of swallowing it.
- Avoid fluoride mouth rinses in children younger than 6 years.

REFERENCES

Higdon, J., Drake, V., & Delage, B. (2014, April 23). *Phosphorus*. Linus Pauling Institute. https://lpi.oregonstate.edu/mic/minerals/phosphorus

National Institutes of Health Office of Dietary Supplements. (n.d.). *Phosphorus.* Retrieved April 24, 2020, from https://ods.od.nih.gov/factsheets/Phosphorus-HealthProfessional/

National Institutes of Health Office of Dietary Supplements. (n.d.). *Magnesium.* Retrieved April 23, 2020, from https://ods.od.nih.gov/factsheets/Magnesium-Consumer/

National Institutes of Health. (2018, July). *Dental caries (tooth decay).* Retrieved April 20, 2020, from https://www.nidcr.nih.gov/research/data-statistics/dental-caries.

Centers for Disease Control and Prevention. (2020, January 15). *Community water fluoridation.* Retrieved April 20, 2020, from https://www.cdc.gov/fluoridation/index.html.

American Academy of Pediatrics. (2009). A pediatric guide to children's oral health. Elk Grove Village, IL: American Academy of Pediatrics, 10.

U.S. National Library of Medicine. (2020, April 9). *Fluoride in diet.* Medline Plus. Retrieved from https://medlineplus.gov/ency/article/002420.htm

Reference Manual of Pediatric Dentistry. (2018). Fluoride Therapy. http://www.aapd.org/media/Policies_Guidelines/G_fluoridetherapy.pdf

Clark, M. B., & Slayton, R. L. (2014). Fluoride use in caries prevention in the primary care setting. Pediatrics, 134(3), 626-633.

VITAMIN D IS ESSENTIAL FOR THE HEALTH OF BONES AND MORE

Vitamin D is unique among the vitamins because we can synthesize most of what we need in our skin. Sunlight is an essential ingredient in this process, so vitamin D is sometimes called the "sunshine vitamin." However, the amount of vitamin D synthesized in the body is often not enough to meet our needs, so many people also need to consume dietary sources.

Metabolism and Functions of Vitamin D

Vitamin D synthesis in the skin begins with the conversion of cholesterol to 7-dehydrocholesterol. Then, in the presence of ultraviolet (UV) rays from sunlight, 7-dehydrocholesterol is converted to vitamin D_3 (also called cholecalciferol), which is transported to the liver by a binding protein.

In dietary sources, vitamin D may be present in the form of vitamin D_3 from animal products or vitamin D_2 (also called ergocalciferol), made by plants, mushrooms, and yeast. Dietary vitamin D_2 and vitamin D_3 are transported to the liver via chylomicrons and then taken up in chylomicron remnants.

Vitamins D_2 and D_3 are both inactive until they undergo two hydroxylations—chemical reactions that add a hydroxyl (-OH) group. The first hydroxylation occurs in the liver, creating calcidiol. This is the circulating form of vitamin D and the form measured in blood to assess a person's vitamin D status. The second hydroxylation occurs in the kidneys and forms calcitriol, the biologically active form of vitamin D. Recall from our discussion of regulation of blood calcium that one of the actions of parathyroid hormone (PTH) is to stimulate enzymes in the kidney that perform this last step in the activation of vitamin D. Active vitamin D increases the absorption of both calcium and phosphorus in the intestine, as well as working with PTH to reduce calcium loss in the urine and stimulate release of calcium and phosphorus from the bone. In these ways, vitamin D plays a critical role in both maintaining blood calcium homeostasis and enhancing the supply of calcium and phosphorus for bone mineralization. Vitamin D deficiency results in poor bone mineralization, with serious consequences in both children and adults, as we'll discuss later on this page.

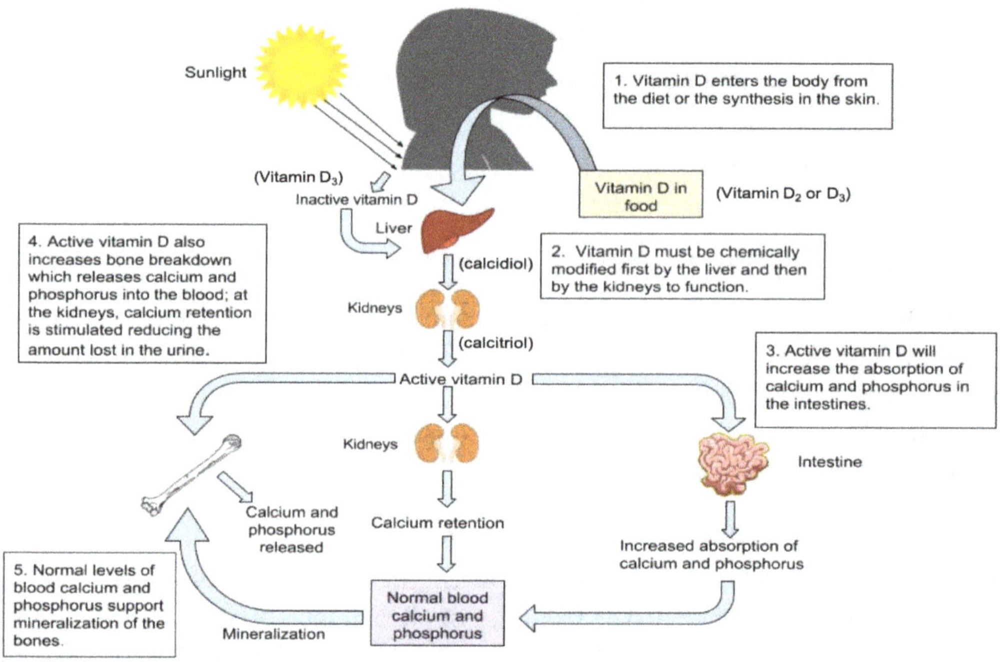

Figure 9.11. Vitamin D can be synthesized in the skin (Vitamin D_3) or provided in the diet (Vitamin D_2 or D_3). It is converted by reactions occurring first in the liver (making calcidiol) and then kidney (making calcitriol, the active form). Once active, vitamin D works in several ways to ensure blood calcium homeostasis and enhance the availability of calcium for bone mineralization.

Beyond its role in bone health, Vitamin-D has many other functions in the body. Cells throughout the body have Vitamin-D receptors in their nuclei, and by binding to these receptors, Vitamin D is thought to regulate the expression of hundreds of genes. Specifically, vitamin D is known to play important roles in regulating cellular differentiation and growth, immunity, insulin secretion, and blood pressure. Studies have found correlations between low circulating vitamin D levels and increased risks of chronic diseases, including cancer, diabetes, cardiovascular disease, and multiple sclerosis. However, it has been difficult to determine if a lack of vitamin D actually contributes to the cause of these diseases, and research in this area is ongoing.

Sunlight As A Source Of Vitamin D

In most people, vitamin D synthesis in the skin provides a significant portion of their body's needs, and a little sun exposure can go a long way. Vitamin D researchers suggest that most people need between 5 and 30 minutes of sun exposure between 10 AM and 3 PM, at least twice per week, in order to synthesize adequate vitamin D. However, any factor that decreases exposure to UV rays can interfere with vitamin D synthesis, including the following:[2]

- Geographic latitude and season. Your location on Earth and the time of the year affects your exposure to UV radiation from the sun. Exposure to UV light is greatest at the equator and declines as you move further north or south. Likewise, in the summer months, the sun is directly overhead for a greater part of the day, so you have more opportunities to synthesize vitamin D. In the winter, the sun stays lower in the sky, day length is shorter, and cloud cover is more likely to block the sun's rays, all of which decrease opportunities to synthesize vitamin D. North of about 35 degrees latitude, vitamin D synthesis is inadequate for at least a few months during the winter, because the sun simply doesn't get high enough in the sky to provide enough UV radiation on earth's surface. For example, in Boston and at the California-Oregon border (42nd parallel north), vitamin D synthesis occurs only from March until October. However, in Los Angeles (34th parallel north), vitamin D synthesis occurs year round. Ozone and air pollution can also block UV rays and decrease vitamin D synthesis.
- Skin pigmentation. Darker skin pigmentation, caused by greater melanin production in the skin, decreases UV light absorption. This helps to protect the skin from damage from UV radiation—a helpful adaptation for those living closer to the equator—but it also reduces synthesis of vitamin D. People with darker skin pigmentation need to spend more time in the sun in order to synthesize the same amount of vitamin D as lighter-skinned people.

- Age. The efficiency of vitamin D synthesis declines with age. In addition, older adults often spend less time outside so may receive less exposure to sunlight.
- Sun-protective behavior. While some UV light exposure is needed to synthesize vitamin D, UV radiation is also carcinogenic, and too much exposure increases the risk of skin cancer. It's wise to protect your skin from UV radiation by applying sunscreen, covering up with clothing and a hat, finding shade, and avoiding sun exposure in the middle of the day. People who are highly vigilant in these sun-protective behaviors or simply aren't able to go outside during the day (UV rays don't penetrate glass) may not get enough UV light for vitamin D synthesis.

With so many factors affecting UV radiation exposure, many people are unable to synthesize enough vitamin D for at least part of the year. Because vitamin D is fat-soluble, liver and adipose storage can supply the body for a while. Beyond that, dietary sources and supplements may be needed to meet the vitamin D requirement.

Dietary Sources Of Vitamin D

Only a few foods are naturally good sources of vitamin D. These include fatty fish such as salmon, tuna, and mackerel, as well as fish liver oil (e.g., cod liver oil). Smaller amounts are found in egg yolks, cheese, and beef liver. Additionally, some mushrooms grown in UV light can be a good source of vitamin D.

Most cow's milk is fortified with vitamin D in the U.S. and Canada, but other dairy products such as ice cream and cheese are not. Fortified orange juice, soymilk and other plant-based beverages, and breakfast cereal can all contribute to dietary intake of vitamin D, although amounts added vary significantly between products.[2]

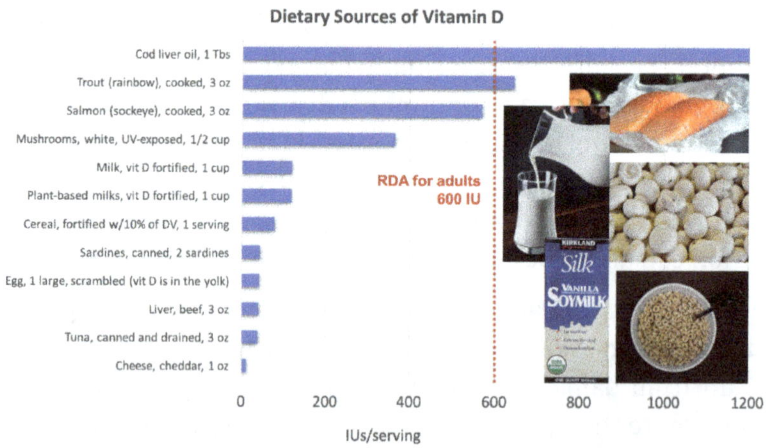

Figure 9.12. Dietary sources of vitamin D. Source: Examples of good sources pictured include salmon, milk, mushrooms, fortified soy milk, and fortified cereal. NIH Office of Dietary Supplements

Both vitamin D_2 and D_3 supplements are also available. Some studies have found D_3 to be more effective at raising circulating vitamin D levels, but others haven't found a difference in efficacy of the two forms.[1,2] Human breast milk doesn't contain adequate vitamin D, so the American Academy of Pediatrics recommends that breastfed infants receive a supplement with 400 IU of vitamin D per day until they are weaned to vitamin D-fortified formula or cow's milk. A vitamin D supplement may also be recommended for older children and adults, depending on dietary intake and sun exposure, but this should be discussed with a healthcare provider.

Vitamin D Deficiency and Toxicity

In children, vitamin D deficiency causes rickets, a disease in which the bones are soft, weak, and deformed. Rickets was very common in the U.S. until the 1930s, when milk processors were asked to add vitamin D to cow's milk. Milk fortification has virtually eliminated rickets from the U.S. and other developed countries. However, rickets does still occur in breastfed infants and children raised on vegan diets who aren't provided with other sources of vitamin D, particularly if they have darker skin pigmentation.[1,3]

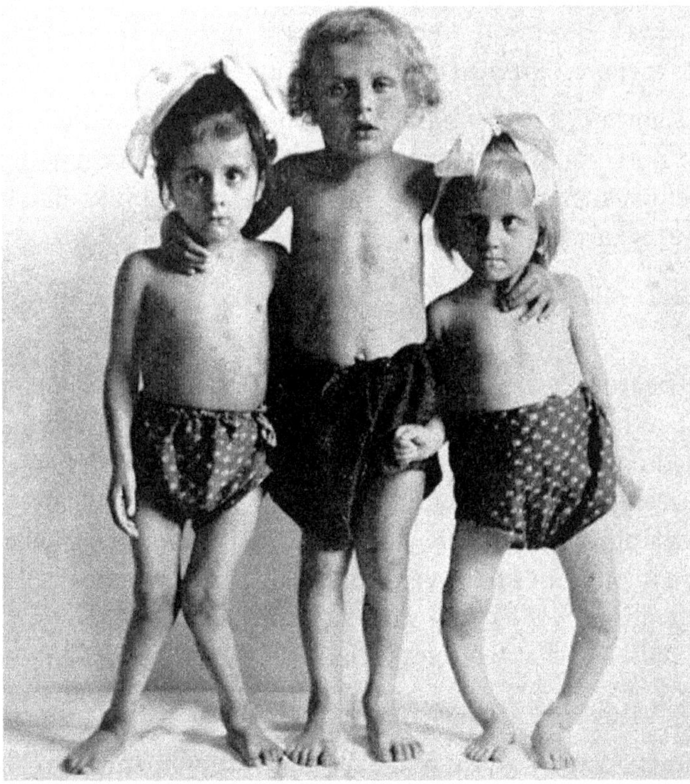

Figure 9.13. Vitamin D deficiency in children causes rickets, a disease in which inadequate vitamin D leads to soft, weak, and deformed bones.

In adults, vitamin D deficiency causes osteomalacia, characterized by softening of bones, reduced bone mineral density, and increased risk of osteoporosis. Because bones are continuously remodelled throughout the lifespan, inadequate vitamin D limits the calcium available to continue to rebuild bone tissue. Vitamin D deficiency can also cause bone pain, as well as muscle weakness and pain, symptoms that can increase the risk of falling and fractures, particularly in older adults. Although vitamin D toxicity is rare, taking excessive amounts of vitamin D in supplement form can lead to hypercalcemia, or high blood calcium. Hypercalcemia can cause kidney damage and calcium deposits to develop in soft tissues such as the kidneys, blood vessels, or other parts of the cardiovascular system. Synthesis of vitamin D from the sun does not cause vitamin D toxicity, because vitamin D synthesis is tightly regulated and decreases if the body has abundant vitamin D.

REFERENCES

Higdon, J., Drake, V., & Delage, B. (2014, April 23). Vitamin D. Linus Pauling Institute. https://lpi.oregonstate.edu/mic/vitamins/vitamin-D#deficiency

National Institutes of Health Office of Dietary Supplements. (n.d.). Office of Dietary Supplements—Vitamin D. Retrieved April 30, 2020, from https://ods.od.nih.gov/factsheets/VitaminD-HealthProfessional/

Lemoine, A., Giabicani, E., Lockhart, V., Grimprel, E., & Tounian, P. (2020). Case report of nutritional rickets in an infant following a vegan diet. Archives De Pediatrie: Organe Officiel De La Societe Francaise De Pediatrie. https://doi.org/10.1016/j.arcped.2020.03.008

ENERGY METABOLISM'S RELATIONSHIP WITH VITAMINS AND MINERALS

Have you ever heard that taking vitamins will give you more energy? Or have you bought a product that claimed it could boost your energy level because it has added vitamins? Based on the knowledge that you've learned in this course, you know that vitamins are not broken down in the body to provide energy. So where does the idea that vitamins give you energy come from? On this page we will provide an overview of the B vitamins and several minerals that are important to the process of energy metabolism in the body, and take a closer look at two of those vitamins (folate and vitamin B_{12}) that have some important implications in our health.

The Role of B Vitamins and Minerals in Energy Metabolism

All of the B Vitamins and several minerals play a role in energy metabolism; they are required as functional parts of enzymes involved in energy release and storage. Many enzymes don't work optimally, or even at all, unless bound to other specific helper molecules, called coenzymes or cofactors. Binding to these molecules promotes optimal conformation and function for their respective enzymes.

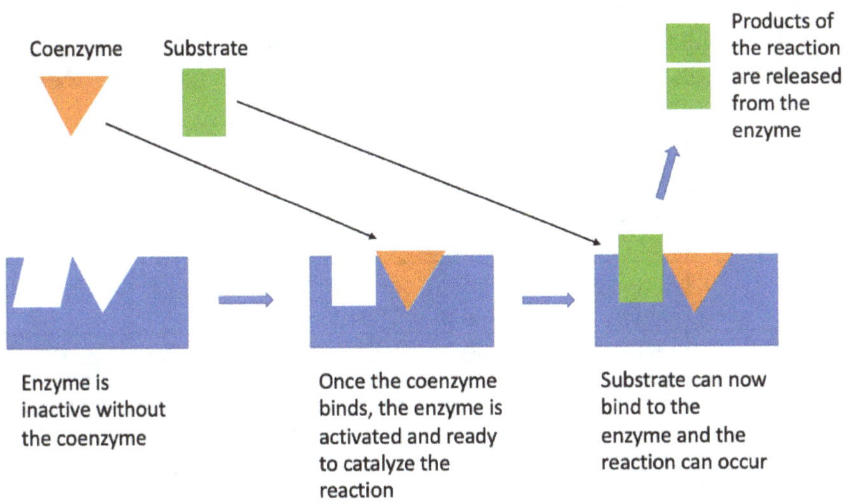

Figure 9.14. Role of a coenzyme assisting in an enzymatic reaction to break down a substrate.

Vitamins that bind to enzymes are referred to as coenzymes—*organic* molecules which are required by enzymes to catalyze a specific reaction. They assist in converting a substrate to an end product. Cofactors are the *inorganic* minerals that assist in these enzymatic reactions. Coenzymes and cofactors are essential in catabolic pathways (*i.e.* breaking down substances) and play a role in many anabolic pathways (*i.e.* building substances). Table 9.1 lists the vitamins and minerals that participate in energy metabolism and their key functions in that process.

Table 9.1. Vitamins and minerals involved in energy metabolism and the role they each play.

Nutrients Involved in Energy Metabolism	
B Vitamins	**Role in Energy Metabolism**
Thiamin (B1)	Assists in glucose metabolism and RNA, DNA, and ATP synthesis
Riboflavin (B2)	Assists in carbohydrate and fat metabolism
Niacin (B3)	Assists in glucose, fat, and protein metabolism
Pantothenic Acid (B5)	Assists in glucose, fat, and protein metabolism, cholesterol and neurotransmitter synthesis
Pyridoxine (B6*)	Assists in the breakdown of glycogen and synthesis of amino acids, neurotransmitters, and hemoglobin
Biotin (B7)	Assists in amino acid synthesis and glucose, fat, and protein metabolism,
Folate (B9)	Assists in the synthesis of amino acids, RNA, DNA, and red blood cells
Cobalamin (B12*)	Protects nerve cells and assists in fat and protein catabolism, folate function, and red blood cell synthesis

Nutrients Involved in Energy Metabolism	
B Vitamins	Role in Energy Metabolism
Minerals	**Role in Energy Metabolism**
Iodine	Assists in metabolism, growth, development, and synthesis of thyroid hormone
Manganese	Assists in carbohydrate and cholesterol metabolism, bone formation, and the synthesis of urea
Sulfur	A component in sulfur-containing amino acids necessary in certain enzymes; a component in thiamin and biotin
Chromium	Assists in carbohydrate, lipid, and protein metabolism, DNA and RNA synthesis
Molybdenum	Assists in metabolism of sulfur-containing amino acids and synthesis of DNA and RNA

*Normally used instead of common names

Do B Vitamins Give You Energy?

Because B vitamins play so many important roles in energy metabolism, it is common to see marketing claims that B vitamins boost energy and performance. This is a myth that is not backed by science. The "feeling" of more energy from energy-boosting supplements stems more from the high amount of added sugars, caffeine, and other herbal stimulants that accompany the high doses of B vitamins in these products. As discussed, B vitamins are needed to support energy metabolism and growth, but taking in more than required does not supply you with more energy. A great analogy of this phenomenon is the gas in your car. Does it drive faster with a half-tank of gas or a full one? It does not matter; the car drives just as fast as long as it has gas. Similarly, depletion of B vitamins will cause problems in energy metabolism, but having more than is required to run metabolism does not speed it up. And because B vitamins are water-soluble, they are not stored in the body and any excess will be excreted from the body, essentially flushing out the added expense of the supplements.

Deficiency and Toxicity

The B vitamins important for energy metabolism are naturally present in numerous foods, and many other foods are enriched with them; therefore, B vitamin deficiencies are rare. Similarly, most of the minerals involved in energy metabolism and listed above are trace minerals that are not frequently deficient in the diet. However, when a deficiency of one of these vitamins or minerals does occur, symptoms can be seen throughout the body because of their relationship to energy metabolism, which happens in all cells of the body. A lack of these vitamins and minerals typically impairs blood health and the conversion of macronutrients into usable energy (i.e., ATP). Deficiency can also lead to an increase in susceptibility to infections, tiredness, lack of energy, and a decrease in concentration.[1] Groups most at risk for a deficiency in any of these micronutrients are people on calorie-limited diets, people with imbalanced or insufficient nutrition, people with eating disorders, and people experiencing extensive levels of physical or emotional stress.[1]

Because of their water-solubility, toxicities of most of these nutrients are also uncommon, as excess intake is often excreted from the body. That doesn't mean taking high doses comes without risks. Large quantities, particularly through supplements, can lead to adverse side effects or cause interactions with medications. For example, too much niacin can cause flushing of the skin or dangerous drops in blood pressure, and a high intake of B_6 can lead to neuropathy. When taking vitamin or mineral supplements, always pay attention to the recommended dietary allowance and avoid exceeding the tolerable upper intake level (UL).

FOLATE

Folate, or vitamin B_9, is a required coenzyme for the synthesis of several amino acids and for making RNA and DNA. Therefore, rapidly dividing cells are most affected by folate deficiency. Red blood cells, white blood cells, and platelets are continuously being synthesized in the bone marrow from dividing stem cells. When folate is deficient, cells cannot divide normally. A consequence of folate deficiency is macrocytic anemia. Macrocytic means "big cell," and anemia refers to fewer red blood cells or red blood cells containing less hemoglobin. Macrocytic anemia is characterized by larger and fewer red blood cells that are less efficient at carrying oxygen to cells. It is caused by red blood cells being unable to produce DNA and RNA fast enough—cells grow but do not divide, making them large in size.

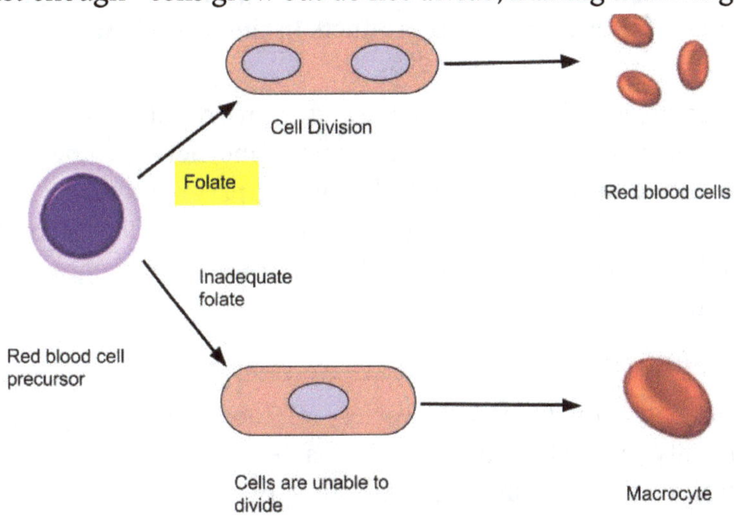

Figure 9.15. Folate and the formation of macrocytic anemia.

Folate is especially essential for the growth and specialization of cells of the central nervous system. Children whose mothers were folate-deficient during pregnancy have a higher risk of neural tube birth defects. Folate deficiency is causally linked to the development of spina bifida, a neural tube defect that occurs in a developing fetus when the spine does not completely enclose the spinal cord. Spina bifida can lead to many physical and mental disabilities (Figure 9.16). In 1998, the U.S. Food and Drug Administration (FDA) began requiring manufacturers to fortify enriched breads, cereals, flours, and cornmeal with folic acid (a synthetic form of folate) to increase the consumption of folate in the American diet and reduce the risk of neural tube defects. Observational studies show that the prevalence of neural tube defects was decreased after the fortification of enriched cereal and grain products with folate compared to before these products were fortified.

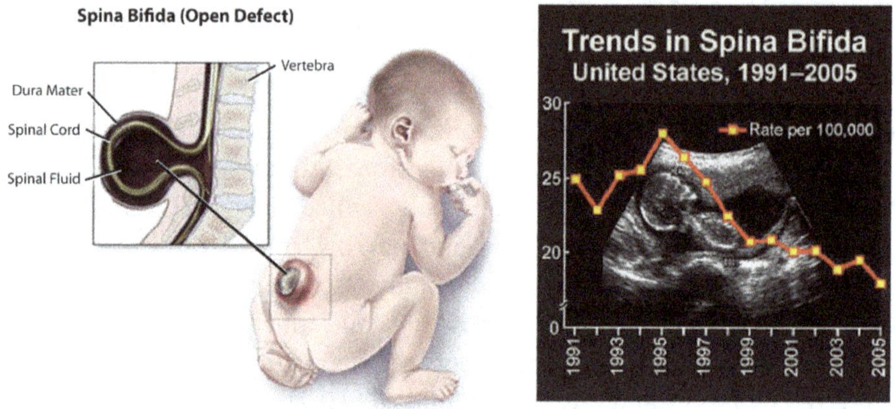

Figure 9.16. Spina bifida (left) is a neural tube defect that can have serious health consequences. The prevalence of cases of spina bifida has decreased significantly with the fortification of cereal and grain products in the United States beginning in 1998.

Additionally, results of clinical trials have demonstrated that neural tube defects are significantly decreased in the offspring of mothers who began taking folic acid supplements one month prior to becoming pregnant and throughout pregnancy. In response to the scientific evidence, the Food and Nutrition Board of the Institute of Medicine (IOM) raised the RDA for folate to 600 micrograms per day for pregnant women.

Dietary Sources of Folate

Folate is found naturally in a wide variety of foods, including vegetables (particularly dark leafy greens), fruits, nuts, beans, legumes, meat, poultry, eggs, and grains. As mentioned previously, folic acid (the synthetic form of folate) is also found in enriched foods such as grains.

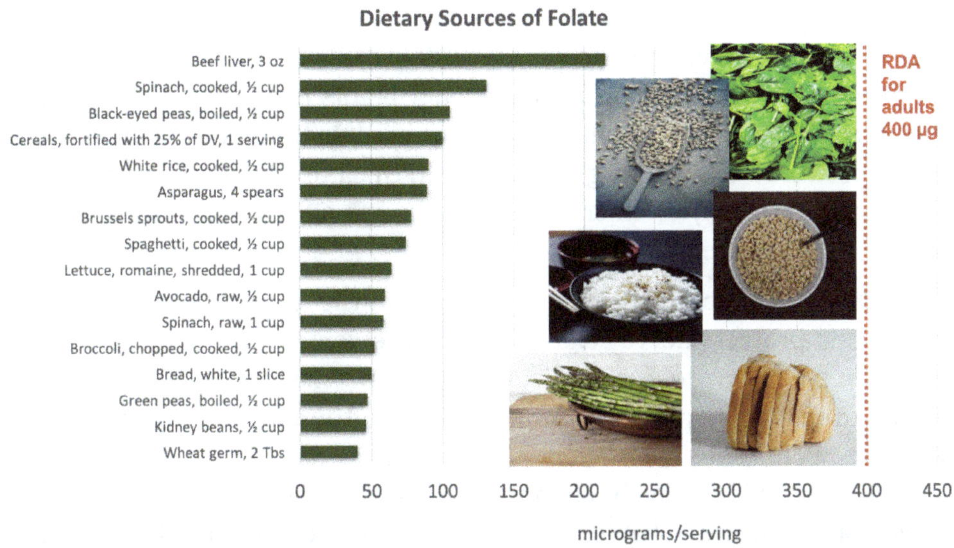

Figure 9.17. Dietary sources of folate. Examples of good sources pictured include spinach, black-eyed peas, fortified cereal, rice, and bread and asparagus. Source: NIH Office of Dietary Supplements

Folate Deficiency and Toxicity

Folate deficiency is typically due to an inadequate dietary intake; however, smoking and heavy, chronic alcohol intake can also decrease absorption, leading to a folate deficiency.[2] The primary sign of a folate deficiency is macrocytic anemia, characterized by large, abnormal red blood cells, which can lead to symptoms of fatigue, weakness, poor concentration, headache, irritability, and shortness of breath. Other symptoms of folate deficiency can include mouth sores, gastrointestinal distress, and changes in the skin, hair and nails. Women with insufficient folate intakes are at increased risk of giving birth to infants with neural tube defects and low intake during pregnancy has been associated with preterm delivery, low birth weight, and fetal growth retardation.[3]

Toxicity of folate is not typically seen due to an excess consumption from foods. However, there is concern regarding a high intake of folic acid from supplements because it could mask a deficiency in vitamin B_{12}. Because folate and vitamin B_{12} deficiencies are manifested by similar anemias, if a person with vitamin B_{12} deficiency is taking a high dose of folic acid, the macrocytic anemia would be corrected while the underlying B_{12} deficiency went undetected, which could result in significant neurological damage. Thus, a tolerable upper intake level (UL) has been established for folate to prevent irreversible neurological damage due to high folic acid intake masking a B_{12} deficiency.[2]

Vitamin B_{12} (COBALAMIN)

Vitamin B_{12} is a unique vitamin because it contains an element (cobalt) and is found almost exclusively in animal products. Neither plants nor animals can synthesize vitamin B_{12}; only bacteria can synthesize it. The vitamin B_{12} found in animal-derived foods was produced by microorganisms within the animals. Animals consume the microorganisms in soil, or microorganisms in the GI tract of animals produce vitamin B_{12} that can then be absorbed.

Vitamin B_{12} helps to prevent the breakdown of the myelin sheath, a cover that surrounds and protects nerve cells. It is also an essential part of coenzymes. It is necessary for fat and protein catabolism, folate coenzyme function, and hemoglobin synthesis. An enzyme requiring vitamin B_{12} is needed by a folate-dependent enzyme to synthesize DNA. Thus, a deficiency in vitamin B_{12} has similar consequences to health as a folate deficiency. In children and adults, vitamin B_{12} deficiency causes macrocytic anemia, and in babies born to cobalamin-deficient mothers there is an increased risk for neural tube defects.

In order for the human body to absorb vitamin B_{12}, the stomach, pancreas, and small intestine must be functioning properly. Cells in the stomach secrete a protein called intrinsic factor that is necessary for vitamin B_{12} absorption, which occurs in the small intestine. Impairment of secretion of this protein either caused by an autoimmune disease or by chronic inflammation of the stomach (such as that occurring in some people with H.pylori infection), can lead to the disease pernicious anemia, a type of macrocytic anemia. Vitamin B_{12} malabsorption is most common in older adults, who may have impaired functioning of digestive organs, a normal consequence of aging.

Vitamin B_{12} and folate play key roles in converting homocysteine, an amino acid found in the blood, to the amino acid methionine. High levels of homocysteine in the blood increases the risk for heart disease. Low levels of vitamin B_{12}, folate, or vitamin B_6 will increase homocysteine levels, thereby increasing the risk of heart disease.

Dietary Sources of Vitamin B_{12}

Vitamin B_{12} is found naturally in animal products such as fish, meat, poultry, eggs, and milk products. Although vitamin B_{12} is not generally present in plant foods, fortified breakfast cereals are also a good source of vitamin B_{12}. Because vitamin B_{12} is only found primarily in animal products, it is important for strict vegetarians who consume no animal products (vegans) to get vitamin B_{12} either through supplements, nutritional yeast, or fortified products like cereals and soy milk. Recent research suggests some plant-based sources like edible algae, mushrooms, and fermented vegetables may contain substantial amounts of vitamin B_{12} as well.[2]

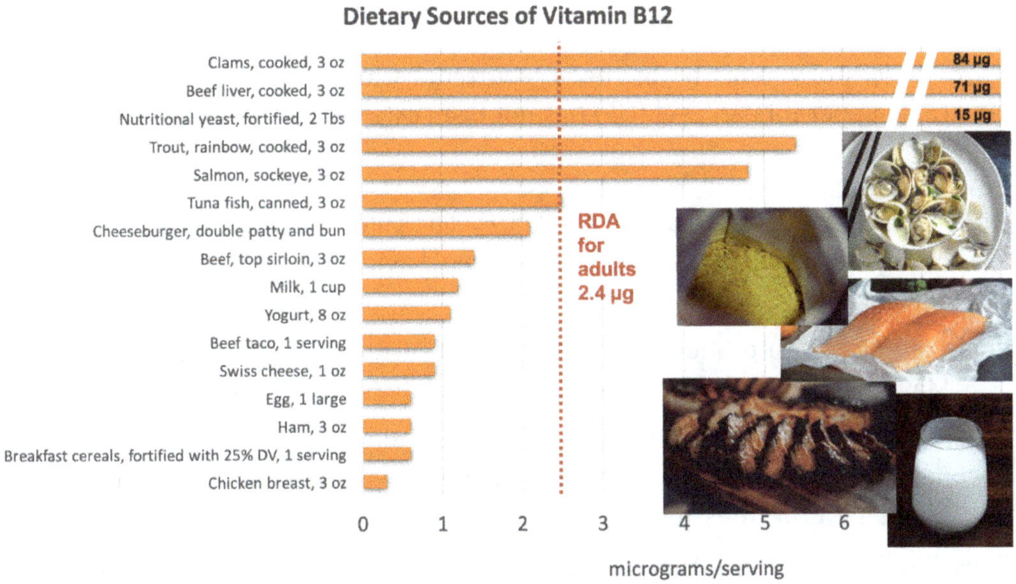

Figure 9.18. Dietary sources of vitamin B12. Examples of good sources pictured include clams, salmon, nutritional yeast, red meat, and milk. Source: NIH Office of Dietary Supplements, www.bragg.com

Vitamin B_{12} Deficiency and Toxicity

When there is a deficiency in vitamin B_{12}, inactive folate (from food) is unable to be converted to active folate and used in the body for the synthesis of DNA. However, folic acid from supplements or fortified foods is available to be used as active folate in the body without vitamin B_{12}. Therefore, if there is a deficiency in vitamin B_{12}, macrocytic anemia may occur. Vitamin B_{12} deficiency also results in nerve degeneration and abnormalities that can often precede the development of anemia. These include a decline in mental function and burning, tingling, and numbness of legs. These symptoms can continue to worsen, and deficiency can be fatal.

The most common cause of vitamin B_{12} deficiency is the condition of inadequate intrinsic factor production that leads to poor vitamin B_{12} absorption, resulting in pernicious anemia. This condition is common in people over the age of 50, because production of intrinsic factor decreases with age. Pernicious anemia is treated by large oral doses of vitamin B_{12} or by putting the vitamin under the tongue (sublingual), where it is absorbed into the bloodstream without passing through the intestine. In patients that do not respond to oral or sublingual treatment, vitamin B_{12} is given by injection. Because vitamin B_{12} is found primarily in foods of animal origin, a strict vegetarian diet can result in cases of vitamin B_{12} deficiency.

Therefore, careful dietary planning to include fortified sources or supplements of vitamin B_{12} is important to prevent deficiency in a vegan diet.

No toxicity of vitamin B_{12} has been reported. Because of the low risk for toxicity, a tolerable upper intake level (UL) has not been established for vitamin B_{12}.[4]

REFERENCES

Huskisson, E., Maggini, S., & Ruf, M. (2007). The role of vitamins and minerals in energy metabolism and well-being. Journal of international medical research, 35(3), 277-289.

Linus Pauling Institute-Micronutrient Information Center. (2020). Folate. Retrieved May 2, 2020 from https://lpi.oregonstate.edu/mic/vitamins/folate

National Institutes of Health Office of Dietary Supplements. (2020, March 11). Folate. Retrieved May 2, 2020, from https://ods.od.nih.gov/factsheets/Folate-HealthProfessional/

Linus Pauling Institute-Micronutrient Information Center. (2020). Vitamin B_{12}. Retrieved May 2, 2020, from https://lpi.oregonstate.edu/mic/vitamins/vitamin-B12

THE ROLE OF VITAMINS AND MINERALS IN BLOOD HEALTH

Blood is essential to life. It transports absorbed nutrients and oxygen to cells, removes metabolic waste products for excretion, and carries molecules, such as hormones, to allow for communication between organs.

Blood is a connective tissue of the circulatory system, made up of four components forming a matrix:

- Red blood cells (or erythrocytes), which transport oxygen to cells
- White blood cells (or leukocytes), which are part of the immune system and help destroy foreign invaders
- Platelets, which are fragments of cells that circulate to assist in blood clotting
- Plasma, which is the fluid portion of the blood and contains proteins that help transport nutrients (e.g., fat-soluble vitamins) and aid in blood clotting.

Maintaining healthy blood, including its continuous renewal, is essential to support its vital functions. Blood health is acutely sensitive to deficiencies in some vitamins and minerals, especially iron and vitamin K.

Iron

Iron is a trace mineral. It is a necessary component of many proteins in the body responsible for functions such as the transport of oxygen, energy metabolism, immune function, and antioxidant defense. Iron is also important in brain development and function.

Iron is essential for oxygen transport because of its role in hemoglobin, a protein in red blood cells that transports oxygen to cells and gives red blood cells their color. Hemoglobin is composed of four globular peptides, each with an iron-containing heme complex in the center (Figure 9.19).

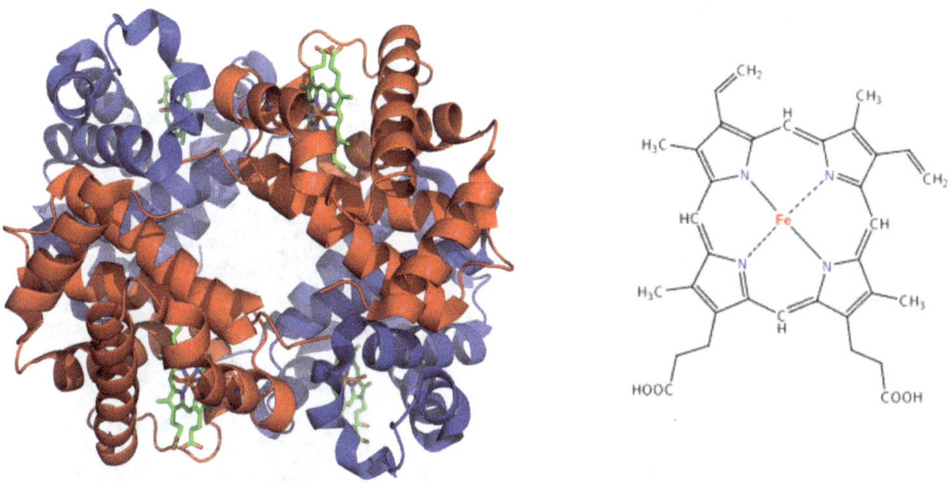

Figure 9.19. The structure of hemoglobin and the heme complex. On the left, the structure of hemoglobin includes four globular peptides (shown in blue and red) and the iron-containing heme groups (shown in green). On the right is a closer view of the heme complex.

The iron in hemoglobin is what binds to oxygen, allowing for transportation to cells. If iron levels are low, hemoglobin is not synthesized in sufficient amounts, and the oxygen-carrying capacity of red blood cells is reduced, resulting in anemia. Iron is also an important part of myoglobin, a protein similar to hemoglobin but found in muscles.

Dietary Sources of Iron

There are two types of iron found in foods: heme iron and non-heme iron.

- Heme iron is iron that is part of the proteins hemoglobin and myoglobin, so it is found only in foods of animal origin, such as meat, poultry, and fish. Heme iron is the most bioavailable form of iron. About 40% of the iron in animal foods is heme iron and 60% is non-heme iron.
- Non-heme iron is the mineral by itself and is not a part of hemoglobin or myoglobin. Non-heme iron can be found in foods from both plants (e.g., nuts, beans, vegetables, and fortified and whole grains) and animals. It

is less bioavailable than heme iron. Consuming vitamin C, meat, poultry, and seafood with non-heme iron increases its bioavailability. For example, eating an orange (a good source of vitamin C) along with your bowl of vegetarian chili will help you to absorb more of the iron from the beans and vegetables. On the other hand, chemicals such as phytates (found in beans and grains) and plant polyphenols (found in fruits, vegetables, some cereals, legumes, tea, coffee, and wine) decrease bioavailability due to binding with iron.[1]

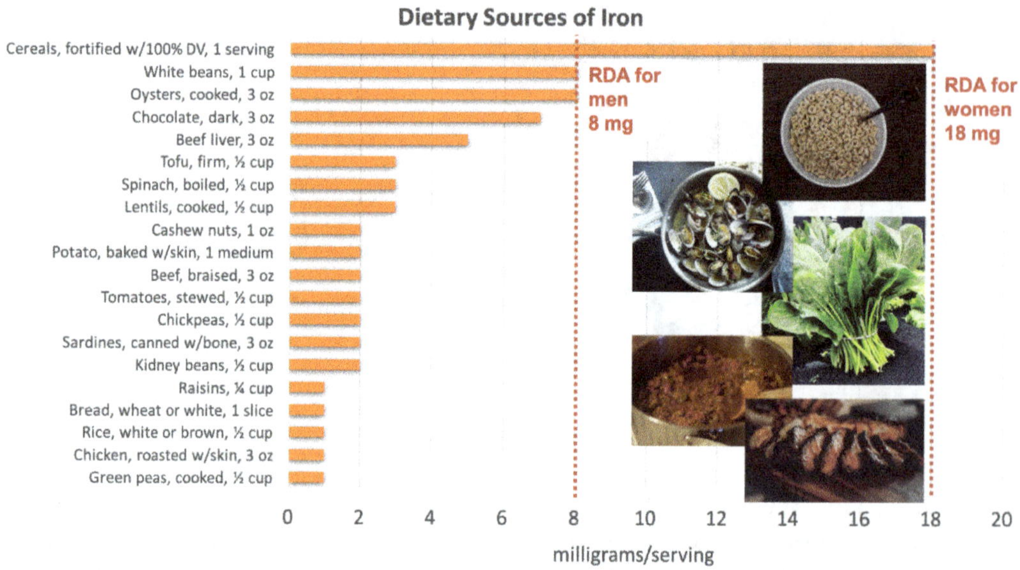

Figure 9.20. Dietary sources of iron. Examples of good sources pictured include fortified cereal, clams, spinach, lentils and red meat. Source: NIH Office of Dietary Supplements

The bioavailability of iron is approximately 14% to 18% from mixed diets and 5% to 12% from vegetarian diets.[2] Bioavailability is influenced by the dietary factors previously mentioned, as well as iron status. The body has no physiological mechanism to excrete iron; therefore, iron balance is tightly regulated by absorption.[2] When iron stores are low, more dietary iron will be absorbed.

The majority of the body's iron needs are not met by dietary sources, but rather by recycling iron within the body (***endogenous*** sources). Ninety percent of daily iron needs are met by recycling iron released from the breakdown of aging cells, mostly red blood cells.[2]

Iron Deficiency and Toxicity

Iron deficiency is one of the most common nutrient deficiencies, affecting more than a third of the world's population.[3] Iron-deficiency anemia is a condition that develops from having insufficient iron levels in the body, resulting in fewer and smaller red blood cells that contain less hemoglobin. As a result, blood carries less oxygen from the lungs to cells. Iron-deficiency anemia has the following signs and symptoms, which are linked to the essential functions of iron in energy metabolism and blood health:

- Fatigue
- Weakness
- Pale skin
- Shortness of breath
- Dizziness
- Swollen, sore tongue
- Abnormal heart rate

Infants, children, adolescents, and women are the populations most at risk worldwide for iron-deficiency anemia.[4] Infants who are premature, low birthweight, or have a mother with iron deficiency are at risk for iron deficiency, because they are born with low iron stores relative to the amount needed for growth and development. Young children and adolescent girls are at risk for iron deficiency due to rapid growth, low dietary intake of iron, as well as heavy menstruation for adolescent girls. In these populations, iron-deficiency anemia can also cause the following signs and symptoms: pica (an intense craving for and ingestion of non-food items such as paper, dirt, or clay), poor growth, failure to thrive, and poor performance in school, as well as mental, motor, and behavioral disorders. Women who experience heavy menstrual bleeding, or who are pregnant, are also at risk for iron inadequacy due to their increased requirements for iron. To give you a better understanding of these risks, it is helpful to look at how much higher the RDA is for women of reproductive age and pregnant women compared to men (Table 9.2).[1] To put this in perspective, 3 ounces of beef contains about 3 milligrams of iron, making it challenging for some women to meet their daily iron requirement.

Table 9.2. A comparison of the RDAs for adult women of reproductive age, pregnancy, and adult men.

Population Group	RDA for Iron
Women of reproductive age, 19-50 years	18 mg/day
Pregnancy, 19- 50 years	27 mg/day
Men, 19-50 years	8 mg/day

Additionally, those who frequently donate blood, as well as people with cancer, heart failure, or gastrointestinal diseases such as Crohn's, are at greater risk for iron inadequacy.[1]

Iron deficiency progresses through three stages:[1]

- Mild depletion of iron stores – No physical symptoms will be present, because hemoglobin levels are not affected, but there will be a decrease in ferritin (a storage form of iron).
- Iron-deficient erythropoiesis (erythrocyte or red blood cell production) – Iron stores are depleted and iron transport is decreased due to a decrease in transferrin (the transport protein for iron), but hemoglobin levels are usually within the normal range.
- Iron-deficiency anemia – Iron stores are exhausted and hemoglobin levels are reduced, resulting in microcytic anemia (small red blood cells) and hypochromic anemia (low color red blood cells).

Healthy adults are at little risk of iron overload from foods, but too much iron from supplements can result in gastric upset, constipation, nausea, vomiting, and abdominal pain.[1] The body excretes little iron; therefore, the potential for toxicity from supplements is a concern. In children, death has occurred from ingesting as little as 200 mg of iron, so it is critical to keep iron supplements out of children's reach. The tolerable upper intake for iron is 45 mg per day for adults, based on the amount that causes gastrointestinal distress.[1]

vitamin k

Vitamin K refers to a group of fat-soluble vitamins that are similar in chemical structure. They act as coenzymes and have long been known to play an essential role in blood coagulation or clotting. Vitamin K is also required for maintaining bone health, as it modifies a protein which is involved in the bone remodeling process.

Dietary Sources of Vitamin K

Vitamin K is found in the highest concentrations in green vegetables such as collard and turnip greens, kale, broccoli, and spinach. Soybean and canola oil are also common sources of vitamin K in the U.S. diet.[5] Additionally, vitamin K can be synthesized by bacteria in the large intestine, but the bioavailability of bacterial vitamin K is unclear.

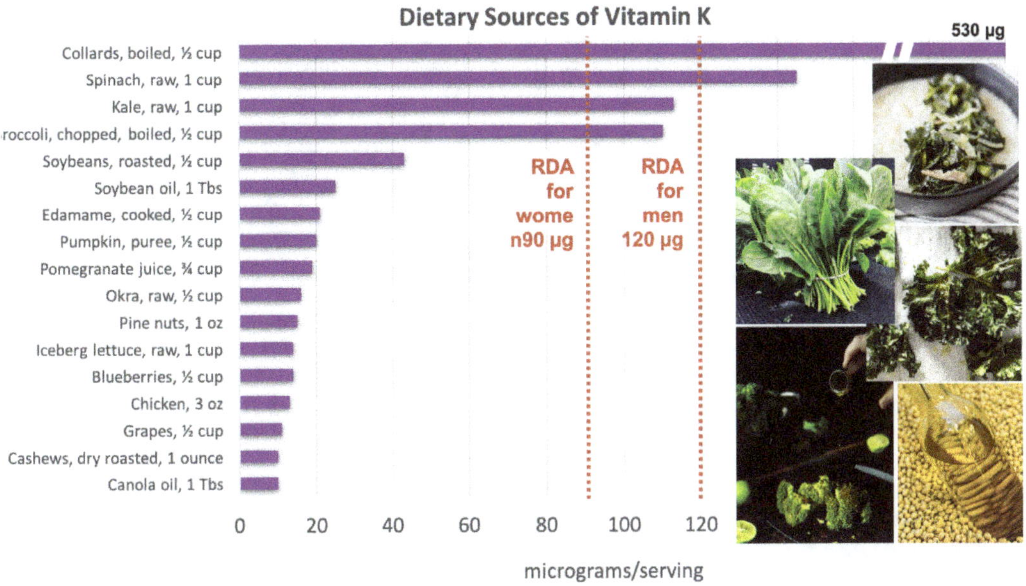

Figure 9.21. Dietary sources of vitamin K. Examples of good sources pictured include chard, spinach, kale, broccoli, and soybean oil. Source: NIH Office of Dietary Supplements

Vitamin K Deficiency and Toxicity

Vitamin K deficiency is rare, as most U.S. diets are adequate in vitamin K.[5] A deficiency in vitamin K causes excessive bleeding (or hemorrhage). When there is damage to a blood vessel that results in bleeding, like a small tear, the body can stop this bleeding through a cascade of reactions. Without adequate vitamin K, blood does not clot properly, and this small bleed can turn into a larger problem, causing excessive bleeding or hemorrhage. People at risk are those who have malabsorption and other gastrointestinal disorders such as celiac disease. Newborns are also at risk for vitamin K deficiency during the first few months of life, as there is poor transfer of vitamin K across the placenta, and breastmilk is also low in vitamin K. Therefore, it has become a routine practice to inject newborns with a single intramuscular dose of vitamin K to prevent hemorrhaging, as bleeding within the skull is especially life-threatening.

The Food and Nutrition Board (FNB) has not established a UL for vitamin K because it has a low potential for toxicity.

UNIT-10

NUTRITION AND PHYSICAL ACTIVITY

NUTRITION AND PHYSICAL ACTIVITY: AN INTRODUCTION

In this course, our focus is on how nutrients support health. However, enjoying good health is about so much more than simply meeting nutrient needs. As we discussed in Unit 1, the World Health Organization (WHO) defines health as "a state of complete physical, mental and social well-being and not merely the absence of disease or infirmity." For most people, being physically active contributes to all three of those elements of health: physical, mental, and social well-being.

Not surprisingly, the more you ask of your body in terms of physical activity, the more attention you'll need to give to fueling it with nutrients. With optimal nourishment, your body will in turn reward you with improved performance and energy levels. Fueling a physically active body is based on the same principles we've discussed throughout this course, but requirements for energy and some nutrients are increased. In this unit, we'll cover the elements of physical fitness, the benefits of physical activity, and recommendations for fueling an active lifestyle and athletic performance.

PHYSICAL FITNESS' ESSENTIAL ELEMENTS AND BENEFITS

Becoming and staying physically fit is an important part of achieving optimal health. A well-rounded exercise program can improve your health in a number of ways. It keeps the heart and lungs strong, strengthens muscles and bones, helps to protect against chronic disease, improves brain and mental health, and can support healthy body weight. There are four main components of physical fitness as it relates to health and function: cardiorespiratory fitness, musculoskeletal fitness, flexibility, and balance. 1 Each component offers specific health benefits, but optimal health requires some degree of balance between all four. (Speed, or the ability to move the body quickly, is also considered a component of physical fitness, but it is less related to health and more relevant to competitive athletes.[1])

Some forms of exercise can help build multiple types of fitness. For example, riding a bicycle not only builds cardiorespiratory endurance; it can also improve muscle strength and endurance. Some forms of yoga and Pilates can also build muscle strength and endurance, along with flexibility and balance. However, building fitness in all four categories generally requires incorporating a range of activities into your regular routine. As you increase physical activity, you will notice that you are able to continue your activity longer and with greater ease.

The Essential Elements Of Physical Fitness

Cardiorespiratory Fitness

Cardiorespiratory fitness is defined as "the ability to perform large-muscle, whole-body exercise at moderate-to-vigorous intensities for extended periods of time," according to the Physical Activity Guidelines for Americans.1 Cardiorespiratory fitness is built by aerobic activity, during which you move your body's large muscle groups in a continuous and rhythmic way, and your heart and breathing rate increase. Examples of aerobic activities include walking, running, swimming, jumping rope, and riding a bike.

Aerobic physical activity can range in level of intensity; for example, brisk walking is considered moderate intensity, and running or jogging is considered vigorous intensity. As intensity increases, so do heart and breathing rates to meet increased demand for oxygen in working muscles. Regular, moderate aerobic activity—about thirty minutes at a time on five days per week—trains the body to deliver oxygen more efficiently, which strengthens the heart and lungs and reduces the risk of cardiovascular disease. Strengthening your heart muscle and increasing the blood volume pumped in each heartbeat boosts your ability to supply your body's cells with oxygen and nutrients and remove carbon dioxide and metabolic wastes. It also leads to a lower resting heart rate for healthy individuals. Aerobic physical activity can also help maintain a healthy weight.

Many aerobic activities also count as bone-strengthening activity, which promotes bone growth and strength by producing a force on the body's bones. Weight-bearing activities like running, walking, and doing jumping jacks are considered bone-strengthening, whereas activities like swimming and biking that put less force on the body's bones are not. Bone-strengthening activities are especially important during childhood and adolescence, when they promote greater bone mass and strength and improved bone structure. In older adults, participating in aerobic and bone-strengthening activities slows the decline in bone density that often occurs with aging.

Musculoskeletal Fitness

Musculoskeletal fitness is defined as "the integrated function of muscle strength, muscle endurance, and muscle power to enable performance."[1] Musculoskeletal fitness is developed and maintained by weight or resistance training, often called anaerobic exercise. Anaerobic exercise consists of short duration, high intensity movements that rely on immediately available energy sources and require little or no oxygen during the activity. This type of training includes lifting weights, working with resistance bands, using body weight (such as for push-ups and pull-ups), and doing work such as carrying heavy loads or vigorous yard work. Building musculoskeletal fitness is not just for athletes and bodybuilders—it's important for children, older adults, and everyone in between. The support that your muscles provide allows you to work, play, and live more efficiently.

Muscle-strengthening activity often involves the use of resistance machines, resistance bands, free weights, or other tools. However, you do not need to pay for a gym membership or expensive equipment to strengthen your muscles. Homemade weights, such as plastic bottles filled with sand, can work just as well. You can also use your own body weight and do push-ups, leg squats, abdominal crunches, planks, and other exercises to build your muscles. What is important is that your training incorporates exercises that together will work all the major muscle groups of the body, including the legs, hips, chest, back, shoulders, arms, and abdomen. To improve muscle strength, do one to three sets of 8 to 12 repetitions per exercise, with enough weight that it is difficult to do another repetition at the end of a set.

If strength training is performed at least twice per week, it can help improve muscular fitness and bone health, as well as manage health conditions like diabetes, arthritis, hypertension, and many others. Strength training can also help to maintain muscle mass during weight loss.[1,2]

Flexibility

Flexibility is the range of motion available to your joints. Many types of physical activity, including dancing, yoga, and martial arts, require significant joint movement, and improving flexibility may allow you to do these activities more easily and fluidly. Even daily activities such as housework and getting in and out of the car are easier with more functional flexibility, especially as we age.

Yoga, tai chi, Pilates, and stretching exercises can increase flexibility. These exercises can improve your range of motion and promote better posture and balance. However, contrary to popular belief, research on stretching has shown that it does not reduce risk of injury or later muscle soreness.[3]

Balance

Balance is the ability to maintain equilibrium during movement or while stationary. Many forms of physical activity and daily living activities require balance. Exercises to improve balance are especially important in older adults, as balance tends to deteriorate with age, putting older adults at greater risk of falls and fractures.[4] Activities that can improve balance include walking backward, standing on one leg, and using a wobble board. Activities such as tai chi and yoga can also improve balance, and especially for older adults, can also count as moderate-intensity aerobic activity.[1]

Physical Activity Benefits And Guidelines

Regular physical activity is one of the best things you can do to achieve optimal health. Individuals who are physically active for 150 minutes per week lower the risk of dying early by 33 percent compared to those who are inactive.[3] The 2018 Physical Activity Guidelines for Americans were issued by the Department of Health and Human Services (HHS) to provide evidence-based guidelines to Americans aged three and older about how to improve health and reduce chronic disease risk through physical activity. The full guidelines are available online at the following link: Physical Activity Guidelines for Americans.4

Key Guidelines for Adults

- Adults should move more and sit less throughout the day. Some physical activity is better than none. Adults who sit less and do any amount of moderate-to-vigorous physical activity gain some health benefits.
- For substantial health benefits, adults should do at least 150 minutes (2 hours and 30 minutes) to 300 minutes (5 hours) a week of moderate-intensity, or 75 minutes (1 hour and 15 minutes) to 150 minutes (2 hours and 30 minutes) a week of vigorous-intensity aerobic physical activity, or an equivalent combination of moderate- and vigorous-intensity aerobic activity. Preferably, aerobic activity should be spread throughout the week.
- Additional health benefits are gained by engaging in physical activity beyond the equivalent of 300 minutes (5 hours) of moderate-intensity physical activity a week.
- Adults should also do muscle-strengthening activities of moderate or greater intensity and that involve all major muscle groups on 2 or more days a week, as these activities provide additional health benefits.

Improving your overall fitness involves sticking with a physical activity program on a regular basis, so that being active becomes part of your routine.

Choosing physical activity that you enjoy will increase the likelihood of sustaining the activity. If you are nervous or unsure about becoming more active, check with your doctor about the best way to get started. The good news is that moderate-intensity activities such as brisk walking are very safe for most people, and the health advantages of becoming active far outweigh the risks. Physical activity can also be social; regularly meeting a friend for a walk or joining a hiking group can make getting active more fun and social, plus help you stick to your plan. Physical activity provides a wealth of benefits—physical, mental, and emotional.

Table 10.1. Physical and emotional benefits of exercise.

Physical Benefits	Mental and Emotional Benefits
Longer life: A regular exercise program can reduce your risk of dying early from heart disease, certain cancers, and other leading causes of death.	**Mood improvement:** Aerobic activity, strength-training, and more contemplative activities such as yoga, all help break cycles of worry and distraction, effectively draining tension from the body.
Healthier weight: Exercise, along with a healthy, balanced eating plan, may help you lose extra weight, maintain weight loss, or prevent excessive weight gain.	**Depression relief:** Exercise can produce antidepressant effects in the body. Studies have shown that physical activity reduces the risk of and helps people cope with the symptoms of depression.

Physical Benefits	Mental and Emotional Benefits
Cardiovascular disease prevention: Being active boosts HDL cholesterol and decreases unhealthy triglycerides, which reduces the risk of cardiovascular diseases.	**Cognitive skills retention:** Regular physical activity can help people maintain thinking, learning, and judgment as they age.
Management of chronic conditions: A regular routine can help to prevent or manage a wide range of conditions and concerns, such as metabolic syndrome, type 2 diabetes, depression, arthritis, and certain types of cancer.	**Better sleep:** A good night's sleep is essential for clear thinking, and regular exercise promotes healthy, sound sleep. It can also help you fall asleep faster and deepen your rest, promoting better mental and emotional wellbeing.
Energy boosts: Regular physical activity can improve muscle tone and strength and provide a boost to your cardiovascular system. When the heart and lungs work more efficiently, you have more energy.	
Strong bones: Research shows that aerobic activity and strength training can slow the loss of bone density that typically accompanies aging.	

The Fitt Principle

One helpful tool for putting together an exercise plan is the FITT. FITT stands for:

- Frequency – how often you exercise
- Intensity – how hard you work during your exercise session
- Time – how long you exercise for
- Type– what kind of exercise you do

You can manipulate the principles of FITT to better meet your exercise goals and to boost your motivation to exercise. You will be more likely to stick to a workout plan that is fun, has flexibility, and works with your lifestyle. By changing up the types of exercise you do, varying the intensity of your workouts, and choosing days of the week and times of the day that work best with your schedule, you can create a plan for success with your exercise goals. As you design your physical activity plan, make sure you find ways to move that you truly enjoy and can sustain and consider the components of the FITT principle to establish more detailed goals and create purpose for your workouts.

REFERENCES:

2018 Physical Activity Guidelines for Americans. US Department of Health and Human Services. Retrieved August 28, 2020, from https://health.gov/paguidelines/second-edition/.

AmericanCollegeofSportsMedicine.(2019). ResistanceTrainingforHealth. Accessed August 28, 2020, from https://www.acsm.org/docs/default-source/files-for-resource-library/resistance-training-for-health.pdf?sfvrsn=d2441c0_2

Bosu, O., Doctor, K., Dixon, K., Smith, L., & Waits, J. B. (2016). Stretching for Prevention of Exercise-Related Injury. American Family Physician, 94(7), 547–547. https://www.aafp.org/pubs/afp/issues/2016/1001/p547.html

Mayo Clinic. (2018). Fitness training: Elements of a well-rounded routine. Accessed August 28, 2020, from https://www.mayoclinic.org/healthy-lifestyle/fitness/in-depth/fitness-training/art-20044792

Centers for Disease Control and Prevention. (2020). Benefits of Physical Activity. Accessed August 28, 2020, from https://www.cdc.gov/physicalactivity/basics/pa-health/index.htm?CDC_AA_refVal=https%3A%2F%2Fwww.cdc.gov%2Fphysicalactivity%2Feveryone%2Fhealth%2Findex.html

SOURCES OF ENERGY FOR EXERCISE

The human body uses carbohydrate, fat, and protein in food and from body stores for energy to fuel physical activity. These essential nutrients are needed regardless of the intensity of the activity you are doing. If you are lying down and reading a book or running a marathon, these macronutrients are always needed in the body. However, in order for these nutrients to be used as fuel for the body, their energy must be transferred into the high energy molecule known as adenosine triphosphate (ATP). ATP is the body's immediate fuel source and can be generated either with aerobic metabolism in the presence of oxygen or anaerobic metabolism without the presence of oxygen. The type of metabolism that is predominately used during physical activity is determined by the availability of oxygen and how much carbohydrate, fat, and protein are used.

Anaerobic and Aerobic Metabolism

Anaerobic metabolism occurs in the cytosol of the muscle cells. As seen in Figure 10.1., a small amount of ATP is produced in the cytosol without the presence of oxygen. Anaerobic metabolism uses glucose as its only source of fuel and produces pyruvate and lactic acid. Pyruvate can then be used as fuel for aerobic metabolism. Aerobic metabolism takes place in mitochondria of the cell and is able to use carbohydrates, protein, or fat as fuel sources. Aerobic metabolism is a much slower process than anaerobic metabolism, but it can produce much more ATP and is the process by which the majority of the ATP in the body is generated.

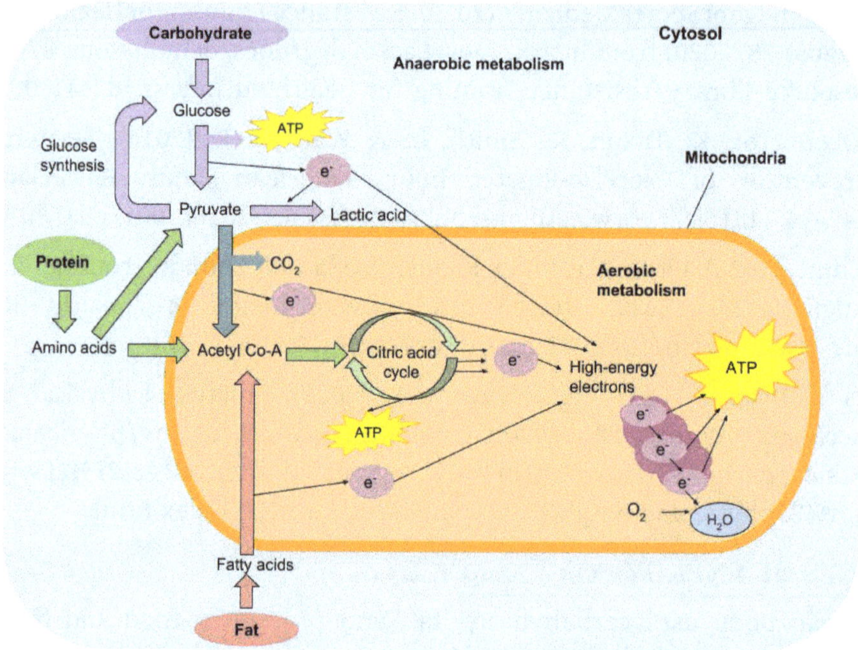

Figure 10.1. Anaerobic vs aerobic metabolism. Note that carbohydrate is the only fuel utilized in anaerobic metabolism, but all three macronutrients can be used for fuel during aerobic metabolism.

Physical Activity Duration and Fuel Use

The respiratory system plays a vital role in the uptake and delivery of oxygen to muscle cells throughout the body. Oxygen is inhaled by the lungs and transferred from the lungs to the blood, where the cardiovascular system circulates the oxygen-rich blood to the muscles. The oxygen is then taken up by the muscles and can be used to generate ATP. When the body is at rest, the heart and lungs are able to supply the muscles with adequate amounts of oxygen to meet the energy needs for aerobic metabolism. However, during physical activity, your muscles need more energy and oxygen. In order to provide more oxygen to the muscle cells, your heart rate and breathing rate will increase. The amount of oxygen that is delivered to the tissues via the cardiovascular and respiratory systems during exercise depend on the duration, intensity and physical conditioning of the individual.

- During the first few steps of exercise, your muscles are the first to respond to the change in activity level. Your lungs and heart do not react as quickly, and during those beginning steps, they can't yet increase the delivery of oxygen. In order for our bodies to get the energy that is needed in these beginning steps, the muscles rely on a small amount of ATP that is stored in resting muscles. The stored ATP is able to provide energy for only a few seconds before it is depleted. Once the stored ATP is just about used up, the body resorts to another high-energy molecule known as creatine phosphate to convert ADP (adenosine diphosphate) to ATP. After about 10 seconds, the stored creatine phosphate in the muscle cells is also depleted as well.

- About 15 seconds into exercise, the stored ATP and creatine phosphate are used up in the muscles. The heart and lungs have still not adapted to the increased oxygen need, so the muscles must begin to produce ATP by anaerobic metabolism (without oxygen). Anaerobic metabolism can produce ATP at a rapid pace but only uses glucose as its fuel source. The glucose is obtained from muscle glycogen. At around 30 seconds, anaerobic pathways are operating at their full capacity, but because the availability of glucose is limited, it cannot continue for a long period of time.
- As your exercise reaches two to three minutes, your heart rate and breathing rate have increased to supply more oxygen to your muscles. Aerobic metabolism is the most efficient way of producing ATP; it produces significantly more ATP for each molecule of glucose than anaerobic metabolism. Although the primary source of ATP in aerobic metabolism is carbohydrates, fatty acids and protein can also be used as fuel to generate ATP.

Figure 10.2. Energy systems used to fuel exercise change with duration of exercise. The ATP-creatine phosphate system is used up within seconds. The short-term and long-term systems kick in and provide energy for exercise as the duration of the workout goes on.

The fuel sources for anaerobic and aerobic metabolism will change depending on the amount of nutrients available and the type of metabolism.

- Glucose may come from blood glucose (which is from dietary carbohydrates, liver glycogen, and glucose synthesis) or muscle glycogen. Glucose is the primary energy source for both anaerobic and aerobic metabolism.
- Fatty acids are stored as triglycerides in muscles, but about 90 percent of stored energy is found in adipose tissue. As low- to moderate-intensity exercise continues using aerobic metabolism, fatty acids become the predominant fuel source for exercising muscles.
- Although protein is not considered a major energy source, small amounts of amino acids are used while resting or doing an activity. The amount of amino acids used for energy metabolism increases if the total energy intake from your diet does not meet your nutrient needs or if you are involved in long endurance exercise. When amino acids are broken down and the nitrogen-containing amine group is removed, the remaining carbon molecule can be broken down into ATP via aerobic metabolism, or it can be used to make glucose. When exercise continues for many hours, amino acid use will increase as an energy source and for glucose synthesis.

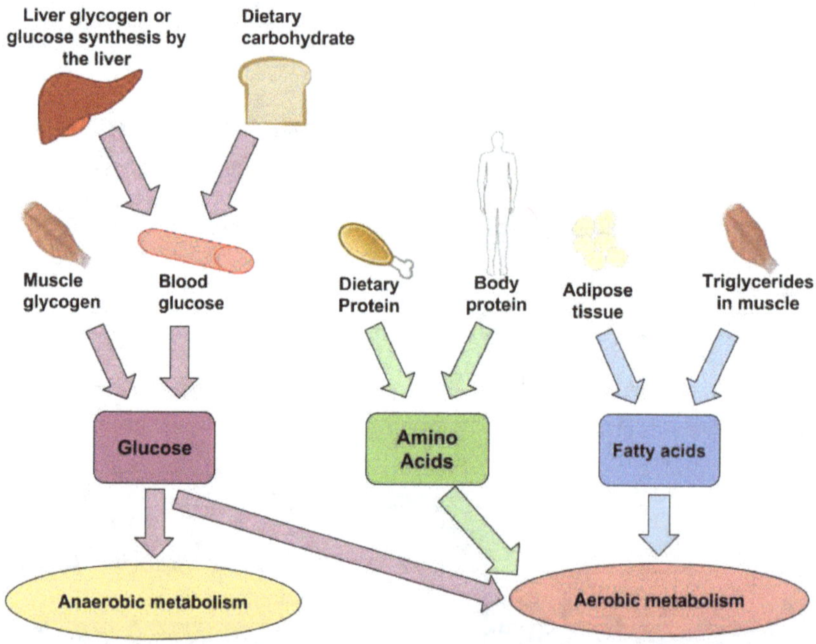

Figure 10.3. Fuel sources for anaerobic and aerobic metabolism. Both dietary sources and body storage of carbohydrates, fat, and protein can all be used to fuel activity. Amount varies depending on duration and intensity of the activity.

Physical Activity Intensity And Fuel Use

Exercise intensity determines the contribution of different fuel sources used for ATP production. Both anaerobic and aerobic metabolism combine during exercise to ensure that the muscles are equipped with enough ATP to carry out the demands placed on them. The contribution from each type of metabolism depends on the intensity of an activity. During low-intensity activities, aerobic metabolism is used to supply enough ATP to muscles. However, during high-intensity activities, more ATP is needed, so the muscles must rely on both anaerobic and aerobic metabolism to meet the body's demands.

Table 10.2. Summary of Fuels Used for Activities of Different Intensities and Durations.

Activity Intensity	Activity Duration	Preferred Fuel	Oxygen Needed?	Activity Example
Very high	30 sec – 3 min	Glucose	No – anaerobic	Sprinting
High	3 min – 20 min	Glucose	Yes – aerobic	Jogging
Low to moderate	>20 min	Fat	Yes – aerobic	Walking

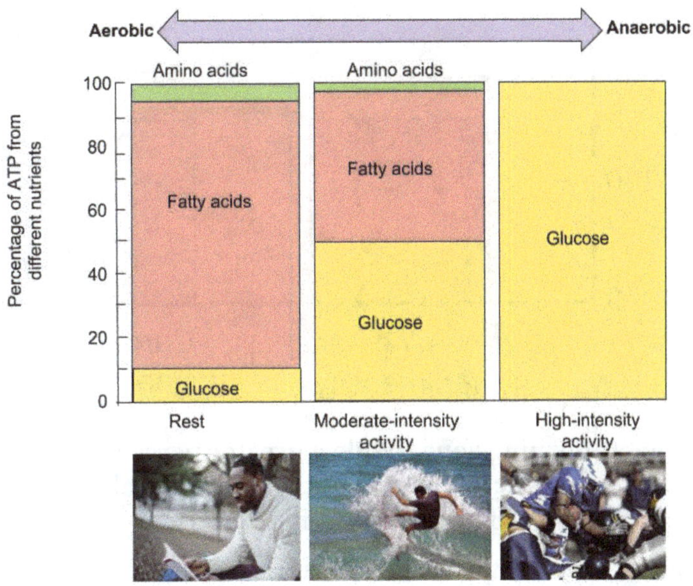

Figure 10.4. The effect of exercise intensity on fuel sources. Anaerobic exercise utilizes only glucose for fuel. As activities become more aerobic, the body can utilize fatty acids and, to a small extent, amino acids, for energy production.

During low-intensity activities, the body will use aerobic metabolism over anaerobic metabolism, because it is more efficient and produces larger amounts of

ATP. Fatty acids are the primary energy source during low-intensity activity. With fat reserves in the body being almost unlimited, low-intensity activities are able to continue for a long time. Along with fatty acids, a small amount of glucose is used as well. Glucose differs from fatty acids, because glycogen storages can be depleted. As glycogen stores are depleted, the glucose supply becomes depleted, and fatigue will eventually set in.

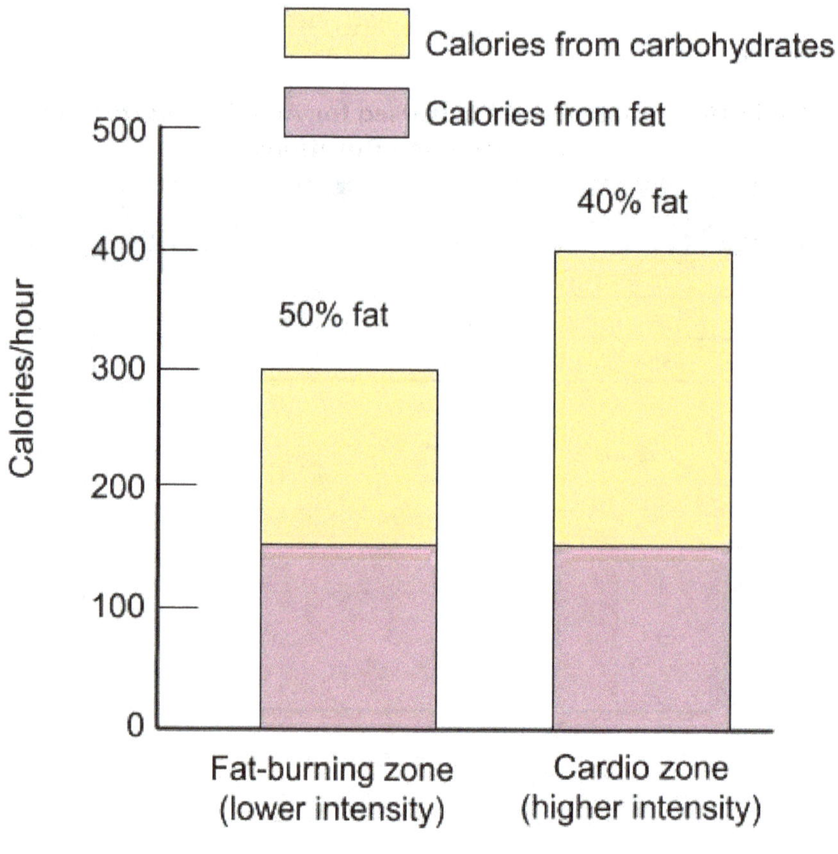

Figure 10.5. The fat-burning zone. While a greater percentage of calories burned in lower intensity exercise come from fat, the overall total calorie burn is greater in higher intensity exercise.

One important clarification about exercise intensity and fuel sources is the concept of the fat-burning zone. Many people think that in order to lose body fat, they should exercise at a lower intensity so that fat is the primary fuel source. The fat-burning zone is typically referred to as a low-intensity aerobic activity that keeps your heart rate between 60 and 69 percent of maximum heart rate. The cardio zone, on the other hand, is a high-intensity aerobic activity that keeps the heart rate between about 70 and 85 percent of maximum heart rate. So which zone do you burn the most fat in? Technically, your body burns a higher percentage of calories from fat during a low-intensity aerobic activity. When you begin a low-intensity activity, about 50% of the calories burned come from fat, whereas in the cardio zone only 40% come from fat. However, this isn't the whole story. High-intensity activity burns more total calories per minute. At this higher rate of energy expenditure, you can burn just as much or more total fat and more total calories as during a lower intensity activity. If weight loss is one of your goals, high-intensity activities will burn more total calories, helping to shift to negative energy balance, and will promote a greater level of fitness. However, the best exercise program is one that is enjoyable, sustainable, and safe for you; if you're just starting out, it's wise to begin with low- to moderate-intensity activities and work your way up from there.

ATHLETE NUTRITIONAL NEEDS

Nutrition is essential to your performance during all types of exercise. As an athlete, the foods consumed in your diet are used to provide the body with enough energy and specific nutrients to fuel an activity and maximize performance. Athletes have different nutritional needs than the general population in order to support their vigorous activity levels in both practice and competition. On this page, we'll explore the specific nutrient needs of athletes and how they differ from the nutrient needs of less active individuals.

Energy Needs

To determine an athlete's nutritional needs, it is important to revisit the concept of energy metabolism. Energy intake is the foundation of an athlete's diet, because it supports optimal body functions, affects intake of macronutrients and micronutrients, and assists in maintaining body composition. Energy needs for athletes increase depending on their energy expenditure. The amount of energy expended during physical activity is contingent on the intensity, duration, and frequency of the exercise. Competitive athletes may need 3,000 to over 5,000 calories daily compared to a typical inactive individual who needs about 2,000 calories per day. Energy needs are also affected by an individual's sex, age, and weight. Weight-bearing exercises, such as running, burn more calories per hour than non-weight-bearing exercises, such as swimming, since weight-bearing exercises require your body to move against gravity. Additionally, men typically burn more calories than women for the same activity, because men have more muscle mass which requires more energy to support and move around.

Macronutrient Needs

The composition of macronutrients in the diet is a key factor in maximizing performance for athletes. As discussed on the previous page, carbohydrates, fat, and protein can all be utilized for energy production during exercise, though the amount utilized of each nutrient varies depending on the intensity and duration of the exercise.

Carbohydrates

Carbohydrates are an important fuel source for the brain and muscle during exercise. Carbohydrate storage in the liver and muscle is relatively limited, and therefore it is important for athletes to regularly consume enough carbohydrates from their diet. Carbohydrate needs should increase about 3-10 g/kg/day depending on the athlete's type and level of training and competition (Table 10.3.).[1]

Table 10.3. Athlete's daily needs for carbohydrate fuel1

Activity Level	Example of Exercise	Increase of Carbohydrate (g/kg of athlete's body weight/day)
Light	Low-intensity or skill-based activities	3-5
Moderate	Moderate exercise program (about 1 hour/day)	5-7
High	Endurance program (about 1-3 hours/day of moderate to high intensity exercise)	6-10
Very high	Extreme commitment (4-5 hours/day of moderate to high intensity exercise)	8-12

Fat

Fat is a necessary component of a healthy diet to provide energy and essential fatty acids and to facilitate the absorption of fat-soluble vitamins. Athletes are recommended to consume the same amount of fat in the diet as the general population, 20 to 35% of their energy intake. Although these recommendations are in accordance with public health guidelines, athletes should individualize their needs based on their training level and body composition goals. Fat intakes below 20% of energy intake will reduce the availability of fat-soluble vitamins and essential fatty acids, especially omega-3 fatty acids.

Protein

Although protein accounts for only about 5% of energy expended, dietary protein is necessary to support metabolic reactions that generate ATP, which rely heavily on proteins such as enzymes and transport proteins. Additional protein also helps muscles with maintenance, growth, and repair. For these reasons, athletes have higher protein needs than the general population. It is recommended that athletes consume 1.2 to 2.0 g/kg/day of protein in order to support these functions. Higher intakes may also be needed for short periods of intense training or when reducing energy intake.[1]

It is important to consume adequate amounts of protein and to understand that the quality of the protein consumed affects the amount needed. Complete protein foods such as meats, dairy, and eggs contain all of the essential amino acids in relative amounts that most efficiently meet the body's needs for growth, maintenance, and repair of muscles. Vegetarian diets contain mostly incomplete protein sources, which have lower digestibility and amino acid patterns that do not match human needs as closely as most animal proteins. To compensate for this, vegetarian athletes need to consume more dietary protein than non-vegetarians and should target the upper end of the recommended protein intake.

In addition to the amount and quality of proteins consumed, timing of protein intake has been shown to impact muscle protein synthesis. Studies show that the synthesis of muscle protein is optimized with high quality protein consumption after exercise, ideally 15 to 25 grams of protein in the early recovery phase after a workout (0-2 hours after exercise). A similar amount of protein should be consumed every 3-5 hours, spread out across the day over multiple meals within the 24 hours post-workout, so that amino acids are always available for optimal protein synthesis.[2] One recommended strategy is to aim for 0.4 g/kg/meal across four meals in order to reach 1.6 g/kg/day for muscle synthesis.[3]

Although athletic training increases protein needs, athletes can meet their protein requirement through high quality food sources, and most do not need to consume protein supplements. Here are some examples of snacks or small meals that contain at least 15 to 25 grams of protein[4]:

- ½ cup of granola plus ¼ cup almonds (16 grams protein)
- 7 oz. Greek yogurt (20 grams protein)
- Peanut butter sandwich (20 grams protein)
- 2 scrambled eggs and 1 cup of milk (20 grams protein)
- 3 oz. canned tuna (½ can) with ½ cup crackers (22 grams protein)
- Turkey and cheese sandwich (32 grams protein)

These whole food options have the benefit of coming packaged with other nutrients, including carbohydrates to replenish glycogen stores, fiber, and micronutrients, and are often less costly than most protein supplements. Like all dietary supplements, protein shakes and other supplements are not well-regulated; some contain unnecessary additives such as sweeteners and herbs, and some have been found to contain unsafe levels of heavy metals like arsenic and mercury. Protein supplements do have the benefit of being convenient and shelf-stable. If you choose to use a protein supplement, look for one certified by a third-party testing organization and with a simple ingredient list.[5]

Micronutrient Needs

Vitamins and minerals are essential for energy metabolism, the delivery of oxygen, protection against oxidative damage, and the repair of body structures. When exercise increases, the amounts of many vitamins and minerals needed are also increased. Currently, there are no special micronutrient recommendations for athletes, but most athletes will meet their needs by consuming a balanced diet that meets their energy needs. Because the energy needs of athletes increase, they often meet their higher need for vitamins and minerals through the additional food they consume to meet energy needs. However, athletes who limit energy intake or utilize extreme weight-loss practices may put themselves at risk for vitamin and mineral deficiencies.

A Word on Dietary Supplements and Ergogenic Aids

Many athletes consider taking dietary supplements or ergogenic aids (i.e., substances used to enhance performance) in an effort to improve performance, increase energy levels, or make up for poor nutrition choices. However, it is important to remember that supplements and ergogenic aids are not regulated, leading to frequent use of false advertising and unsubstantiated claims by the supplement industry. Athletes must be careful not only in deciphering the claims of products, but also in researching their safety and efficacy, particularly in relation to any rules and regulations that govern the sport in which the athlete participates. Very few supplements that claim to have ergogenic benefits have sound evidence to back up those claims, and in some situations, consuming them could be dangerous. Most athletes can meet their nutrition needs without added supplements. Athletes who have nutrition concerns should consult with a sports dietitian or other sport science professional to make sure their individual needs are met safely.

Water and Electrolyte Needs

During exercise, being appropriately hydrated contributes to performance. Water is needed to cool the body, transport oxygen and nutrients, and remove waste products from the muscles. Water needs are increased during exercise due to the extra water losses through evaporation and sweat. Dehydration can occur when there are inadequate water levels in the body and can be very hazardous to the health of an individual. As the severity of dehydration increases, the exercise performance of an individual will begin to decline (see Figure 10.6). It is important to continue to consume water before, during, and after exercise to avoid dehydration as much as possible.

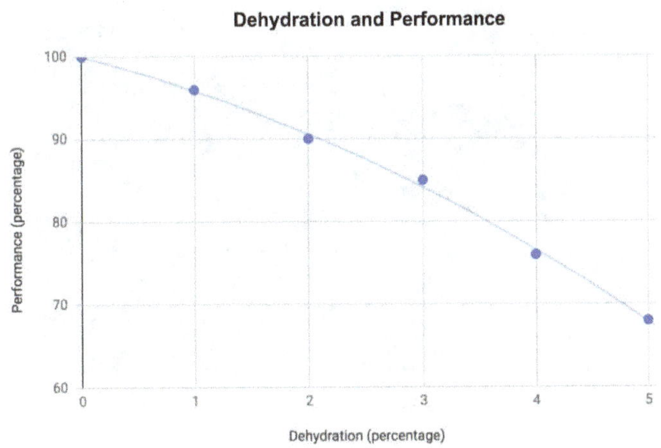

Figure 10.6. The effect of dehydration on exercise performance.

During exercise, thirst is not a reliable short-term indicator of the body's water needs, as it typically is not enough to replace the water lost. Even with constant replenishing of water throughout a workout, it may not be possible to drink enough water to compensate for the losses. Dehydration occurs when water loss is so significant that total blood volume decreases, which leads to a reduction in oxygen and nutrients transported to the muscle cells. A decreased blood volume also reduces blood flow to the skin and the production of sweat, which can increase body temperature. As a result, the risk of heat-related illnesses such as heat exhaustion or heat stroke, increases. The external temperature during exercise can also play a role in the risk of heat-related illnesses. As the external temperature increases, it becomes more difficult for the body to dissipate heat. As humidity also increases, the body is unable to cool itself through evaporation.

Hyponatremia and Sports Drinks

Sweating during exercise helps our bodies to stay cool. Sweat consists of mostly water, but it also causes losses of sodium, potassium, calcium and magnesium. During most types of exercise, the amount of sodium lost is very small, and drinking water after a workout will replenish the sodium in the body. However, during long endurance exercises, such as a marathon or triathlon, sodium losses are larger and must be replenished. If water is replenished without sodium, the sodium already in the body will become diluted. These low levels of sodium in the blood will cause a condition known as hyponatremia. When sodium levels in the blood are decreased, water moves into cells through osmosis, which causes swelling. Accumulation of fluid in the lungs and the brain can cause serious, life-threatening conditions such as seizure, coma, and death (see Unit 9).

In order to avoid hyponatremia, athletes should increase their consumption of sodium in the days leading up to an event and consume sodium-containing sports drinks during their race or event. A well-concocted sports drink contains sugar, water, and sodium in the correct proportions so that hydration is optimized. The sugar is helpful in maintaining blood-glucose levels needed to fuel muscles, the water keeps an athlete hydrated, and the sodium enhances fluid absorption and replaces some of that lost in sweat. The American College of Sports Medicine states that the goal of drinking fluids during exercise is to prevent dehydration, which compromises performance and endurance.

Homemade Sports Drink

Note: The nutrition profile of commercial sports drinks is 50 to 70 calories per 8 ounces, with about 110 milligrams of sodium. Following is a simple recipe that offers this profile, but at a much lower cost than expensive store-bought brands—without additives, colors, or preservatives.

Ingredients:

- ¼ cup (50 g) sugar
- ¼ teaspoon salt
- ¼ cup (60 ml) water
- ¼ cup (60 ml) orange juice (not concentrate) plus 2 tablespoons lemon juice
- 3 ½ cups (840 ml) cold water

Method

- In the bottom of a pitcher, dissolve the sugar and salt in the hot water.
- Add the juice and the remaining water; chill.
- Quench that thirst!
- **Yield:** 1 quart (1 L)
- **Nutrition Information**: 200 total calories; 50 calories per 8 ounces (240 ml); 12 g carbohydrate; 110 mg sodium

The hydration goal for obtaining optimal endurance and performance is to replace what is lost, not to over-hydrate. Perspiration rates are variable and dependent on many factors including body composition, humidity, temperature, and type of exercise. A person's sweat rate can be approximated by measuring weight before and after exercise—the difference in weight will be the amount of water weight you lost.

Who Needs Sports Drinks?

Scientific studies show that, under certain circumstances, consuming sports drinks (instead of plain water) during high-intensity exercise lasting longer than one hour significantly enhances endurance, and some evidence also indicates it enhances performance. There is no consistent evidence that drinking sports drinks instead of plain water enhances endurance or performance in individuals exercising less than one hour at a time and at low to moderate intensities. Children and adult athletes exercising for more than one hour at high-intensity (tennis, rowing, rugby, soccer, etc.) may benefit from consuming a sports drink rather than water. However, consuming sports drinks provides no benefit over water to endurance, performance, or exercise recovery for those exercising less than an hour. In fact, as with all other sugary drinks containing few to no nutrients, they are only another source of calories. Drinking sports drinks when you are doing no exercise at all is not recommended.

REFERENCES:

Thomas, D. T., Erdman, K. A., & Burke, L. M. (2016). Position of the Academy of Nutrition and Dietetics, Dietitians of Canada, and the American College of Sports Medicine: nutrition and athletic performance. Journal of the Academy of Nutrition and Dietetics, 116(3), 501-528.

Kerksick, C. M., Arent, S., Schoenfeld, B. J., Stout, J. R., Campbell, B., Wilborn, C. D., ... & Willoughby, D. (2017). International society of sports nutrition position stand: nutrient timing. Journal of the International Society of Sports Nutrition, 14(1), 1-21.

Schoenfeld, B. J., & Aragon, A. A. (2018). How much protein can the body use in a single meal for muscle-building? Implications for daily protein distribution. Journal of the International Society of Sports Nutrition, 15(1), 1-6.

U.S. Department of Agriculture. (n.d.). FoodData Central. Retrieved September 9, 2020, from https://fdc.nal.usda.gov/

Gelsomin, E. (2020, March 9). The scoop on protein powder. Harvard Health Blog. https://www.health.harvard.edu/blog/the-scoop-on-protein-powder-2020030918986

UNIT-11

THE IMPACT OF NUTRITION OVER THE LIFESPAN

INTRODUCTION

Throughout all of the seasons of life, food is a basic human need. That's just as true for a newborn infant filling her tiny tummy with breast milk as it is for her mother, whose body is performing the physiological feat of synthesizing milk for her baby. And it's equally true for her grandparents, who are preparing a meal to share with the new family once they come home from the hospital. Food nourishes our bodies in every stage of life, but eating together is also a way of nurturing our social and emotional connections with one another.

Yet, as we grow and change, so do our nutrient needs, and every stage of life brings unique challenges. In this unit, we'll cover nutritional considerations in several key stages: pregnancy and breastfeeding, infancy, toddlerhood, childhood, adolescence, and older adulthood.

PREGNANCY AND LACTATIONAL NUTRITION

Becoming a parent is a huge commitment, and for all its many rewards, it's also costly to have a baby. And that cost comes not just in time and money. For mothers, having a baby also represents a nutritional and metabolic investment. Throughout pregnancy and infancy, a new human grows and develops, and this feat is made possible in large part because of the nutrients that the mother provides. Nutrient requirements are greater in pregnancy and lactation to support maternal health, the healthy growth and development of the fetus, and the production of milk to nourish the infant.

Physiology Of Pregnancy

Human pregnancy lasts for approximately 40 weeks (when counted from the first day of a woman's last menstrual period) and is roughly divided into thirds, or trimesters. Dramatic changes occur throughout pregnancy, including in the earliest days, sometimes before women realize they're pregnant. Therefore, adequate nutrition is vital for women who are trying to conceive or may become pregnant. In addition, women who are either underweight or obese before becoming pregnant can find it more difficult to get pregnant and face greater risks of pregnancy complications. Therefore, it's recommended to establish healthy eating and exercise habits and work towards a healthy weight before pregnancy.1 Nutrition before conception is also important for fathers; studies have shown that sperm quality is better in fathers who have a healthy body weight, consume adequate amounts of folate and omega-3 fatty acids, and follow a healthy dietary pattern, such as a Mediterranean diet emphasizing seafood, poultry, whole grains, legumes, and fruits and vegetables.2

Pregnancy begins when the sperm fertilizes the egg, and the single cell formed from this union begins to divide and differentiate. During the first few weeks of development, the cells of the uterine lining provide nutrients to the developing embryo. Between the 4th and 12th weeks, the developing placenta gradually takes over the role of feeding the embryo, which is called a fetus beginning in the 9th week of pregnancy. The placenta forms from tissues deriving from both the fetus and the mother, and this new organ becomes the interface between the two. The placenta provides nutrition and respiration, handles waste from the fetus, and produces hormones important to maintaining the pregnancy. The blood of the fetus and mother do not mix in the placenta, but they come close enough that nutrients (including glucose, amino acids, fatty acids, vitamins, and minerals) and oxygen can pass from the mother's blood to the fetus, and carbon dioxide and waste products can be passed from the fetus to the mother.

The first trimester is a key period for the formation of embryonic and fetal organs. During the second and third trimesters, the fetus continues to grow and develop, with final development of organs such as the brain, lungs, and liver continuing in the last few weeks of pregnancy. Infants born before 37 weeks of pregnancy are considered preterm, and because they have not completed fetal development, are at increased risk for a range of health problems. Advances in neonatal care have greatly improved outcomes for these babies born too soon, but in most cases, the best scenario is for the fetus to have the full term of pregnancy to develop in the protective environment of the uterus, receiving nutrients through the placenta.

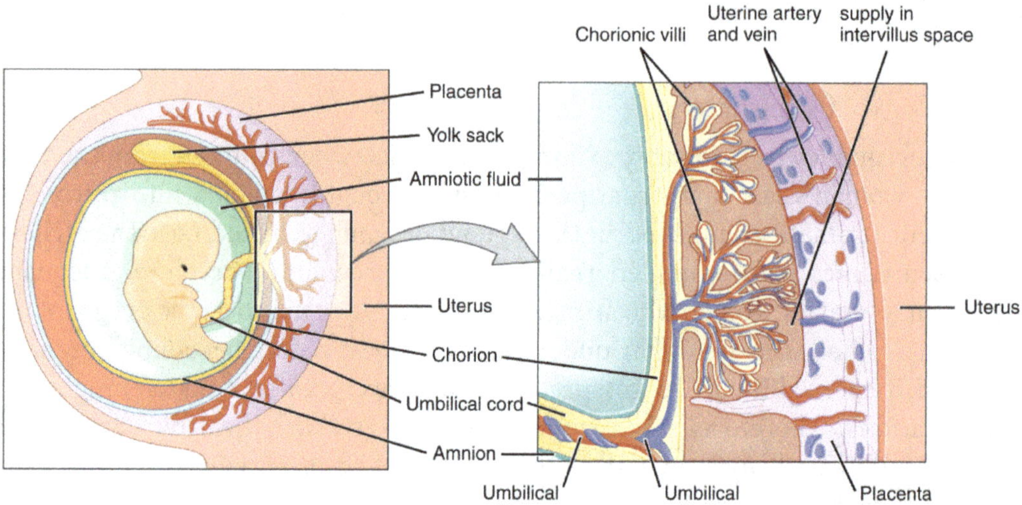

Figure 11.1. Cross-section of the placenta. In the placenta, the maternal and fetal bloodstreams do not mix, but nutrients, gasses, and waste products can diffuse between them.

Nutrient Requirements During Pregnancy

Pregnant women need more calories, macronutrients, and micronutrients than they did before pregnancy. However, the increase in nutrient requirements is relatively greater than the increase in caloric needs, emphasizing the importance of a nutrient-dense diet. A dietary pattern focused on vegetables, fruits, whole grains, nuts, legumes, fish, and vegetable oils, and lower in red and processed meats, refined grains, and added sugars is associated with a reduced risk of pregnancy complications such as gestational diabetes and hypertension.[5] There is nothing revolutionary about this dietary pattern—it's what is recommended for everyone! However, it's even more beneficial during pregnancy, as it promotes both maternal and fetal health.

Energy Intake And Weight Gain

During the first trimester, energy requirements are generally not increased, so women should consume about the same number of calories as they did before pregnancy. As fetal growth ramps up, energy requirements increase by about 340 calories per day in the second trimester and 450 calories per day in the third trimester. This is just an average; individual energy requirements vary depending on factors such as activity level and body weight before pregnancy. Weight gain is a normal part of pregnancy. The growth of the fetus accounts for about 6 to 8 pounds of weight gain by the end of the pregnancy. Much of the rest comes from the development and expansion of tissues and fluids to support the pregnancy, including the placenta, uterus, breasts, amniotic fluid, blood, and maternal body fluids. In addition, women starting pregnancy at a normal weight should gain about 8 to 10 pounds of body fat and protein during pregnancy, in part to prepare for lactation.

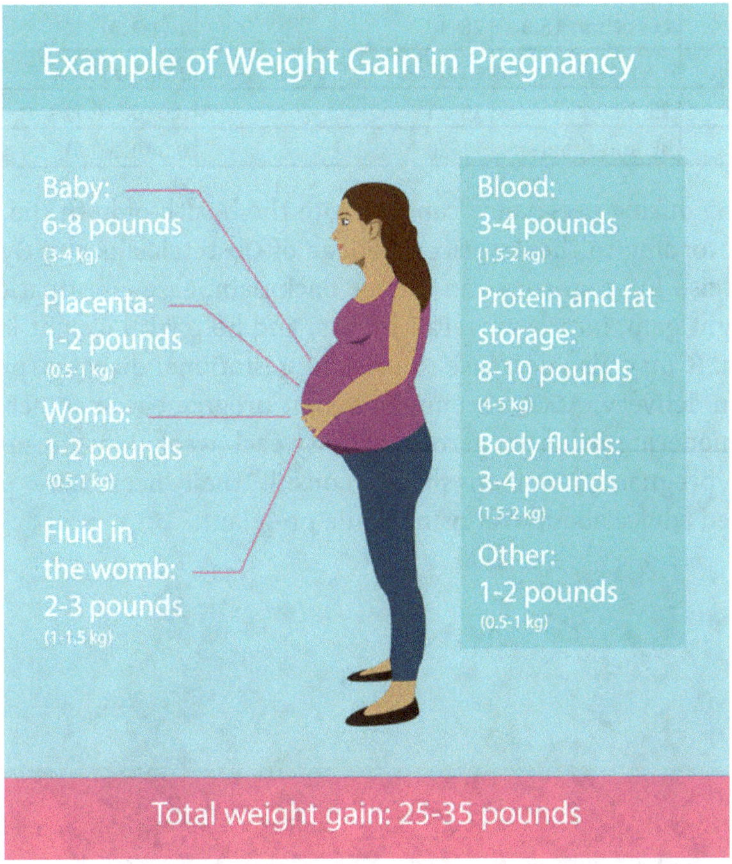

Figure 11.2. Components of weight gain in healthy pregnant women with normal BMI before pregnancy. Weight gain comes not just from the baby, but from many different body systems changing to support the pregnancy.

The Institute of Medicine recommends different amounts of weight gain in pregnancy depending on pre-pregnancy BMI. Women who were underweight before pregnancy need to gain more, and those who were overweight or obese before pregnancy need to gain less. Gaining too little weight in pregnancy can compromise fetal growth, leading to the baby being born too small, which can cause increased risk of illness, difficulty feeding, and developmental delays. Gaining too much weight in pregnancy can cause the baby to be too big at birth, leading to birth complications and increasing the risk of needing a cesarean birth. It can also make it more difficult to lose weight after pregnancy.[8]

Table 11.1. Recommended weight gain during pregnancy depends on prepregnancy BMI.

Prepregnancy Weight Category	Body Mass Index (BMI)	Recommended Range of Total Weight Gain (lb)	Recommended Rates of Weight Gain in the 2nd and 3rd Trimesters (lb/wk)
Underweight	Less than 18.8	28-40	1 (1-1.3)
Normal weight	18.5-24.9	25-35	1 (0.8-1)
Overweight	25-29.9	15-25	0.6 (0.5-0.7)
Obese	30 and greater	11-20	0.5 (0.4-0.6)

Exercising during pregnancy can promote the health of both the mother and the baby. According to the American College of Obstetricians and Gynecologists, regular exercise in pregnancy can reduce back pain, ease constipation, promote healthy weight gain, improve overall fitness, and help with weight loss after the baby is born. It may also decrease the risk of gestational diabetes, preeclampsia, and cesarean delivery. ACOG recommends that pregnant women get at least 150 minutes of moderate-intensity aerobic activity each week. Most forms of exercise are safe in pregnancy; women should consult their healthcare provider for individualized advice about exercising while pregnant.

Macronutrient Requirements

The Acceptable Macronutrient Distributions Ranges (AMDR) for macronutrients are the same for all healthy adults, pregnant or not, with about 45 to 65 percent of calories coming from carbohydrates, 20 to 35 percent from fats, and 10 to 35 percent from protein. As energy intake increases, a pregnant woman needs more of each of these macronutrients. The RDA for carbohydrates increases from 130 grams per day for non-pregnant adults to 175 grams per day for pregnant women. This level of carbohydrate intake provides energy for fetal development and ensures adequate glucose for both the mother's and the fetus's brain. The recommended fiber intake in pregnancy, expressed as an AI, is the same as for all adults: 14 grams of fiber per 1,000 calories consumed. As caloric intake increases during pregnancy, so should fiber intake, emphasizing the importance of choosing whole food sources of carbohydrates.

Additional protein is also needed during pregnancy. Protein builds muscle and other tissues, enzymes, antibodies, and hormones in both the mother and the fetus, as well as supporting increased blood volume and the production of amniotic fluid. The RDA for protein during pregnancy is 1.1 grams per kg body weight per day, coming to about 71 grams per day for an average woman—roughly 25 grams more than needed before pregnancy. There is not a specific RDA for fat during pregnancy. Fats should continue to make up 25 to 35 percent of daily caloric intake, providing energy and essential fatty acids (linoleic acid and alpha-linolenic acid), as well as helping with fat-soluble vitamin absorption. The omega-3 polyunsaturated fatty acids DHA and EPA become more important during pregnancy and lactation, because they are essential for brain and eye development of the fetus and infant. The Dietary Guidelines for Americans recommend that pregnant women consume 8 to 12 ounces of seafood each week, in part to provide omega-3 fatty acids. Fish with high levels of mercury should be avoided; these include king mackerel, marlin, orange roughy, shark, swordfish, tilefish, and bigeye tuna.

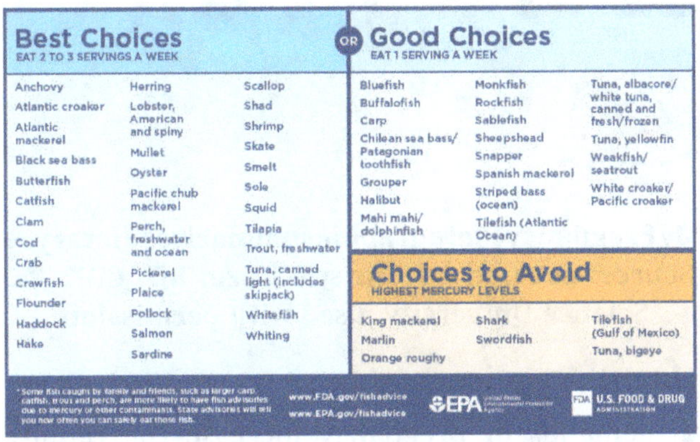

Figure 11.3. Advice from the EPA and FDA about choosing safe fish to consume during pregnancy and breastfeeding.

Consider the following recommendations for healthy eating during pregnancy, as summarized in the My Pregnancy Plate graphic:

- Choose large portions of a variety of non-starchy vegetables such as leafy greens, broccoli, carrots, bell peppers, tomatoes, mushrooms, and cabbage.
- Choose a variety of whole fruits, limiting juice and dried fruit.
- Aim for 2 to 3 servings of nonfat or 1 percent milk or yogurt (unsweetened or slightly sweetened).
- Choose protein sources like poultry, beans, nuts, eggs, tofu, and cheese. Aim for 8 to 12 ounces of low-mercury seafood each week.
- Choose fiber-rich sources of carbohydrates, including whole grains, legumes, and starchy vegetables such as sweet potatoes and squash.
- Drink mainly water, decaf coffee, or tea. (Caffeine is discussed below).

Figure 11.4. The My Pregnancy Plate graphic summarizes dietary recommendations for pregnancy. Source: Created by Christie Naze, RD, CDE, ©Oregon Health & Science University, used with permission.

Vitamin and Mineral Requirements

The physiological demands of pregnancy increase the requirement for many vitamins and minerals. These requirements can be met from food sources, but obstetricians also generally recommend that women take a prenatal supplement while trying to conceive and during pregnancy. This ensures that nutrient requirements are met, while also providing a little peace of mind if pregnancy-related nausea and vomiting limit dietary variety and quality.[1]

The following table compares the recommended intake levels of some vitamins and minerals to the levels needed in pregnancy.

Table 11.2. Recommended Micronutrient intakes during pregnancy. Sources: The American College of Obstetricians and Gynecologists and the NIH Office of Dietary Supplements.

Nutrient	RDA/AI, Nonpregnant Adult Women	RDA/AI, Pregnant Adult	Importance
Vitamin A	700 mcg	770 mcg	**Forms healthy skin and eyesight; helps with bone growth**
Vitamin B6	1.3 mg	1.9 mg	**Helps form red blood cells; helps the body metabolize macronutrients**
Vitamin B12	2.4 mcg	2.6 mcg	**Maintains nervous system; helps form red blood cells**
Vitamin C	75 mg	85 mg	**Promotes healthy gums, teeth, and bones**
Vitamin D	600 IU	600 IU	**Builds fetal bones and teeth; promotes healthy eyesight and skin**
Folate	400 mcg	600 mcg	**Helps prevent neural tube defects; supports growth and development of fetus and placenta**
Calcium	1,000 mg	1,000 mg	**Builds strong bones and teeth**
Iron	18 mg	27 mg	**Helps red blood cells deliver oxygen to fetus**
Iodine	150 mcg	220 mg	**Essential for healthy brain development**
Choline	425 mg	450 mg	**Important for development of fetal brain and spinal cord**

Among the micronutrients, folate, iron, and iodine deserve a special mention. Folate is essential for the growth and specialization of cells of the central nervous system (see Unit 9). Mothers who are folate-deficient during pregnancy have a higher risk of having a baby with a neural tube birth defect such as spina bifida. Folic acid fortification of grains in the U.S. has helped to raise folate intake in the general population and has reduced the incidence of neural tube defects. However, dietary intake is often not adequate to meet the requirement of 600 mcg folate per day in pregnancy, and the American College of Obstetricians and Gynecologists recommends that pregnant women take an additional 400 mcg of folic acid each day, an amount usually included in prenatal supplements.[13] The neural tube closes by day 28 of pregnancy, before a woman may realize she is pregnant, so it's important to consume adequate folate while trying to conceive. Because about 45 percent of pregnancies in the U.S. are unplanned, a folic acid supplement is a good idea for anyone who may become pregnant.

Iron intake is important because of the increase in blood volume during pregnancy. Iron is an essential component of hemoglobin, the protein responsible for oxygen transport in blood, so adequate iron intake supports oxygen delivery to both maternal and fetal tissues (see Unit 9). Good dietary sources of iron include meat, poultry, seafood, nuts, legumes, and fortified or whole grain cereals. Iron should also be included in a prenatal supplement or taken separately. Iodine is essential for fetal brain development, but recent data suggests that many pregnant women in the U.S. do not consume enough iodine to meet their increased requirement.15 Much of the iodine in the typical American diet comes from dairy products and iodized salt. However, iodine intake in the U.S. has declined in recent decades as more people watch their intake of table salt and/or switch to kosher or sea salt, which aren't iodized. In addition, processed foods are generally made with non-iodized salt, and the intake of processed foods has increased. Meanwhile, the popularity of dairy products has declined. Most people in the U.S. still consume enough iodine to meet their requirement, but pregnant women may be at risk for iodine deficiency because of their increased need for this mineral. Iodine deficiency during pregnancy can cause miscarriage, stillbirth, and major neurodevelopmental deficits and growth retardation in the fetus.16 Unfortunately, many prenatal vitamins do not contain iodine, so it's worth checking the label to ensure that iodine is included. The micronutrients involved with building the skeleton—vitamin D, calcium, phosphorus, and magnesium—are crucial during pregnancy to support fetal bone development. Although the levels are the same as those for nonpregnant women, many women do not typically consume adequate amounts and should make an extra effort to meet those needs.

As always, it's important to read supplement labels carefully, with the aim of choosing a prenatal supplement that contains close to the RDA or AI for micronutrients and avoiding those that exceed the UL, unless under the specific direction of a healthcare provider. In particular, both vitamin A and zinc consumed in excessive amounts can cause birth defects. Beta-carotene is typically used as the vitamin A source in prenatal supplements, because unlike vitamin A, it doesn't cause birth defects (see Unit 8).

Foods And Other Substances To Avoid

It's not just nutrients that can cross the placenta. Other substances such as alcohol, nicotine, cannabinoids (from cannabis), and both prescription and recreational drugs can also pass from mother to fetus. Exposure to these substances can have lasting and detrimental effects on the health of the fetus. For this reason, pregnant women are advised to avoid using alcohol, tobacco, cannabis, and recreational drugs during pregnancy. Medical providers can help pregnant women quit using these substances and advise them on the safety of specific medications needed during pregnancy.

Some substances are so detrimental that a woman should avoid them even if she suspects that she might be pregnant. For example, consumption of alcoholic beverages results in a range of abnormalities that fall under the umbrella of fetal alcohol spectrum disorders. They include learning and attention deficits, heart defects, and abnormal facial features (Figure 11.5). Alcohol enters the fetus's bloodstream via the umbilical cord and can slow fetal growth, damage the brain, or even result in miscarriage. The effects of alcohol are most severe in the first trimester, when the organs are developing. There is no known safe amount of alcohol in pregnancy.

Craniofacial features associated with fetal alcohol syndrome

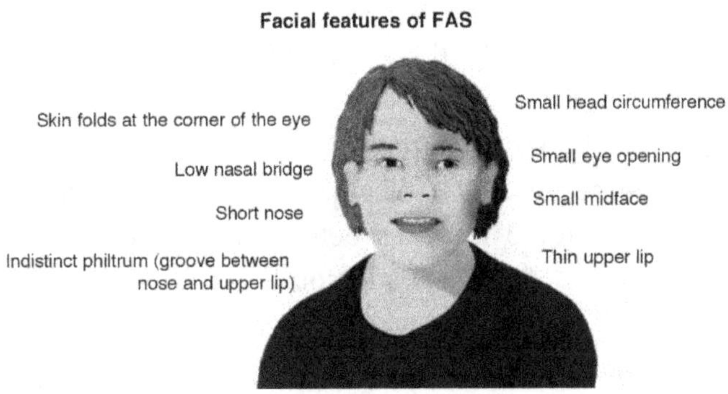

Figure 11.5. Craniofacial features associated with fetal alcohol syndrome.

Caffeine use should be limited to less than 200 milligrams per day, or the equivalent of about two cups of coffee. Consumption of greater amounts is linked to miscarriage and preterm birth. Keep in mind that caffeine is also found in other sources such as chocolate, energy drinks, soda, tea, and some over-the-counter pain and headache medications.

It's also important to pay special attention to avoiding foodborne illness during pregnancy, as it can cause major health problems for both the mother and the developing fetus. For example, the foodborne illness caused by the bacteria Listeria monocytogenes, called listeriosis, can cause miscarriage, stillbirth, and fetal or newborn meningitis. According to the CDC, pregnant women are ten times more likely to become ill with this disease than nonpregnant adults, likely because they have a dampened immune response. Other common foodborne illnesses, such as those caused by Salmonella and E. coli, can also be very serious in pregnant women.

Foods more likely to be contaminated with foodborne pathogens should be avoided by pregnant women to decrease their chances of infection. These include the following:

- Unpasteurized dairy products such as soft cheeses
- Raw or smoked seafood
- Hot dogs and deli meats (or heat to 165° before eating)
- Paté and other meat spreads
- Undercooked or raw meat, poultry, and eggs
- Raw sprouts
- Raw dough
- Unpasteurized juice and cider

Following standard food safety practices can also go a long way towards preventing foodborne illness, during pregnancy or anytime. The CDC offers this summary of food safety tips:

- COOK. Use a food thermometer to ensure that foods are cooked to a safe internal temperature: 145°F for whole beef, pork, lamb, and veal (allowing the meat to rest for 3 minutes before carving or consuming), 160°F for ground meats, and 165°F for all poultry, including ground chicken and turkey.
- CLEAN. Wash your hands after touching raw meat, poultry, and seafood. Also wash your work surfaces, cutting boards, utensils, and grill before and after cooking.
- CHILL. Keep your refrigerator below 40°F and refrigerate foods within 2 hours of cooking (1 hour during the summer heat).
- SEPARATE. Germs from raw meat, poultry, seafood, and eggs can spread to produce and ready-to-eat foods unless you keep them separate. Use different cutting boards to prepare raw meats and any food that will be eaten without cooking.

Maternal Health During Pregnancy

Especially during the first trimester, it's very common for women to feel nauseous, often leading to vomiting, and to be very sensitive to smells. This so-called "morning sickness" is misnamed, because it can make women feel awful any time of day—morning, noon, and night. Nausea and vomiting during pregnancy affects as many as four out of five pregnancies. Symptoms can range from mild nausea to a much more severe illness called ***hyperemesis gravidarum***, which causes relentless vomiting and affects between 0.3 and 3 percent of all pregnancies. Most women find that their nausea and vomiting subsides by the end of the first trimester, but in rare and unfortunate cases, it can continue until the baby is born.

Dietary changes can help alleviate nausea and vomiting. Here are some suggestions:

- Choose foods that are low in fat and easily digestible, such as those in the BRATT diet (bananas, rice, applesauce, toast, and tea). The goal is simply to find foods that can be tolerated. It is reassuring to remember that pregnancy does not require extra calories during the first trimester, and a prenatal supplement can help meet micronutrient needs.
- **An empty stomach can worsen feelings of nausea**, so try eating a few crackers or dry toast before getting out of bed in the morning to avoid moving around with an empty stomach. It may also help to eat five or six small meals per day and to eat small snacks such as crackers, fruits, and nuts throughout the day.
- Many pregnant women experiencing nausea and vomiting find meat unappetizing. **Try other sources of protein** such as dairy foods (e.g., milk, yogurt, ice cream), nuts and seeds (as well as nut butters), and protein powders and shakes.
- **Ginger can help settle your stomach**; it's available in capsule, candy, and tea form. Ginger ale, if made with real ginger, might also be helpful.

If nausea and vomiting remain unmanageable after making these dietary changes, medications may be necessary. Hyperemesis gravidarum, though rare, can be very serious. Constant vomiting can lead to malnutrition, weight loss, dehydration, and electrolyte imbalance. Hospitalization may be required so that women can receive fluids and electrolytes through an intravenous line, and sometimes a feeding tube is necessary. Getting help for this condition can be a frustrating process of trial-and-error, and unfortunately, there are still many unknowns about its causes and effective treatment options.

As pregnancy progresses, another common complaint is **heartburn**, caused by **gastroesophageal reflux.** This is caused by the upward, constrictive pressure of the growing uterus on the stomach, as well as decreased peristalsis in the GI tract. (See Unit 3 for management strategies.)

About 6 percent of pregnancies in the U.S. are affected by **gestational diabetes** (see Unit 4).[24] This is a type of diabetes that develops during pregnancy in women who didn't previously have diabetes. Gestational diabetes is managed by monitoring blood glucose levels, eating a healthy diet, and exercising regularly. Sometimes, insulin injections are needed. If blood glucose levels aren't well controlled in the mother, the fetus will also have high blood glucose levels. This can cause the baby to grow too big, leading to a greater chance of birth complications and increased likelihood of needing a cesarean birth.

REFERENCES

American College of Obstetricians and Gynecologists. (2019). Prepregnancy Counseling. Obstetrics & Gynecology, 133(1), e78-89.

Nassan, F. L., Chavarro, J. E., & Tanrikut, C. (2018). Diet and men's fertility: Does diet affect sperm quality? Fertility and Sterility, 110(4), 570–577. https://doi.org/10.1016/j.fertnstert.2018.05.025

Betts, J. G., Young, K. A., Wise, J. A., Johnson, E., Poe, B., Kruse, D. H., Korol, O., Johnson, J. E., Womble, M., & DeSaix, P. (2013). Embryonic Development. In Anatomy and Physiology. OpenStax. https://openstax.org/books/anatomy-and-physiology/pages/28-2-embryonic-development

CDC. (2019). Preterm Birth. Reproductive Health. https://www.cdc.gov/reproductivehealth/maternalinfanthealth/pretermbirth.htm

Raghavan, R., Dreibelbis, C., Kingshipp, B. L., Wong, Y. P., Abrams, B., Gernand, A. D., Rasmussen, K. M., Siega-Riz, A. M., Stang, J., Casavale, K. O., Spahn, J. M., & Stoody, E. E. (2019). Dietary patterns before and during pregnancy and maternal outcomes: A systematic review. The American Journal of Clinical Nutrition, 109(Supplement_1), 705S-728S. https://doi.org/10.1093/ajcn/nqy216

Kominiarek, M. A., & Rajan, P. (2016). Nutrition Recommendations in Pregnancy and Lactation. The Medical Clinics of North America, 100(6), 1199–1215. https://doi.org/10.1016/j.mcna.2016.06.004

Institute of Medicine, Food and Nutrition Board. (2005). Dietary Reference Intakes for Energy, Carbohydrate, Fiber, Fat, Fatty Acids, Cholesterol, Protein, and Amino Acids (Macronutrients). The National Academies Press.

American College of Obstetricians and Gynecologists. (2013). Weight Gain During Pregnancy. Committee Opinion No. 548. Reaffirmed 2020. Obstetrics & Gynecology, 121, 210–212.

American College of Obstetricians and Gynecologists. (n.d.). Exercise During Pregnancy. Patient Resources. Retrieved August 13, 2020, from https://www.acog.org/en/Patient Resources/FAQs/Pregnancy/Exercise During Pregnancy

Lowensohn, R. I., Stadler, D. D., & Naze, C. (2016). Current Concepts of Maternal Nutrition. Obstetrical & Gynecological Survey, 71(7), 413–426. https://doi.org/10.1097/OGX.0000000000000329

Hibbeln, J. R., Spiller, P., Brenna, J. T., Golding, J., Holub, B. J., Harris, W. S., Kris-Etherton, P., Lands, B., Connor, S. L., Myers, G., Strain, J. J., Crawford, M. A., & Carlson, S. E. (2019). Relationships between seafood consumption during pregnancy and childhood and neurocognitive development: Two systematic reviews. Prostaglandins, Leukotrienes, and Essential Fatty Acids, 151, 14–36. https://doi.org/10.1016/j.plefa.2019.10.002

U.S. Department of Agriculture and U.S. Department of Health and Human Services. (2020). Dietary Guidelines for Americans, 2020-2025, 9th Edition. Retrieved from https://www.dietaryguidelines.gov/

American College of Obstetricians and Gynecologists. (n.d.). Nutrition During Pregnancy. Retrieved August 9, 2020, from https://www.acog.org/en/Patient Resources/FAQs/Pregnancy/Nutrition During Pregnancy

Finer, L. B., & Zolna, M. R. (2016). Declines in Unintended Pregnancy in the United States, 2008–2011. New England Journal of Medicine, 374(9), 843–852. https://doi.org/10.1056/NEJMsa1506575

Council on Environmental Health, Rogan, W. J., Paulson, J. A., Baum, C., Brock-Utne, A. C., Brumberg, H. L., Campbell, C. C., Lanphear, B. P., Lowry, J. A., Osterhoudt, K. C., Sandel, M. T., Spanier, A., & Trasande, L. (2014). Iodine deficiency, pollutant chemicals, and the thyroid: New information on an old problem. Pediatrics, 133(6), 1163–1166. https://doi.org/10.1542/peds.2014-0900

National Institutes of Health Office of Dietary Supplements. (n.d.). Iodine—Fact Sheet for Health Professionals [Office of Dietary Supplements]. Retrieved August 13, 2020, from https://ods.od.nih.gov/factsheets/Iodine-HealthProfessional/

Patel, A., Lee, S. Y., Stagnaro-Green, A., MacKay, D., Wong, A. W., & Pearce, E. N. (2019). Iodine Content of the Best-Selling United States Adult and Prenatal Multivitamin Preparations. Thyroid: Official Journal of the American Thyroid Association, 29(1), 124–127. https://doi.org/10.1089/thy.2018.0386

American College of Obstetricians and Gynecologists. (n.d.). Tobacco, Alcohol, Drugs, and Pregnancy. Retrieved August 13, 2020, from https://www.acog.org/en/Patient Resources/FAQs/Pregnancy/Tobacco Alcohol Drugs and Pregnancy

CDC. (2017, June 29). Prevent Listeria Infections. Centers for Disease Control and Prevention. https://www.cdc.gov/listeria/risk-groups/pregnant-women.html

Foodsafety.gov, A. S. for P. (2019, April 28). People at Risk: Pregnant Women [Text]. FoodSafety.Gov. https://www.foodsafety.gov/people-at-risk/pregnant-women

Callahan, A. (2018, August 3). What Causes Morning Sickness? The New York Times. https://www.nytimes.com/2018/08/03/well/what-causes-morning-sickness.html

American College of Obstetricians and Gynecologists. (n.d.). Morning Sickness: Nausea and Vomiting of Pregnancy. Patient Resources. Retrieved August 13, 2020, from https://www.acog.org/en/Patient Resources/FAQs/Pregnancy/Morning Sickness Nausea and Vomiting of Pregnancy

American College of Obstetricians and Gynecologists. (2018). ACOG Practice Bulletin No. 189: Nausea And Vomiting Of Pregnancy. Obstetrics & Gynecology, 131(1), e15. https://doi.org/10.1097/AOG.0000000000002456

Deputy, N. P., Kim, S. Y., Conrey, E. J., & Bullard, K. M. (2018). Prevalence and Changes in Preexisting Diabetes and Gestational Diabetes Among Women Who Had a Live Birth—United States, 2012–2016. MMWR. Morbidity and Mortality Weekly Report, 67. https://doi.org/10.15585/mmwr.mm6743a2

CDC. (2019, May 30). Gestational Diabetes. Centers for Disease Control and Prevention. https://www.cdc.gov/diabetes/basics/gestational.html

EARLY NUTRITION FOR INFANTS

Infancy is a time of dramatic change and development. In the first year of life, babies triple their body weight and develop from tiny bundles whose daily activities involve eating, sleeping, and creating dirty diapers to toddlers well on their way to walking, talking, and feeding themselves. The first 1,000 days of life, beginning at conception and continuing through toddlerhood, also represent the most active period of brain development in the lifespan, laying the foundation and establishing neural networks to support cognitive, motor, and social-emotional skills throughout life. All of these critical developmental processes are supported through sensitive caregiving, a safe home environment, and of course, a healthy diet.[1] Nutrient requirements on a per-kilogram body weight basis are higher during infancy than in any other stage in the lifespan.

Early Infancy: Meeting Nutrient Needs Through Milk

At birth and continuing through the first 4 to 6 months of life, breast milk, infant formula, or some combination of the two should be the sole source of nutrition for infants. This is because young infants' gastrointestinal tracts aren't yet ready to process more complex foods, and they lack the oral motor skills to swallow solid foods safely. Breast milk is uniquely adapted to meet the nutrient needs of young infants, and infant formula is also designed and regulated to ensure that it is safe and provides adequate nutrition. Any other substitute, including cow's milk, goat's milk, plant-based beverages such as soy milk, homemade infant formula, or watered down formula, should be avoided. These do not provide the right balance of nutrients to meet infants' requirements and can cause serious problems such as damage to the intestines or kidneys.[2]

CHOOSING BREASTFEEDING OR FORMULA-FEEDING

The World Health Organization and the American Academy of Pediatrics, as well as many other health organizations, recommend that infants be exclusively breastfed (only receiving breast milk, with no formula or other foods) for about the first 6 months of life. Breast milk is considered the optimal source of nutrition for infants, but it also contains many other bioactive molecules, including immunoglobulins (or antibodies), hormones, enzymes, growth factors, and protective proteins.[3] In addition, breast milk contains special carbohydrates, called human milk oligosaccharides, that are indigestible to infants (they lack the enzymes to break them down) but help to establish a healthy gut microbiome by serving as a food source for friendly bacteria and binding up harmful bacteria.[4]

Breastfeeding is beneficial to babies in many ways; it reduces a baby's risk of gastrointestinal, respiratory, and ear infections, and it may also protect babies from sudden infant death syndrome (SIDS), eczema, and asthma. It is also associated with a small increase in IQ and a reduced risk of obesity later in childhood. In addition, breastfeeding benefits the health of mothers; it's associated with a reduced risk of hypertension, type 2 diabetes, and breast and ovarian cancer.[3] However, it's important to note that most of the data on benefits of breastfeeding come from observational studies, and these studies have many confounding factors. Women who breastfeed tend to have higher incomes, be more educated, be older in age, and are more likely to be white compared with those who don't breastfeed. This means breastfed babies are often born with more advantages beyond how they are fed, so it can be difficult to separate correlation from causation in studies of infant feeding outcomes.[5]

Despite the challenges in breastfeeding research, breastfeeding has many well-established benefits, and most mothers intend to breastfeed. In 2017, 84% of babies born in the U.S. were breastfed at birth, but just 58% were still breastfeeding at 6 months of age, and 26% were exclusively breastfed through 6 months.[6] These statistics represent the many challenges women face with breastfeeding. Although breastfeeding is natural, it doesn't always come naturally or easily. Breastfeeding challenges can often be addressed with the help of an experienced professional such as a lactation consultant, but this support is not easy for all women to access. Women may also find it hard to establish and continue breastfeeding if they have to return to work just a few weeks or months postpartum, a scenario common in the U.S., the only developed country that has no national policy for paid parental leave. There are also a few circumstances in which babies should not or can not be breastfed. In the U.S., it's recommended that women infected with HIV do not breastfeed, because the virus can pass to infants in milk. Infants born with rare metabolic disorders may not be able to metabolize breast milk and need to receive special formulas. There are also some medications which, if taken by the mother, are not safe for breastfeeding infants. In these cases, mothers may be advised to formula-feed.[3]

Many women find breastfeeding enjoyable and a wonderful way to bond with their baby. For others, breastfeeding can be a struggle for one or more of the reasons discussed above, and that struggle can overshadow a mother's relationship with her baby. For all of these reasons, the choice to breastfeed or formula-feed, or to feed some of both, is complex and individual. Parents should be supported whatever their decision, and they can be confident that infant formula is a safe, nutritionally-adequate option.

Nutrient Composition of Breast Milk

The first milk, called colostrum, is produced immediately after birth and continues for the first two to five days after the arrival of the baby. Colostrum is yellowish in color, thicker than mature breast milk, and produced in small quantities. It is low in fat and easily digestible, yet rich in protein, fat-soluble vitamins, and minerals. Colostrum is also a concentrated source of immunoglobulins that pass from the mother to the baby and provide immune protection to the newborn. The stomach capacity of newborns is small, so they only consume a teaspoon or two of colostrum per feeding in the first few days of life, but they need to feed often. Frequent, on-demand feeding (whenever the baby is hungry, not on a schedule) helps to promote full milk production.

After a couple of days, colostrum is replaced by transitional milk, which is produced in much greater volume and lasts through 7 to 14 days postpartum. Compared with colostrum, transitional milk has more fat and lactose, and less protein and immunoglobulins, and it is also more calorie-dense. Finally, women begin to produce mature milk and will continue to make this type of milk through the end of lactation. Mature milk contains about 87% water, 4% fat, 1% protein, and 7% lactose. Together, these meet infants' macronutrient and caloric requirements.

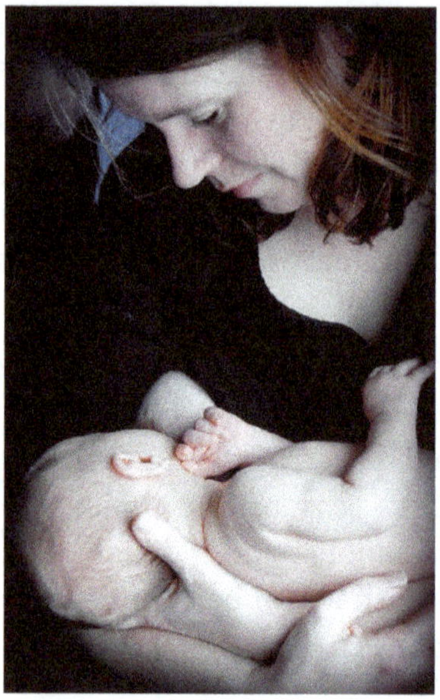

However, there is variability even within mature milk. In a given feeding, the milk secreted at the beginning of the feeding (called foremilk) is thinner and higher in lactose than the milk at the end of the feeding (called hindmilk). The higher levels of fat in the hindmilk helps to ensure the baby's energy needs are met.[9] Milk can also vary from day to night; nighttime milk is higher in fat and the sleep-promoting hormone melatonin. As breastfeeding continues beyond 6 or 7 months, the levels of some vitamins and minerals begin to decline. This is around the time that babies begin to eat some solid foods, so foods can help to fill in nutritional gaps, while breast milk continues to be an important source of nutrients.

Breast milk provides enough of all of the micronutrients that young infants need with two main exceptions: vitamin K and vitamin D. For this reason, newborns should receive a vitamin K shot soon after birth; otherwise, vitamin K deficiency can lead to serious bleeding disorders such as hemorrhage (see Unit 9). The American Academy of Pediatrics also recommends that breastfed newborns be given a vitamin D supplement beginning in the first few days of life and continuing until they are weaned to formula or cow's milk, both of which are fortified with adequate vitamin D. (Cow's milk can be given beginning at 12 months of age.)[11]

Breast milk is also low in iron, although what little is there is absorbed very efficiently. Newborns are born with a certain amount of iron absorbed from their mothers during pregnancy, and they utilize this stored iron—in addition to that provided in breast milk—to meet their iron requirement in early infancy. However, stored iron is depleted by around 4 months of age, so the American Academy of Pediatrics recommends that exclusively breastfed infants begin taking an iron supplement at this age and continuing until they are eating substantial amounts of iron-rich solid foods, such as meat or iron-fortified cereals. Iron deficiency remains a significant problem, with 11% of 1-year-olds in the U.S. estimated to be iron-deficient.[11] Iron is essential for brain development, and iron deficiency may cause lasting developmental deficits.

Nutrition For Breastfeeding Mothers

Breastfeeding mothers have nutrient requirements similar to women in the third trimester of pregnancy. After all, they're continuing to provide nutrients for their babies through their milk, and milk production requires energy, macronutrients, micronutrients, and water. Breastfeeding women have a remarkable ability to make enough milk to meet their babies' nutrient needs even without an optimal diet, but eating well supports maternal health and energy levels during this demanding time.

Breastfeeding increases energy requirements by about 450 to 500 calories per day, part of which can be supplied by using adipose stored in pregnancy and part of which should be supplied from greater caloric intake. In general, breastfeeding mothers can eat a wide variety of foods, and they don't usually need to avoid or restrict any specific foods. However, as in pregnancy, they should continue to avoid high-mercury fish. Consuming low-mercury fish 2 to 3 times per week provides the omega-3 fatty acids DHA and EPA, which pass into breast milk and support brain and eye development for the breastfeeding infant.[13] There is no need to avoid common food allergens, such as peanut or dairy; avoiding these foods while breastfeeding has not been shown to reduce babies' chances of developing food allergies.[14] Sometimes women find that certain foods, such as garlic or spicy foods, are associated with fussiness or gas in their infants, and they may experiment with avoiding those foods. However, most infants don't have a problem tolerating these foods. In fact, flavors from the mother's diet pass into breast milk, so researchers hypothesize that when mothers eat a wide variety of foods and flavors while breastfeeding, their children grow up to be more adventurous eaters.

In addition to consuming a nutrient-dense diet, obstetricians sometimes recommend that breastfeeding mothers continue taking a prenatal supplement to ensure that their micronutrient needs are met. Breastfeeding mothers should also be sure to drink plenty of fluids to support milk production.

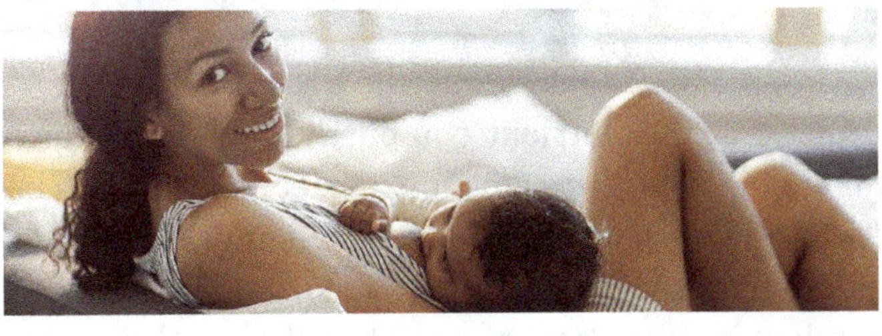

Similar to pregnancy, substances consumed by the breastfeeding mother can pass to the infant in her milk. However, how much passes into the milk depends on the type of substance and the timing of consumption relative to breastfeeding. For example, it's considered safe for a breastfeeding mother to have an alcoholic drink so long as she waits at least two hours before breastfeeding, because by that point, most of the alcohol will have cleared her bloodstream and will not pass into her milk. Caffeine lasts longer in the blood but is considered safe in moderation, with a limit of about 300 milligrams per day.[13] Cannabinoids, the chemicals found in cannabis, are fat-soluble and remain in a mother's bloodstream and body tissues for much longer. Tetrahydrocannabinol (THC), the main psychoactive chemical in cannabis, has been detected in breast milk as much as 6 weeks after a mother used cannabis.[16] Although it's not clear how these chemicals affect babies when consumed in breast milk, medical organizations agree it's best to avoid using cannabis while breastfeeding.

The National Institutes of Health's LactMed database is a great resource for information about the safety of drugs, medications, and supplements during breastfeeding.

Nutrient composition of infant formula

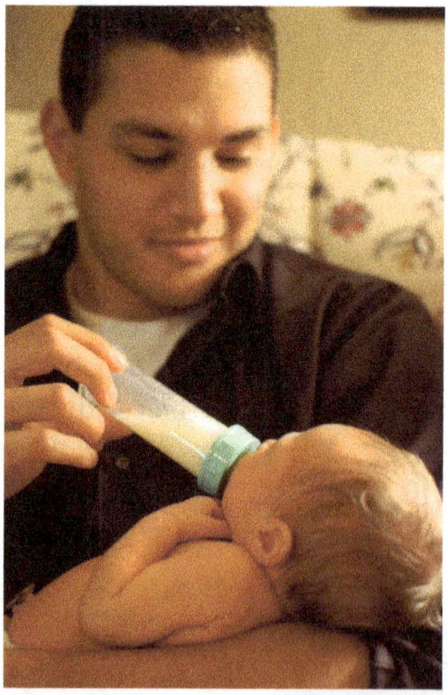

To a certain extent, the nutrient composition of infant formula is modeled after that of human milk, and although it by no means duplicates breast milk, it is a safe and effective substitute. Formula is made from ingredients such as cow's milk, soy, vegetable oils, and corn syrup. These may be combined in ways to mimic the overall macronutrient composition of breast milk, but the content of individual amino acids, fatty acids, and sugars can vary somewhat. In addition, some of these nutrients are less digestible, so to compensate, formulas tend to have higher levels of some nutrients, such as protein, compared with breast milk. Formula contains more of some micronutrients, such as iron, vitamin D, and vitamin K, so deficiencies of these vitamins and minerals are more common in breastfed infants if they don't receive appropriate supplementation or solid foods when the time comes.[17]

Infant formula also does not contain most of the bioactive molecules found in breast milk, although formula companies are beginning to add versions of some of these molecules. For example, many formulas now include some type of indigestible sugar molecule intended to act as a prebiotic to feed healthy gut bacteria, similar to human milk oligosaccharides. At this point, however, the evidence that such ingredients are beneficial to infants is not very convincing. However, novel ingredients like these are often used as marketing tools, with labels touting vague structure-function claims such as "brain-boosting" and "immune-supporting." Consumers should know that there is often little evidence that "designer ingredients" in infant formula make them healthier for babies. All infant formulas are required by law to be safe and meet the nutrient requirements of infants, and in most cases, basic store brand formulas cost less and are just as good as other products on the shelf. It may be true that infant formula can't replicate the complexity of breast milk, but it has a very strong track record of safety, and infants can grow and thrive with formula-feeding.[17]

REFERENCES

Schwarzenberg, S. J., Georgieff, M. K., & Nutrition, C. O. (2018). Advocacy for Improving Nutrition in the First 1000 Days to Support Childhood Development and Adult Health. Pediatrics, 141(2). https://doi.org/10.1542/peds.2017-3716

Korioth, T. (2020). Don't feed homemade formula to babies; seek help instead. AAP News. https://www.aappublications.org/news/2019/02/25/homemadeformulapp022519

American Academy of Pediatrics Section on Breastfeeding. (2012). Breastfeeding and the Use of Human Milk. Pediatrics, 129(3), e827-41. https://doi.org/10.1542/peds.2011-3552

Bode, L. (2015). The functional biology of human milk oligosaccharides. Early Human Development, 91(11), 619–622. https://doi.org/10.1016/j.earlhumdev.2015.09.001

American Academy of Pediatrics Committee on Nutrition. (2014). Breastfeeding. In Pediatric Nutrition (7th ed., pp. 41–59). American Academy of Pediatrics.

CDC. (2019, December 31). 2018 Breastfeeding Report Card. Centers for Disease Control and Prevention. https://www.cdc.gov/breastfeeding/data/reportcard.htm

U.S. Department of Health and Human Services. (2011). The surgeon general's call to action to support breastfeeding. U.S. Department of Health & Human Services, Office of the Surgeon General. https://www.surgeongeneral.gov/library/calls/breastfeeding/calltoactiontosupportbreastfeeding.pdf

Chzhen, Y., Gromada, A., & Rees, G. (2019). Are the world's richest countries family friendly? Policy in the OECD and EU. Florence.

Martin, C. R., Ling, P.-R., & Blackburn, G. L. (2016). Review of Infant Feeding: Key Features of Breast Milk and Infant Formula. Nutrients, 8(5). https://doi.org/10.3390/nu8050279

Italianer, M. F., Naninck, E. F. G., Roelants, J. A., van der Horst, G. T. J., Reiss, I. K. M., Goudoever, J. B. van, Joosten, K. F. M., Chaves, I., & Vermeulen, M. J. (2020). Circadian Variation in Human Milk Composition, a Systematic Review. Nutrients, 12(8). https://doi.org/10.3390/nu12082328

American Academy of Pediatrics. (2016). Vitamin D & Iron Supplements for Babies: AAP Recommendations. HealthyChildren.org. Retrieved September 3, 2020, from https://www.healthychildren.org/English/ages-stages/baby/feeding-nutrition/Pages/Vitamin-Iron-Supplements.aspx

Cusick, S. E., & Georgieff, M. K. (2016). The Role of Nutrition in Brain Development: The Golden Opportunity of the "First 1000 Days." The Journal of Pediatrics, 175, 16–21. https://doi.org/10.1016/j.jpeds.2016.05.013

CDC. (2020, February 10). Diet considerations for breastfeeding mothers. Centers for Disease Control and Prevention. https://www.cdc.gov/breastfeeding/breastfeeding-special-circumstances/diet-and-micronutrients/maternal-diet.html

Abrams, E. M., & Chan, E. S. (2019). It's Not Mom's Fault: Prenatal and Early Life Exposures that Do and Do Not Contribute to Food Allergy Development. Immunology and Allergy Clinics of North America, 39(4), 447–457. https://doi.org/10.1016/j.iac.2019.06.001

Forestell, C. A. (2017). Flavor Perception and Preference Development in Human Infants. Annals of Nutrition & Metabolism, 70 Suppl 3, 17–25. https://doi.org/10.1159/000478759

LactMed. (2020). Cannabis. In Drugs and Lactation Database. National Library of Medicine (US). http://www.ncbi.nlm.nih.gov/books/NBK501587/

American Academy of Pediatrics Committee on Nutrition. (2014). Formula Feeding of Term Infants. In Pediatric Nutrition (7th ed., pp. 61–81). American Academy of Pediatrics.

Nutrition in Later Infancy and Toddlerhood

In early infancy, nutrition choices are relatively simple (though not necessarily easy!). When the baby is hungry, it's time to breastfeed or prepare a bottle. But in later infancy and toddlerhood, a baby's food horizons expand. This is an exciting period of learning about foods and how to eat with the rest of the family.

Introducing Solid Foods

The World Health Organization recommends that babies begin eating some solid foods at 6 months while continuing to breastfeed. Other health organizations offer more flexible advice, recommending that solid foods be introduced sometime between 4 and 6 months, depending on the baby's development, interest in eating solids, and family preferences. Regardless, most babies aren't ready to eat solid foods before 4 months, and starting too soon may increase the risk of obesity. Yet it's also important not to start solids too late, as beyond 6 months, breast milk alone can't support a baby's nutrient requirements.[1]

However, as babies begin to eat solids, breast milk or formula continue to be the nutritional foundation of the diet. This period is also called complementary feeding, because solid foods are meant to complement the nutrients provided by breast milk or formula. Between 6 and 12 months, babies gradually eat more solid foods and less milk so that by 12 months, formula is no longer needed. Breastfeeding mothers may choose to wean at 12 months or continue breastfeeding as long as she and the baby like.

Babies should be developmentally ready to eat solids before trying their first foods. A baby ready for solids should be able to do the following:[1]

- Sit up without support (e.g., in a high chair or lap)
- Open mouth for a spoonful of food and swallow it without gagging or pushing it back out
- Reach for and grasp food or toys and bring them to his or her mouth.

The American Academy of Pediatrics recommends beginning with iron-rich foods such as pureed meat or iron-fortified cereal (e.g., rice cereal, oatmeal), as iron is usually the most limiting nutrient at this age, particularly for exclusively breastfed babies. Once these foods are introduced, others can gradually be added to the diet, introducing one at a time to keep an eye out for allergic reactions. Work up to a variety of foods from all of the food groups, as babies are willing to try just about anything at this stage, and this is an opportunity for them to learn about different flavors. You can also gradually increase texture, from pureed to mashed food, then lumpy foods to soft finger foods. By 12 months, most babies can eat most of the foods at the family table, with some modifications to avoid choking hazards.

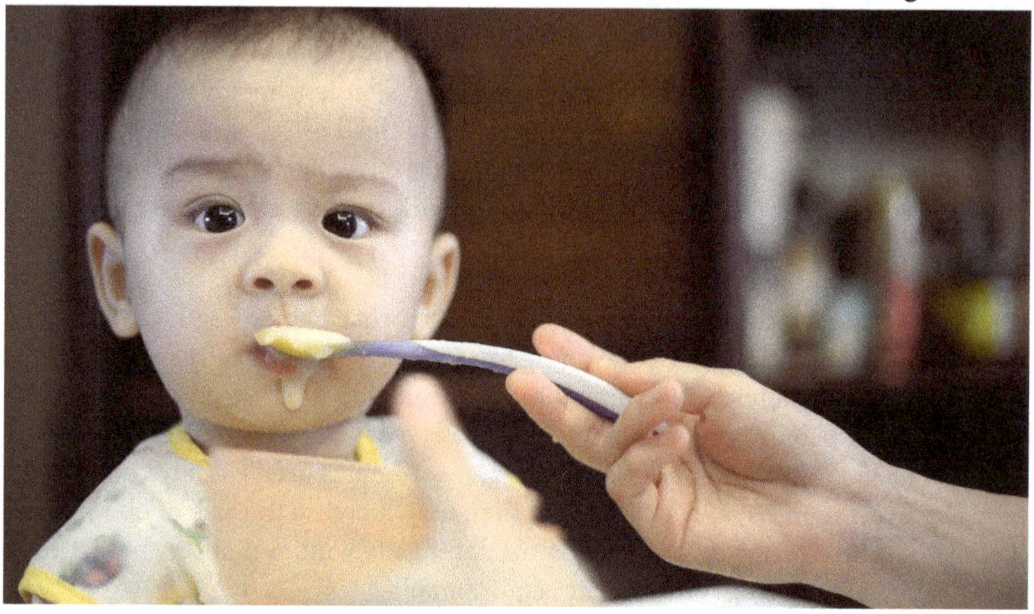

When choosing good complementary foods, there are three main goals: (1) to meet nutrient requirements; (2) to introduce potentially allergenic foods; and (3) to support your baby in learning to eat many different flavors and textures. Parents should be sure to include the following:[1,2]

- Good sources of iron and zinc, as both minerals can be limiting for breastfed infants. Good sources include meat, poultry, fish, and iron-fortified cereal. Beans, whole grains, and green vegetables add smaller amounts of iron.
- Adequate fat to support babies' rapid growth and brain development. Good sources include whole fat yogurt, avocado, nut butters, and olive oil for cooking vegetables. Fish is also a great food for babies, because it provides both iron and fat, and it's a good source of omega3- fatty acids like DHA and EPA, which support brain development.
- A variety of vegetables, fruits, and whole grains, so that your baby learns to like many different tastes and textures. It may take babies and toddlers 8 to 10 exposures of a new food before they learn to like it, so don't be discouraged if your baby doesn't like some foods right away.

There is no need to avoid giving your baby common food allergens, such as peanut, egg, dairy, fish, shellfish, wheat, soy, or tree nuts. In fact, studies indicate that introducing at least some of these foods during the first year can prevent food allergies from developing. The evidence is strongest for peanut allergy. A randomized controlled trial published in 2015, called the Learning Early About Peanut Allergy (LEAP) study, showed that in infants considered high-risk for food allergies, feeding peanut products beginning between 4 and 11 months reduced peanut allergy by 81 percent, compared with waiting to give peanuts until age 5.[3] Similarly, early introduction of egg seems to protect children from developing an egg allergy.[4,5] With any new food, keep an eye out for symptoms of an allergic reaction, such as hives, vomiting, wheezing, and difficulty breathing.[6]

FOODS TO AVOID IN THE FIRST YEAR

There are only a few foods that should be avoided in the first year. These include the following:

- Cow's milk can't match the nutrition provided by breast milk or formula and can cause intestinal bleeding in infants. However, dairy products such as yogurt and cheese are good choices for babies who have started solids. Babies can eat other dairy products, like yogurt and cheese, and cow's milk can be added to the diet at 12 months.
- Plant-based beverages such as soy and rice milk aren't formulated for infants, lack key nutrients, and often have added sugar.
- Juice and sugar-sweetened beverages have too much sugar. Whole fruit in a developmentally-appropriate form (pureed, mashed, chopped, etc.) is a better choice.
- Honey may contain botulism, which can make infants very ill.
- Unpasteurized dairy products or juices, and raw or undercooked meats or eggs, which can be contaminated with harmful foodborne pathogens.
- Added sugar and salt should be kept to a minimum so that your baby learns to like many different flavors and doesn't develop preferences for very sweet or salty foods.
- Choking hazards such as whole nuts, grapes, popcorn, hot dogs, and hard candies should be avoided.

Responsive Feeding and Infant Growth

Regardless of whether infants are fed breast milk or formula, and continuing when they start solid foods, it's important that caregivers use a responsive feeding approach. Responsive feeding is grounded in 3 steps:

- The child signals hunger and satiety. This may occur through vocalizations (e.g., crying, talking), actions (e.g., pointing at food, or turning away when full), and facial expressions.

- The caregiver recognizes the cues and responds promptly and appropriately. For example, if the baby seems hungry, he or she is offered food promptly. If the baby turns his or her head or pushes away the breast, bottle, or an offered bite of food, the caregiver does not pressure the baby to eat more. The child experiences a predictable response to his or her signals.

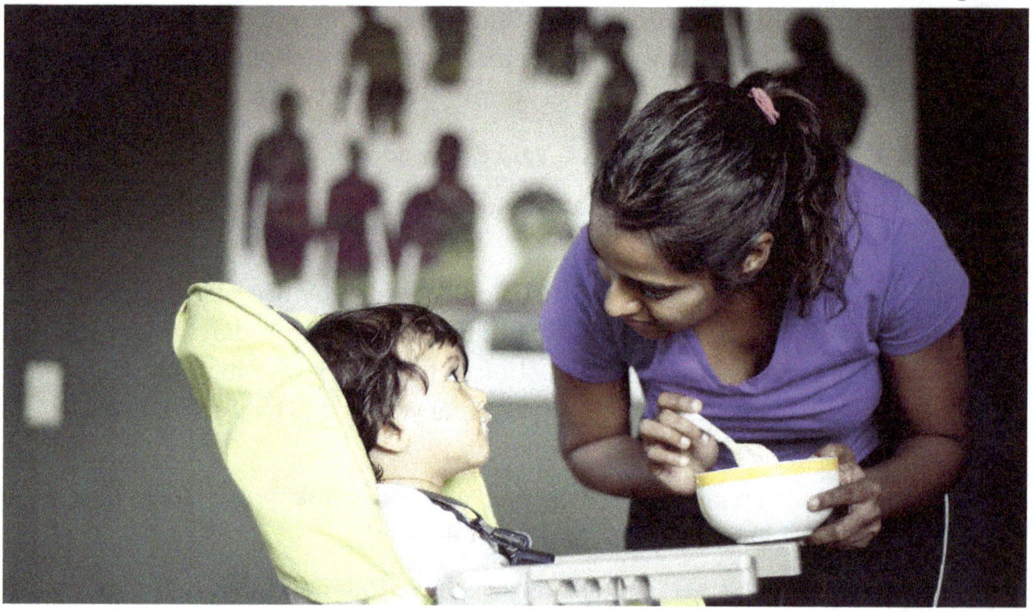

With breastfeeding, responsive feeding simply means feeding the baby when he or she signals hunger, and the baby usually turns away, spits out the nipple, or falls asleep when full. With bottle-feeding, whether feeding breast milk or formula in a bottle, it's a little trickier. It's human nature to want the baby to finish the bottle that you've prepared, but a responsive feeding approach means that you let the baby decide when he or she has had enough. Pressure to eat more can cause the baby to grow too fast in infancy, which is correlated with becoming overweight or obese later in childhood. When feeding solid foods, the same responsive feeding principles apply, although solids should be offered at predictable meal- and snack-times to avoid constant grazing throughout the day. If babies are offered appropriate, nutrient-dense foods, and fed responsively, parents generally don't need to worry about serving sizes or amounts eaten. They can trust that their babies will eat what they need.[2]

The best way to determine if children are getting enough food to eat is to track their growth. Pediatricians and other healthcare providers do this by measuring a child's weight, length, and head circumference at each check-up and plotting their measurements periodically on growth charts from the World Health Organization. Growth charts allow you to compare your child's growth to a population of other healthy children. Sometimes parents worry that if their child is in the 15th percentile for weight, that means she's not growing well, but this isn't the case. Children come in different shapes and sizes, and some grow faster than others. Thus, a child in the 15th percentile for weight may be a bit smaller than average, but when it comes to body size, the goal is not to be average or above average. The goal is to grow steadily and predictably in a way that is healthy for that individual child. If a child who was previously in the 15th percentile was suddenly measured at the 5th percentile or 50th percentile, that might indicate a health or nutrition problem that warrants further evaluation.

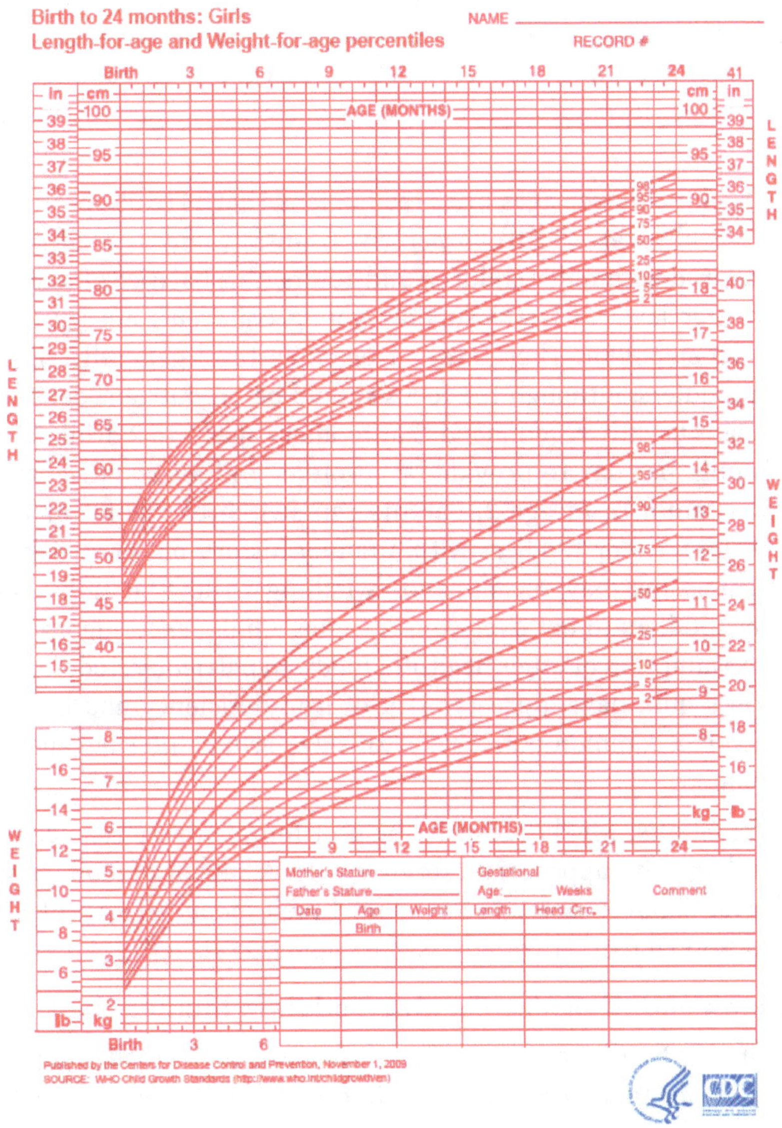

Figure 11.6. WHO growth chart for girls from birth to 24 months.

Feeding Toddlers

Toddlerhood represents a stage of growing independence for children. They gain the physical abilities to feed themselves confidently, and their growing language skills mean they can verbalize food preferences more clearly. Gradually, through exposure and experience, they learn to eat foods more and more like the rest of their family.[8]

In the toddler years, it's important to shift from a mindset of feeding "on demand," which is appropriate for infants, to one of predictable structure, with sit-down meals and snacks (usually three meals and two to three snacks each day). This prevents constant grazing and means that children come to the table hungry, ready to enjoy a nourishing meal. As much as you can, sit down to meals together so that your toddler learns that part of the joy of eating is enjoying time with loved ones.[2,8]

The AMDRs for children ages 1 to 3 recommend that 45 to 65 percent of calories come from carbohydrate, 30 to 40 percent from fat, and 5 to 20 percent from protein. Compared with older children and adults, this balance of macronutrients includes a higher level of fat to support young children's energy demands for growth and development. Therefore, fat or cholesterol generally should not be restricted in toddlers, although the focus should be on nutrient-dense sources of fat. Pediatricians usually recommend that toddlers ages 1 to 2 drink 2 to 3 cups of whole cow's milk per day to provide fat, protein, and micronutrients, including calcium and vitamin D. At age 2, parents can switch to low-fat or nonfat milk to reduce fat intake. For toddlers with a family history or other risk factors for obesity, pediatricians may recommend switching to low-fat milk sooner. It is important for toddlers to not over-consume cow's milk, as filling up on milk will reduce the consumption of other healthful foods. In particular, toddlers who drink too much cow's milk have a greater risk of iron deficiency and iron deficiency anemia, which is a common nutrient for this age group and can cause deficits in brain development.

Just as for adults, MyPlate can be helpful for planning balanced meals for children 2 and up, with appropriate serving sizes. A ballpark recommendation for serving sizes for children ages 2 to 6 is about 1 tablespoon per year of age for each food, with additional food provided based on appetite.

Other Recommendations for Feeding Toddlers Include the Following:

- Continuing to offer a variety of foods from all of the food groups, including a mix of vegetables and fruits of different colors, tastes, and textures.
- Include whole grains and protein sources, such as poultry, fish, meats, tofu, or legumes in most meals and snacks.

- Limit salty foods and sugary snacks for health reasons, and so that your child doesn't come to expect these tastes in foods.
- By 12 to 15 months, wean toddlers from a bottle, transitioning to giving milk at meals in a cup. Prolonged bottle use tends to promote overconsumption of milk and can cause dental caries, particularly when toddlers fall asleep with a bottle.
- Continue to take care with choking hazards, as many choking incidents happen in children younger than 4. Common choking hazards include hot dogs, hard candy, nuts, seeds, whole grapes, raw carrots, apples, popcorn, marshmallows, chewing gum, sausages, and globs of peanut butter. Ensuring that children are sitting down when eating can help to prevent choking accidents.
- Stick to cow's milk and water as main beverage choices. Juice can be enjoyed occasionally in small servings (<0.5 cups/day) but is high in sugar, and whole fruit provides more nutrition. Plant-based beverages such as soy milk can be used in the case of a dairy allergy, lactose intolerance, or strong dietary preference, but be aware that these can be high in sugar and may not offer the same nutrients as cow's milk, so check labels carefully. Flavored cow's milk, soda, sports drinks, energy drinks, and caffeinated beverages should be avoided.
- Most children can get all of the nutrients they need from their diet, even if it seems that their intake is variable and they are somewhat picky about their choices. Pediatricians may prescribe a fluoride supplement for children living in areas with low fluoride levels in drinking water. They may also recommend a vitamin D supplement for children who do not consume adequate levels in their diet.

REFERENCES:

American Academy of Pediatrics Committee on Nutrition. (2014). Complementary Feeding. In Pediatric Nutrition (7th ed., pp. 123–139). American Academy of Pediatrics.

Perez-Escamilla, R., Segura-Perez, S., & Lott, M. (2017). Feeding Guidelines for Infants and Young Toddlers: A Responsive Parenting Approach. Healthy Eating Research. http://healthyeatingresearch.org

Du Toit, G., Roberts, G., Sayre, P. H., Bahnson, H. T., Radulovic, S., Santos, A. F., Brough, H. A., Phippard, D., Basting, M., Feeney, M., Turcanu, V., Sever, M. L., Gomez Lorenzo, M., Plaut, M., & Lack, G. (2015). Randomized Trial of Peanut Consumption in Infants at Risk for Peanut Allergy. New England Journal of Medicine, 0(0), null. https://doi.org/10.1056/NEJMoa1414850

Burgess, J. A., Dharmage, S. C., Allen, K., Koplin, J., Garcia-Larsen, V., Boyle, R., Waidyatillake, N., & Lodge, C. J. (2019). Age at introduction to complementary solid food and food allergy and sensitization: A systematic review and meta-analysis. Clinical and Experimental Allergy: Journal of the British Society for Allergy and Clinical Immunology, 49(6), 754–769. https://doi.org/10.1111/cea.13383

Greer, F. R., Sicherer, S. H., Burks, A. W., Nutrition, C. O., & Immunology, S. on A. A. (2019). The Effects of Early Nutritional Interventions on the Development of Atopic Disease in Infants and Children: The Role of Maternal Dietary Restriction, Breastfeeding, Hydrolyzed Formulas, and Timing of Introduction of Allergenic Complementary Foods. Pediatrics, 143(4). https://doi.org/10.1542/peds.2019-0281

American Academy of Pediatrics. (2018). Food Allergies in Children. HealthyChildren.Org. Retrieved September 9, 2020, from https://www.healthychildren.org/English/healthy-living/nutrition/Pages/Food-Allergies-in-Children.aspx

American Academy of Pediatrics. (2015). How to Read a Growth Chart: Percentiles Explained. HealthyChildren.Org. Retrieved September 3, 2020, from https://www.healthychildren.org/English/health-issues/conditions/Glands-Growth-Disorders/Pages/Growth-Charts-By-the-Numbers.aspx

HOW TO RAISE HEALTHY EATERS

"Raising a healthy eater takes years. Children learn bite by bite, food by food, meal by meal. The goal of raising a healthy eater is to help your child grow up with positive eating attitudes and behaviors; it is not to get him to eat his peas for tonight's supper."

Ellyn Satter, MS, MSSW, RDN

As we discussed previously, what we feed kids is important to ensure that they are meeting their nutrient requirements. But, just as important as WHAT we feed kids, is HOW we feed kids. The structure and environment that parents provide when feeding not only impacts nutrition, but it can also affect weight and behavioral problems. Families who regularly eat together have children who eat more fruits and vegetables, have healthier weights, and are less likely to use drugs, alcohol, or tobacco when they are older.[1]

Family meals have many nutritional, social, and emotional benefits. In this section, we will discuss best practices for providing family meals and raising healthy eaters.

The Division of Responsibility

The gold standard for feeding children is Ellyn Satter's Division of Responsibility (sDOR). Ellyn Satter is a dietitian, family therapist, and internationally recognized authority on eating and feeding. The Division of Responsibility outlines the optimal relationship between parent and child when it comes to feeding: the parent determines the *what, where, and when* of feeding, and the child chooses *how much* to eat and *whether* to eat from the foods provided.[2]

The Parents' Jobs with Feeding:

- Provide structured, sit-down meals and snacks at predictable times
- Decide what foods will be offered at meals and snacks
- Make mealtimes pleasant
- Teach table manners
- Only offer water between meal and snack times; no other foods or beverages
- Be considerate of food preferences, but don't cater to likes and dislikes
- Trust your child knows how much food to eat and will grow into the body right for them

The Child's Jobs with Feeding:

- Eat the amount of food he or she needs
- Learn to eat a variety of foods that are offered at family meals
- Have good table manners
- Grow into the body that is right for him or her

The division of responsibility is built on the trust that children know when they are hungry and full, that they want to eat, and that they want to learn to eat grown-up foods. Parents can best support their child's eating by providing the structure of family sit-down meals and snacks and modeling relaxed, enjoyable eating. The division of responsibility is an authoritative approach, providing structure and limits, but allowing the child autonomy within those limits. At meals and snack times, parents promote children's independence by allowing them to pick and choose foods from what has been made available and deciding how much of each to eat.

As your child learns to eat grown-up foods, every day will look slightly different—some days eating a lot and other days eating very little. Don't cross the division of responsibility and pressure your child to eat more or less of something; this kind of pressure backfires. Pressure can be positive: praise, rewards, bribing, making special food, playing games, or talking about nutrition to encourage kids to eat more of nutritious foods or less of "bad" foods. Pressure can also be negative: restricting, threatening, punishing, shaming, or withholding dessert or fun activities. Instead of pressure, follow the division of responsibility in feeding.

Eating Competence

Eating competence is an evidence-based model that defines the interrelated spectrum of eating attitudes and behaviors.[3] The model is based on the principle that internal cues of hunger, appetite, and satiety are reliable and can be used to inform food selection and guide energy balance and body weight. Satter breaks eating competence down into 4 basic components: (1) eating attitudes; (2) food acceptance attitudes and skills; (3) internal regulation attitudes and skills; and (4) contextual attitudes and skills for providing family meals.

According to Ellyn Satter, "Eating Competence is being positive, comfortable, and flexible with eating as well as matter-of-fact and reliable about getting enough to eat of enjoyable food. Even though they don't worry about what and how much to eat, competent eaters do better nutritionally, are more active, sleep better, and have better lab tests. They are more self-aware and self-accepting, not only with food, but in all ways. To be a competent eater, be relaxed, self-trusting, and joyful about eating, and take good care of yourself with food."

Current nutrition advice often focuses on avoidance of foods and reliance on outside indicators, such as diet plans, to guide eating. Eating competence is the opposite. It encourages us to seek food that is enjoyable and to let internal processes such as hunger and satiety to guide eating.

Eating competence comprises both permission and discipline.5

- Permission is choosing foods that are enjoyable and eating those foods in amounts that are satisfying, based on hunger, appetite, and satiety.
- Discipline is providing the structure of regular meals and sit-down snacks, and paying attention to internal regulators while eating.

In "Secrets of Feeding a Healthy Family: How to Eat, How to Raise Good Eaters, How to Cook," Satter discusses how the eating competence model is built on trust: "Trust in our love of food and good eating; trust in following our inclinations to eat the food we like in amounts that are satisfying; trust that taking time to enjoy eating is time well-spent; trust that taking pleasure in eating supports being healthy; trust that behaving in such a self-respecting way is legitimate."

Today's nutrition advice is often based on control—the opposite of trust. Control is looking to outside instruction for *what* and *how much* to eat. It is sticking to a strict macro or calorie count; it is eating food because it is good for you (even though it is not appealing); it is eating a defined amount of food that is not related to hunger and satiety; and it is restricting food. Control often means restrained eating, and restrained eaters actually consume *more* food when exposed to forbidden foods.[6] Also, when food choices are externally dictated, especially when promoting negative energy balance, the body's physiological and psychological defense mechanisms are activated, which can lead to gaining excess weight and accumulating excess fat after food restriction.

Parents should consider their own eating competence as they work to build eating competence in their children. Do you feel positive about eating and about food? Do you enjoy eating a variety of foods and trying new foods? Do you trust your internal regulators of hunger, appetite, and satiety to guide how much you eat? Do you have structured meals and snacks? It's much more difficult to raise eating competent children if the adults in the family do not model a healthy relationship with food. Many new parents find that having children inspires them to work towards greater eating competence for themselves.

Meal planning is a good place to start when working on building eating competence. Tips for meal planning:[8]

- Start with what your family is currently eating, and cluster those foods into meals and snacks.
- Try to provide food from each of the food groups for meals, and from 2 to 3 groups for snacks (dairy, fruit, vegetable, protein, and grain). Make sure they are foods you like and enjoy.
- Include bread or similar food like rice at every meal. Bread is always an easy-to-like food that family members can choose when they aren't excited about other options. Pair familiar foods with unfamiliar foods and favorite foods with not-so-favorite foods.
- Include fat in food preparation to make foods enjoyable. For meals to be sustainable, they must be satisfying.
- Let everyone choose what tastes good to them from what is provided on the table.

The Picky Eater

All young children can be picky when it comes to food and refuse to try new foods; this is normal behavior. It is only a concern when children get stuck with being picky. Follow the division of responsibility and adjust your expectations. Steer clear of the following traps that can limit food acceptance:

- Catering the menu to only child-friendly foods or favorite foods
- Making a separate meal when your child complains about what is offered
- Asking your child what they want to eat
- Pressuring your child to eat
- Offering food outside of scheduled meal and snack time

The biggest goal with a finicky eater is to not make eating an issue, and to have mealtime be pleasant. Provide the structure of scheduled meals and snacks, and the permission for your child to choose from what is offered, with no pressure.

REFERENCES:

Academy of Nutrition and Dietetics. (2019). Family Meals: Small Investment Big Payoff. Retrieved from https://www.eatright.org/food/nutrition/eating-as-a-family/family-meals–small-investment–big-payoff

Ellyn Satter Institute. (2019). Raise a healthy child who is a joy to feed. Retrieved from https://www.ellynsatterinstitute.org/how-to-feed/the-division-of-responsibility-in-feeding/

Lohse B, Satter E, Horacek T, GebreselassieT, Oakland MJ. Measuring Eating Competence: psychometric properties and validity of the ecSatter Inventory. J Nutr Educ Behav. 2007;39 (suppl):S154-S166.

Ellyn Satter Institute. (2019). The joy of eating: Being a competent eater. Retrieved from https://www.ellynsatterinstitute.org/how-to-eat/the-joy-of-eating-being-a-competent-eater/

Satter, E. (2008). Eat as Much as You Want. In Secrets of Feeding a Healthy Family: How To Eat, How To Raise Good Eaters, How To Cook. (p. 29). Madison, WI: Kelcy Press.

Stirling LJ, Yeomans MR. Effect of exposure to a forbidden food on eating in restrained and unrestrained women. Int J Eat Disord. 2004; 35:59-68.

Keys A, Brozek J, Henschel A, Mickelsen O, Taylor H. The Biology of Human Starvation. Minneapolis: University of Minnesota Press; 1950.

Satter, E. (2008). Feed Yourself Faithfully. In Secrets of Feeding a Healthy Family: How To Eat, How To Raise Good Eaters, How To Cook. (p. 48). Madison, WI: Kelcy.

HOW TO RAISE HEALTHY EATERS

Puberty marks the beginning of adolescence, the time between childhood and young adulthood. The DRI recommendations divide adolescence into two age groups: 9 through 13 years, and 14 through 18 years. The onset of puberty brings a number of changes, including the development of the reproductive organs, the onset of menstruation in females, growth spurts, and changing body composition. Fat usually assumes a larger percentage of change in body weight in girls, while teenage boys experience greater muscle and bone development. All of these changes should be supported with sound nutrition.

NUTRIENT NEEDS IN ADOLESCENCE

Energy and Macronutrients

Adequate energy intake is necessary to support the dramatic growth that takes place during adolescence. For ages 9 to 13, girls should consume about 1,400 to 2,200 calories per day, and boys should consume 1,600 to 2,600 calories per day. For ages 14 to 18, girls should consume about 1,800 to 2,400 calories per day, and boys should consume about 2,000 to 3,200 calories per day. Calorie needs vary based on activity level. The extra energy required for physical development during the teenage years should be obtained primarily from nutrient-dense foods instead of empty-calorie foods, to support adequate nutrient intake and a healthy body weight.

For children and adolescents ages 4 through 18, the AMDR for carbohydrates is 45 to 65 percent of daily calories, and most of these calories should come from high-fiber foods such as whole grains. The AMDR for protein is 10 to 30 percent of daily calories, and lean proteins, such as meat, poultry, fish, beans, nuts, and seeds are excellent ways to meet protein needs. The AMDR for fat is 25 to 35 percent of daily calories. The focus should be on unsaturated plant fats to prevent chronic diseases.

Micronutrients

Micronutrient recommendations for adolescents are mostly the same as for adults, though children this age need more of certain minerals. The most important micronutrients for adolescents are calcium, vitamin D, vitamin A, and iron.

- Calcium levels increase to 1,300 mg/day during adolescence to support bone growth and prevent osteoporosis later in life. Low-fat dairy products and foods fortified with calcium, such as breakfast cereals and orange juice, are excellent sources of calcium.
- Iron needs increase for adolescent girls with the onset of menstruation (15 mg/day for ages 9 to 13 and 18 mg/day for ages 14 to 18). Adolescent boys also need additional iron for the development of lean body mass (11 mg/day for ages 14-18).
- Vitamin A is critical to support the rapid development and growth that happens during adolescence. Adequate fruit and vegetable intake meets vitamin A needs.

COMMON NUTRITION-RELATED HEALTH CONCERNS IN ADOLESCENCE

Disordered Eating

Eating disorders involve extreme behaviors related to food and exercise. They encompass a group of conditions marked by under-eating or overeating, as discussed in Unit 7. Eating disorders stem from stress, low self-esteem, and other psychological and emotional issues. They are most prevalent among adolescent girls but have been increasing among adolescent boys in recent years. Because eating disorders often lead to malnourishment, adolescents with eating disorders are deprived of the crucial nutrients their still-growing bodies need. Girls with anorexia experience nutritional and hormonal problems that negatively influence peak bone density, and therefore may be at increased risk for osteoporosis and fracture throughout life.[1] It is important for parents to watch for signs and symptoms of these disorders, including sudden weight loss, lethargy, vomiting after meals, and the use of appetite suppressants. Eating disorders can lead to serious complications or even be fatal if left untreated. Treatment includes cognitive, behavioral, and nutritional therapy.

Obesity

Children need adequate caloric intake for growth, and it is important not to impose very restrictive diets. However, exceeding caloric requirements on a regular basis can lead to childhood obesity, which has become a major problem worldwide. According to the CDC National Center for Health Statistics, the prevalence of obesity was 18.4% for youth ages 6 to11 and 20.6% for youth ages 12 to 19 in 2106.[2]

There are a number of factors that may contribute to this problem, including:

- Early life factors, such as lack of breastfeeding support
- Larger portion sizes
- Limited access to nutrient-rich foods
- Increased access to fast foods and vending machines
- Declining physical education programs in schools
- Insufficient physical activity and a sedentary lifestyle
- Media messages encouraging the consumption of unhealthy foods

Children who suffer from obesity are more likely to become overweight or obese adults. Obesity has a profound effect on self-esteem, energy, and activity level. Even more importantly, it is a major risk factor for a number of diseases later in life, including cardiovascular disease, Type 2 diabetes, stroke, hypertension, and certain cancers.[3]

One major contributing factor to childhood obesity is the consumption of added sugars, especially in the form of sugar sweetened beverages.[4] Added sugars include not only sugar added to food at the table, but also ingredients in items such as bread, cookies, cakes, pies, jams, and soft drinks. In addition, sugars are often "hidden" in items added to foods after they're prepared, such as ketchup, salad dressing, and other condiments. According to the National Center for Health Statistics, young children and adolescents consume an average of 362 calories per day from added sugars, or about 16% of daily calories, 10% more than what the Dietary Guidelines for Americans recommends.[5] Adolescent boys (ages 12 to19 years) have the greatest intake of added sugar, averaging 442 calories. The primary offenders are processed and packaged foods, along with soda and other beverages. These foods are not only high in sugar, they are also light in terms of nutrients and often take the place of healthier options.

If a child gains weight inappropriate to growth, parents and caregivers should nurture eating competence and follow the division of responsibility as previously discussed in this unit. In addition, it is extremely beneficial to increase a child's physical activity and limit sedentary activities, such as watching television, playing video games, or surfing the Internet. Programs to address childhood obesity can include behavior modification, exercise counseling, psychological support or therapy, family counseling, and family meal-planning advice. For most, the goal is not weight loss, but rather allowing height to catch up with weight as the child continues to grow. Rapid weight loss is not recommended for preteens or younger children due to the risk of deficiencies and stunted growth.

Nutritionally Vulnerable

One of the psychological and emotional changes that takes place during this life stage includes the desire for independence as adolescents develop individual identities apart from their families. One way that teenagers assert their independence is by choosing what to eat. They have their own money to purchase food and tend to eat more meals away from home. Too many poor choices can make young people nutritionally vulnerable.

At this life stage, young people still need the structure of family meals. Evidence shows that eating family meals is associated with nutritional benefits, including eating a diet with more fruits, vegetables, fiber, and micronutrients, and less fried food, soda, and saturated and trans fat.

REFERENCES

National Institute of Health: Osteoporosis and Related Bone Diseases National Resource Center. (2018). What People With Anorexia Nervosa Need To Know About Osteoporosis. Retrieved from https://www.bones.nih.gov/health-info/bone/osteoporosis/conditions-behaviors/anorexia-nervosa

Hales, C.M., Carroll, M.D., Fryar, C.D., Ogden, C.L. (2017). Prevalence of Obesity Among Adults and Youth: United States, 2015–2016. National Center for Health Statistics. NCHS Data Brief, No. 288. Retrieved from https://www.cdc.gov/nchs/data/databriefs/db288.pdf

World Health Organization. (2017). Obesity and Overweight Fact Sheet. Retrieved from http://www.who.int/mediacentre/factsheets/fs311/en/.

Keller, A., & Bucher Della Torre, S. (2015). Sugar-Sweetened Beverages and Obesity among Children and Adolescents. A Review of Systematic Literature Reviews. Childhood obesity (Print), 11(4), 338–346. https://doi.org/10.1089/chi.2014.0117

Ervin R.B., Kit B.K., Carroll M.D. (2012). Consumption of Added Sugar among US Children and Adolescents, 2005–2008. National Center for Health Statistics. NCHS Data Brief, No. 87. Retrieved from http://www.cdc.gov/nchs/data/databriefs/db87.pdf.

Gillman MW, Rifas-Shiman SL, Frazier AL, et al. Family dinner and diet quality among older children and adolescents. Arch Fam Med. 2000;9:235-240. doi: 10.1001/archfami.9.3.235.

OLDER ADULTS' NUTRITION

The senior years are the period from age 51 until the end of life. A number of physiological and emotional changes take place during this life stage, and as they age, older adults can face a variety of health challenges. Blood pressure rises, and the immune system may have more difficulty battling invaders and infections. The skin becomes thinner and more wrinkled and may take longer to heal after injury. Older adults may gradually lose an inch or two in height. And short-term memory might not be as keen as it once was. However, many older adults remain in relatively good health and continue to be active into their golden years. Good nutrition is important to maintaining health later in life. In addition, the fitness and nutritional choices made earlier in life set the stage for continued health and happiness.

As noted in Unit 1, Dietary Reference Intakes (DRIs) vary based on age. Beginning at age 51, nutrient requirements for adults change in order to fit the nutritional issues and health challenges that older people face. Because the process of aging affects nutrient needs, some requirements for nutrients decrease as a person ages, while requirements for other nutrients increase. On this page, we will take a look at the changing nutrient requirements for older adults as well as some special concerns for the aging population.

Nutrient Needs In Older Adults energy and macronutrients

Due to reductions in lean body mass and metabolic rate, older adults have lower calorie needs than younger adults. The energy requirements for people ages 51 and older are 1,600 to 2,200 calories for women and 2,000 to 2,800 calories for men, depending on activity level. The decrease in physical activity that is typical of older adults also influences nutrition requirements. The AMDRs for carbohydrates, protein, and fat remain the same from middle age into old age. Older adults should substitute more unrefined carbohydrates, such as whole grains, for refined ones. Fiber is especially important in preventing constipation and diverticulitis, which is more common as people age, and it may also reduce the risk of colon cancer. Protein should be lean, and healthy fats, such as omega-3 fatty acids, are a part of any good diet.

Micronutrients

The recommended intake levels of several micronutrients are increased in older adulthood, while others are decreased. A few nutrient changes to note include the following:

- To slow bone loss, the recommendations for calcium increase from 1,000 milligrams per day to 1,200 milligrams per day for both men and women.
- Also to help protect bones, vitamin D recommendations increase from 600 IU to 800 IU per day for men and women.
- Vitamin B_6 recommendations rise to 1.7 milligrams per day for older men and 1.5 milligrams per day for older women to help lower levels of homocysteine and protect against cardiovascular disease.
- Due to a decrease in the production of stomach acid, which can lead to an overgrowth of bacteria in the small intestine and decrease absorption of vitamin B_{12}, older adults need an additional 2.4 micrograms per day of B_{12} compared to younger adults.
- For elderly women, higher iron levels are no longer needed post-menopause, and recommendations decrease from 18 milligrams per day to 8 milligrams per day.

Common Health Concerns In Older Adults

Older adults may face serious health challenges in their later years, many of which have ties to nutrition.

- Increased occurrence of cancer, heart disease, and diabetes
- Loss of hormone production, bone density, muscle mass, and strength, as well as changes in body composition (increase of fat deposits in the abdominal area, increasing the risk for type 2 diabetes and cardiovascular disease)
- Increased occurrence of dementia, resulting in memory loss, agitation and delusions
- Decreased kidney function, becoming less effective in excreting metabolic products such as sodium, acid, and potassium, which can alter water balance and hydration status
- Decreased immune function, resulting in more susceptibility to illness
- Increased risk for neurological disorders and psychological conditions (e.g., depression), influencing attitudes toward food, along with the ability to prepare or ingest food

- Dental problems can lead to difficulties with chewing and swallowing, which in turn can make it hard to maintain a healthy diet
- Lower efficiency in the absorption of vitamins and minerals
- Being either underweight or overweight

Nutrition Concerns for Older Adults

Dietary choices can help improve health during this life stage and address some of the nutritional concerns that many older adults face. In addition, there are specific concerns related to nutrition that affect adults in their later years. They include medical problems, such as disability and disease, which can impact diet and activity level.

Sensory Issues

At about age 60, taste buds begin to decrease in size and number. As a result, the taste threshold is higher in older adults, meaning that more of the same flavor must be present to detect the taste. Many elderly people lose the ability to distinguish between salty, sour, sweet, and bitter flavors. This can make food seem less appealing and decrease appetite. Intake of foods high in sugar and sodium can increase due to an inability to discern those tastes. The sense of smell also decreases, which impacts attitudes toward food. Sensory issues may also affect digestion, because the taste and smell of food stimulates the secretion of digestive enzymes in the mouth, stomach, and pancreas.

Dysphagia

Some older adults have difficulty getting adequate nutrition because of the disorder dysphagia, which impairs the ability to swallow. Stroke, which can damage the parts of the brain that control swallowing, is a common cause of dysphagia. Dysphagia is also associated with advanced dementia because of overall brain function impairment. To assist older adults suffering from dysphagia, it can be helpful to alter food consistency. For example, solid foods can be pureed, ground, or chopped to allow more successful and safe swallowing. This decreases the risk of aspiration, which occurs when food flows into the respiratory tract and can result in pneumonia. Typically, speech therapists, physicians, and dietitians work together to determine the appropriate diet for dysphagia patients.

Obesity in Old Age

Similar to other life stages, obesity is a concern for the elderly. Adults over age 60 are more likely to be obese than young or middle-aged adults. Reduced muscle mass and physical activity mean that older adults need fewer calories per day to maintain a normal weight. Being overweight or obese increases the risk for potentially fatal conditions that can afflict older adults, particularly cardiovascular disease and type 2 diabetes. Obesity is also a contributing factor for a number of other conditions, including arthritis.

For older adults who are overweight or obese, dietary changes to promote weight loss should be combined with an exercise program to protect muscle mass. This is because dieting reduces muscle as well as fat, which can exacerbate the loss of muscle mass due to aging. Although weight loss among the elderly can be beneficial, it is best to be cautious and consult with a healthcare professional before beginning a weight loss program.

The Anorexia of Aging

In addition to concerns about obesity among senior citizens, being underweight can be a major problem. A condition known as the anorexia of aging is characterized by poor food intake, which results in dangerous weight loss. This major health problem among the elderly leads to a higher risk for immune deficiency, frequent falls, muscle loss, and cognitive deficits. It is important for health care providers to examine the causes for anorexia of aging, which can vary from one individual to another. Understanding why some elderly people eat less as they age can help healthcare professionals assess the risk factors associated with this condition. Decreased intake may be due to disability or the lack of a motivation to eat. Also, many older adults skip at least one meal each day. As a result, some elderly people are unable to meet even reduced energy needs.

Nutrition interventions for anorexia of aging should focus primarily on a healthy diet. Remedies can include increasing the frequency and variety of meals and adding healthy, high-calorie foods (such as nuts, potatoes, whole-grain pasta, and avocados) to the diet. The use of flavor enhancements with meals and oral nutrition supplements between meals may help to improve caloric intake.[1] Health care professionals should consider a patient's habits and preferences when developing a nutrition treatment plan. After a plan is in place, patients should be weighed on a weekly basis until they show improvement.

Vision Problems

Many older people suffer from vision loss and other vision problems. Age-related macular degeneration is the leading cause of blindness in Americans over age sixty.[2] This disorder can make food planning and preparation extremely difficult, and people who suffer from it often depend on caregivers for their meals. Self-feeding also may be difficult if an elderly person cannot see their food clearly. Friends and family members can help older adults with shopping and cooking. Food-assistance programs for older adults (such as Meals on Wheels) can also be helpful. Diet may also help to prevent macular degeneration. Consuming colorful fruits and vegetables increases the intake of lutein and zeaxanthin, two antioxidants that provide protection for the eyes.

Longevity and Nutrition

Bad habits and poor nutrition have an accrual effect. The foods you consume in your younger years will impact your health as you age, from childhood into the later stages of life. As a result, good nutrition today means optimal health tomorrow. Therefore, it is best to start making healthy choices from a young age and maintain them as you mature. However, research suggests that adopting good nutritional choices later in life, during the 40s, 50s, and even the 60s, may still help to reduce the risk of chronic disease as you grow older.[3]

Even if past nutrition and lifestyle choices were not aligned with dietary guidelines, older adults can still do a great deal to reduce their risk of disability and chronic disease. There are a number of changes middle-aged adults can implement, even after years of unhealthy choices. Choices include eating more dark, green, leafy vegetables, choosing lean sources of protein such as lean meats, poultry, fish, beans, and nuts, and engaging in moderate physical activity for at least thirty minutes per day, several days per week. The resulting improvements will go a long way toward providing greater protection against falls and fractures, and helping to ward off cardiovascular disease and hypertension, among other chronic conditions.

Nutrition Strategies for Older Adults

What can make it harder for me to eat healthy as I age?

Some changes that can happen as you age can make it harder for you to eat healthy. These include changes in your

- Home life, such as suddenly living alone or having trouble getting around
- Health, which can make it harder for you to cook or feed yourself
- Medicines, which can change how food tastes, make your mouth dry, or take away your appetite
- Income, which means that you may not have as much money for food
- Sense of smell and taste
- Problems chewing or swallowing your food

What can I do if I am having trouble eating healthy?

Sometimes health issues or other problems can make it hard to eat healthy. Here are some tips that might help:

- If you are **tired of eating alone**, try organizing some potluck meals or cooking with a friend. You can also look into having some meals at a nearby senior center, community center, or religious facility.
- If you are **having trouble chewing**, see your dentist to check for problems
- If you are **having trouble swallowing**, try drinking plenty of liquids with your meal. If that does not help, check with your health care provider. A health condition or medicine could be causing the problem.
- If you're **having trouble smelling and tasting** your food, try adding color and texture to make your food more interesting
- If you **aren't eating enough**, add some healthy snacks throughout the day to help you get more nutrients and calories
- If an **illness is making it harder for you to cook or feed yourself**, check with your health care provider. He or she may recommend an occupational therapist, who can help you find ways to make it easier.

[4]U.S. National Library of Medicine (2020, June 23). *Nutrition for Older Adults*. MedlinePlus. Retrieved August 18, 2020, from https://medlineplus.gov/nutritionforolderadults.html

Figure 11.7. Nutrition strategies for older adults

REFERENCES

Cox, N. J., Ibrahim, K., Sayer, A. A., Robinson, S. M., & Roberts, H. C. (2019). Assessment and treatment of the anorexia of aging: A systematic review. Nutrients, 11(1), 144. https://www.ncbi.nlm.nih.gov/pmc/articles/PMC6356473/

Ehrlich, R., Harris, A., Kheradiya, N. S., Winston, D. M., Ciulla, T. A., & Wirostko, B. (2008). Age-related macular degeneration and the aging eye. Clinical interventions in aging, 3(3), 473.

Rivlin, R. S. (2007). Keeping the young-elderly healthy: Is it too late to improve our health through nutrition? The American journal of clinical nutrition, 86(5), 1572S-1576S.

U.S. National Library of Medicine (2020, June 23). Nutrition for Older Adults. MedlinePlus. Retrieved August 18, 2020, from https://medlineplus.gov/nutritionforolderadults.html

GLOSSARY

Absorption: The process of nutrients traveling from the lumen of the GI tract to the blood or lymph after digestion.

Acceptable macronutrient distribution ranges (AMDR): The calculated range of how much energy from carbohydrate, fat, and protein is recommended for a healthy diet.

Accessory organs: Organs that are not part of the intestinal tract itself, but have ducts that deliver digestive juices into the tract to help aid in digestion; includes the salivary glands, liver, gallbladder, and pancreas.

Acetyl CoA: An important molecule for many metabolic pathways, including the Krebs Cycle in aerobic metabolism.

Active vitamin A: Another term for retinol found in animal-derived foods.

Added sugars: Concentrated sweeteners that are added as ingredients to foods to make them sweeter; they decrease a food's nutrient density.

Adenosine triphosphate (ATP): The energy-containing molecule found in the cells of all animals and humans; powers cellular work.

Adequate: Providing sufficient amounts of calories and each essential nutrient, as well as fiber.

Adequate Intake (AI): Nutrient recommendations set when there is not enough evidence to establish an RDA; based on observing healthy people and seeing how much of the nutrient in question they are consuming.

Aerobic Activity: Continuous exercise (lasting more than 2 minutes) that increases heart and breathing rate (e.g., walking, jogging, biking) and primarily relies on energy generated through aerobic metabolism.

Aerobic Metabolism: The metabolic pathways that require oxygen to generate ATP for cells.

Air Displacement Plethysmography (ADP): A non-invasive, quick, but more expensive tool to estimate body composition by measuring air displacement when a person sits in an enclosed chamber.

Albumin: Butterfly-shaped protein; has many functions in the body including maintaining fluid and acid-base balance and transporting molecules.

Allergen: A foreign substance that causes an immune response in the body.

Alpha-Linolenic Acid: An essential omega-3 fatty acid.

Amino Acids: The building blocks of protein.

Amphiphilic: Molecules that are both water- and fat-soluble.

Anabolic Pathways: The synthesis of larger molecules, which requires energy.

Anaerobic Exercise: Activities that consist of short duration, high intensity movements that rely on immediately available energy sources and require little or no oxygen during the activity.

Anaerobic Metabolism: The metabolic system that can generate ATP without oxygen.

Anaphylaxis: A severe allergic reaction involving more than one organ system (e.g., a rash coupled with difficulty breathing).

Anemia: A condition in which oxygen-carrying capacity of red blood cells is reduced, often from a reduction in hemoglobin.

Anion: A negatively-charged electrolyte (e.g., chloride).

Anorexia Nervosa: An eating disorder in which a person obsesses about their weight and the food that they eat, resulting in extreme nutrient inadequacy and eventually organ malfunction.

Anorexia of Aging: A condition that is characterized by poor food intake, which results in dangerous weight loss in older adults.

Antibiotic Resistance: When bacteria develop resistance to antibiotics, and the antibiotics no longer work to kill the bacteria causing infections.

Antibodies: Proteins that circulate in blood, recognize harmful intruders like bacteria and viruses, and surround and destroy them.

Antioxidants: Molecules that can donate an electron to stabilize and neutralize free radicals.

Apolipoproteins: Found on the surface of lipoproteins; both fat- and water-soluble.

Appetite: The psychological desire to eat.

Aspiration: Occurs when food flows into the respiratory tract and can result in pneumonia.

Atherosclerosis: The narrowing of arteries due to buildup of plaque.

Atoms: The fundamental unit of matter; the smallest unit of an element.

Authorized Health Claims: Claims that have stronger scientific evidence to back them than qualified health claims.

Autoimmune Diseases: A disease where the immune system produces antibodies that attack and damage the body's own tissues.

Balance: An element of physical fitness that is the ability to maintain equilibrium during movement or while stationary.

Balanced: Including a combination of foods from the different food groups.

Basal Metabolic Rate (BMR): The energy expended by the body when at rest, to fuel the behind-the-scenes activities required to sustain life (e.g., respiration, circulation); largest component of energy expenditure.

Behavioral Weight Loss Interventions: Interventions that help individuals develop skills that support healthy lifestyle and body weight.

Best Weight: The weight that a person can achieve and maintain while living their healthiest and happiest life.

Beta-Carotene: A type of carotenoid; a precursor to active vitamin A.

Bile: A chemical made by the liver and stored in the gallbladder; acts as an emulsifier, which allows fat droplets to mix with the watery digestive juices in the small intestine.

Binge-Eating Disorder: An eating disorder characterized by periodic overeating with a feeling of loss of control over eating but not accompanied with fasting, purging, or compulsive exercise.

Binge-Eating Disorder (BED): An eating disorder characterized by periodic overeating with a feeling of loss of control over eating but not accompanied with fasting, purging, or compulsive exercise.

Bioavailability: The amount of a substance that is absorbed from the intestine into the bloodstream.

Bioelectric Impedance Analysis (BIA): A simple, non-invasive, quick tool that estimates body composition by sending a small amount of electricity through the body.

Biological Macromolecules: The raw materials used to build living organisms; formed when atoms of carbon, hydrogen, oxygen, and nitrogen bond with each other in unique and varied ways.

Blinding: technique to prevent bias in intervention studies, where either the research team, the subject, or both don't know what treatment the subject is receiving.

Body Composition: The proportion of a person's body mass that is made up of water, lean mass, fat, and other components.

Body Mass Index (BMI): A simple formula expressing the ratio of body weight and height; an inexpensive screening tool used in clinical and research settings to estimate body size and health risk.

Bolus: What food is referred to once it has been chewed and moistened.

Bone Mineral Density (BMD): A test that can detect osteoporosis and predict the risk of bone fracture.

Bran: The outer skin of a wheat kernel; contains antioxidants, B vitamins, and fiber.

Brush Border: Another name for microvilli, because their appearance resembles the bristles on a brush.

Bulimia Nervosa: An eating disorder characterized by episodes of eating large amounts of food followed by purging, which is accomplished by vomiting and with the use of laxatives and diuretics.

Calcidiol: A precursor to the active form of vitamin D; the circulating form of vitamin D and the form measured in blood to assess a person's vitamin D status.

Calcitonin: A hormone secreted by the thyroid gland; decreases blood calcium levels.

Calcitriol: Activated vitamin D; increases the absorption of calcium in the intestine and works with PTH to release calcium from bone and reduce calcium loss in urine.

Carbohydrates: Macromolecules composed of carbon, hydrogen, and oxygen; major fuel source for all cells of the body.

Cardiorespiratory Fitness: An element of physical fitness developed by perform large-muscle, whole-body exercise at moderate-to-vigorous intensities for extended periods of time.

Carotenoid: Brightly-colored yellow, orange, and red pigments synthesized by plants.

Case-Control Studies: Research that compares a group of cases and controls, looking for differences between the two groups that might explain their different health outcomes.

Catabolic Pathways: The breakdown of complex molecules into simpler molecules to generate energy.

Cation: A positively-charged electrolyte (e.g., sodium).

Celiac Disease: An autoimmune disorder; the body has an abnormal immune reaction to gluten; results in damage to the villi, decreasing the surface area for nutrient absorption.

Cell Membrane: A thin covering around the cell that separates the internal and external environments.

Cells: The smallest structural and functional units of all living things.

Cellular Differentiation: The process by which cells change from stem cells to more specialized cells with specific structure and function.

Cellular Respiration: A key pathway of energy metabolism occurring in cells of aerobic organisms; results in the production of ATP.

Cellulose: One of the most common types of fiber; the main component in plant cell walls.

Chemical Bond: The attractive force between atoms; contains energy.

Chemical Digestion: The breakdown of macronutrients into their chemical building blocks (e.g., starch into glucose) with the aid of enzymes.

Cholecalciferol: Also known as vitamin D3, this type of vitamin D is made by the skin when exposed to UV light and found in animal-derived foods.

Cholecystokinin (CCK): A hormone that is secreted in response to nutrients in the gut, especially fat and protein, and signals satiety; aids in nutrient digestion by inhibiting food intake and stimulating pancreatic secretions, gall bladder contractions, and intestinal motility.

Cholesterol: The most well-known sterol; only found in animal fats.

Cholesterol Esters: Cholesterol molecules with a fatty acid attached; part of the structure of lipoproteins.

Chylomicrons: A type of lipoprotein that serves as a transport vehicle for lipids absorbed from the small intestine into lymph and blood.

Chyme: A semiliquid mass of partially digested food and gastric juices.

Chymotrypsin: An enzyme made by the pancreas; facilitates the chemical breakdown of proteins in the small intestine.

Coagulation: Also known as clotting; the process by which blood changes from a liquid to a semi-liquid or gel to form a clot.

Coenzymes: Organic molecules required by enzymes to catalyze a specific reaction.

Cofactors: Inorganic minerals that assist in enzymatic reactions.

Cohort Studies: Research that follows a group of people (a cohort) over time, measuring factors such as diet and health outcomes.

Collagen: A protein important to the strength and structure of muscles, bones, tendons, ligaments, connective tissue, and skin.

Colostrum: Breast milk produced in small quantities for the first two to five days after the birth of the baby; low in fat, easily digestible, rich in protein, fat-soluble vitamins, and minerals, as well as immunoglobulins that provide immune protection to the newborn.

Compact Bone: Also known as cortical bone, this dense bone tissue surrounds all spongy bone and makes up approximately 80 percent of the adult skeleton.

Complementary Feeding: The time in later infancy when solid foods are added to the baby's diet, while breastmilk and/or formula continue to be the nutritional foundation of the diet.

Complementary Proteins: Two or more incomplete protein sources that can be combined to make a complete protein.

Complete Protein Source: A food that contains all of the essential amino acids in adequate amounts needed by the body.

Complete Proteins: A high-quality protein that contains all 9 essential amino acids; found in animal foods and soy and quinoa.

Complex Carbohydrates: Large carbohydrates that contain many sugar units, also called polysaccharides; include starch, glycogen, and fiber.

Conditionally Essential Amino Acids; Amino acids that must be obtained in the diet in certain situations when more are needed than the body can synthesize.

Confounding Factors: Factors that can affect the outcome in question.

Connective Tissue: A tissue that connects, supports, and/or binds other tissues in the body.

Constipation: Infrequent bowel movements (less than 3 times per week); stools that are hard, dry, or lumpy, and often painful to pass.

Control: A group that is used as a comparison to other experimental groups within a research study.

Correlations: Relationships between two factors (e.g., nutrition and health).

Covalent Bond: The strongest, most stable type of chemical bond in the biological world, made from the sharing of electrons.

Creatine Phosphate: A high-energy molecule that can be used to generate ATP for cells during the first 10 seconds of an activity.

Crohn's Disease: A chronic inflammatory disease that can affect any part of the GI tract.

Cross-Sectional Studies: Research that collects information about a population of people at one point in time.

Cytoplasm: The cellular fluid within a cell.

Daily Value (DV): An approximate recommendation for daily intake for a nutrient, developed by the FDA for use on food labels; allows consumers to see how much of a nutrient is provided by a serving of a food relative to about how much they need each day.

Deamination: A process that removes nitrogen from amino acids before they are used to synthesize ATP, glucose, or fat.

Dehydration: A significant water loss in the body, such that the body doesn't have enough fluids to function properly.

Denaturation: When the three-dimensional structure of a protein is unfolded due to a change in the environment (e.g., acid, heat); results in loss of protein function.

Diabetes: A chronic disease in which the body isn't able to regulate blood glucose; includes type 1, type 2, and gestational diabetes.

Diarrhea: Loose, watery stools.

Diet Culture: Beliefs continuously promoted by diet companies, media, and influencers that center the idea that being thin is the same as being healthy, and that anyone who applies themselves by eating the right foods in the right amounts can achieve that ideal.

Dietary Fiber: Non-digestible carbohydrates that are found in plants.

Dietary Habits: Routines such as what a person eats, how much a person eats, how frequently meals are consumed, and how often a person eats out.

Dietary Reference Intakes (DRI): A set of recommendations developed by the National Academies of Sciences, Engineering, and Medicine to describe the amounts of specific nutrients and energy that people should consume in order to stay healthy.

Dietary Supplement: A product that is intended to supplement the diet and can be taken by mouth; contains one or more dietary ingredients (e.g., vitamins, minerals, herbs or other botanicals, amino acids, or enzymes).

Dietitian: A healthcare professional who has registered credentials (requiring a bachelor's or master's degree in dietetics, completion of a dietetic internship, passing a national exam, and maintenance of registration through ongoing continuing education) and can provide nutritional care in the areas of health and wellness.

Disaccharides: A simple carbohydrate that contains two sugar units; includes maltose, sucrose, and lactose.

Division of Responsibility: The gold standard approach to feeding children, defining the optimal relationship between parent and child when it comes to feeding: the parent determines the what, where, and when of feeding, and the child chooses how much to eat and whether to eat from the foods provided.

Docosahexaenoic Acid (DHA): An important omega-3 fatty acid that can be synthesized in the body; helps lower blood triglycerides and blood pressure, reduce inflammation, prevent blood clot formation, and promote normal growth and development in infants.

Double-Blind: Neither the research team nor the subject know what treatment the subject is receiving.

Dual Energy X-Ray Absorptiometry: dual energy X-ray absorptiometry (DXA) One of the most accurate but more expensive methods of measuring body composition; this method scans the body with low-dose X-ray beams to determine fat, muscle, and bone mass.

Dual-Energy X-Ray Absorptiometry (DXA): One of the most accurate but more expensive methods of measuring body composition; this method scans the body with low-dose X-ray beams to determine fat, muscle, and bone mass.

Duodenum: The first segment of the small intestine.

Dysphagia: Difficulty swallowing.

Eating Competence: An evidence-based model developed by Ellyn Satter; based on the principle that internal cues of hunger, appetite, and satiety are reliable and can be used to inform food selection and guide energy balance and body weight.

Eicosanoids: A large family of important signaling molecules, with roles such as regulating inflammation.

Eicosapentaenoic Acid (EPA): An important omega-3 fatty acid that can be synthesized in the body; it helps lower blood triglycerides and blood pressure, reduce inflammation, prevent blood clot formation, and promote normal growth and development in infants.

Elastin: A protein that gives connective tissue its elasticity and flexibility.

Electrolytes: Substances that dissociate into charged ions when dissolved in water (e.g., sodium chloride (NaCl) dissociates into sodium (Na^+) and chloride (Cl^-) in water).

Electrons: Small, negative particles that are found outside of the nucleus of an atom in regions called shells.

Element: Substance made entirely from one specific type of atom.

Empty Calories: Calories from solid fats and/or added sugars, adding few other nutrients; make foods less nutrient dense.

Emulsifiers: Chemicals that allow fat to mix with watery liquids.

Endoplasmic Reticulum: An organelle that processes and packages proteins and lipids for transport.

Endoscopy: A procedure in which a camera is inserted into the GI tract to visualize the interior.

Endosperm: The largest part of the wheat kernel; provides energy in the form of starch to support reproduction.

Energy Balance: When energy intake equals energy expenditure; weight should remain stable.

Energy Metabolism: The process of converting food into energy.

Energy-Yielding: Nutrients that provide energy to the body; include carbohydrates, proteins, and fats.

Enriched: Food ingredients with added nutrients; usually refined grains that have lost naturally-occurring nutrients during processing.

Enterocytes: Absorptive cells that line the small intestine.

Enzymes: Proteins that help speed up or facilitate chemical reactions in the body; they bring together two compounds to react, without undergoing any changes themselves.

Epithelial Tissue: A tissue that lines and protects organs.

Ergocalciferol: Also known as vitamin D2, this type of vitamin D is made by plants, mushrooms, and yeast.

Ergogenic Aids: Substances used to enhance performance.

Esophagus: A muscular tube that transports food from the mouth to the stomach.

Essential Amino Acids: Amino acids that must be obtained in the diet because they can't be synthesized by the body in sufficient amounts.

Essential Fatty Acids: Fatty acids that can't be synthesized in the body and must be consumed in the diet.

Estimated Average Requirement (EAR): The amount of a nutrient that meets the requirements of 50 percent of people within a group of the same life stage and sex.

Estimated Energy Requirement (EER): An estimate of how many calories a person needs to consume, on average, each day to stay healthy, based on their age, sex, height, weight, and physical activity level.

Excretion: The final step of digestion; undigested materials are removed from the body.

Exercise-Related Activity Thermogenesis (EAT): Planned, structured, and repetitive physical activity with the objectives such as improving health or fitness or having fun (e.g., strength training, sports).

Extracellular Fluid (ECF): The fluid outside of cells; includes both blood plasma and interstitial fluid.

Fat-Soluble Vitamins: Vitamins that dissolve in fat; include vitamins A, D, E, and K.

Fatty Acids: Long chains of carbon and hydrogen molecules with an acid (-COOH) at one end.

Feces: Undigested materials that are stored in the anus until they are excreted from the body.

Ferritin: A storage form of iron.

Fiber: A polysaccharide made by plants to provide protection and structural support.

FITT: A helpful tool for putting together an exercise plan that includes considering the frequency, intensity, time, and type of exercise.

Flexibility: An element of physical fitness developed through stretching and other activities, enhancing the ability of joints to move through the full range of motion.

Fluorosis: A condition caused by excessive intake of fluoride; characterized by mottling (i.e., white speckling) and pitting of the teeth.

Folic Acid: A synthetic form of folate.

Food: A substance, usually derived from plants, animals, or fungi, that is consumed and provides nutrients.

Food Allergy: The result of the immune system mistakenly identifying a food protein as an invasive threat; usually causes immediate symptoms, such as rash, swollen face or throat, or difficulty breathing.

Food Deserts: Areas where healthy foods simply aren't available or easily accessible.

Food Insecurity: Having inconsistent access to enough food for an active, healthy life.

Food Intolerance: Occurs when a person has difficulty digesting a specific food or nutrient (due to missing a specific digestive enzyme); symptoms include unpleasant GI symptoms such as gas, bloating, flatulence, cramping, and diarrhea.

Food Security: Having consistent access to enough food for an active, healthy life.

Free Radical: An atom or group of atoms with an unpaired electron; they are unstable, highly reactive, and can be damaging in excess.

Frequency: A component of the FITT acronym that describes how often you exercise.

Fructose: A monosaccharide that is one of the sweetest sugars; found in fruits, vegetables, honey, and high fructose corn syrup.

Galactose: A monosaccharide that is rarely found in food alone, but found in dairy products bonded to glucose.

Gallbladder: An accessory organ located behind the liver; stores, concentrates, and secretes bile.

Gastric Lipase: An enzyme produced by cells of the stomach; aids in the chemical breakdown of triglycerides.

Gastroesophageal Reflux (GER): When the acidic chyme in the stomach escapes back into the esophagus.

Gastroesophageal Reflux Disease (GERD): A disease diagnosis when heartburn occurs more than twice a week.

Gene Expression: The process by which the instructions contained in DNA are used to synthesize a product such as a protein.

Germ: The embryo of the seed, which can sprout into a new plant; contains B vitamins, protein, minerals, and healthy fats.

Gestational Diabetes: A type of diabetes that develops during pregnancy in women who did not previously have diabetes.

Ghrelin: Known as the "hunger hormone"; produced in the stomach and communicates hunger to the brain.

Gi Tract: A one-way tube about 25 feet in length, beginning at the mouth and ending at the anus; where digestion takes place.

Glucagon: A hormone made by the pancreas and released when blood glucose is low; it causes glycogen in the liver to break down, results in raising blood glucose concentrations.

Gluconeogenesis: The synthesis of new glucose molecules from amino acids.

Glucose: A monosaccharide that is a product of photosynthesis and an important fuel source for the body; found in fruits and vegetables, honey, corn syrup, and high fructose corn syrup.

Gluten: A group of proteins found in wheat, rye, and barley; in people with celiac disease, gluten causes antibodies to attack the cells lining the small intestine.

Glycerol: The three-carbon backbone of triglycerides.

Glycogen: A polysaccharide made up of long, branched chains of glucose; a storage form of carbohydrate in animals.

Glycolysis: A cycle that occurs in the cytoplasm of the cell and plays a central role in the production of energy through anaerobic metabolism.

Golgi Apparatus: An organelle that distributes macromolecules like proteins and lipids around the cell.

Health: The state of complete physical, mental, and social well-being and not merely the absence of disease or infirmity.

Health at Every Size (HAES): An approach that aims to decrease our culture's obsession with body size and weight, decrease weight discrimination and stigma, and instead promote size acceptance and inclusivity.

Health Claims: Statements on food packaging that link the food or a component in the food to reducing the risk of a disease.

Health Equity: The fair and just opportunity for all to be as healthy as possible.

Heartburn: A burning, often painful, sensation in the chest or throat; caused by gastroesophageal reflux.

Helicobacter Pylori (H. pylori): Bacteria that cause peptic ulcers.

Heme Iron: Iron that is part of the proteins hemoglobin and myoglobin; most bioavailable form of iron, found only in animal foods.

Hemoglobin: A protein in red blood cells that transports oxygen to cells and gives red blood cells their color.

Hemorrhage: Excessive bleeding.

High-Density Lipoproteins (HDL): A type of lipoprotein that picks up cholesterol from the body's cells and returns it to the liver for disposal; sometimes called "good cholesterol," because high blood levels are associated with a lower risk of cardiovascular disease.

Homeostasis: The body's ability to maintain equilibrium or a steady state, such as maintaining blood glucose concentration.

Homocysteine: An amino acid found in blood.

Hormones: Chemical messengers that are produced by endocrine glands and travel in the blood to target cells to initiate a specific reaction or cellular process.

Hunger: The physiological need to eat.

Hydrochloric Acid: An acid that is a component of gastric juices; creates an acidic environment in the stomach, killing bacteria and aiding in protein digestion.

Hydrogenation: The process of adding hydrogen to the carbon-carbon double bonds of a fatty acid, thus making it more saturated.

Hydrolysis: The splitting of one molecule into two with the addition of water; the main chemical reaction in digestion.

Hydroxyapatite: Tiny crystals made from inorganic minerals that form around collagen fibers to provide strength to bones.

Hydroxylations: Chemical reactions that add a hydroxyl (-OH) group to a compound.

Hypercalcemia: High blood calcium.

Hyperglycemia: High blood sugar.

Hyperkeratosis: Thickening of the outer layer of the skin due to overproduction of the protein keratin; can be a symptom of vitamin A deficiency.

Hyperlipidemia: An elevated level of lipids, including triglycerides and cholesterol, in the blood.

Hypernatremia: Elevated blood sodium concentration.

Hypertension: High blood pressure.

Hypochromic Anemia: A type of anemia characterized by low color red blood cells.

Hypoglycemia: Low blood sugar.

Hyponatremia: Low blood sodium concentration.

Hypothalamus: A region in the brain responsible for coordinating many basic functions to maintain homeostasis in the body, including control of hunger and satiety.

Ileum: The final segment of the small intestine, where the majority of absorption occurs.

Immunoglobulin E (IgE): A type of antibody produced by the immune system in response to a specific substance.

In Vitro Study: Experiments that are conducted outside of living organisms, within flasks, dishes, plates, or test tubes.

In Vivo: Research that is conducted in living organisms, such as rats and mice.

Incomplete Protein Source: A food that does not contain all of the essential amino acids in adequate amounts needed by the body.

Incomplete Proteins: A lower quality protein that does not contain all 9 essential amino acids in proportions needed to support growth and health; found in plant foods.

Inflammatory Bowel Disease (IBD): Includes two types of disorders — ulcerative colitis and Crohn's disease.

Information Literacy: The ability to find information, evaluate whether it is accurate and useful, and apply it effectively.

Ingestion: The first process of digestion; the entry of food into the GI tract through the mouth.

Inorganic: Nutrients that do not contain both carbon and hydrogen; can not be created or destroyed.

Insoluble Fiber: A type of fiber that does not typically dissolve in water and helps prevent constipation; lignin, cellulose, and hemicellulose are common types of insoluble fibers, and food sources include wheat bran, vegetables, fruits, and whole grains.

Insulin: A hormone made by the pancreas and released when blood glucose is high; it results in lowering blood glucose concentrations.

Insulin Resistance: A condition where cells stop responding to insulin.

Intensity: A component of the FITT acronym that describes how hard you work during your exercise session.

Intermediate-Density Lipoproteins (IDL): A type of lipoprotein that is created as triglycerides are removed from VLDL.

Interstitial Fluid (IF): The fluid that surrounds cells.

Intervention Studies: Research that includes some type of treatment or change imposed by the researchers; sometimes called experimental studies or clinical trials.

Intracellular Fluid (ICF): The fluid contained within cells.

Intrinsic Factor: A protein secreted in the stomach that is necessary for vitamin B_{12} absorption.

Iron-Deficiency Anemia: A condition that develops from having insufficient iron levels in the body, resulting in fewer and smaller red blood cells that contain less hemoglobin.

Irritable Bowel Syndrome (IBS): A type of functional GI disorder that is caused by a disruption in the signals between the brain and gut; symptoms include abdominal pain, bloating, the feeling of not being able to finish a bowel movement, as well as diarrhea or constipation or both, often in cycles.

Junk Science: Untested or unproven claims or ideas, usually meant to push an agenda, sell a product, or promote special interests.

Keratin: A strong, fibrous protein; an important component of skin, hair, and nails.

Ketoacidosis: A condition which results from too many ketones accumulating in the blood, resulting in the blood being too acidic.

Ketone Bodies: Another name for ketones.

Ketones: Compounds that are made when fatty acid breakdown is high and glucose is limited.

Ketosis: The accumulation of ketones in the blood.

Krebs Cycle: A cycle that occurs in the mitochondria of cells and plays a central role in the production of energy through aerobic metabolism.

Kwashiorkor: A protein deficiency characterized by swelling of the feet and abdomen, poor skin health, poor growth, low muscle mass, and liver malfunction.

Lactase: An enzyme produced by the enterocytes; breaks lactose into its building blocks, glucose and galactose.

Lactose: A disaccharide made of a glucose molecule bonded to a galactose molecule; found in dairy products like milk, yogurt, and cheese.

Lactose Intolerance: A common food intolerance; not enough of the enzyme lactase is produced to effectively digest the milk sugar lactose.

Large Intestine: The part of the GI tract that lies between the small intestine and the anus; water absorption occurs here.

Leptin: A hormone produced by adipose tissue; its production increases as fat stores increase, and it communicates to the brain to suppress hunger and increase energy expenditure.

Lingual Lipase: An enzyme produced by cells on the tongue; begins the chemical breakdown of triglycerides.

Linoleic Acid: An essential omega-6 fatty acid.

Lipases: A group of enzymes that facilitate the chemical breakdown of triglycerides.

Lipid Panel: A standard blood test that reports total cholesterol, LDL, HDL, and triglycerides.

Lipids: A family of organic compounds that are mostly insoluble in water; the three main types are triglycerides, sterols, and phospholipids.

Lipoprotein Lipase: An enzyme that sits on the surface of cells that line the blood vessels; it breaks down triglycerides, allowing fatty acids and glycerol to enter nearby cells.

Lipoproteins: Transport vehicles for moving water-insoluble lipids around the body.

Liver: An accessory organ located just under the rib cage on the right side of the abdomen; produces bile and helps get rid of toxins.

Low-Density Lipoproteins (LDL): A type of lipoprotein that delivers cholesterol to the body's cells; sometimes called "bad cholesterol," because high blood LDL is a risk factor for atherosclerosis.

Lumen: The interior space of the GI tract where digestion and absorption occur.

Lysosomes: An organelle that breaks down macromolecules and destroys foreign invaders.

Lysozyme: An enzyme secreted in the saliva; attacks the walls of bacteria, causing them to rupture.

Macrocytic Anemia: A type of anemia characterized by larger and fewer red blood cells; caused by folate deficiency.

Macronutrients: Nutrients that are needed in large amounts and include carbohydrates, lipids, and proteins.

Major Minerals: Minerals required by the body in amounts greater than 100 milligrams per day.

Malnutrition: A lack of proper nutrition, which can be caused by not getting enough or getting too much food or nutrients.

Maltase: An enzyme produced by the enterocytes; breaks maltose into two glucose molecules.

Maltose: A disaccharide made of two glucose molecules bonded together; found in sprouted grains.

Marasmus: A protein and calorie deficiency characterized by an extreme emaciated appearance, poor skin health, poor growth, and increased risk of infection.

Mastication: Chewing; increases the surface area of the food and allows for food to be broken into small enough pieces to be swallowed safely.

Matter: Anything that has mass and takes up space.

Mechanical Digestion: The physical process of making food particles smaller to increase both surface area and mobility without changing the chemical nature of the food (e.g., mastication and peristalsis).

Meta-Analysis: A type of systematic review that combines data from multiple studies and uses statistical methods to summarize it, as if creating a mega-study from many smaller studies.

Metabolic Health: The body's ability to maintain normal homeostasis and effectively regulate measures like blood pressure, blood lipids, and blood glucose.

Metabolically Healthy Obese (MHO): metabolically obese normal weight (MONW)

Micelles: Structures that consist of bile salts clustered around the products of fat digestion; aid in absorption of fats into enterocytes.

Microcytic Anemia: A type of anemia characterized by small red blood cells.

Micronutrients: Nutrients required by the body in smaller amounts; include all of the essential minerals and vitamins.

Microvilli: Hair-like projections that line the enterocytes' cell membrane to increase surface area.

Minerals: Inorganic elements classified according to how much the body requires.

Mitochondria: An organelle often called the powerhouse of the cell; generates usable energy for the cell from energy-yielding nutrients.

Modeling: Dismantling of bone tissue at one site and building up at another, changing the shape of the bone.

Moderation: Avoiding extremes, neither too much nor too little of any one food or nutrient.

Molecules: A group of two or more atoms held together by chemical bonds.

Monosaccharides: The smallest of the carbohydrates, containing just one sugar unit; include glucose, fructose, and galactose.

Monounsaturated Fatty Acid: A fatty acid with one double bond.

Mouth: The oral cavity where ingestion and mastication occur.

Muscle: A tissue that contracts to provide movement and support.

Musculoskeletal Fitness: An element of physical fitness that is developed through the integrated function of muscle strength, muscle endurance, and muscle power to enable performance.

Mutual Supplementation: The process of combining complementary proteins.

Myelin Sheath: A cover that surrounds and protects nerve cells and allows electrical impulses to transmit quickly and efficiently.

Myoglobin: A protein similar to hemoglobin but found in muscles; provides oxygen to muscles.

Naturally-Occurring Sugars: Sugars found naturally in foods such as fruits, veggies, and dairy; come packaged with fiber, vitamins, and minerals.

Negative Energy Balance: When energy intake is less than energy expenditure; usually results in weight loss.

Negative Nitrogen Balance: When the amount of excreted nitrogen is greater than that consumed, meaning that the body is breaking down protein to meet its demands.

Nerve: A tissue that responds and reacts to signals in the environment.

Neutrons: Small, neutral particles that are found in the nucleus of an atom.

Night Blindness: A condition that makes it difficult to see in low-light conditions.

Nitrogen Balance: When the amount of nitrogen consumed equals the amount of nitrogen excreted.

Non-Celiac Gluten Sensitivity (NCGS): A diagnosis when people test negative for celiac disease but still believe that consuming gluten is causing symptoms.

Non-Diet Approaches: Approaches that focus on establishing a healthy relationship with food and more body acceptance and positivity regardless of shape and size (e.g: Satter Eating Competence Model, Health at Every Size).

Non-Exercise Activity Thermogenesis (NEAT): Energy expenditure for unstructured and unplanned activities (e.g., daily-living activities like cleaning the house); also includes the energy required to maintain posture and spontaneous movements such as fidgeting and pacing.

Non-Heme Iron: The mineral form of iron by itself, not a part of hemoglobin or myoglobin; found in foods from both plants and animals and is less bioavailable than heme iron.

Nonessential Amino Acids: Amino acids that are not required in the diet because the body can synthesize them.

Nonsteroidal Anti-Inflammatory Drugs (NSAIDs): A classification of drugs such as aspirin or ibuprofen; long-term use is the second leading cause of peptic ulcers.

Nucleus: A membrane-bound organelle within the cell that contains genetic material (DNA).

Nutrient Claims: Statements regulated by the FDA that provide straight-forward information about the level of a nutrient or calories in the food, such as "fat-free," "low calorie," or "reduced sodium."

Nutrient Density: A measure of the amount of nutrients provided by a food (especially vitamins, minerals, fiber, protein, and healthy fats) relative to the calories it contains.

Nutrients: Chemical molecules that are found in foods; required by our bodies to maintain life and support growth and health.

Nutrition: The study of how food affects the health of the body.

Nutritionist: A title commonly used to imply expertise in nutrition, but this term has no legal regulation; anyone can call themselves a nutritionist with or without proper training.

Obesity: Abnormal or excessive fat accumulation that presents a risk to health, defined by The World Health Organiztion.

Obesity Paradox: The finding that overweight or mild obesity based on BMI can be associated with better health outcomes.

Obesogenic Environments: Built environments that promote weight gain by encouraging food intake and limiting physical activity.

Observational Studies: In nutrition, research that is conducted by collecting information on people's dietary patterns or nutrient intake to look for associations with health outcomes. Observational studies do not give participants a treatment or intervention; instead, they look at what they're already doing and see how it relates to their health.

Omega-3 Fatty Acids: Fatty acids with the first double bond at the third carbon From The Omega End.

Omega-6 Fatty Acids: Fatty acids with the first double bond at the sixth carbon from the omega end.

Organ: A group of similar tissues arranged in a specific manner to perform a specific physiological function (e.g., the heart, the lungs).

Organ System: A group of two or more organs that work together to perform a specific physiological function (e.g., the digestive system, the central nervous system).

Organelles: Tiny organs within the cell that perform a specific task for the cell.

Organic: Complex nutrients that can be made by living organisms from many elements (especially carbon, hydrogen, oxygen, and sometimes nitrogen).

Organism: The highest level of organization; a complete living system capable of conducting all of life's biological processes.

Orthorexia: An extreme, unhealthy fixation on healthy or "clean" eating.

Osmosis: The diffusion of fluid through a semipermeable membrane toward higher solute concentration.

Ossification: The process of building new bone.

Osteoblasts: The cells that are responsible for building new bone.

Osteoclast: The cells that are responsible for the breakdown of bone.

Osteomalacia: A disease caused by a vitamin D deficiency in adults and characterized by softening of bones, reduced bone mineral density, and increased risk of osteoporosis.

Osteopenia: Low bone mass, which can lead to osteoporosis if untreated.

Osteoporosis: A bone disease that occurs when bone density or bone mass decreases, becoming thinner, more porous, and more susceptible to breaking.

Oxidation: The loss of electrons during a reaction.

Oxidative Stress: Free radical-induced damage that can contribute to disease.

Pancreas: An accessory organ located behind the stomach; produces and secretes pancreatic juices, which contain bicarbonate that neutralizes the acidity of the stomach-derived chyme and enzymes that further break down proteins, carbohydrates, and lipids.

Pancreatic Amylase: An enzyme secreted from the pancreas into the small intestine; continues the chemical breakdown of starch to smaller glucose chains and maltose.

Pancreatic Lipases: Enzymes produced by the pancreas; chemically break down triglycerides in the small intestine.

Parathyroid Glands: Four pea-sized glands located at the back of the thyroid gland; secrete parathyroid hormone.

Parathyroid Hormone (PTH): A hormone produced by the parathyroid glands; increases blood calcium levels.

Peak Bone Mass: The point when bones have reached their maximum strength and density.

Peer-Reviewed Manuscripts: Scientific papers that are reviewed by other experts in the field, who were not directly involved in the research, before publication.

Pepsin: An enzyme found in gastric juices; aids in the chemical breakdown of proteins.

Peptic Ulcers: Sores on the tissues lining the esophagus, stomach, or duodenum.

Peptide Bond: A special chemical bond between two amino acids.

Peristalsis: Sequential, alternating waves of contraction and relaxation of the smooth muscles in the GI tract; acts to propel food along the GI tract.

Pernicious Anemia: A type of macrocytic anemia caused by vitamin B12 malabsorption due to lack of intrinsic factor.

Phospholipids: A lipid that is both water- and fat-soluble due to the hydrophilic phosphate "head," and the hydrophobic lipid "tail."

Photosynthesis: A biological process that captures energy originating from sunlight and converts it to glucose that all organisms use to power daily functions.

Physical Activity: As a component of total energy expenditure, includes both exercise-related activity thermogenesis (EAT) and non-exercise activity thermogenesis (NEAT); contributes anywhere from 15 to 30 percent of energy expenditure.

Phytochemicals: The compounds found in plants that give them their smell, taste, and color; some have been shown to affect human health.

Pica: An intense craving for and ingestion of non-food items such as paper, dirt, or clay.

Placebo: A "fake" treatment that contains no active ingredients, such as a sugar pill.

Placebo Effect: The beneficial effect that results from a subject's belief in a treatment, not from the treatment itself.

Placenta: An organ that develops during pregnancy; provides nutrition and respiration, handles waste from the fetus, and produces hormones important to maintaining the pregnancy.

Plasma: The fluid component of the blood.

Platelets: Fragments of cells that circulate and assist in blood clotting.

Polyunsaturated Fatty Acid: A fatty acid with two or more double bonds.

Positive Energy Balance: When energy intake is greater than energy expenditure; usually results in weight gain.

Positive Nitrogen Balance: When the amount of excreted nitrogen is less than what is consumed, such as during pregnancy or growth in childhood, times the body requires more protein to build new tissues.

Prediabetes: A condition that involves insulin resistance, but not full-blown type 2 diabetes.

Preformed: preformed or active vitamin A

Preformed Vitamin A: Another term for retinol found in animal-derived foods.

Primary Structure: One-dimensional polypeptide chain of amino acids.

Prospective: Looking into the future.

Proteases: Enzymes that aid in the chemical breakdown of proteins in the small intestine.

Protein Folding: The third major step of protein synthesis; the amino acid chain folds into three-dimensional shapes.

Proteins: Macromolecules composed of chains of amino acids, which are simple subunits made of carbon, oxygen, hydrogen, and nitrogen.

Protons: Small, positive particles that are found in the nucleus of an atom.

Provitamin: A substance that can be converted into the active form of a vitamin (e.g., beta-carotene is a provitamin to active vitamin A).

Pyruvate: A 3-carbon molecule that is the end product of glycolysis.

Qualified Health Claims: Claims that have some evidence to support them, but not as much, so there's less certainty that these claims are true.

quaternary structure: Multiple folded polypeptides called subunits that have combined to make one larger functional protein.

Randomized Controlled Trial (RCT): The gold standard for intervention studies, because the research involves a control group and participants are randomized.

Recommended Daily Allowances (RDA): Nutrient recommendations that are set to meet the needs of the vast majority (97 to 98 percent) of the target healthy population.

Rectum: The last part of the GI tract; serves as a temporary holding area for feces.

Red Blood Cells: Also known as erythrocytes; transport oxygen to cells and remove carbon dioxide from cells.

refined grains: Grains and grain products made only from the endosperm.

Remodeling: Degrading and building up of bone tissue at the same location.

Resorption: The breakdown of bone.

Respiration: Breathing; taking in oxygen and removing carbon dioxide.

Responsive Feeding: An approach to feeding that is responsive to the child's cues; the caregiver recognizes child cues and responds promptly and appropriately.

Retinol: The form of vitamin A found in animal-derived foods.

Retrospective: Looking at what happened in the past.

Rhodopsin: A pigment in the eye that is especially important to vision in low-light conditions.

Ribosomes: An organelle that assembles proteins based on the instructions of DNA.

Rickets: A disease caused by vitamin D deficiency in children and characterized by soft, weak, and deformed bones.

Saliva: A mixture of water, enzymes, and other chemicals secreted from the salivary glands into the mouth.

Salivary Amylase: A digestive enzyme produced by the salivary glands; starts the chemical breakdown of starch or amylose.

Salivary Glands: The glands that make and secrete saliva.

Salt-Sensitive: A physiological trait that causes blood pressure to increase with a high-sodium diet.

Satiety: The feeling of being full.

Satter Eating Competence Model: A model developed by Ellyn Satter; focuses on eating attitudes, food acceptance, regulation of food intake and body weight, and management of the eating context.

Saturated Fatty Acid: Fatty acids in which each carbon is bonded to two hydrogen atoms, with single bonds between the carbons.

Scientific Method: An organized process of inquiry used in nutritional science, and every other science; made up of a cyclical process of steps including observation/question, hypothesis, experiment, analysis, and conclusion.

Scurvy: A disease caused by vitamin C deficiency.

Secondary Structure: Polypeptide chain that has folded into two-dimensional simple coils (also called helices) and sheets.

Segmentation: Consists of localized contractions of circular muscle of the GI tract; isolates small sections of the intestine, moving contents back and forth while continuously subdividing, breaking up, and mixing the contents; mixes food with digestive juices and facilitates absorption.

Selectively Permeable: Allowing some substances but not others to pass through freely (e.g., cell membranes allow water to cross, but other substances require special transport proteins, channels, and often energy).

Simple Carbohydrates: Small carbohydrate molecules made up of just one (monosaccharides) or two (disaccharides) sugar units.

Single-Blind Study: Either the research team or the subject know what treatment is being given, but not both.

Skinfold Test: A simple, non-invasive, and low-cost way to assess fat mass; calipers are used to measure the thickness of skin on three to seven different parts of the body, and these numbers are then entered into a conversion equation.

Small Intestine: A part of the GI tract that lies between the stomach and the large intestine; where most digestion and nutrient absorption occurs.

Social Determinants Of Health: Economic and social circumstances, such as poverty and racism, that impact health.

Soluble Fiber: A type of fiber that dissolves in water and helps to decrease blood glucose spikes and lower blood cholesterol levels; pectins and gums are common types of soluble fibers, and good food sources include oat bran, barley, nuts, seeds, beans, lentils, peas, and some fruits and vegetables.

Solutes: A dissolved substance in a solution.

Speed: The ability to move the body quickly.

Spina Bifida: A neural tube defect that occurs in a developing fetus when the spine does not completely enclose the spinal cord; caused by folate deficiency.

Spongy Bone: Also known as trabecular bone, this porous, lattice-like bone tissue makes up about 20 percent of the adult skeleton and is found at the ends of long bones, in the cores of vertebrae, and in the pelvis.

Starch: A polysaccharide made up of long chains of glucose; a storage form of carbohydrate in plants.

Sterols: Lipids that have a multi-ring structure.

Stomach: An expansion of the GI tract that links the esophagus to the first part of the small intestine; aids in both mechanical and chemical breakdown of food.

Structure-Function Claims: Vague statements about nutrients playing some role in health processes; not regulated by the FDA.

Subcutaneous Fat: Fat stored just underneath the skin.

Sucrase: An enzyme produced by the enterocytes; breaks sucrose into its building blocks, glucose and fructose.

Sucrose: A disaccharide made of a glucose molecule bonded to a fructose molecule; found in fruits, vegetables, table sugar, maple syrup, and honey.

Sugar Alcohols: A type of sugar substitute; include sorbitol, mannitol, lactitol, erythritol, and xylitol.

Sugar Substitutes: Artificial, non-nutritive, high-intensity, or low-calorie sweeteners.

Sweat Rate: The amount of fluids lost through sweat during exercise; it is calculated by measuring weight before and after exercise and is useful for determining hydration needs.

Systematic Review: Researchers formulate a research question and then systematically and independently identify, select, evaluate, and synthesize all high-quality evidence from previous research that relates to the research question.

Tertiary Structure: Polypeptide chain that has folded into a three-dimensional organized shape based on the interactions of the amino acids.

Tetany: A condition in which muscles can't relax, and instead become stiff and contract involuntarily; caused by calcium deficiency.

The Division Of Responsibility: The gold standard approach to feeding children, defining the optimal relationship between parent and child when it comes to feeding: the parent determines the what, where, and when of feeding, and the child chooses how much to eat and whether to eat from the foods provided.

Thermic Effect Of Food (TEF): The energy needed to digest, absorb, and store the nutrients in foods; accounts for 5 to 10 percent of total energy expenditure.

Thyroid Gland: A small, butterfly-shaped gland located at the base of the neck; secretes hormones that influence metabolism.

Time: A component of the FITT acronym that describes how long you exercise for.

Tissue: A group of many similar cells that share a common structure and work together to perform a specific function.

Tolerable Upper Intake Level (UL): The highest level of continuous intake of a particular nutrient that may be taken without causing health problems.

Total Energy Expenditure (TEE): The sum of daily caloric expenditure; includes basal metabolic rate (BMR), thermic effect of food (TEF), and physical activity.

Trace Minerals: Minerals required by the body in amounts of 100 milligrams or less per day.

Trans Fatty Acid: A fatty acid where the hydrogen atoms are bonded on opposite sides of the carbon chain, resulting in a more linear structure.

Transcription: The first major step of protein synthesis; a process in which the genetic code of DNA is copied into messenger RNA.

Transferrin: The transport protein for iron.

Translation: The second major step of protein synthesis; information on messenger RNA is translated into building a protein.

Triglycerides: The main form of lipids in the body and in foods; made up of three fatty acids bonded to a glycerol backbone.

Trypsin: An enzyme that facilitates the chemical breakdown of protein in the small intestine; activates other protein-digesting enzymes.

Type: A component of the FITT acronym that describes what kind of exercise you do.

Type 1 Diabetes: An autoimmune disease in which the cells of the pancreas that create insulin are destroyed.

Type 2 Diabetes: The most common type of diabetes, occurring when cells stop responding to insulin; strongly associated with abdominal obesity.

Ulcerative Colitis: A chronic inflammatory disease specific to the large intestine and rectum.

Unsaturated Fatty Acids: Fatty acids that have one or more points of unsaturation, or double bonds, between the carbons.

Variety: Consuming different foods within each of the food groups on a regular basis.

Vegans: A strict vegetarian who consumes no animal products.

Very-Low-Density Lipoprotein (VLDL): A type of lipoprotein that is made in the liver and delivers triglycerides to the body's cells.

Villi: Tiny, finger-like projections that cover the lining of the small intestine to increase surface area.

Visceral Fat: Fat surrounding vital organs and stored deep within the abdominal cavity.

Vitamins: Essential, non-caloric, organic micronutrients that are required for many bodily functions.

Waist Circumference: A measurement of waist size, taken just above the hip bone and level with the belly button and used to assess abdominal fat; waist circumferences greater than 40 inches for men and 35 inches for women are associated with greater health risks.

Waist-To-Hip Ratio: The ratio of waist and hip circumference (at its widest part), used to estimate abdominal fat; abdominal obesity is defined by the World Health Organization as waist-to-hip ratio above 0.90 for males and 0.85 for females.

Water: One of the most vital nutrients; composed of two hydrogens and one oxygen per molecule of water.

Water-Soluble Vitamins: Vitamins that dissolve in water; include vitamin C and all of the B vitamins.

Weight Bias: Negative attitudes, beliefs, assumptions and judgments toward individuals based on their body size.

Weight Stigma: The discriminatory acts targeted towards individuals who live in larger bodies; a result of weight bias.

White Blood Cells: Also known as leukocytes; immune cells that help destroy foreign invaders.

Whole Grains: Grains and grain products made from the entire grain seed, including the bran, germ, and endosperm.

Xerophthalmia: Abnormal dry eyes and clouded vision, usually caused by vitamin A deficiency.

INDEX

R

S

T

U

V

X

Y

www.ingramcontent.com/pod-product-compliance
Lightning Source LLC
LaVergne TN
LVHW062212200926
848484LV00006B/1378